Recent Advances and Perspectives in Deoxynivalenol Research

Special Issue Editor
Marc Maresca

MDPI • Basel • Beijing • Wuhan • Barcelona • Belgrade

MDPI

Special Issue Editor
Marc Maresca
Aix Marseille Université
France

Editorial Office
MDPI AG
St. Alban-Anlage 66
Basel, Switzerland

This edition is a reprint of the Special Issue published online in the open access journal *Toxins* (ISSN 2072-6651) from 2013–2014 (available at: http://www.mdpi.com/journal/toxins/special_issues/deoxynivalenol).

For citation purposes, cite each article independently as indicated on the article page online and as indicated below:

Author 1; Author 2. Article title. *Journal Name*. **Year**. Article number/page range.

First Edition 2017

ISBN 978-3-03842-470-3 (Pbk)
ISBN 978-3-03842-471-0 (PDF)

Photo courtesy of Marc Maresca

Table of Contents

About the Special Issue Editor

Marc Maresca was born on 25 October 1974. He is an assistant professor at the Institute of Molecular Sciences of Marseille (ISM2, UMR-CNRS 7313). Research performed at this multidisciplinary institute aims at combining different techniques (cell biology, microbiology, toxicology, nuclear magnetic resonance, mass spectrometry, chromatography, chemometrics, peptide synthesis) in order to understand biological mechanisms (eukaryotic and prokaryotic systems). Dr Marc MARESCA is specialized in using and characterizing in vitro cell models for toxicity studies, including human intestine epithelium and brain cell models as demonstrated by his publications in both subjects. Part of his research is on food-contaminants and their toxicity, including pesticides and mycotoxins (particularly deoxynivalenol but also ochratoxin, patulin and fumonisin). He has published major articles on the effects of deoxynivalenol and its derivatives on the human gut. He also has expertise in antibiotic development (particularly antimicrobial peptides) and works to understand their mechanisms of action (pore-forming activity, safety of antibiotics to human and animal cells).

Preface to "Recent Advances and Perspectives in Deoxynivalenol Research"

Life is a battle and thus living organisms have developed strategies to win this war. Among the different strategies employed by micro-organisms to dominate their habitat is the production of toxins including bacteria and fungi and their use as bioweapons. Mycotoxins are secondary metabolites produced by molds that play such a role.

For many years, one of these mycotoxins, the food-associated trichothecene Deoxynivalenol (DON or vomitoxin) has attracted the attention of scientists. This is due, in part, to its high prevalence in animal/human food and feed products, as demonstrated through the successful use of urinary biomarkers confirming the exposure of humans to substantial doses of this toxin. DON is also one of the most hazardous mycotoxins; it affects the functions of nerve, endocrine, immune and intestinal cells. In addition to its toxicity to animal cells (this could be considered as collateral damage), DON is also known to affect plant cell functions; such effects certainly play a role during the colonization of wheat and cereals by DON-producing fungi such as Fusarium species. The toxicity of DON seems to depend on the presence of an epoxide function which allows its binding to ribosomes, causing the so-called "ribotoxic stress" effect, and the activation of specific kinases (including PKR and MAP kinases), eventually leading to the inhibition of the protein synthesis and to cell death. Due to its ability to activate PKR and MAP kinases, DON also acts as a proinflammatory signal at low doses whereas higher doses are immunosuppressive due to cellular toxicity. In animals, as well as affecting systemic and intestinal immunity, DON also impacts the functions of the brain and endocrine cells, causing anorexia and vomiting. Food not only contains native toxin, but also large amounts of plant and fungal derivatives of DON (including the fungal metabolites 3 and 15 acetyl-DON (3 and 15ADON) and the plant derivative 3-O-glucoside-DON (D3G)) and possibly, although no study has yet confirmed it, of animal derivatives (i.e., 3 and 15-glucuronide DON) potentially present in meat and animal-derived products. New DON derivatives were also recently found in plants and food products, including DON-oligoglycosides, DON-glutathione, DON-S-Cysteine, DON-S-Cysteinyl-glycine, and DON-sulfonate. Although previous research has shed light on the mechanisms of action of DON, important questions remain. For example, little is known about the ability of the fungi to transmit from the soil to the cereals, and about the levels of DON and DON metabolites in different plant tissues during natural and experimental contamination. Data on the effects of DON and its metabolites on plant cells are also scarce. Similarly, how DON enters the cells (animal or plant cells) and how it binds/acts on ribosomes is not perfectly characterized. Finally, if ribosomes are the only target of DON, how the toxin could activate different kinases, depending on the toxin dose, remains a mystery. We hope that some of these questions will be answered in this Special Issue that focuses on one of the most studied and relevant food-associated mycotoxins.

Marc Maresca

Special Issue Editor

toxins

Review

From the Gut to the Brain: Journey and Pathophysiological Effects of the Food-Associated Trichothecene Mycotoxin Deoxynivalenol

Marc Maresca

Aix Marseille Université, CNRS, iSm2 UMR 7313, Marseille 13397, France; m.maresca@univ-amu.fr;
Tel.: +33-491-288-445; Fax: +33-491-284-440

Received: 25 February 2013; in revised form: 11 April 2013; Accepted: 12 April 2013; Published: 23 April 2013

Abstract: Mycotoxins are fungal secondary metabolites contaminating food and causing toxicity to animals and humans. Among the various mycotoxins found in crops used for food and feed production, the trichothecene toxin deoxynivalenol (DON or vomitoxin) is one of the most prevalent and hazardous. In addition to native toxins, food also contains a large amount of plant and fungal derivatives of DON, including acetyl-DON (3 and 15ADON), glucoside-DON (D3G), and potentially animal derivatives such as glucuronide metabolites (D3 and D15GA) present in animal tissues (e.g., blood, muscle and liver tissue). The present review summarizes previous and very recent experimental data collected *in vivo* and *in vitro* regarding the transport, detoxification/metabolism and physiological impact of DON and its derivatives on intestinal, immune, endocrine and neurologic functions during their journey from the gut to the brain.

Keywords: deoxynivalenol; mycotoxin; trichothecene; detoxification; intestinal absorption; intestine; brain; endocrine; glial cells; immune cells

---

1. Introduction

Deoxynivalenol (DON, vomitoxin) belongs to a family of mycotoxins called trichothecenes. Trichothecenes (including T-2 toxin, nivalenol, DON and satratoxins) are structurally related molecules produced by fungi of *Fusarium* and *Stachybotrys* species [1]. They are small sesquiterpenoids all having in common an epoxide group at position 12–13 that is critical for their toxicity [2–5] (Figure 1). It has been proposed that the epoxide group allows them to bind to ribosomes, a mechanism known as the ribotoxic stress effect, leading to the activation of various protein kinases, the modulation of gene expression, the inhibition of protein synthesis and cell toxicity [5–8].

Analyses of the occurrence of DON in food and feed matrices have demonstrated that DON is one of the most prevalent food-associated mycotoxins, particularly in cereals and cereal-derived products [9]. In the US, 73% and 92% of wheat and corn samples, respectively, were found positive for DON [10]. In Europe, a large-scale collaborative study conducted on more than 40,000 food samples has shown that DON was present in 57% of all samples, with a percentage of positive samples varying depending of the country (*i.e.*, from 15% to 100% for Belgium and France, respectively) and at levels ranging from 91 to 5000 μg/kg [11]. Similarly, another study conducted on 82 feed matrices in Europe has demonstrated that 67 of them were contaminated with DON, 52 samples being highly contaminated with levels of DON ranging from 74 to 9528 μg/kg [12]. DON is, moreover, resistant to high temperature (up to 350 °C), thereby making it stable during processing and cooking, leading to its persistence throughout the food chain [13].

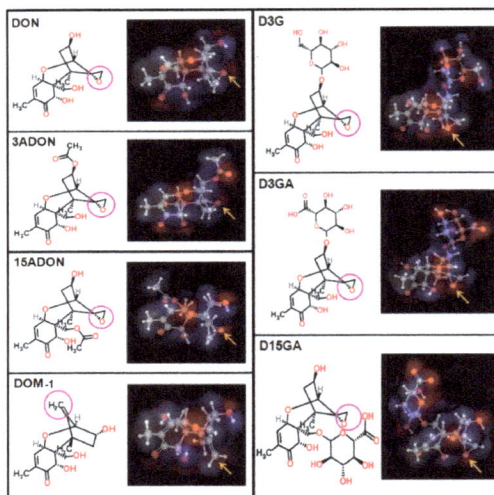

Figure 1. Chemical structure of DON and its major derivatives. DON and its derivatives were drawn using Marvin software. Images on the right show an electrostatic map of the molecules, the blue color indicating positive region, the red color indicating negative region and the gray color indicating neutral region. The purple circles on the left images and yellow arrows on the right images indicate the position of the epoxide or de-epoxide function in DON and its derivatives.

In addition to its prevalence, DON is one of the most hazardous food-associated mycotoxins [4,7,8,13–16]. A provisional maximum tolerable daily intake (PMTDI) for DON of 1 μg/kg of body weight and per day has been proposed by the Joint FAO/WHO Expert Committee on Food Additives (JECFA) [14]. The ingestion of DON has been associated with alterations of the intestinal, immune and nervous systems, thus leading, in cases of acute exposure, to illnesses characterized by vomiting, anorexia, abdominal pain, diarrhea, malnutrition, headache and dizziness [4,7,8,17]. Toxicity of DON relies on its ability to cross the biological barriers (*i.e.*, the intestinal and blood-brain barriers) and to affect the functions and viability of the cells forming such organ systems.

The present paper compiles experimental data collected *in vivo* and *in vitro* regarding: (i) the transport of DON and DON derivatives from the gut to the brain; (ii) their detoxification; and (iii) their impact on the animal and human physiology.

2. Transport and Metabolism of DON

2.1. Structure and Physicochemical Properties of DON and Its Derivatives

Food and feed are contaminated both by native DON and its derivatives. The structure and partition coefficient (log*D*) of DON and its metabolites are given in Figure 1, Figure 2. The major derivatives of DON correspond to metabolites formed either by fungi (*i.e.*, the acetylated derivatives: 3- and 15-acetyl-DON or 3ADON and 15ADON), plants (*i.e.*, 3-*O*-glucoside-DON or D3G), animals (*i.e.*, glucuronic acid derivatives: DON-3 and DON-15-glucuronide or D3GA and D15GA) or bacteria (*i.e.*, the de-epoxide diene derivatives of DON: DOM-1) [18,19]. Various studies have shown that food contains large amounts of DON metabolites, mainly the fungal and plant derivatives 3/15ADON and D3G, with up to 75% of the total amount of DON corresponding to DON metabolites [19]. In addition, although no studies confirm it, animal derivatives of DON (*i.e.*, D3/15GA) may be theoretically present in food originated from animal tissues and blood. The amount of DON metabolites has not been considered in the regulatory limits fixed by food agencies for DON due to the lack of data regarding their absorption and toxicity [19].

Calculation of the partition coefficient demonstrates that metabolic modifications of DON lead to important changes in the polarity of the molecule (Figure 2). LogD of DON is −0.97 at pH 7, suggesting a polar behavior. The presence of an acetyl moiety in the fungal metabolites 3ADON and 15ADON or the absence of the oxygen linked to the epoxide function in the bacterial diene metabolite DOM-1, result in a decrease in the polarity of the molecule compared to the native toxin (logD values of the metabolites being less negative than the one of DON with a value at neutral pH of −0.35 and −0.53 for DOM-1 and 3/15ADON, respectively).

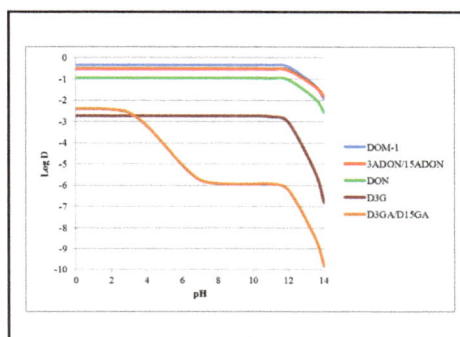

Figure 2. LogD values of DON and its derivatives. LogD values of DON and its derivatives at various pH values were calculated using Marvin software.

Conversely, the presence of a glucoside or a glucuronide moiety in the plant and animal metabolites D3G, D3GA and D15GA leads to an increase in their polarity compared to DON (their logD values being more negative than the one of the native toxin with a value at neutral pH of −2.74 and −5.75 for D3G and D3/15GA, respectively). As discussed below, increase or decrease in the polarity of DON metabolites may affect their ability to enter the cells and thus to be absorbed by the intestine and/or to cause cell toxicity.

2.2. Cell Entry of DON and Its Derivatives

No studies have been conducted to characterize the exact mechanism of the cell entry of DON, with only speculations being possible at present (Figure 3). One possibility is that cell entry of DON does not occur at all and that cellular effects of DON described in part 3.1, such as activations of various kinases, rely on its interaction with membrane receptors/proteins activating such signal pathways. Although the direct effect of DON on membrane proteins could not be ruled out, data support the idea that at least a part of DON enters the cells, *i.e.*, the fact that: (i) DON binds to intracellular ribosomes; and (ii) DON is substrate of intracellular detoxification enzymes (see part 2.4.). Studies using intestinal cells have shown that the cell entry of DON does not saturate when the extracellular concentration of toxin increases, suggesting that its entry takes place through a passive diffusion mechanism [20,21]. An important question is how DON, with its logD value of −0.97 at neutral pH that makes it behave like a polar molecule, could diffuse across the cell membrane. Based on the fact that only molecules bearing a logD value close to zero or positive are able to enter the cells through lipid diffusion [22], the ability of DON to enter the cells through such a mechanism is theoretically low to nil. This leaves only two possibilities for DON to cross the cell membrane: (i) a diffusion through an uncharacterized membrane-associated passive transporter; and/or (ii) a bulk phase endocytosis/pinocytosis process (Figure 3) [23,24].

3

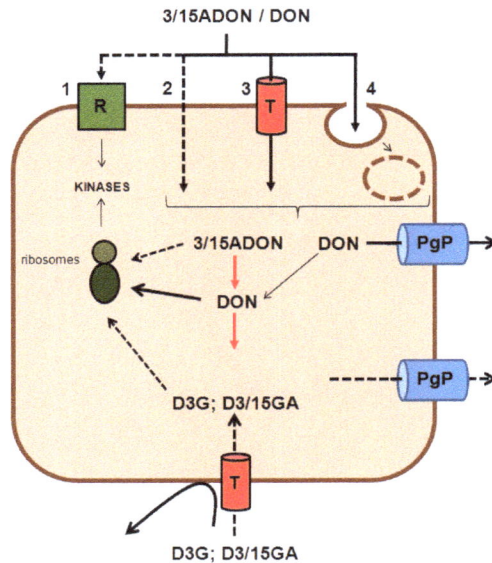

Figure 3. Cell entry of DON and its derivatives. Although highly unlikely, cellular effects of DON could rely on its ability to directly bind membrane receptor(s) (R) (**1**). However, the fact that DON interacts with ribosomes and is substrate of intracellular detoxification enzymes rather suggests that DON enters the cells. Cell entry of DON and its acetylated derivatives (3/15ADON) could take place through membrane diffusion across lipids (**2**), through a membrane transporter (T) (**3**) or through bulk phase endocytosis/pinocytosis (**4**) Once in the cell, 3/15ADON could be transformed in DON by intracellular carboxyl-esterases. DON (and possibly 3/15ADON) reacts then with ribosomes to cause cell effects. Detoxification of DON involves the production of glucuronide-metabolites by UDP-glucuronosyltransferases. In addition, P-glycoproteins (PgP) are responsible for the efflux/excretion of DON and possibly of its derivatives. The absence of cell effects of D3G and D3/15GA suggests either that: (i) these derivatives do not cross the cell membrane (**5**); or (ii) that they efficiently enter the cell but do not bind to ribosomes (**6**), the first hypothesis being more likely. Dashed lines/arrows and full lines/arrows indicate unlikely/hypothetical and likely mechanisms, respectively.

As for DON, no data exists regarding the mechanism of cell entry of DON derivatives. As explained in Section 3.1, alterations in their ability to enter the cells and/or to bind to ribosomes/receptors may explain the difference of cell toxicity and toxicokinetics observed for DON derivatives compared to the native toxin. One can suppose that DON derivatives with log*D* values closer to zero (*i.e.*, DOM-1 and 3/15ADON) may have higher ability to diffuse across the lipids of the membrane. Conversely, glucoside and glucuronide metabolites of DON (D3G, D3/15GA) bearing bigger molecular masses and more polarity would have a reduced ability to enter the cells through lipid diffusion. Similarly, modifications of DON (size, polarity) may also theoretically affect the ability of DON derivatives to interact with membrane transporters if such transporters are involved. Future studies should help identify the mechanism(s) that permit the entry of DON and its derivatives into the cells.

2.3. Bacterial Transformation and Intestinal Absorption of DON and Its Derivatives

Metabolism of DON in plants and fungi has been fully described recently [18,19] and thus will not be elaborated upon in the present review focusing on modifications of DON by bacteria, animals and humans.

The first phase of the intoxication by DON and its derivatives corresponds to their passage through the gut wall, such transport being possibly affected by bacterial metabolism. The intestinal tract of animals and humans contains vast amounts of bacteria forming the commensal microbiota that lives in symbiosis with the host. At present, the microbiota could be considered as an additional organ system, playing important roles in the maturation of the intestinal and immune systems, in the nutrition of the host, and finally in the protection of the host against pathogenic micro-organisms and hazardous chemicals/xenobiotics, including DON and its derivatives [18,25–27].

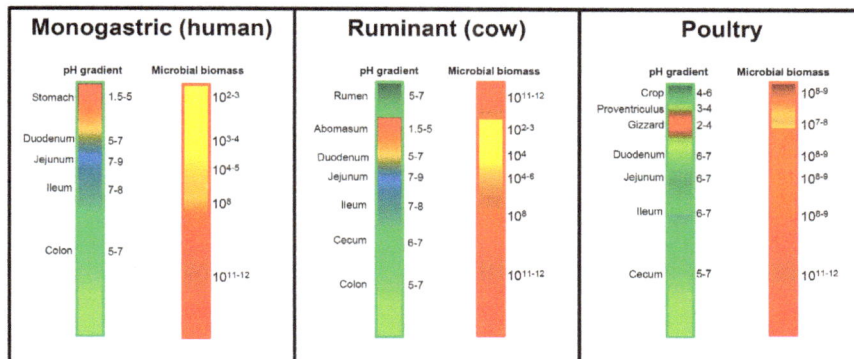

Figure 4. Regional pH and bacterial densities in the digestive tract. pH and bacterial density (per mL of intestinal fluid content) of the different segments of the digestive tract of humans, ruminants and poultry are indicated in the figure. Values were obtained from publications [27–31].

The efficiency of the intestinal absorption and metabolism of DON greatly varies between animal species due in part to the localization of bacteria along their intestine in relation to regional pH (Figure 4) [27–31]. On this basis, animals could be divided into two groups: (i) those with a high bacterial content located both before and after the small intestine such as polygastric animals (*i.e.*, ruminants that have bacteria in their rumen and in their colon) and birds (including poultry that have bacteria in their crop and in their cecum); and (ii) those with high bacterial content located only after the small intestine, in their colon, such as most of the monogastric species (including humans, pigs and rodents). Localization of the gut bacteria prior or after the small intestine has a major effect on the bioavailability of ingested DON and its metabolites (Figure 5, Figure 6).

In monogastric animals, large amounts of ingested DON can cross the intestinal epithelium and reach the blood compartment (Figure 5). For example, in pigs, 54% to 89% of the ingested toxin is absorbed *in vivo* after acute and chronic oral exposure to DON, respectively [32], possible explanations for the higher oral bioavailability of DON after chronic exposure are discussed in the following. After oral intoxication of pigs, DON starts to appear in the plasma after 30 min and its serum concentration reaches a peak value within three to four hours post-ingestion, thereby suggesting a fast and efficient absorption of the toxin through the proximal small intestine [32–34]. Accordingly, *in vitro* experiments conducted with intestinal segments from pigs have shown that the intestinal absorption of DON takes place mainly through the jejunum [35]. Similarities between the human and pig intestines (also in terms of DON effects as described in part 3.2. suggest that humans could also efficiently absorb ingested DON.

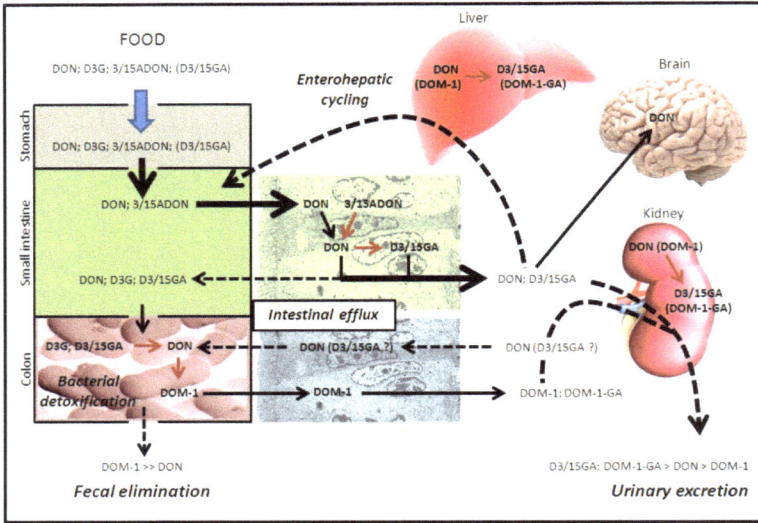

Figure 5. Intestinal absorption, detoxification and excretion of DON and its derivatives in monogastric species (e.g., humans/pigs/rodents). Humans and monogastric animals are exposed to DON and DON derivatives through the ingestion of contaminated food. Details are given in the text (parts 2.3 and 2.4). DOM-1-GA corresponds to glucuronide derivatives of DOM-1. Red arrows indicate transformation of DON or DON derivatives, dashed arrows indicate excretion/elimination mechanisms.

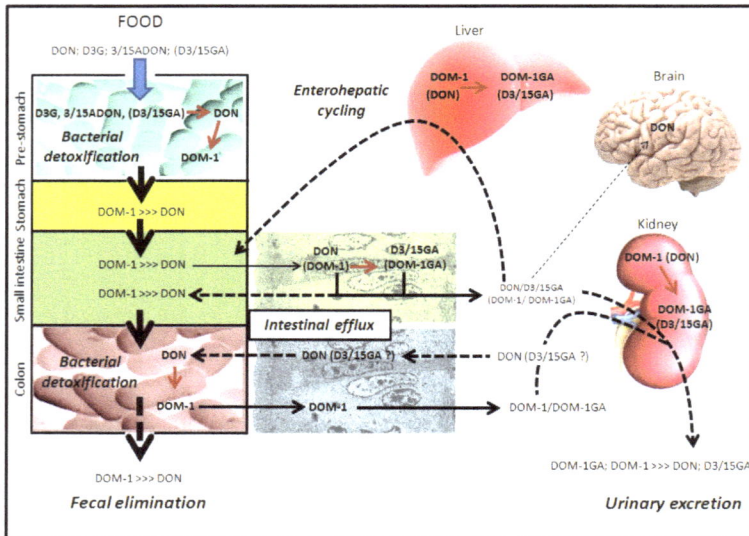

Figure 6. Intestinal absorption, detoxification and excretion of DON and its derivatives in ruminants and poultry. Poultry and polygastric animals are exposed to DON and DON derivatives through the ingestion of contaminated food. Details are given in the text (parts 2.3. and 2.4.). DOM-1-GA corresponds to glucuronide derivatives of DOM-1. Red arrows indicate transformation of DON or DON derivatives, dashed arrows indicate excretion/elimination mechanisms.

Even though *in vivo* toxicokinetic data of DON in humans are not available, the intestinal absorption of DON by humans has been elegantly evaluated using an *in vitro* model of human intestinal epithelial cells (IEC), *i.e.*, the Caco-2 cell line [20]. Caco-2 cells have been used for decades to mimic the human IEC and to study the intestinal absorption of drugs and toxins, this cell line giving apparent permeability coefficient (Papp) values predictive of the *in vivo* oral bioavailability of molecules in humans [36]. Authors showed that DON has a Papp value of 5.02×10^{-6} cm/s in Caco-2, corresponding to a potential *in vivo* oral bioavailability of 50% to 60% in human, a value in accordance with the *in vivo* experiments conducted on pigs. The Papp of DON was also measured *in vitro* in poultry, giving even higher values (*i.e.*, 1.7×10^{-5} cm/s, corresponding to a predicted *in vivo* oral bioavailability of 70%) [37]. As discussed below, the observed oral bioavailability of DON in poultry is much lower (*i.e.*, around 20%) than expected, in relation to the presence of intestinal bacteria able to transform DON in DOM-1 before the small intestine in birds. Few data are available regarding the mechanism of intestinal absorption of DON. Experiments conducted *in vitro* with Caco-2 cells [20] or with intestinal segments from poultry [37] showed that the intestinal absorption of DON does not saturate but is proportionally dependent on the extracellular concentration of DON, thus suggesting that the intestinal absorption of DON takes place through passive transcellular and/or paracellular diffusion. As explained in part 2.2., based on its ability to target intracellular ribosomes and to be substrate of intracellular detoxification enzymes, at least a part of the intestinal absorption of DON has to take place through transcellular transport. Although the relative contribution of transcellular and paracellular transport in DON absorption has not been evaluated yet, contribution of the paracellular mechanism may massively increase and become predominant in the case of alterations of intestinal permeability. Decreases in the intestinal tightness are observed in various conditions such as inflammatory bowel disease (IBD) (including Crohn's disease), intestinal infections by viruses, or pathogenic bacteria and exposure to DON or other mycotoxins [14,38–44]. This could explain the observed higher absorption of DON in pigs chronically exposed to oral DON compared to acute exposure [32].

The rank order of sensitivity of animals to ingested DON is pigs > poultry/ruminants [4]. As mentioned above, intestinal explants from poultry and pigs possess a similar ability to intestinally absorb DON, suggesting that the difference in their sensitivity to ingested DON does not rely on their ability to intestinally absorb DON. In fact, the sensitivity of animals to oral DON relies on the localization of the intestinal bacteria in their gut in relation to their ability to generate 9,12-diene DON or DOM-1, the non-toxic de-epoxide derivative of DON [18,45].

The presence of high bacterial contents that are able to convert toxic DON into its non-toxic de-epoxide metabolite DOM-1 before the small intestine in ruminants (rumen-associated bacteria) and poultry (crop-associated bacteria) massively decreases the amount of native DON reaching the small intestine, making such animal species almost insensitive to oral intoxication by DON (Figure 6) [4]. For example, only a small amount of the ingested DON reaches the small intestine as native toxin in poultry and sheep (19.3% and 7.5% of the ingested DON being found in the blood of intoxicated poultry and sheep, respectively) [46–48]. Similarly, in cows, 16% of ingested toxin reaches the small intestine [49] and only 1% crosses the intestinal wall to reach the blood [50].

In monogastric animals, due to the high absorption of DON by the small intestine, bacterial transformation of DON in DOM-1 could only be possible if a part of the ingested DON reaches the colon and/or in the case of intestinal/hepatic excretion of absorbed DON (Figure 5). This explains why only a low percentage of ingested DON is found in the feces of monogastric animals as DOM-1, with most of the ingested DON being eliminated in the urine as glucuronide-DON, DON, glucuronide-DOM-1 and DOM-1 (Figure 5) [51,52].

No studies have looked at the intestinal absorption of DOM-1 in animals or humans. However, based on the fact that DOM-1 is only formed by intestinal bacteria in the gut lumen and that a percentage of ingested DON is found in urine as DOM-1, we could suppose that DOM-1 formed by intestinal bacteria is efficiently absorbed by the gut (Figure 5, Figure 6) [53].

Not all bacteria are able to transform DON in DOM-1 [18]. In pigs, it has been demonstrated that only chronic oral exposure to DON leads to the formation of DOM-1 by the microbiota [54]. Experiments conducted with human feces coming from five volunteers showed that only one spontaneously possesses bacteria able to transform DON in DOM-1 [53]. Taken together, experiments conducted with pigs and humans suggest that naive intestinal bacteria naturally do not possess the ability to detoxify DON and that pre-exposure of the microbiota to DON induces the appearance of the bacterial detoxification activity, either through the induction of the expression of particular enzymes and/or the selection of particular detoxifying bacterial species [18]. Initially, aerobic bacteria were thought to be unable to form DOM-1 as they rather transform DON in 3-epi-DON and 3-keto-4-DON, both having an intact epoxide function [18,55]. However, recent data suggested that some soil bacteria are also able to form DOM-1 both in aerobic or anaerobic condition [56]. It has to be noted that although some bacteria and micro-organisms were initially thought to be able to totally mineralize DON, data suggest that adsorption of the toxin to the cell wall and bacterial uptake are in fact responsible for the disappearance of the toxin from the medium [18], with such adsorption certainly playing an important role in the neutralization of DON by the intestinal bacteria.

Fungal (*i.e.*, 3/15ADON) and plant (*i.e.*, D3G) metabolites of DON are also present in food and could thus be absorbed by the intestine and/or metabolized by intestinal bacteria (Figure 5, Figure 6). In addition, although no published studies describe it, animal derivatives of DON (*i.e.*, D3/15GA) could also be theoretically present in animal-derived food (animal tissues, blood) and thus be ingested by humans or animals. In pigs, the ingestion of 3ADON leads to the appearance of DON (58%) and DON metabolites (glucuronide-DON and DOM-1 (42%)) but not of 3ADON in the blood [52]. This result suggests either that: (i) 3ADON is not directly absorbed by IEC but requires its initial transformation into DON by gut bacteria or by luminal intestinal lipases before its absorption; or (ii) 3ADON is directly absorbed by IEC that transform it intracellularly into DON before its release in the blood. Luminal intestinal lipases and microbial esterases/lipases could theoretically cut the acetyl moieties of the fungal metabolites 3ADON and 15ADON to release DON in the intestinal lumen [57,58]. Similarly, IEC possess intracellular carboxylesterases (CES) [59] potentially able to transform absorbed 3ADON and 15ADON into DON. An *in vitro* study has shown that isolated IEC are sensitive to 3/15ADON [60], proving that these derivatives could be directly absorbed by the IEC without the requirement of intestinal lipases or microbial lipases/esterases. The relative contribution of intestinal lipases, bacteria and IEC in the metabolism of acetyl-DON may depend of the animal species. In monogastric animals, the significance of the transformation of 3/15ADON by colonic bacteria is limited due to their high absorption by IEC [60], suggesting a major role of intestinal lipases and/or epithelial CES in that case (Figure 5). To date, direct evidence of the transformation by intestinal CES of 3/15ADON in DON are unfortunately missing. In ruminants and poultry, bacterial de-acetylation of 3/15ADON could happen prior the small intestine and may theoretically impact their bioavailability.

Little is known regarding the intestinal absorption and bacterial metabolism of the more polar metabolites of DON, *i.e.*, D3G, D3GA and D15GA. On the basis of the hypothesis that DON enters IEC by lipid diffusion, such polar metabolites should have lower intestinal absorption efficiency compared to the native toxin. Similarly, one could suppose that addition of glucoside or glucuronide moiety would impact the interaction of DON with its membrane transporter, affecting their cell entry through this mechanism (see part 2.2.). No data exists regarding the intestinal absorption of D3/15GA, but the lack of toxicity of these metabolites suggests that their oral bioavailability is certainly low to nil [61]. Similarly, nothing is known regarding the bacterial metabolism of D3/15GA. One could suspect that bacterial beta-glucuronidases would certainly transform them into DON with or without consequences, depending if the transformation occurs prior to or after the small intestine. As with D3/15GA, D3G is unable to cause toxicity [62]. Data have, however, shown that intestinal bacteria are able to transform D3G into DON through the hydrolysis of its glucoside moiety [18,19,53,63,64]. Interestingly, the bacterial activity leading to the transformation of D3G in DON is spontaneously present in the feces and does not seem to require its induction as observed for the transformation of

DON in DOM-1 [53]. In monogastric animals, the transformation of D3G into DON after the small intestine does not allow the absorption of the released DON since most of the toxin remains in the feces, suggesting that D3G is not hazardous at least for these animals due to its limited intestinal absorption (Figure 5) [64]. Again, the situation may be totally different in ruminants and poultry where the transformation of D3G in DON would take place before the small intestine, potentially allowing the absorption of the released toxin.

2.4. Metabolism and Excretion of DON and Its Derivatives by the Animals

The ingestion of native DON and its derivatives leads to the presence of a native toxin in the body of intoxicated animals. As with many xenobiotics, DON is then subject to detoxification and excretion (Figure 5, Figure 6).

Transport studies using Caco-2 cells have demonstrated that human IEC have the ability to apically excrete DON [21]. Whereas the apical (AP) to basolateral (BL) transport of DON by human IEC is insensitive to transporter's inhibitors, its BL to AP excretion is sensitive to P-glycoprotein inhibitors, particularly inhibitors of the multidrug resistance-associated protein (MRP-2) transporter [21]. In addition to reduce the net absorption of ingested DON by IEC of the small intestine, net excretion by IEC of the colon may account for the total excretion of DON (and possibly D3/15GA) by the body. Detoxification of ingested xenobiotics generally takes place in the IEC, the liver and the kidneys. Detoxification of DON certainly starts in IEC, directly after its intestinal absorption. Although the detoxification metabolite DOM-1 is present in the blood of animals orally intoxicated with DON, as mentioned above, the transformation of DON in DOM-1 is not related to animal detoxification as it occurs in the intestinal lumen and corresponds to bacterial detoxification followed by the intestinal absorption of DOM-1 [45,65]. Body detoxification of DON mostly involves the formation of glucuronide metabolites (mainly D3GA and D15GA) by UDP-glucuronosyltransferases. Such metabolites are less toxic than the parental toxin due to their lower log$D$ value (Figure 2) making them less efficient at crossing the cell membrane and/or at binding to ribosomes [61]. The amount of glucuronide-DON formed greatly differs, depending of the animal species used. Thus, in sheep, glucuronide metabolites correspond to 75% of the systemic DON [46], whereas in pigs, the percentage of glucuronide metabolites varies from 5% to 58%, depending if the animals were exposed to DON or 3ADON, respectively [33,52]. This suggests that in addition to the animal species used, the form of the ingested toxin, either native or conjugated, also impacts its detoxification, at least in pigs. The precise site of the formation of the glucuronide-DON is not characterized at present, intestinal, liver and kidney cells being theoretically able to form glucuronide metabolites. Liver microsomes extracted from animals and humans have been shown to be able to transform DON in glucuronide-DON, mainly D3GA and D15GA [66,67]. Experiments conducted with intestinal or renal microsomes are not available to confirm that these tissues could also detoxify DON. However, the fact that in sheep the amount of glucuronide-DON formed is higher after oral exposure to DON (75%) than after intravenous injection (21%) suggests that IEC are responsible for a large part of the formation of such detoxification products in case of oral intoxication [46].

Regarding the excretion of DON, it seems that most of the toxin is excreted in the urine as glucuronide-DON, glucuronide-DOM-1, DON and DOM-1 (Figure 5, Figure 6). A study in humans has shown that 91% of the DON excreted in urine is glucuronide-DON, D15GA being predominantly found [68,69]. In pigs orally intoxicated by DON, 68% of the toxin is excreted in the urine as unchanged DON and glucuronide-DON, the remaining being mostly eliminated in feces (20%) as DOM-1 and DON (80% and 20% of total DON in feces, respectively) [32,34]. Similarly, oral exposure of pigs to 3ADON leads to the elimination of the toxin mainly in urine (up to 80%) as DON and glucuronide-DON with only low amounts of toxin (2%) being present in the feces as DON and DOM-1 (48% and 52%, respectively) [52]. As mentioned previously, intoxication of rats with D3G does not require body detoxification since most of the D3G is not absorbed; 3.7% of the toxin is eliminated in the urine as DON, glucuronide-DON and D3G, and all the rest is eliminated in feces as DON and DOM-1 [64].

Mechanism(s) of excretion of DON/DOM-1 and glucuronide-DON/DOM-1 are unknown at present and could involve both glomerular filtration of the metabolites present in the blood and their excretion through *P*-glycoproteins expressed by intestinal, renal or hepatic epithelial cells, as demonstrated for DON in IEC [21].

Overall, excretion of DON and its detoxification metabolites is quite efficient, half of plasmatic DON being eliminated after 6 h in pigs and in sheep [34,46]. The fast elimination of DON suggests a low binding of DON and its metabolites to serum albumin, at least in animals. Accordingly, the *in vitro* toxicity of DON is not modified by the presence of bovine serum albumin (BSA) in the medium, contrary to the toxicity of ochratoxin A, another mycotoxin with a strong affinity for serum albumin and a longer plasmatic half-life (up to 840 h) [70,71]. It has to be noted that a recent study described the interaction of DON with human serum albumin [72], indicating that in humans, DON could possibly have a longer plasmatic half-life.

2.5. Transport of DON through the Blood-Brain Barrier (BBB)

As described in Section 3.4, DON is able to cause alterations of the brain functions. Although part of these alterations could be attributed to peripheral effects, data suggest that neurologic effects of DON rely in part to the direct action of DON on brain cells. This requires the crossing of the blood-brain barrier (BBB) by the toxin. The BBB is formed by the close apposition of endothelial and glial cells, forming a selective barrier controlling the passage of molecules from the plasma to the cerebro-spinal fluid (CSF) [73]. *In vivo* studies have shown that DON crosses the BBB in various animal models. DON transport across the BBB occurs rapidly, native toxins being detected in the brain of animals within a few minutes (2 to 60 min, depending of the animal species) after exposure [74]. The ability of DON to cross the BBB depends on the animal species. In pigs, 25% to 30% of the plasmatic DON is found in the CSF, the toxin having a CSF half-life similar to its plasmatic one and being still detectable in the CSF 20 h after the intoxication [74]. In mice, the BBB crossing of DON is lower, the concentration of DON in the brain corresponding to 10% of the plasmatic concentration [75]. Finally, in sheep, only 5% of the plasmatic DON crosses the BBB [74]. Transport of DON across the BBB of other animal species and humans has not yet been evaluated, though it would not be surprising that some animal species may have higher or lower BBB permeability to DON. The best example of variation of BBB permeability for a specific molecule between animal species comes from another family of mycotoxins: the fumonisins. Indeed, BBB permeability to these toxins ranges from low/nil (for mice) to high (for horses) [76,77]. Importantly, it was demonstrated that perturbations of the BBB permeability, caused by LPS-induced neuro-inflammation, increase the brain accumulation of fumonisins in mice [78]. At present, no studies have looked at the effects of neuro-inflammation and perturbations of the BBB on the brain accumulation of DON. Similarly, nothing is known regarding the mechanism(s) responsible for the transport (absorption and excretion) of DON across the BBB. Regarding the ability of DON derivatives to enter the brain, only native toxin is found in the CSF of intoxicated animals, suggesting that neither DOM-1, nor D3/15GA are able to cross the BBB [74].

3. Pathophysiological Effects of DON

3.1. Cellular Effects of DON

As with other trichothecenes, DON is able to cause cellular effects through its ability to target ribosomes and to cause ribotoxic stress [4–8]. Trichothecenes have all in common an epoxide group at position 12–13 critical for their action on ribosomes, explaining why the de-epoxide diene metabolite DOM-1 is non-toxic [2–5].

Binding of DON to the ribosomes could occur through the reaction of the epoxide moiety of DON with the nucleotides forming ribosomal RNA (rRNA) [6–8]. Nucleotides contain amine groups potentially able to react with epoxide (Figure 7) [79]. At present, nothing is known regarding the precise nature of the chemical reaction(s) allowing DON to bind to rRNA. Surprisingly, no report has been made of the interaction of DON with other nucleotide-containing molecules, such as mRNA or DNA. Aflatoxins (AFL), another family of mycotoxins with an epoxide function after their metaboliation by CYP450, selectively bind to guanine and cytosine residues present both in DNA and RNA [80]. This suggests—if the absence of reaction between DNA/mRNA and DON is confirmed—either that nucleotides from rRNA have a particular spatial organization allowing their specific interaction with DON, or that the real target of DON in rRNA is not nucleotides.

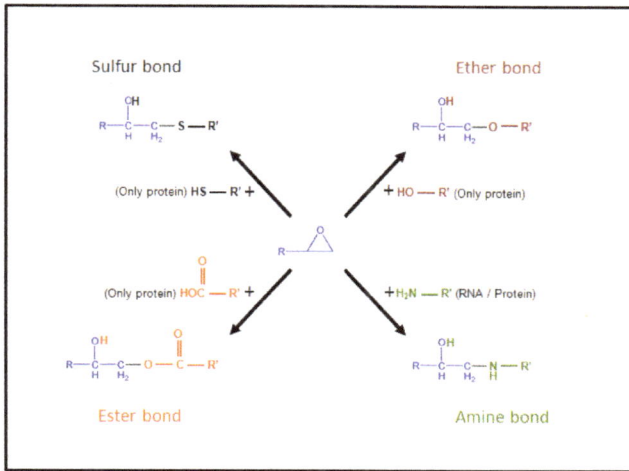

Figure 7. Chemical reactivity of the epoxide moiety. Epoxide moiety of DON could theoretically react with nucleophile functions present on the puric/pyrimidic bases of the nucleotides forming nucleic acid (DNA and RNA) such as amine group and/or on the side chain of amino acids forming the proteins such as: amine, hydroxyl, carboxyl and thiol groups.

Proteins possess amine, carboxyl, thiol and hydroxyl groups all potentially able to react with the epoxide function of DON (Figure 7), as demonstrated for AFL that forms adduct with the amine function of the lateral chain of lysine residues in serum albumin [81]. Thus, one could imagine that the binding of DON to rRNA takes place through its interaction with rRNA-associated protein(s). However, in that case again, it is not clear why DON only targets such rRNA-associated protein(s) and not other cellular proteins. One could suppose that only rRNA possesses both the correct spatial organization and chemical functions, present either on nucleotides or rRNA-associated proteins, allowing their selective interaction with DON. What we know is that the binding of DON to rRNA causes their cleavage and the activation of various cellular signaling pathways affecting cell functions and potentially leading to cell apoptosis (Figure 8) [6–8,82,83]. Signal pathways activated by DON correspond to the one generally activated by ribotoxins and ribotoxic stress, including: (i) two rRNA associated protein kinases, *i.e.*, the double-stranded RNA (dsRNA)-activated protein kinase (PKR) and the hematopoietic cell kinase (Hck); and (ii) the MAP kinases (p38, ERK1/2, JNK), affecting the expression of proteins involved in the innate immunity (through NFκB activation) and apoptosis (through p53) [6–8].

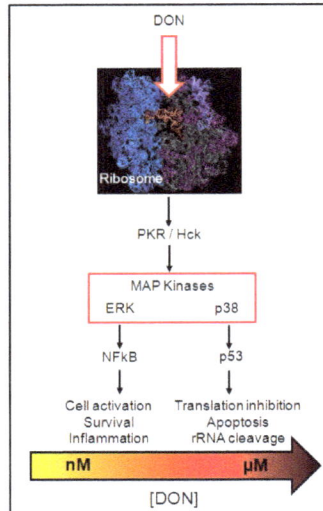

Figure 8. Cell effects of DON. Effects of DON on cell signal pathways in macrophages. Top image shows the organization of eukaryotic ribosome. The small subunit (40S) on the left contains an RNA molecule (cyan) and 20 proteins (dark blue); the large subunit (60S) on the right contains two RNA molecules (grey and slate) and more than 30 proteins (magenta). The image also shows a transfer RNA (orange) bound to the active site of the ribosome.

Initially, it was proposed that the binding of DON to rRNA cause their cleavage that in turn activate PKR and Hck, leading to the downstream activation of MAP kinases, NFκB and apoptosis pathways [6]. But recent elegant work from Pestka's group has shown that, in fact, apoptosis activation is not the consequence but is rather the cause of the rRNA cleavage through the activation of caspases and RNases [82,83]. In accordance with this hypothesis, the authors showed that signaling pathway activation occurs in minutes after exposure to low or high doses of DON, whereas rRNA cleavage appears only after hours of exposure to high doses. In addition, authors demonstrated that inhibitors of signal pathways and apoptosis inhibit rRNA cleavage caused by DON, thus definitively proving that rRNA cleavage is the consequence of cell signaling induced by DON and not the opposite.

The actual hypothesis regarding DON effects is that after its cell entry, DON binds to rRNA through the interaction of its epoxide moiety with functional group(s) present on the nucleotides and/or rRNA-associated proteins (such as PKR and Hck) leading to the rapid activation of the rRNA-associated protein kinases PKR and Hck which then activate MAP kinases, the type of MAP kinases activated being different depending of the doses of DON used. Thus, in macrophages, low doses (nM) activate preferentially ERK, causing cell survival and gene expression, whereas high doses (μM) activate p38 leading to apoptosis, rRNA cleavage and protein synthesis inhibition (Figure 8).

At low doses, DON has been showed to regulate the expression of various genes involved in the innate immunity and the inflammatory reactions through selective transcription, mRNA stabilization and translational regulation [7,8,82,83]. In addition to PKR, Hck, MAP kinases and NFκB other proteins participate in the transcriptional/translational effects of DON, including the HuR/Elav-like RNA binding protein 1, the CCAAT/enhancer-binding protein (CHOP) homologous protein, the peroxisome proliferator-activated receptor γ (PPARγ), the early growth response gene 1 (EGR-1), the activating transcription factor 3 (ATF3), the histone methylase, and GRP78/BiP [84–89]. It will not be surprising that additional signaling proteins participate in DON effects. Accordingly, a recent study from Pestka's group showed that DON affects the phosphorylation of 188 proteins, including proteins involved in transcription, epigenetic modulation, cell cycle, RNA processing, translation, ribosome biogenesis, cell differentiation and cytoskeleton organization [90].

Few studies have looked at the cell effects of DON derivatives. It has been known for decades that the loss of the epoxide moiety leads to the absence of cell effects of DOM-1 due to its inability to bind to ribosome independently of its cell entry [18]. More polar derivatives of DON (D3/15GA and D3G) are also non-toxic due either to their inability to cross the cell membrane and/or to bind to ribosomes, the relative participation of each event being unknown at present [61,62]. Toxicity of the less polar derivative (3/15ADON) depends of the organ and animal species, differences in their ability to enter the cells and/or to bind to ribosomes compared to DON possibly being involved. Using lymphocytes, 3ADON and 15ADON were found less toxic than DON [91–93]. Conversely, using pig IEC and intestinal explants, Oswald's group ranked the toxicity of DON and its acetylated derivatives as follows: 3ADON < DON < 15ADON [60]. *In vivo* experiments on mice confirmed the higher intestinal toxicity of 15ADON compared to DON in case of ingestion but not after i.p. injection, suggesting a particular sensitivity of the intestinal epithelium to acetyl-DON [94]. Based on the fact that ribosomes are theoretically the same in all cells, the difference of sensitivity of lymphocytes and IEC to acetyl-DON suggests that acetylation may affect the cell entry of DON derivatives, IEC being more efficient at transporting acetyl-DON than lymphocytes. Another attractive hypothesis would be that DON and 3/15ADON have a similar ability to enter the cells, but that only DON, and not 3/15ADON, binds to ribosomes. In that case, the higher sensitivity of IEC compared to lymphocytes could rely on the higher ability of IEC to transform 3/15ADON into DON through CES activity.

Overall, according to their ability to enter the cells, only DON and 3/15ADON have been shown to affect the functions of intestinal, immune and brain cells; DON effects on these systems are interconnected as described below.

3.2. Impacts of DON on the Intestinal Functions

Intestinal epithelial cells (IEC) are the first target of DON in case of natural exposure through ingestion of contaminated food. Whereas only IEC of the small intestine are exposed apically to ingested DON, IEC of the small intestine and colon are potentially exposed basolaterally to systemic DON that has crossed the intestinal wall to reach the blood compartment. Numerous studies have demonstrated that DON impacts IEC functions (Figure 9 and Table 1) (for review: [14,16,95]).

DON alters the proliferation and viability of animal and human IEC. In human IEC, the inhibition of the cell proliferation is observed at low doses (IC_{50} = 1–5 µM), cytotoxic effects being observed at higher doses (30–40 µM) [39,96]. Similarly, high doses of DON (IC_{50} = 10–50 µM) cause cell toxicity and apoptosis in rat and pig IEC [97–100]. Importantly, studies conducted on pig IEC have shown that the status of the cells (undifferentiated *versus* differentiated) and the site of DON exposure (apical or basolateral) affect its toxicity, DON being more toxic to undifferentiated IEC and when added basolaterally [97–99]. Whereas the higher sensitivity of undifferentiated cells compared to differentiated cells (10-times more sensitive) may be explained by differences in their cell cycle, no formal reasons explain the higher susceptibility of cells exposed basolaterally compared to apically (4-times difference), especially on the basis of the supposed passive diffusion entry of DON in the cells. No studies have been conducted to see if IEC absorb/accumulate more DON when the toxin is added basolaterally, explaining such a difference. Experiments need to be performed to understand how the site of exposure affects so much the effect of DON on IEC both in terms of toxicity and gene expression [97,98,101].

In vitro and *in vivo* experiments have also shown that DON inhibits the intestinal absorption of nutrients (at least glucose and amino acids) by human [39] and animal IEC [37,102–104]. The sodium-glucose dependent transporter (SGLT-1) activity is particularly sensitive to DON inhibition with an IC_{50} of 10 µM [39]. In addition to nutritional consequences, inhibition of SGLT-1 could explain the diarrhea associated with the ingestion of DON, since this transporter is responsible for the daily absorption of 5 L of water by the gut [105]. How DON causes inhibition of SGLT1 and other nutrient transporters is unknown at present, this inhibition being possibly related either to non-specific effects such as protein synthesis inhibition or ATP depletion, or to specific modulation of the expression/membrane targeting/activity of the transporters. According to the second hypothesis, activation of MAP kinases in IEC by proinflammatory signals causes the inhibition of the activity of membrane inserted SGLT-1 without affecting its expression [106,107].

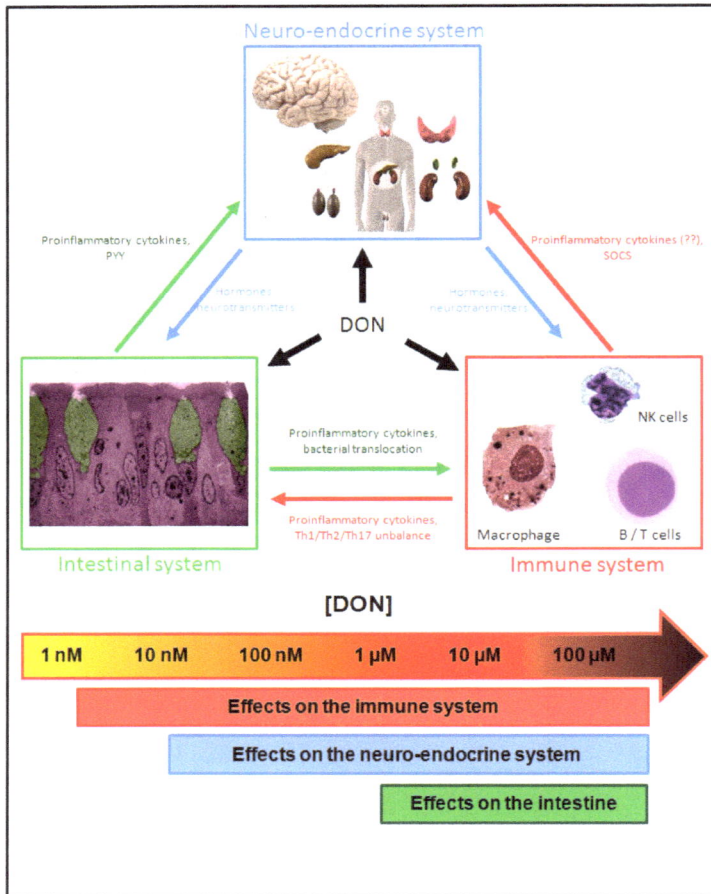

Figure 9. Effects of DON on the intestinal, immune and neuro-endocrine systems. Effects of DON on the intestinal, immune and neuro-endocrine systems are explained in the text. Doses at which the effects occur are schematically indicated at the bottom of the figure. It appears that the order of sensitivity of the systems is as follow: immune > neuro-endocrine > intestinal (Intestinal microscopy image courtesy of Cendrine Nicoletti).

Based on the ability of DON to activate MAP kinases in IEC, it would not be surprising that DON inhibits glucose absorption through such a mechanism. Finally, although no data are available to support such a hypothesis, one could speculate that D3G could act as a competitive inhibitor of SGLT-1 through its glucoside moiety.

Table 1. Risk evaluation of the exposure to DON for humans. Risks associated to DON exposure in humans were evaluated using the doses required to cause physiological alterations and PMTDI/higher exposure of the human population. Intestinal, blood and CSF concentrations of DON were calculated on the basis of a human adult weighting 70 kg, having a small intestinal volume of 1 L, a blood volume of 5 L and assuming that humans behave like pigs regarding blood and CSF concentration of DON [14,74]. The physiological alterations occurring at doses of DON with a safety factor inferior to 30 compared to the PMTDI/highest exposure-related concentrations are indicated in red.

Organ/system affected	Effect	Doses required	Times the PMTDI-related concentrations	Times the highest dose-related concentrations
INTESTINE	Inhibition of the cell proliferation	1–5 µM	4.8–23	2–10
	Increase in β-Defensin expression	2 µM	10	3.9
PMTDI-related intestinal concentration = 210 nM	Decrease in nutrient absorption	10 µM	48	19
	Decrease in mucin expression	10 µM	48	19
	Increase in intestinal pathogenic Th17	10 µM	48	19
	Increase in bacterial translocation	10 µM	48	19
Highest dose-related intestinal concentration = 504 nM	Increase in IL-8 secretion	1–20 µM	48–95	2–39
	Modification of the microbiota	20 µM	95	39
	Increase in intestinal permeability	10–50 µM	48–245	19–99
	Cytotoxicity	>30 µM	142	59
	Decrease in IL-8 secretion	>30 µM	142	59
	Increase in IgA secretion	500 µM	2 380	992
IMMUNE SYSTEM	Increase in lymphocyte proliferation	1–30 nM	0.6–20	0.3–8
	Activation of macrophages	1–100 nM	0.6–66	0.3–28
PMTDI-related blood concentration = 1.5 nM	Increase in proinflammatory cytokines	0.1–1 µM	66–666	28–277
	Decrease in lymphocyte proliferation	>100 nM	>66	>28
Highest dose-related blood concentration = 3.6 nM	Inhibition of NK cells	>150 nM	>100	>41
	Apoptosis of macrophages	>300 nM	>200	>83
	Apoptosis of lymphocytes	10 µM	6666	2777
ENDOCRINE SYSTEM	Steroid perturbations	0.3–3 µM	200–2 000	83–833
	Increase in insulin secretion	1.44 µM	960	400
	Decrease in IGF-1/IGFALS	1.8 µM	1200	500
PMTDI-related blood concentration = 1.5 nM	Increase in secretion of PYY	7 µM	4666	1944
Highest dose-related blood concentration = 3.6 nM				
BRAIN PMTDI-related brain concentration = 0.45 nM	Feed refusal	1.5–75 nM	3–166	1.4–69
	Activation of microglia	10–100 nM	22–222	9–92
	Inhibition of glutamate uptake	50 nM	111	46
	Cell death/Inhibition of microglia	>300 nM	>666	>277
	Vomiting	1.2 µM	2666	1111
	Direct neuroinflammation	1.5 µM	3333	1388
Highest dose-related brain concentration = 1.08 nM	Cell death of astrocytes	31 µM	68,888	28,703

In addition to directly affecting the activity of nutrient transporters, DON also affects the permeability of the intestinal epithelium through modulation of the tight junction complexes ($IC_{50} = 10$ to 50 µM) [20,39,40,60,97,98,100,108]. Studies have demonstrated that activation of MAP kinases (particularly ERK) by DON affects the expression and cellular localization of proteins forming or being associated with the tight junctions such as claudins, ZO-1, resulting in an increase in the paracellular permeability of the intestine [20,60,97,98,108]. Acetylated DON derivatives are also able to affect the tight junctions through activation of the MAP kinases pathway, a direct correlation existing between their ability to activate MAP kinases and to open tight junctions, with the following order 3ADON < DON < 15ADON [60]. In addition to affecting nutrient absorption and causing intestinal inflammation [14,40], the increase in paracellular intestinal permeability may explain why animals chronically exposed to the toxin have higher DON oral bioavailability [32]. Interestingly, others in addition to us reported that incubation of IEC with doses of DON not able to affect the tight junctions causes a transcellular bacterial translocation across the intestinal epithelium suggesting a possible role of DON as risk factor for inflammatory bowel diseases (IBD) and intestinal bacterial infections [14,40,99,109]. In addition to opening tight junctions and promoting bacterial translocation,

DON also modifies the production of the intestinal mucus. Thus, *in vivo* studies with pigs have shown that ingestion of DON causes a decrease in the number of goblet cells and in the production of mucus [110,111], potentially explaining the observed perturbations of the microbiota in pigs exposed to DON [14,112]. Accordingly, we have preliminary *in vitro* data showing that DON modifies the expression/production of mucins by human cells exposed to 10 μM of toxin (personal communication).

Finally, innate immunity related to IEC is also affected by DON both directly (through the activation of signal pathways by the toxin) and indirectly (through the crossing of luminal bacterial antigens caused by the bacterial translocation, mucus alteration and the opening of the tight junctions) [14,40]. Thus, DON (1 to 20 μM) affects the expression of proteins involved in the epithelial innate immunity, including COX-2 and β-defensins [113–115]. Similarly, numerous studies using animal and human cells have demonstrated that DON stimulates the expression and secretion of interleukin-8 (IL-8), a chemoattractant cytokine causing the recruitment/activation of circulating immune cells and thus potentially participating indirectly in the central effects of DON in terms of feed refusal and emesis. Induction of the intestinal inflammation by DON takes place through the activation of PKR/Hck/MAP kinases/NFκB pathways [40,86,89,99,114,116,117]. Study with human IEC has shown that DON has a biphasic effect on the secretion of IL-8, low doses of toxin (1 to 25 μM, non-cytotoxic) causing a massive increase in the secretion of IL-8, whereas higher doses (50 to 100 μM, cytotoxic) inhibit it [40]. Similarly, as described for immune cells, low doses (10–20 μM) of DON potentate the effects of pro-inflammatory molecules such as cytokines or bacteria components (flagella, LPS) on intestinal IL-8 secretion, whereas higher doses of DON inhibit it [40,116]. Taken together, such a biphasic effect explains why DON acts: (i) as a proinflammatory toxin leading to intestinal inflammation at low doses; and (ii) as an inhibitor of the intestinal immunity leading to higher susceptibility of animals to intestinal infections at higher doses [14,40,118,119].

Taken together, the opening of tight junctions, the increase in the bacterial translocation, and the decrease in the mucus production caused by DON may promote the passage across the intestinal epithelium of xenobiotics (pharmaceutics, pesticides, others mycotoxins), harmful molecules (prion, bacterial toxins, alimentary allergens) and pathogenic micro-organisms (bacteria, fungi, viruses) present in food and water.

As detailed below, in addition to local effects, the alterations by DON of the intestinal functions, including epithelial innate immunity, may have consequences on the systemic immunity (part 3.3.) and on the brain functions (part 3.4.).

3.3. Impacts of DON on the Immune Functions

The second organ system targeted by DON once the toxin has crossed the intestinal epithelium is the immune system. *In vivo* and *in vitro* studies have shown that immune cells (including macrophages, B and T lymphocytes and natural killer (NK) cells) are very sensitive to DON and its toxic derivatives (3/15ADON), exposure to the toxin leading either to immunostimulatory/inflammatory or immunosuppressive effects depending of the dose, as demonstrated with IEC (Figure 9 and Table 1) [5–8,12,93,120,121].

Due to their ability to phagocytose pathogens, to present antigens and to secrete cytokines regulating B/T cells functions, monocytes/macrophages are critical in the immune system as they link together the innate and the acquired immune responses [122]. Macrophages are highly sensitive to DON exposure. Stimulation of macrophages with low doses of DON (nM range) causes their activation, the secretion of inflammatory cytokines such as IL-1β, IL-2, IL-4, IL-5, IL-6 and TNFα and the expression of intracellular proteins involved in the innate immunity such as COX-2 and iNOS through the selective activation of ERK, NFκB and activator protein-1 (AP-1) [5–8,123–125]. In addition to its direct stimulatory effect, DON at low doses also potentates the stimulatory effects of cytokines/bacterial components on macrophages [124,125]. In parallel to macrophage activation, low doses of DON also affect their ability to phagocytose and to kill bacteria, leading either to a decrease or an increase in the phagocytosis depending of the type of bacteria used in the assay [109,126]. As shown with IEC, higher doses of DON (µM range) possess suppressive effects on macrophage activations (cytokine secretion, phagocytosis, bacterial killing) and induce their apoptosis [124,125,127] such deleterious effects certainly contributing to the observed increase in the susceptibility to infection of animals exposed to DON [119,127,128]. As mentioned in part 3.1., it has to be noted that both macrophage activation and apoptosis induced by DON depend on the type of MAP kinases activated, *i.e.*, ERK for the survival/activation signal and p38 for the inhibition/pro-apoptotic signal [6–8]. It is interesting to note that macrophages are the most sensitive cells regarding DON toxicity, such cells being 10 to 100-fold more sensitive compared to other cell types, including fibroblasts, lymphocytes, IEC or astrocytes. Hypotheses could only be formulated regarding such differences, the higher sensitivity of macrophages to DON toxicity relying either on: (i) a potential and unproved higher ability of DON to enter/accumulate in these cells; and/or (ii) on a specific activation of JAK/STAT pathway leading to apoptosis in these cells [129]. In addition to impacting the innate immunity, alterations of the macrophage functions by DON also affect the acquired immune response. Thus, the decrease in the phagocytosis/bacterial killing and cytokine production induced by DON may inhibit the ability of macrophages to play their role as antigen-presenting cells (APC) and to activate B and T cells. Accordingly, macrophage perturbation was proposed to play a role in the aberrant production of IgA by the B cells of the intestinal Peyer's patches [130,131].

Independently of the alterations of the macrophages, DON also affects the proliferation and functions of lymphocytes, including B, T and NK cells.

Natural killer (NK) cells are effector lymphocytes of the innate immunity playing an important role in the immune surveillance against tumors and microbial infections [132]. Low doses of DON (150–300 nM) are able to inhibit the activity of human NK cells suggesting that DON exposure could indirectly favor the emergence of tumors through a decrease in the immune vigilance associated to NK cells, at least in humans [93].

DON also affects lymphocytes of the acquired immunity (B and T cells). At high doses (superior to 10 µM), DON causes the apoptosis of lymphocytes, leading to immuno-suppression, increased susceptibility to infection, reactivation of latent infections and decreased vaccine efficiency [6–8,119,128, 133,134]. At lower doses, DON has a biphasic effect on the mitogen-induced proliferation of human and animal lymphocytes, 1 to 30 nM of toxin stimulating the proliferation, whereas 100 to 600 nM of DON suppress it [134,135]. At low doses (nM), DON also increases the expression of cytokines by lymphocytes, including IL-2, IL-4, IL-6, IL-8 and TNFα [136]. Alterations of the lymphocyte proliferation and of the secretion of particular cytokines may explain the imbalance in the Th1/Th17/Th2 immune responses observed after intoxication of the animals with DON. In mice, DON exposure results in a parallel suppression and stimulation of the systemic Th1 and Th2 immune responses, respectively [119]. Similarly, exposure of intestinal explants from pigs to DON at 10 µM causes a profound alteration of the intestinal Th17 immune response with a selective increase in the expression of genes associated to the pathogenic/inflammatory Th17 cells (*i.e.*, IL-23A, IL-22, IL-21) without affecting the expression of the genes associated to the regulatory/protective Th17 cells (*i.e.*, the anti-inflammatory cytokine IL-10 and TGF-β) [117]. Modification of the secretion of cytokines by T cells and macrophages located in the Peyer's patches could also explain how DON modifies the production of antibodies by the B cells, the exposure to DON being characterized by an increase in the production of IgA and a parallel decrease in the production of IgM and IgG [130,131,134,137]. Importantly, part of the IgA produced after exposure to DON reacts with self-antigens and gut bacteria as observed in IBD [138]. Based on the ability of DON to cause intestinal and immune alterations mimicking the one found in IBD, we proposed in 2010 that DON could play a role in such diseases, our hypothesis being now defended by others and, more importantly, being confirmed by the recent work conducted on pigs by Oswald's group showing the activation of intestinal pathogenic Th17 at 10 µM of DON [14,117].

The effects of DON derivatives on immune cells have been studied. As observed with other cell systems, DOM-1 and glucuronide-DON have been found non-toxic to immune cells [61,139], no studies having tested the effect of D3G. Regarding acetyl-DON derivatives, it has been shown

17

that 3ADON and 15ADON are less toxic than DON to human and mouse lymphocytes [91–93,120], difference compared to DON in their ability to enter the cells and/or to bind to ribosomes potentially explaining it (see part 3.1.).

In addition to affecting the immunity, alterations of the immune cells by DON could affect the intestinal and the brain functions. Indeed, local activation of intestinal immune cells by DON could reinforce the direct proinflammatory effect of DON on IEC through a vicious circle in which IEC and immune cell-mediated inflammations potentate each other as described in IBD [14]. In addition, intestinal and systemic production of cytokines could affect the endocrine system and the brain functions and thus participate in the growth retardation, feed refusal and emesis caused by DON as explained below.

3.4. Impacts of DON on the Brain and Endocrine Functions

Studies have demonstrated that DON affects the nervous and the endocrine systems (Figure 9 and Table 1).

Regarding the endocrine perturbations, it was shown that DON (at 0.3–3 µM) modifies the gene expression, viability and synthesis/secretion of steroid hormones by human adrenocortical cells, causing an increase in the secretion of progesterone and a parallel decrease in the production of testosterone, estradiol and cortisol [140]. Stimulatory effect of DON on the secretion of progesterone was furthermore confirmed in animals, such endocrine perturbation potentially leading to reproductive toxicity [141,142]. Systemic inflammation induced by nanomolar doses of DON also causes the production of suppressors of cytokine signaling (SOCS) able to inhibit the induction by the growth hormone of the hepatic secretion of IGF-1 and IGF acid labile subunit (IGFALS) eventually resulting in growth retardation [143,144]. Finally, DON increases the secretion of insulin and of the gut satiety hormone peptide YY (PYY), two hormones with anorexic action [145,146]. Importantly, antagonist of the PYY receptor partially prevents the anorexigenic effect of DON, showing that PYY plays a role in the anorexia induced by DON [146].

In addition to endocrine perturbations, DON causes perturbations of brain cells. As mentioned in Section 2.5, part of the plasmatic DON is able to cross the BBB to directly act on neurons and glial cells forming the brain [74,75]. An *in vitro* study conducted on brain cells isolated from newborn rats has shown that DON affects the viability and functions of astrocytes and microglial cells [147]. The sensitivity of astrocytes to DON toxicity is similar to the one observed with epithelial cells or lymphocytes (IC_{50} of 31 µM on the cell viability). Microglial cells, in accordance with their origin (monocytes) are much more sensitive to DON toxicity with an IC_{50} of 259 nM on cell survival (more than 100-fold difference compared to astrocytes). Whether or not the higher sensitivity of microglia to DON toxicity relies on JAK/STAT pathway activation as observed for monocytes/macrophages [129] remains to be determined. In addition to affect their viability, DON is also able to modify the functions of glial cells. DON has a biphasic effect on the microglia-associated neuro-inflammation [147]. At doses inferior or equal to 100 nM, DON potentates the neuro-inflammation caused by LPS in terms of iNOS induction and TNF-α secretion. Conversely, at doses superior to 300 nM, DON dose-dependently inhibits the neuro-inflammation induced by LPS certainly through a general cytotoxic effect of DON on microglia [147]. We also found that DON, at doses not causing toxicity to astrocytes, inhibits their ability to reabsorb the excitatory neurotransmitter glutamate through EAAT1/2 transporters (IC_{50} = 50 nM, total inhibition at 1 µM) [147]. Surprisingly and contrarily to what we found with another mycotoxin, ochratoxin A [148], such inhibition is associated to a massive increase in the membrane expression of EAAT1/2 through an unidentified mechanism. Inhibition induced by DON of the glutamate uptake by astrocytes may have major consequences since this activity prevents neuronal damage caused by high excitotoxic extracellular glutamate concentrations [149] and that perturbation of glutamate clearance by astrocytes could also contribute to brain tumor progression [150], pain hypersensitivity [151] and to alterations in learning and memory consolidation [152]. Although very interesting, these *in vitro* data showing the perturbation of glial cells by DON now need to be confirmed by *in vivo* studies.

In vivo studies have shown that DON affects the activity of brain neurons, particularly in relation to anorexia and emesis; exposure of pigs to 10–75 or >150 µg of DON/kg BW (body weight)/day causing partial/total feed refusal or vomiting, respectively (for review: [153]). Importantly, higher doses of DON are required in mice, *i.e.*, 0.5 to 5 mg/kg of BW causing anorexia, suggesting that pigs are more sensitive to brain effects than mice [153]. This could be related to the higher ability of DON to cross the BBB in pigs compared to mice (30% *versus* 10% of the plasmatic DON reaching the CSF in pigs and mice, respectively [74,75]) and/or to a higher sensitivity of pigs to emetic/anorexigenic stimuli, the important question being whether or not humans are closer to pigs or mice regarding the brain effects of DON. Regarding the mechanism involved in feed behavior effects of DON, it was first shown that emesis and anorexia induced by DON rely on central serotoninergic activities, as demonstrated for other emetic molecules [154–156]. An *in vivo* study conducted on rats next identified a role of neurons from the area postrema in the DON-induced conditioned taste aversion [157]. More recently, *in vivo* studies proved that oral exposure to DON at 1 mg/kg of BW and at 6 to 25 mg/kg of BW in pigs and mice, respectively, activates central anorexigenic neurocircuitries, including POMC and nesfatin-1 neurons present in specific area of the brain controlling the food intake and the vomiting [158–160]. Furthermore, it was demonstrated that, in addition to systemic/peripheral inflammation, DON also causes a central neuro-inflammation with an increased expression of proinflammatory molecules in the brain, including IL-1β and TNF-α and the anorexigenic prostaglandin PGE2 synthesized by mPGES-1 [159]. Although it was initially proposed that central and/or peripheral inflammation may cause the DON-induced anorexia as observed with LPS [161], *in vivo* data do not support such a hypothesis. Indeed, inhibition of the TNF-α signaling does not affect DON-induced anorexia [162]. Similarly, the section of the vagus nerve known to prevent the anorexigenic effect of peripheral inflammation induced by LPS does not affect DON-induced brain activation [158]. Finally, DON still causes anorexia in mPGES-1 knock-out mice that are resistant to anorexia induced by LPS, showing that peripheral and central inflammations caused by DON are not involved in the DON-induced anorexia and that, although LPS and DON activate a similar brain area, they use different mechanisms to do so [159]. At present, the exact mechanism involved in DON-induced anorexia is still a mystery. One could speculate, based on an antagonist study, that the intestinal secretion of PYY induced by DON is totally responsible for its anorexigenic action [146]. However, the fact that the direct injection of DON in the CSF leads to activation of the anorexigenic neurons and to anorexia rule out such a hypothesis. Accordingly, although peripheral secretion of PYY could play a role, DON-induced anorexia certainly also depends on the central effect of the toxin independently of its neuro-inflammatory effect [159]. We could propose that DON either activates neurons directly involved in feed refusal and/or affects glial cells regulating anorexigenic neuronal circuitries [163]. Future studies should help confirm such a hypothesis.

Not a lot of studies have looked at the brain effects of DON derivatives. 3ADON and 15ADON possess similar anorexic effects compared to DON, potentially in accordance with the fast and efficient conversion of such derivatives in DON when they enter the body [164]. The absence of DOM-1 and D3/15GA in the CSF of intoxicated animals suggests that such metabolites are not able to cross the BBB or to enter the brain [74].

4. Conclusions: Global View of the Effects of DON and Risk Assessment for Humans Exposed to DON

In vivo and in vitro studies have demonstrated that DON is able to alter the functions of the gut, the immune system, the endocrine system and the brain, modifications of each system happening at specific doses of DON and potentially affecting the functions of the others (Figure 9 and Table 1). DON-induced perturbations of the intestinal functions and of the intestinal immunity are observed at micromolar doses. Although the intestine is thus the less sensitive organ system regarding DON toxicity, we have to remember that the intestine is also the organ system exposed to the higher doses of DON, making DON-induced perturbation of the gut likely in case of ingestion of the toxin. In addition to affecting the gut functions, intestinal effects of DON also lead to alteration of the systemic immunity and of the endocrine/brain systems through the release of proinflammatory cytokines and of gut-associated hormones, such as the anorexigenic hormone PYY. Alterations of the immune system observed at nanomolar to micromolar concentrations of DON, in addition to affecting the immunity, may impact the intestinal and the neuro-endocrine functions through a vicious cycle, as observed in IBD or in the case of the peripheral inflammation caused by LPS [14,161]. Finally, perturbations of the neuro-endocrine system, in addition to causing modifications of the behavior including appetite, in turn affect the gut and the immune system functions through the release of neuro-endocrine mediators. Importantly, DON-induced inflammation of the intestine and brain could increase the permeability of the intestinal and blood-brain barriers and thus increase the crossing of these barriers by DON (and others toxins), ultimately affecting its bioavailability and its toxicity.

The use of highly innovative and promising methods based on the measurement of exposure biomarkers has shown that humans are significantly exposed to DON and its derivatives [165–167]. Although DON-induced perturbations have been demonstrated both *in vivo* and *in vitro*, one major question remains: are the doses causing such alterations realistic? To be as straightforward as possible, are the doses causing intestinal/immune/neuro-endocrine effects susceptible to be reached in humans exposed to food contaminated by DON? To address this question, we compare in Table 1: (i) the concentrations of DON potentially found in the intestinal lumen, the blood and the CSF, based on its provisional maximum tolerable daily intake (PMTDI) and the higher range of exposure in adult and children to DON obtained from the Joint FAO/WHO Expert Committee on Food Additives (JECFA) [14]; to (ii) the doses of DON required to cause alterations in the gut, the immune/endocrine system or the brain. Concentrations of DON in the intestinal lumen, the blood and the CSF have been estimated using the PMTDI/higher exposure of DON and assuming that: (i) a human adult has a body weight of 70 kg, a global small intestinal volume of 1 L (considering the net intake/secretion (around 9 L) and absorption (around 8 L) of water by the gut), and a blood volume of 5 L; and that (ii) toxicokinetics data obtained with pigs orally exposed to DON could be extrapolated to humans [14,74]. On the basis of a PMTDI of 1 µg/kg of BW per day for DON, toxin concentrations should be: 210, 1.5 and 0.45 nM in the intestinal lumen, the blood and the CSF, respectively. On the basis of the worldwide higher exposure in adult and children to DON obtained from the Joint FAO/WHO Expert Committee on Food Additives (JECFA) (0.78 to 2.4 µg/kg of BW per day), DON concentrations would reach maximal values of 504, 3.6 and 1.08 nM in the intestine, the blood and the CSF, respectively.

From the analysis of Table 1, it clearly appears that, as suggested by others and us [14,168], DON represents a risk to human health based on the presence of a low safety factor (inferior to 30) between the doses of DON affecting cell functions and the doses of DON susceptible to be present in relation to its actual PMTDI. The risk concerns mainly the intestinal and immune systems and the brain; DON effects on the endocrine system are being unlikely to be observed in humans exposed to DON at doses close to its PMTDI. Importantly, the risk could be even higher than supposed since the toxicokinetic profile (intestinal absorption, detoxification, excretion, BBB crossing) and/or the cellular effects of DON could be affected by factors not considered in our calculation. This includes: (i) the concomitant presence in food of others xenobiotics and toxins such as drugs, heavy metals, pesticides, bacterial/plant toxins or others mycotoxins [169]; and (ii) the exposure of particular populations to DON, including: vegans/macrobiotics, children and patients suffering from bacterial/viral infection, renal/hepatic diseases, IBD, compromised immunity, neurological disorders or cancers, these populations being at higher risk regarding DON effects [14,170].

Taken together, such observations should alert food agencies and potentially lead to the reevaluation of the actual PMTDI for DON, particularly as new DON metabolites have been found in plants and food products, including DON-oligoglycosides, DON-glutathione, DON-*S*-Cysteine, DON-*S*-Cysteinyl-glycine, DON-sulfonate. Such derivatives represent new "masked" toxins not yet considered in the total intake of DON and for which few or no data are available regarding their intestinal transformation/absorption and their cellular toxicity [139,171,172].

Acknowledgments: Author would like to thanks Gilles Iacazio, Michel Mafféi, Alain Archelas and Fabien Graziani for helpful discussions and Josette Perrier as well as Marius Réglier, for their support. Author also would like to thanks Cendrine Nicoletti that provided microscopy images.

Conflicts of Interest: The author declares no conflict of interest.

References

1. Pitt, J.I. Toxigenic fungi: Which are important? *Med. Mycol.* **2000**, *38*, 17–22.
2. Ueno, Y.; Nakajima, M.; Sakai, K.; Ishii, K.; Sato, N. Comparative toxicology of trichothec mycotoxins: Inhibition of protein synthesis in animal cells. *J. Biochem.* **1973**, *74*, 285–296.
3. Cundliffe, E.; Cannon, M.; Davies, J. Mechanism of inhibition of eukaryotic protein synthesis by trichothecene fungal toxins. *Proc. Natl. Acad. Sci. USA* **1974**, *71*, 30–34. [CrossRef]
4. Rotter, B.A.; Prelusky, D.B.; Pestka, J.J. Toxicology of deoxynivalenol (vomitoxin). *J. Toxicol. Environ. Health* **1996**, *48*, 1–34.
5. Pestka, J.J.; Zhou, H.R.; Moon, Y.; Chung, Y.J. Cellular and molecular mechanisms for immune modulation by deoxynivalenol and other trichothecenes: Unraveling a paradox. *Toxicol. Lett.* **2004**, *153*, 61–73. [CrossRef]
6. Pestka, J.J. Mechanisms of deoxynivalenol-induced gene expression and apoptosis. *Food Addit. Contam. Part A* **2008**, *25*, 1128–1140. [CrossRef]
7. Pestka, J.J. Deoxynivalenol: Mechanisms of action, human exposure, and toxicological relevance. *Arch. Toxicol.* **2010**, *84*, 663–679. [CrossRef]
8. Pestka, J.J. Deoxynivalenol-induced proinflammatory gene expression: Mechanisms and pathological sequelae. *Toxins* **2010**, *2*, 1300–1317. [CrossRef]
9. Streit, E.; Schatzmayr, G.; Tassis, P.; Tzika, E.; Marin, D.; Taranu, I.; Tabuc, C.; Nicolau, A.; Aprodu, I.; Puel, O.; *et al.* Current situation of mycotoxin contamination and co-occurrence in animal feed—Focus on Europe. *Toxins* **2012**, *4*, 788–809. [CrossRef]
10. Canady, R.; Coker, R.; Rgan, S.; Krska, R.; Kuiper-Goodman, T.; Olsen, M. Deoxynivalenol safety evaluation of certain mycotoxins in food. *WHO Food Addit. Ser.* **2001**, *47*, 420–555.
11. Schothorst, R.C.; van Egmond, H.P. Report from SCOOP task 3.2.10 "collection of occurrence data of *Fusarium* toxins in food and assessment of dietary intake by the population of EU member states"—Subtask: Trichothecenes. *Toxicol. Lett.* **2004**, *153*, 133–143. [CrossRef]
12. Monbaliu, S.; van Poucke, C.; Detavernier, C.; Dumoulin, F.; van de Velde, M.; Schoeters, E.; van Dyck, S.; Averkieva, O.; van Peteghem, C.; de Saeger, S. Occurrence of mycotoxins in feed as analyzed by a multi-mycotoxin LC-MS/MS Method. *J. Agric. Food Chem.* **2010**, *58*, 66–71. [CrossRef]
13. Pestka, J.J.; Smolinski, A.T. Deoxynivalenol: Toxicology and potential effects on humans. *J. Toxicol. Environ. Health B* **2005**, *8*, 39–69. [CrossRef]
14. Maresca, M.; Fantini, J. Some food-associated mycotoxins as potential risk factors in humans predisposed to chronic intestinal inflammatory diseases. *Toxicon* **2010**, *56*, 282–294. [CrossRef]
15. Sobrova, P.; Adam, V.; Vasatkova, A.; Beklova, M.; Zeman, L.; Kizek, R. Deoxynivalenol and its toxicity. *Interdiscip. Toxicol.* **2010**, *3*, 94–99.
16. Pinton, P.; Guzylack-Piriou, L.; Kolf-Clauw, M.; Oswald, I.P. The effect on the intestine of some fungal toxins: The trichothecenes. *Curr. Immunol. Rev.* **2012**, *8*, 193–208. [CrossRef]
17. Bryden, W.L. Mycotoxins in the food chain: Human health implications. *Asia Pac. J. Clin. Nutr.* **2007**, *16*, 95–101.
18. Karlovsky, P. Biological detoxification of the mycotoxin deoxynivalenol and its use in genetically engineered crops and feed additives. *Appl. Microbiol. Biotechnol.* **2011**, *91*, 491–504. [CrossRef]
19. Berthiller, F.; Crews, C.; Dall'Asta, C.; de Saeger, S.; Haesaert, G.; Karlovsky, P.; Oswald, I.P.; Seefelder, W.; Speijers, G.; Stroka, J. Masked mycotoxins: A review. *Mol. Nutr. Food Res.* **2013**, *57*, 165–186. [CrossRef]
20. Sergent, T.; Parys, M.; Garsou, S.; Pussemier, L.; Schneider, Y.J.; Larondelle, Y. Deoxynivalenol transport across human intestinal Caco-2 cells and its effects on cellular metabolism at realistic intestinal concentrations. *Toxicol. Lett.* **2006**, *164*, 167–176. [CrossRef]

21. Videmann, B.; Tep, J.; Cavret, S.; Lecoeur, S. Epithelial transport of deoxynivalenol: Involvement of human *P*-glycoprotein (ABCB1) and multidrug resistance-associated protein 2 (ABCC2). *Food Chem. Toxicol.* **2007**, *45*, 1938–1947. [CrossRef]

22. Riley, R.J.; Martin, I.J.; Cooper, A.E. The influence of DMPK as an integrated partner in modern drug discovery. *Curr. Drug Metab.* **2002**, *3*, 527–550. [CrossRef]

23. Větvicka, V.; Fornůsek, L. Limitations of transmembrane transport in drug delivery. *Crit. Rev. Ther. Drug Carrier Syst.* **1988**, *5*, 141–170.

24. Dobson, P.D.; Lanthaler, K.; Oliver, S.G.; Kell, D.B. Implications of the dominant role of transporters in drug uptake by cells. *Curr. Top. Med. Chem.* **2009**, *9*, 163–181. [CrossRef]

25. Chow, J.; Lee, S.M.; Shen, Y.; Khosravi, A.; Mazmanian, S.K. Host-bacterial symbiosis in health and disease. *Adv. Immunol.* **2010**, *107*, 243–274. [CrossRef]

26. Barnett, A.M.; Roy, N.C.; McNabb, W.C.; Cookson, A.L. The interactions between endogenous bacteria, dietary components and the mucus layer of the large bowel. *Food Funct.* **2012**, *3*, 690–699.

27. Walter, J.; Ley, R. The human gut microbiome: Ecology and recent evolutionary changes. *Annu. Rev. Microbiol.* **2011**, *65*, 411–429.

28. Frey, J.C.; Pell, A.N.; Berthiaume, R.; Lapierre, H.; Lee, S.; Ha, J.K.; Mendell, J.E.; Angert, E.R. Comparative studies of microbial populations in the rumen, duodenum, ileum and faeces of lactating dairy cows. *J. Appl. Microbiol.* **2010**, *108*, 1982–1993.

29. Smith, H.W. Observations on the flora of the alimentary tract of animals and factors affecting its composition. *J. Pathol. Bacteriol.* **1965**, *89*, 95–122. [CrossRef]

30. Farner, D.S. The hydrogen ion concentration in avian digestive tracts. *Poultr. Sci.* **1942**, *21*, 445–450. [CrossRef]

31. Ao, T.; Cantor, A.H.; Pescatore, A.J.; Pierce, J.L. *In vitro* evaluation of feed-grade enzyme activity at pH levels simulating various parts of the avian digestive tract. *Anim. Feed Sci. Technol.* **2008**, *140*, 462–468. [CrossRef]

32. Goyarts, T.; Dänicke, S. Bioavailability of the Fusarium toxin deoxynivalenol (DON) from naturally contaminated wheat for the pig. *Toxicol. Lett.* **2006**, *163*, 171–182. [CrossRef]

33. Prelusky, D.B.; Hartin, K.E.; Trenholm, H.L.; Miller, J.D. Pharmacokinetic fate of 14C-labeled deoxynivalenol in swine. *Fundam. Appl. Toxicol.* **1988**, *10*, 276–286. [CrossRef]

34. Dänicke, S.; Valenta, H.; Döll, S. On the toxicokinetics and the metabolism of deoxynivalenol (DON) in the pig. *Arch. Anim. Nutr.* **2004**, *58*, 169–180. [CrossRef]

35. Avantaggiato, G.; Havenaar, R.; Visconti, A. Evaluation of the intestinal absorption of deoxynivalenol and nivalenol by an *in vitro* gastrointestinal model, and the binding efficacy of activated carbon and other adsorbent materials. *Food Chem. Toxicol.* **2004**, *42*, 817–824. [CrossRef]

36. Yee, S. *In vitro* permeability across Caco-2 cells (colonic) can predict *in vivo* (small intestinal) absorption in man—Fact or myth. *Pharm. Res.* **1997**, *14*, 763–766. [CrossRef]

37. Awad, W.A.; Aschenbach, J.R.; Setyabudi, F.M.; Razzazi-Fazeli, E.; Böhm, J.; Zentek, J. *In vitro* effects of deoxynivalenol on small intestinal D-glucose uptake and absorption of deoxynivalenol across the isolated jejunal epithelium of laying hens. *Poult. Sci.* **2007**, *86*, 15–20.

38. Maresca, M.; Mahfoud, R.; Pfohl-Leszkowicz, A.; Fantini, J. The mycotoxin ochratoxin A alters intestinal barrier and absorption functions but has no effect on chloride secretion. *Toxicol. Appl. Pharmacol.* **2001**, *176*, 54–63. [CrossRef]

39. Maresca, M.; Mahfoud, R.; Garmy, N.; Fantini, J. The mycotoxin deoxynivalenol affects nutrient absorption in human intestinal epithelial cells. *J. Nutr.* **2002**, *132*, 2723–2731.

40. Maresca, M.; Yahi, N.; Younès-Sakr, L.; Boyron, M.; Caporiccio, B.; Fantini, J. Both direct and indirect effects account for the pro-inflammatory activity of enteropathogenic mycotoxins on the human intestinal epithelium: Stimulation of interleukin-8 secretion, potentiation of interleukin-1beta effect and increase in the transepithelial passage of commensal bacteria. *Toxicol. Appl. Pharmacol.* **2008**, *228*, 84–92. [CrossRef]

41. Dean, P.; Maresca, M.; Schüller, S.; Phillips, A.D.; Kenny, B. Potent diarrheagenic mechanism mediated by the cooperative action of three enteropathogenic *Escherichia coli*-injected effector proteins. *Proc. Natl. Acad. Sci. USA* **2005**, *103*, 1876–1881.

42. Guttman, J.A.; Finlay, B.B. Tight junctions as targets of infectious agents. *Biochim. Biophys. Acta* **2009**, *1788*, 832–841. [CrossRef]

43. Schulzke, J.D.; Günzel, D.; John, L.J.; Fromm, M. Perspectives on tight junction research. *Ann. N.Y. Acad. Sci.* **2012**, *1257*, 1–19. [CrossRef]

44. Takahashi, A.; Kondoh, M.; Suzuki, H.; Watari, A.; Yagi, K. Pathological changes in tight junctions and potential applications into therapies. *Drug Discov. Today* **2012**, *17*, 727–732. [CrossRef]

45. Worrell, N.R.; Mallett, A.K.; Cook, W.M.; Baldwin, N.C.; Shepherd, M.J. The role of gut micro-organisms in the metabolism of deoxynivalenol administered to rats. *Xenobiotica* **1989**, *19*, 25–32. [CrossRef]

46. Prelusky, D.B.; Veira, D.M.; Trenholm, H.L. Plasma pharmacokinetics of the mycotoxin deoxynivalenol following oral and intravenous administration to sheep. *J. Environ. Sci. Health B* **1985**, *20*, 603–624. [CrossRef]

47. Prelusky, D.B.; Veira, D.M.; Trenholm, H.L.; Hartin, K.E. Excretion profiles of the mycotoxin deoxynivalenol, following oral and intravenous administration to sheep. *Fundam. Appl. Toxicol.* **1986**, *6*, 356–363.

48. Osselaere, A.; Devreese, M.; Goossens, J.; Vandenbroucke, V.; de Baere, S.; de Backer, P.; Croubels, S. Toxicokinetic study and absolute oral bioavailability of deoxynivalenol, T-2 toxin and zearalenone in broiler chickens. *Food Chem. Toxicol.* **2013**, *51*, 350–355.

49. Dänicke, S.; Matthäus, K.; Lebzien, P.; Valenta, H.; Stemme, K.; Ueberschär, K.H.; Razzazi-Fazeli, E.; Böhm, J.; Flachowsky, G. Effects of Fusarium toxin-contaminated wheat grain on nutrient turnover, microbial protein synthesis and metabolism of deoxynivalenol and zearalenone in the rumen of dairy cows. *J. Anim. Physiol. Anim. Nutr.* **2005**, *89*, 303–315. [CrossRef]

50. Prelusky, D.B.; Trenholm, H.L.; Lawrence, G.A.; Scott, P.M. Nontransmission of deoxynivalenol (vomitoxin) to milk following oral administration to dairy cows. *J. Environ. Sci. Health B* **1984**, *19*, 593–609. [CrossRef]

51. Eriksen, G.S.; Pettersson, H.; Johnsen, K.; Lindberg, J.E. Transformation of trichothecenes in ileal digesta and faeces from pigs. *Arch. Tierernahr.* **2002**, *56*, 263–274. [CrossRef]

52. Eriksen, G.S.; Pettersson, H.; Lindberg, J.E. Absorption, metabolism and excretion of 3-acetyl DON in pigs. *Arch. Tierernahr.* **2003**, *57*, 335–345.

53. Gratz, S.W.; Duncan, G.; Richardson, A.J. Human fecal microbiota metabolize deoxynivalenol and deoxynivalenol-3-glucoside and may be responsible for urinary de-epoxy deoxynivalenol. *Appl. Environ. Microbiol.* **2013**, *79*, 1821–1825. [CrossRef]

54. Hedman, R.; Pettersson, H. Transformation of nivalenol by gastrointestinal microbes. *Arch. Tierernahr.* **1997**, *50*, 321–329. [CrossRef]

55. Sato, I.; Ito, M.; Ishizaka, M.; Ikunaga, Y.; Sato, Y.; Yoshida, S.; Koitabashi, M.; Tsushima, S. Thirteen novel deoxynivalenol-degrading bacteria are classified within two genera with distinct degradation mechanisms. *FEMS Microbiol. Lett.* **2012**, *327*, 110–117. [CrossRef]

56. Islam, R.; Zhou, T.; Young, J.C.; Goodwin, P.H.; Pauls, K.P. Aerobic and anaerobic de-epoxydation of mycotoxin deoxynivalenol by bacteria originating from agricultural soil. *World J. Microbiol. Biotechnol.* **2012**, *28*, 7–13. [CrossRef]

57. Udell, M.N.; Dewick, P.M. Metabolic conversions of trichothecene mycotoxins: De-esterification reactions using cell-free extracts of Fusarium. *Z. Naturforsch. C* **1989**, *44*, 660–668.

58. Jaeger, K.E.; Reetz, M.T. Microbial lipases form versatile tools for biotechnology. *Trends Biotechnol.* **1998**, *16*, 396–403. [CrossRef]

59. Hänninen, O.; Lindström-Seppä, P.; Pelkonen, K. Role of gut in xenobiotic metabolism. *Arch. Toxicol.* **1987**, *60*, 34–36. [CrossRef]

60. Pinton, P.; Tsybulskyy, D.; Lucioli, J.; Laffitte, J.; Callu, P.; Lyazhri, F.; Grosjean, F.; Bracarense, A.P.; Kolf-Clauw, M.; Oswald, I.P. Toxicity of deoxynivalenol and its acetylated derivatives on the intestine: Differential effects on morphology, barrier function, tight junction proteins, and mitogen-activated protein kinas. *Toxicol. Sci.* **2012**, *130*, 180–190. [CrossRef]

61. Wu, X.; Murphy, P.; Cunnick, J.; Hendrich, S. Synthesis and characterization of deoxynivalenol glucuronide: Its comparative immunotoxicity with deoxynivalenol. *Food Chem. Toxicol.* **2007**, *45*, 1846–1855.

62. Poppenberger, B.; Berthiller, F.; Lucyshyn, D.; Sieberer, T.; Schuhmacher, R.; Krska, R.; Kuchler, K.; Glössl, J.; Luschnig, C.; Adam, G. Detoxification of the Fusarium mycotoxin deoxynivalenol by a UDP-glucosyltransferase from *Arabidopsis thaliana*. *J. Biol. Chem.* **2003**, *278*, 47905–47914. [CrossRef]

63. Dall'erta, A.; Cirlini, M.; Dall'asta, M.; del Rio, D.; Galaverna, G.; Dall'asta, C. Masked mycotoxins are efficiently hydrolyzed by human colonic microbiota releasing their aglycones. *Chem. Res. Toxicol.* **2013**, *26*, 305–312. [CrossRef]

64. Nagl, V.; Schwartz, H.; Krska, R.; Moll, W.D.; Knasmüller, S.; Ritzmann, M.; Adam, G.; Berthiller, F. Metabolism of the masked mycotoxin deoxynivalenol-3-glucoside in rats. *Toxicol. Lett.* **2012**, *213*, 367–373. [CrossRef]

65. Côté, L.M.; Buck, W.; Jeffery, E. Lack of hepatic microsomal metabolism of deoxynivalenol and its metabolite, DOM-1. *Food Chem. Toxicol.* **1987**, *25*, 291–295. [CrossRef]

66. Maul, R.; Warth, B.; Kant, J.S.; Schebb, N.H.; Krska, R.; Koch, M.; Sulyok, M. Investigation of the hepatic glucuronidation pattern of the Fusarium mycotoxin deoxynivalenol in various species. *Chem. Res. Toxicol.* **2012**, *25*, 2715–2717.

67. Uhlig, S.; Ivanova, L.; Fæste, C.K. Enzyme-assisted synthesis and structural characterization of the 3-, 8- and 15-glucuronides of deoxynivalenol. *J. Agric. Food Chem.* **2013**, *61*, 2006–2012. [CrossRef]

68. Turner, P.C.; Hopton, R.P.; White, K.L.; Fisher, J.; Cade, J.E.; Wild, C.P. Assessment of deoxynivalenol metabolite profiles in UK adults. *Food Chem. Toxicol.* **2011**, *49*, 132–135. [CrossRef]

69. Warth, B.; Sulyok, M.; Fruhmann, P.; Berthiller, F.; Schuhmacher, R.; Hametner, C.; Adam, G.; Fröhlich, J.; Krska, R. Assessment of human deoxynivalenol exposure using an LC-MS/MS based biomarker method. *Toxicol. Lett.* **2012**, *211*, 85–90. [CrossRef]

70. Chu, F.S. Interaction of ochratoxin A with bovine serum albumin. *Arch. Biochem. Biophys.* **1971**, *147*, 359–366. [CrossRef]

71. Hagelberg, S.; Hult, K.; Fuchs, R. Toxicokinetics of ochratoxin A in several species and its plasma-binding properties. *J. Appl. Toxicol.* **1989**, *9*, 91–96.

72. Li, Y.; Wang, H.; Jia, B.; Liu, C.; Liu, K.; Qi, Y.; Hu, Z. Study of the interaction of deoxynivalenol with human serum albumin by spectroscopic technique and molecular modelling. *Food Addit. Contam. Part A* **2013**, *30*, 356–364. [CrossRef]

73. Nico, B.; Ribatti, D. Morphofunctional aspects of the blood-brain barrier. *Curr. Drug Metab.* **2012**, *13*, 50–60. [CrossRef]

74. Prelusky, D.B.; Hartin, K.E.; Trenholm, H.L. Distribution of deoxynivalenol in cerebral spinal fluid following administration to swine and sheep. *J. Environ. Sci. Health B* **1990**, *25*, 395–413.

75. Pestka, J.J.; Islam, Z.; Amuzie, C.J. Immunochemical assessment of deoxynivalenol tissue distribution following oral exposure in the mouse. *Toxicol. Lett.* **2008**, *178*, 83–87. [CrossRef]

76. Diaz, G.J.; Boermans, H.J. Fumonisin toxicosis in domestic animals: A review. *Vet. Hum. Toxicol.* **1994**, *36*, 548–555.

77. Foreman, J.H.; Constable, P.D.; Waggoner, A.L.; Levy, M.; Eppley, R.M.; Smith, G.W.; Tumbleson, M.E.; Haschek, W.M. Neurologic abnormalities and cerebrospinal fluid changes in horses administered fumonisin B1 intravenously. *J. Vet. Intern. Med.* **2004**, *18*, 223–230.

78. Osuchowski, M.F.; He, Q.; Sharma, R.P. Endotoxin exposure alters brain and liver effects of fumonisin B1 in BALB/c mice: Implication of blood brain barrier. *Food Chem. Toxicol.* **2005**, *43*, 1389–1397.

79. Swenberg, J.A.; Lu, K.; Moeller, B.C.; Gao, L.; Upton, P.B.; Nakamura, J.; Starr, T.B. Endogenous *versus* exogenous DNA adducts: Their role in carcinogenesis, epidemiology, and risk assessment. *Toxicol. Sci.* **2011**, *120*, S130–S145. [CrossRef]

80. Bedard, L.L.; Massey, T.E. Aflatoxin B1-induced DNA damage and its repair. *Cancer Lett.* **2006**, *241*, 174–183. [CrossRef]

81. Sabbioni, G.; Skipper, P.L.; Büchi, G.; Tannenbaum, S.R. Isolation and characterization of the major serum albumin adduct formed by aflatoxin B1 *in vivo* in rats. *Carcinogenesis.* **1987**, *8*, 819–824. [CrossRef]

82. He, K.; Zhou, H.R.; Pestka, J.J. Mechanisms for ribotoxin-induced ribosomal RNA cleavage. *Toxicol. Appl. Pharmacol.* **2012**, *265*, 10–18.

83. He, K.; Zhou, H.R.; Pestka, J.J. Targets and intracellular signaling mechanisms for deoxynivalenol-induced ribosomal RNA cleavage. *Toxicol. Sci.* **2012**, *127*, 382–390. [CrossRef]

84. Yang, G.H.; Li, S.; Pestka, J.J. Down-regulation of the endoplasmic reticulum chaperone GRP78/BiP by vomitoxin (Deoxynivalenol). *Toxicol. Appl. Pharmacol.* **2000**, *162*, 207–217. [CrossRef]

85. Yang, H.; Park, S.H.; Choi, H.J.; Moon, Y. Epithelial cell survival by activating transcription factor 3 (ATF3) in response to chemical ribosome-inactivating stress. *Biochem. Pharmacol.* **2009**, *77*, 1105–1115. [CrossRef]

86. Moon, Y.; Yang, H.; Lee, S.H. Modulation of early growth response gene 1 and interleukin-8 expression by ribotoxin deoxynivalenol (vomitoxin) via ERK1/2 in human epithelial intestine 407 cells. *Biochem. Biophys. Res. Commun.* **2007**, *362*, 256–262. [CrossRef]

87. Ansari, K.I.; Hussain, I.; Das, H.K.; Mandal, S.S. Overexpression of human histone ethylase MLL1 upon exposure to a food contaminant mycotoxin, deoxynivalenol. *FEBS J.* **2009**, *276*, 3299–3307. [CrossRef]

88. Choi, H.J.; Yang, H.; Park, S.H.; Moon, Y. HuR/ELAVL1 RNA binding protein modulates interleukin-8 induction by muco-active ribotoxin deoxynivalenol. *Toxicol. Appl. Pharmacol.* **2009**, *240*, 46–54. [CrossRef]

89. Park, S.H.; Choi, H.J.; Yang, H.; Do, K.H.; Kim, J.; Moon, Y. Repression of peroxisome proliferator-activated receptor gamma by mucosal ribotoxic insult-activated CAAT/enhancer-binding protein homologous protein. *J. Immunol.* **2010**, *185*, 5522–5530. [CrossRef]

90. Pan, X.; Whitten, D.A.; Wu, M.; Chan, C.; Wilkerson, C.G.; Pestka, J.J. Global protein phosphorylation dynamics during deoxynivalenol-induced ribotoxic stress response in the macrophage. *Toxicol. Appl. Pharmacol.* **2013**, *268*, 201–211. [CrossRef]

91. Forsell, J.H.; Pestka, J.J. Relation of 8-ketotrichothecene and zearalenone analog structure to inhibition of mitogen-induced human lymphocyte blastogenesis. *Appl. Environ. Microbiol.* **1985**, *50*, 1304–1307.

92. Bondy, G.S.; McCormick, S.P.; Beremand, M.N.; Pestka, J.J. Murine lymphocyte proliferation impaired by substituted neosolaniols and calonectrins—Fusarium metabolites associated with trichothecene biosynthesis. *Toxicon* **1991**, *29*, 1107–1113.

93. Berek, L.; Petri, I.B.; Mesterházy, A.; Téren, J.; Molnár, J. Effects of mycotoxins on human immune functions *in vitro*. *Toxicol. in Vitro* **2001**, *15*, 25–30. [CrossRef]

94. Forsell, J.H.; Jensen, R.; Tai, J.H.; Witt, M.; Lin, W.S.; Pestka, J.J. Comparison of acute toxicities of deoxynivalenol (vomitoxin) and 15-acetyldeoxynivalenol in the B6C3F1 mouse. *Food Chem. Toxicol.* **1987**, *25*, 155–162.

95. Bouhet, S.; Oswald, I.P. The intestine as a possible target for fumonisin toxicity. *Mol. Nutr. Food Res.* **2007**, *51*, 925–931. [CrossRef]

96. Instanes, C.; Hetland, G. Deoxynivalenol (DON) is toxic to human colonic, lung and monocytic cell lines, but does not increase the IgE response in a mouse model for allergy. *Toxicology* **2004**, *204*, 13–21. [CrossRef]

97. Diesing, A.K.; Nossol, C.; Panther, P.; Walk, N.; Post, A.; Kluess, J.; Kreutzmann, P.; Dänicke, S.; Rothkötter, H.J.; Kahlert, S. Mycotoxin deoxynivalenol (DON) mediates biphasic cellular response in intestinal porcine epithelial cell lines IPEC-1 and IPEC-J2. *Toxicol. Lett.* **2011**, *200*, 8–18. [CrossRef]

98. Diesing, A.K.; Nossol, C.; Dänicke, S.; Walk, N.; Post, A.; Kahlert, S.; Rothkötter, H.J.; Kluess, J. Vulnerability of polarised intestinal porcine epithelial cells to mycotoxin deoxynivalenol depends on the route of application. *PLoS One* **2011**, *6*, e17472. [CrossRef]

99. Vandenbroucke, V.; Croubels, S.; Martel, A.; Verbrugghe, E.; Goossens, J.; van Deun, K.; Boyen, F.; Thompson, A.; Shearer, N.; de Backer, P.; *et al.* The mycotoxin deoxynivalenol potentiates intestinal inflammation by *Salmonella typhimurium* in porcine ileal loops. *PLoS One* **2011**, *6*, e23871. [CrossRef]

100. Bianco, G.; Fontanella, B.; Severino, L.; Quaroni, A.; Autore, G.; Marzocco, S. Nivalenol and deoxynivalenol affect rat intestinal epithelial cells: A concentration related study. *PLoS One* **2012**, *7*, e52051.

101. Diesing, A.K.; Nossol, C.; Ponsuksili, S.; Wimmers, K.; Kluess, J.; Walk, N.; Post, A.; Rothkötter, H.J.; Kahlert, S. Gene regulation of intestinal porcine epithelial cells IPEC-J2 is dependent on the site of deoxynivalenol toxicological action. *PLoS One* **2012**, *7*, e34136. [CrossRef]

102. Awad, W.A.; Razzazi-Fazeli, E.; Böhm, J.; Zentek, J. Influence of deoxynivalenol on the D-glucose transport across the isolated epithelium of different intestinal segments of laying hens. *J. Anim. Physiol. Anim. Nutr.* **2007**, *91*, 175–180.

103. Awad, W.A.; Razzazi-Fazeli, E.; Böhm, J.; Zentek, J. Effects of B-trichothecenes on luminal glucose transport across the isolated jejunal epithelium of broiler chickens. *J. Anim. Physiol. Anim. Nutr.* **2008**, *92*, 225–230. [CrossRef]

104. Hunder, G.; Schümann, K.; Strugala, G.; Gropp, J.; Fichtl, B.; Forth, W. Influence of subchronic exposure to low dietary deoxynivalenol, a trichothecene mycotoxin, on intestinal absorption of nutrients in mice. *Food Chem.Toxicol.* **1991**, *9*, 809–814.

105. Meinild, A.; Klaerke, D.A.; Loo, D.D.; Wright, E.M.; Zeuthen, T. The human Na$^+$-glucose cotransporter is a molecular water pump. *J. Physiol.* **1998**, *508*, 15–21.

106. Amador, P.; García-Herrera, J.; Marca, M.C.; de La Osada, J.; Acín, S.; Navarro, M.A.; Salvador, M.T.; Lostao, M.P.; Rodríguez-Yoldi, M.J. Inhibitory effect of TNF-alpha on the intestinal absorption of galactose. *J. Cell. Biochem.* **2007**, *101*, 99–111.

107. Amador, P.; Marca, M.C.; García-Herrera, J.; Lostao, M.P.; Guillén, N.; de La Osada, J.; Rodríguez-Yoldi, M.J. Lipopolysaccharide induces inhibition of galactose intestinal transport in rabbits *in vitro*. *Cell. Physiol. Biochem.* **2008**, *22*, 715–724. [CrossRef]

108. Pinton, P.; Nougayrède, J.P.; del Rio, J.C.; Moreno, C.; Marin, D.E.; Ferrier, L.; Bracarense, A.P.; Kolf-Clauw, M.; Oswald, I.P. The food contaminant deoxynivalenol, decreases intestinal barrier permeability and reduces claudin expression. *Toxicol. Appl. Pharmacol.* **2009**, *237*, 41–48.

109. Vandenbroucke, V.; Croubels, S.; Verbrugghe, E.; Boyen, F.; de Backer, P.; Ducatelle, R.; Rychlik, I.; Haesebrouck, F.; Pasmans, F. The mycotoxin deoxynivalenol promotes uptake of *Salmonella typhimurium* in porcine macrophages, associated with ERK1/2 induced cytoskeleton reorganization. *Vet. Res.* **2009**, *40*, 64. [CrossRef]

110. Obremski, K.; Zielonka, L.; Gajecka, M.; Jakimiuk, E.; Bakuła, T.; Baranowski, M.; Gajecki, M. Histological estimation of the small intestine wall after administration of feed containing deoxynivalenol, T-2 toxin and zearalenone in the pig. *Pol. J. Vet. Sci.* **2008**, *11*, 339–345.

111. Bracarense, A.P.; Lucioli, J.; Grenier, B.; Drociunas Pacheco, G.; Moll, W.D.; Schatzmayr, G.; Oswald, I.P. Chronic ingestion of deoxynivalenol and fumonisin, alone or in interaction, induces morphological and immunological changes in the intestine of piglets. *Br. J. Nutr.* **2012**, *107*, 1776–1786. [CrossRef]

112. Waché, Y.J.; Valat, C.; Postollec, G.; Bougeard, S.; Burel, C.; Oswald, I.P.; Fravalo, P. Impact of deoxynivalenol on the intestinal microflora of pigs. *Int. J. Mol. Sci.* **2009**, *10*, 1–17.

113. Krishnaswamy, R.; Devaraj, S.N.; Padma, V.V. Lutein protects HT-29 cells against Deoxynivalenol-induced oxidative stress and apoptosis: Prevention of NF-kappaB nuclear localization and down regulation of NF-kappaB and Cyclo-Oxygenase-2 expression. *Free Radic. Biol. Med.* **2010**, *49*, 50–60. [CrossRef]

114. Van de Walle, J.; During, A.; Piront, N.; Toussaint, O.; Schneider, Y.J.; Larondelle, Y. Physio-pathological parameters affect the activation of inflammatory pathways by deoxynivalenol in Caco-2 cells. *Toxicol. in Vitro* **2010**, *24*, 1890–1898. [CrossRef]

115. Wan, M.L.; Woo, C.S.; Allen, K.J.; Turner, P.C.; El-Nezami, H. Modulation of porcine β-defensins 1 and 2 upon individual and combined Fusarium toxin exposure in a swine jejunal epithelial cell line. *Appl. Environ. Microbiol.* **2013**, *79*, 2225–2232. [CrossRef]

116. Van de Walle, J.; Romier, B.; Larondelle, Y.; Schneider, Y.J. Influence of deoxynivalenol on NF-kappaB activation and IL-8 secretion in human intestinal Caco-2 cells. *Toxicol. Lett.* **2008**, *177*, 205–214. [CrossRef]

117. Cano, P.M.; Seeboth, J.; Meurens, F.; Cognie, J.; Abrami, R.; Oswald, I.P.; Guzylack-Piriou, L. Deoxynivalenol as a new factor in the persistence of intestinal inflammatory diseases: An emerging hypothesis through possible modulation of Th17-mediated response. *PLoS One* **2013**, *8*, e53647.

118. Hara-Kudo, Y.; Sugita-Konishi, Y.; Kasuga, F.; Kumagai, S. Effects of deoxynivalenol on *Salmonella* enteritidis infection. *Mycotoxins* **1996**, *42*, 51–55.

119. Li, M.; Cuff, C.F.; Pestka, J. Modulation of murine host response to enteric reovirus infection by the trichothecene deoxynivalenol. *Toxicol. Sci.* **2005**, *87*, 134–145. [CrossRef]

120. Atkinson, H.A.; Miller, K. Inhibitory effect of deoxynivalenol, 3-acetyldeoxynivalenol and zearalenone on induction of rat and human lymphocyte proliferation. *Toxicol. Lett.* **1984**, *23*, 215–221. [CrossRef]

121. Miller, K.; Atkinson, H.A. The *in vitro* effects of trichothecenes on the immune system. *Food Chem. Toxicol.* **1986**, *24*, 545–549. [CrossRef]

122. Virelizier, J.L.; Arenzana-Seisdedos, F. Immunological functions of macrophages and their regulation by interferons. *Med. Biol.* **1985**, *63*, 149–159.

123. Sugita-Konishi, Y.; Pestka, J.J. Differential upregulation of TNF-alpha, IL-6, and IL-8 production by deoxynivalenol (vomitoxin) and other 8-ketotrichothecenes in a human macrophage model. *J. Toxicol. Environ. Health A* **2001**, *64*, 619–636. [CrossRef]

124. Döll, S.; Schrickx, J.A.; Dänicke, S.; Fink-Gremmels, J. Deoxynivalenol-induced cytotoxicity, cytokines and related genes in unstimulated or lipopolysaccharide stimulated primary porcine macrophages. *Toxicol. Lett.* **2009**, *184*, 97–106. [CrossRef]

125. Sugiyama, K.; Muroi, M.; Tanamoto, K.; Nishijima, M.; Sugita-Konishi, Y. Deoxynivalenol and nivalenol inhibit lipopolysaccharide-induced nitric oxide production by mouse macrophage cells. *Toxicol. Lett.* **2010**, *192*, 150–154. [CrossRef]

126. Ayral, A.M.; Dubech, N.; le Bars, J.; Escoula, L. *In vitro* effect of diacetoxyscirpenol and deoxynivalenol on microbicidal activity of murine peritoneal macrophages. *Mycopathologia* **1992**, *120*, 121–127. [CrossRef]

127. Waché, Y.J.; Hbabi-Haddioui, L.; Guzylack-Piriou, L.; Belkhelfa, H.; Roques, C.; Oswald, I.P. The mycotoxin deoxynivalenol inhibits the cell surface expression of activation markers in human macrophages. *Toxicology* **2009**, *262*, 239–244. [CrossRef]

128. Li, M.; Harkema, J.R.; Cuff, C.F.; Pestka, J.J. Deoxynivalenol exacerbates viral bronchopneumonia induced by respiratory reovirus infection. *Toxicol. Sci.* **2007**, *95*, 412–426.

129. Wang, X.; Liu, Q.; Ihsan, A.; Huang, L.; Dai, M.; Hao, H.; Cheng, G.; Liu, Z.; Wang, Y.; Yuan, Z. JAK/STAT pathway plays a critical role in the proinflammatory gene expression and apoptosis of RAW264.7 cells induced by trichothecenes as DON and T-2 toxin. *Toxicol. Sci.* **2012**, *127*, 412–424. [CrossRef]

130. Yan, D.; Zhou, H.R.; Brooks, K.H.; Pestka, J.J. Potential role for IL-5 and IL-6 in enhanced IgA secretion by Peyer's patch cells isolated from mice acutely exposed to vomitoxin. *Toxicology* **1997**, *122*, 145–158.

131. Yan, D.; Zhou, H.R.; Brooks, K.H.; Pestka, J.J. Role of macrophages in elevated IgA and IL-6 production by Peyer's patch cultures following acute oral vomitoxin exposure. *Toxicol. Appl. Pharmacol.* **1998**, *148*, 261–273. [CrossRef]

132. Vivier, E.; Tomasello, E.; Baratin, M.; Walzer, T.; Ugolini, S. Functions of natural killer cells. *Nat. Immunol.* **2008**, *9*, 503–510.

133. Pinton, P.; Accensi, F.; Beauchamp, E.; Cossalter, A.M.; Callu, P.; Grosjean, F.; Oswald, I.P. Ingestion of deoxynivalenol (DON) contaminated feed alters the pig vaccinal immune responses. *Toxicol. Lett.* **2008**, *177*, 215–222. [CrossRef]

134. Oswald, I.P.; Marin, D.E.; Bouhet, S.; Pinton, P.; Taranu, I.; Accensi, F. Immunotoxicological risk of mycotoxins for domestic animals. *Food Addit. Contam.* **2005**, *22*, 354–360. [CrossRef]

135. Taranu, I.; Marina, D.E.; Burlacu, R.; Pinton, P.; Damian, V.; Oswald, I.P. Comparative aspects of *in vitro* proliferation of human and porcine lymphocytes exposed to mycotoxins. *Arch. Anim. Nutr.* **2010**, *64*, 383–393. [CrossRef]

136. Meky, F.A.; Hardie, L.J.; Evans, S.W.; Wild, C.P. Deoxynivalenol-induced immunomodulation of human lymphocyte proliferation and cytokine production. *Food Chem. Toxicol.* **2001**, *39*, 827–836. [CrossRef]

137. Pestka, J.J. Deoxynivalenol-induced IgA production and IgA nephropathy-aberrant mucosal immune response with systemic repercussions. *Toxicol. Lett.* **2003**, *140–141*, 287–295. [CrossRef]

138. Rasooly, L.; Pestka, J.J. Vomitoxin-induced dysregulation of serum IgA, IgM and IgG reactive with gut bacterial and self antigens. *Food Chem. Toxicol.* **1992**, *30*, 499–504. [CrossRef]

139. Dänicke, S.; Hegewald, A.K.; Kahlert, S.; Kluess, J.; Rothkötter, H.J.; Breves, G.; Döll, S. Studies on the toxicity of deoxynivalenol (DON), sodium metabisulfite, DON-sulfonate (DONS) and de-epoxy-DON for porcine peripheral blood mononuclear cells and the Intestinal Porcine Epithelial Cell lines IPEC-1 and IPEC-J2, and on effects of DON and DONS on piglets. *Food Chem. Toxicol.* **2010**, *48*, 2154–2162. [CrossRef]

140. Ndossi, D.G.; Frizzell, C.; Tremoen, N.H.; Fæste, C.K.; Verhaegen, S.; Dahl, E.; Eriksen, G.S.; Sørlie, M.; Connolly, L.; Ropstad, E. An *in vitro* investigation of endocrine disrupting effects of trichothecenes deoxynivalenol (DON), T-2 and HT-2 toxins. *Toxicol. Lett.* **2012**, *214*, 268–278. [CrossRef]

141. Medvedova, M.; Kolesarova, A.; Capcarova, M.; Labuda, R.; Sirotkin, A.V.; Kovacik, J.; Bulla, J. The effect of deoxynivalenol on the secretion activity, proliferation and apoptosis of porcine ovarian granulosa cells *in vitro*. *J. Environ. Sci. Health B* **2011**, *46*, 213–219. [CrossRef]

142. Kolesarova, A.; Capcarova, M.; Maruniakova, N.; Lukac, N.; Ciereszko, R.E.; Sirotkin, A.V. Resveratrol inhibits reproductive toxicity induced by deoxynivalenol. *J. Environ. Sci. Health A* **2012**, *47*, 1329–1334.

143. Amuzie, C.J.; Pestka, J.J. Suppression of insulin-like growth factor acid-labile subunit expression—A novel mechanism for deoxynivalenol-induced growth retardation. *Toxicol. Sci.* **2010**, *113*, 412–421. [CrossRef]

144. Voss, K.A. A new perspective on deoxynivalenol and growth suppression. *Toxicol. Sci.* **2010**, *113*, 281–283. [CrossRef]

145. Szkudelska, K.; Szkudelski, T.; Nogowski, L. Short-time deoxynivalenol treatment induces metabolic disturbances in the rat. *Toxicol. Lett.* **2002**, *136*, 25–31. [CrossRef]

146. Flannery, B.M.; Clark, E.S.; Pestka, J.J. Anorexia induction by the trichothecene deoxynivalenol (vomitoxin) is mediated by the release of the gut satiety hormone peptide YY. *Toxicol. Sci.* **2012**, *130*, 289–297. [CrossRef]

147. Razafimanjato, H.; Benzaria, A.; Taïeb, N.; Guo, X.J.; Vidal, N.; di Scala, C.; Varini, K.; Maresca, M. The ribotoxin deoxynivalenol affects the viability and functions of glial cells. *Glia* **2011**, *59*, 1672–1683. [CrossRef]

148. Razafimanjato, H.; Garmy, N.; Guo, X.J.; Varini, K.; di Scala, C.; di Pasquale, E.; Taïeb, N.; Maresca, M. The food-associated fungal neurotoxin ochratoxin A inhibits the absorption of glutamate by astrocytes

through a decrease in cell surface expression of the excitatory amino-acid transporters GLAST and GLT-1. *Neurotoxicology* **2010**, *31*, 475–484.

149. Wang, D.D.; Bordey, A. The astrocyte odyssey. *Prog. Neurobiol.* **2008**, *86*, 342–367.

150. Varini, K.; Benzaria, A.; Taïeb, N.; di Scala, C.; Azmi, A.; Graoudi, S.; Maresca, M. Mislocalization of the exitatory amino-acid transporters (EAATs) in human astrocytoma and non-astrocytoma cancer cells: Effect of the cell confluence. *J. Biomed. Sci.* **2012**, *19*, 10. [CrossRef]

151. Ren, K.; Dubner, R. Neuron-glia crosstalk gets serious: Role in pain hypersensitivity. *Curr. Opin. Anaesthesiol.* **2008**, *21*, 570–579. [CrossRef]

152. Gibbs, M.E.; Hutchinson, D.; Hertz, L. Astrocytic involvement in learning and memory consolidation. *Neurosci. Biobehav. Rev.* **2008**, *32*, 927–944.

153. Bonnet, M.S.; Roux, J.; Mounien, L.; Dallaporta, M.; Troadec, J.D. Advances in deoxynivalenol toxicity mechanisms: The brain as a target. *Toxins* **2012**, *4*, 1120–1138.

154. Fitzpatrick, D.W.; Boyd, K.E.; Wilson, L.M.; Wilson, J.R. Effect of the trichothecene deoxynivalenol on brain biogenic monoamines concentrations in rats and chickens. *J. Environ. Sci. Health B* **1988**, *23*, 159–170. [CrossRef]

155. Prelusky, D.B.; Yeung, J.M.; Thompson, B.K.; Trenholm, H.L. Effect of deoxynivalenol on neurotransmitters in discrete regions of swine brain. *Arch. Environ. Contam. Toxicol.* **1992**, *22*, 36–40.

156. Prelusky, D.B. A study on the effect of deoxynivalenol on serotonin receptor binding in pig brain membranes. *J. Environ. Sci. Health B* **1996**, *31*, 1103–1117.

157. Ossenkopp, K.P.; Hirst, M.; Rapley, W.A. Deoxynivalenol (vomitoxin)-induced conditioned taste aversions in rats are mediated by the chemosensitive area postrema. *Pharmacol. Biochem. Behav.* **1994**, *47*, 363–367. [CrossRef]

158. Girardet, C.; Bonnet, M.S.; Jdir, R.; Sadoud, M.; Thirion, S.; Tardivel, C.; Roux, J.; Lebrun, B.; Wanaverbecq, N.; Mounien, L.; *et al.* The food-contaminant deoxynivalenol modifies eating by targeting anorexigenic neurocircuitry. *PLoS One* **2011**, *6*, e26134. [CrossRef]

159. Girardet, C.; Bonnet, M.S.; Jdir, R.; Sadoud, M.; Thirion, S.; Tardivel, C.; Roux, J.; Lebrun, B.; Mounien, L.; Trouslard, J.; *et al.* Central inflammation and sickness-like behavior induced by the food contaminant deoxynivalenol: A PGE2-independent mechanism. *Toxicol. Sci.* **2011**, *124*, 179–191. [CrossRef]

160. Gaigé, S.; Bonnet, M.S.; Tardivel, C.; Pinton, P.; Trouslard, J.; Jean, A.; Guzylack, L.; Troadec, J.D.; Dallaporta, M. c-Fos immunoreactivity in the pig brain following deoxynivalenol intoxication: Focus on NUCB2/nesfatin-1 expressing neurons. *Neurotoxicology* **2013**, *34*, 135–149. [CrossRef]

161. Langhans, W. Signals generating anorexia during acute illness. *Proc. Nutr. Soc.* **2007**, *66*, 321–330. [CrossRef]

162. Pestka, J.J.; Zhou, H.R. Effects of tumor necrosis factor type 1 and 2 receptor deficiencies on anorexia, growth and IgA dysregulation in mice exposed to the trichothecene vomitoxin. *Food Chem. Toxicol.* **2002**, *40*, 1623–1631. [CrossRef]

163. Dallaporta, M.; Bonnet, M.S.; Horner, K.; Trouslard, J.; Jean, A.; Troadec, J.D. Glial cells of the nucleus tractus solitarius as partners of the dorsal hindbrain regulation of energy balance: A proposal for a working hypothesis. *Brain Res.* **2010**, *1350*, 35–42.

164. Wu, W.; Flannery, B.M.; Sugita-Konishi, Y.; Watanabe, M.; Zhang, H.; Pestka, J.J. Comparison of murine anorectic responses to the 8-ketotrichothecenes 3-acetyldeoxynivalenol, 15-acetyldeoxynivalenol, fusarenon X and nivalenol. *Food Chem. Toxicol.* **2012**, *50*, 2056–2061. [CrossRef]

165. Turner, P.C.; Hopton, R.P.; Lecluse, Y.; White, K.L.; Fisher, J.; Lebailly, P. Determinants of urinary deoxynivalenol and de-epoxy deoxynivalenol in male farmers from Normandy, France. *J. Agric. Food Chem.* **2010**, *58*, 5206–5212.

166. Turner, P.C.; Flannery, B.; Isitt, C.; Ali, M.; Pestka, J. The role of biomarkers in evaluating human health concerns from fungal contaminants in food. *Nutr. Res. Rev.* **2012**, *25*, 162–179. [CrossRef]

167. Warth, B.; Sulyok, M.; Fruhmann, P.; Mikula, H.; Berthiller, F.; Schuhmacher, R.; Hametner, C.; Abia, W.A.; Adam, G.; Fröhlich, J.; *et al.* Development and validation of a rapid multi-biomarker liquid chromatography/tandem mass spectrometry method to assess human exposure to mycotoxins. *Rapid Commun. Mass Spectrom.* **2012**, *26*, 1533–1540. [CrossRef]

168. Muri, S.D.; van der Voet, H.; Boon, P.E.; van Klaveren, J.D.; Brüschweiler, B.J. Comparison of human health risks resulting from exposure to fungicides and mycotoxins via food. *Food Chem. Toxicol.* **2009**, *47*, 2963–2974. [CrossRef]

169. Mattsson, J.L. Mixtures in the real world: The importance of plant self-defense toxicants, mycotoxins, and the human diet. *Toxicol. Appl. Pharmacol.* **2007**, *223*, 125–132. [CrossRef]
170. Wild, C.P.; Gong, Y.Y. Mycotoxins and human disease: A largely ignored global health issue. *Carcinogenesis* **2010**, *31*, 71–82. [CrossRef]
171. Kluger, B.; Bueschl, C.; Lemmens, M.; Berthiller, F.; Häubl, G.; Jaunecker, G.; Adam, G.; Krska, R.; Schuhmacher, R. Stable isotopic labelling-assisted untargeted metabolic profiling reveals novel conjugates of the mycotoxin deoxynivalenol in wheat. *Anal. Bioanal. Chem.* **2012**. [CrossRef]
172. Zachariasova, M.; Vaclavikova, M.; Lacina, O.; Vaclavik, L.; Hajslova, J. Deoxynivalenol oligoglycosides: New "masked" fusarium toxins occurring in malt, beer, and breadstuff. *J. Agric. Food Chem.* **2012**, *60*, 9280–9291. [CrossRef]

Section 1:
Prevalence, Exposure and Remediation of DON and Its Derivatives

toxins

MDPI

Article

Rapid Analysis of Deoxynivalenol in Durum Wheat by FT-NIR Spectroscopy

Annalisa De Girolamo *, Salvatore Cervellieri, Angelo Visconti and Michelangelo Pascale

Institute of Sciences of Food Production, National Research Council of Italy (ISPA-CNR), Via G. Amendola 122/O, 70126 Bari, Italy; salvatore.cervellieri@ispa.cnr.it (S.C.); angelo.visconti@ispa.cnr.it (A.V.); michelangelo.pascale@ispa.cnr.it (M.P.)

* Author to whom correspondence should be addressed; annalisa.degirolamo@ispa.cnr.it;
Tel.: +39-080-592-9351; Fax: +39-080-592-9374.

External Editor: Marc Maresca

Received: 14 July 2014; in revised form: 25 October 2014; Accepted: 27 October 2014; Published: 6 November 2014

Abstract: Fourier-transform-near infrared (FT-NIR) spectroscopy has been used to develop quantitative and classification models for the prediction of deoxynivalenol (DON) levels in durum wheat samples. Partial least-squares (PLS) regression analysis was used to determine DON in wheat samples in the range of <50–16,000 µg/kg DON. The model displayed a large root mean square error of prediction value (1,977 µg/kg) as compared to the EU maximum limit for DON in unprocessed durum wheat (*i.e.*, 1,750 µg/kg), thus making the PLS approach unsuitable for quantitative prediction of DON in durum wheat. Linear discriminant analysis (LDA) was successfully used to differentiate wheat samples based on their DON content. A first approach used LDA to group wheat samples into three classes: A (DON ≤ 1,000 µg/kg), B (1,000 < DON ≤ 2,500 µg/kg), and C (DON > 2,500 µg/kg) (LDA I). A second approach was used to discriminate highly contaminated wheat samples based on three different cut-off limits, namely 1,000 (LDA II), 1,200 (LDA III) and 1,400 µg/kg DON (LDA IV). The overall classification and false compliant rates for the three models were 75%–90% and 3%–7%, respectively, with model LDA IV using a cut-off of 1,400 µg/kg fulfilling the requirement of the European official guidelines for screening methods. These findings confirmed the suitability of FT-NIR to screen a large number of wheat samples for DON contamination and to verify the compliance with EU regulation.

Keywords: deoxynivalenol; FT-NIR; rapid method; wheat; LDA; PLS

---

1. Introduction

Deoxynivalenol (DON), also known as vomitoxin, is a type B trichothecene mycotoxin. It is one of the major secondary metabolites produced by fungi of the *Fusarium* genus, mainly *Fusarium graminearum* and *Fusarium culmorum*, and occurs predominantly in grains, such as wheat, maize, barley, oats and rye [1]. DON inhibits the synthesis of DNA, RNA and proteins and has a hemolytic effect on erythrocytes. DON can cause feed refusal, vomiting, reduced weight gain, diarrhea, hemorrhage, skin lesions, growth depression and immunosuppression, which have a negative impact on human and animal health [2–6]. In order to protect consumers from exposure to DON through the consumption of cereal-based food products, the European Commission has set maximum permitted levels for DON ranging from 200 µg/kg for processed cereal-based food for infants and young children up to 1,750 µg/kg for unprocessed durum wheat, maize and oats [7].

Analytical methods for rapid, sensitive and accurate determination of DON in foods and feeds are highly demanded for exposure risk assessment studies and to enforce regulatory requirements issued by governments and international organizations. A number of analytical methods, such as gas chromatography (GC) with electron-capture or mass spectrometric (MS) detection and high-performance liquid chromatography (HPLC) based on UV or MS detection have been developed to quantitatively measure DON concentration in cereals and derived products [8–10]. Although these traditional methods are sensitive and accurate, most of them involve expensive and time-consuming steps, including sample cleanup and detection, being unsuitable for screening purposes. Large amounts of cereals are processed in the food and feed industry each year, and frequent checks are required to verify the compliance of raw materials with regulation, resulting in a large number of samples to be analyzed. Factors, like promptness and low cost of analysis, minimal sample preparation and environmentally-friendly methods, are of paramount importance to rapidly respond to the demands of the market. In the last few decades, a variety of rapid methods based on competitive enzyme-linked immunosorbent assays (ELISA) or on novel technologies, including lateral flow devices (LFD), membrane-based flow-through enzyme immunoassay, fluorescence polarization (FP) immunoassay, molecularly imprinted polymers (MIP) and surface plasmon resonance (SPR) biosensors, have been reported for the rapid analysis of DON [8,11,12]. However, these methods are destructive, require an extraction step and, in some cases, a clean-up procedure. Recently, a rapid, easy to-perform and non-invasive method using an electronic nose based on metal oxide sensors to distinguish the quality of durum wheat samples based on the content of DON has been reported [13]. Infrared spectroscopy (IR) has gained wide acceptance within food and feed analysis as a rapid analytical tool that requires minimal or no sample preparation, and in contrast with traditional chromatographic analysis, it does not require reagents, nor does it produce chemical waste. Near-infrared (NIR) or mid-infrared (MIR) spectroscopy techniques, both in combination or not with Fourier-transform (FT), are commonly used in a remarkably wide range of applications for the analysis of moisture, oil, fiber, starch, lipids, protein, yeast and bacteria in agricultural products [14,15]. In recent years, the potential of using IR spectroscopy for the detection of mycotoxins, including DON, ochratoxin A, fumonisins and aflatoxins, and mycotoxigenic fungal contamination in cereals and cereal products has been also demonstrated [16–37]. Among mycotoxins, the most investigated one was DON, mainly in *Fusarium*-damaged wheat kernels and ground wheat [16,18–20,22–24,34] and to a lesser extent in maize, barley and oat [17,21,32,33]. The majority of these IR methods for the determination of DON in wheat is based on NIR or UV-Vis-NIR spectroscopy. The feasibility of using FT-NIR spectroscopy for the qualitative and quantitative prediction of DON in unprocessed ground durum and common wheat was reported for the first time by our research group [19]. Performance results suggested the use of FT-NIR as a sorting tool for screening purposes; however, both qualitative and quantitative models merited further implementation in a larger study involving more wheat samples with a homogeneous distribution of DON levels around the EU maximum limit (*i.e.*, 1,750 µg/kg) for unprocessed durum wheat in order to make the model more robust and reliable [19]. Moreover, the classification model included both common and durum wheat samples and used a cut-off level (300 µg/kg) far from the EU maximum limit (ML) to distinguish the two classes of wheat samples [19]. FT-NIR spectroscopy for DON determination in whole wheat kernels was also reported by some authors [23,34]. The advantages of FT include improvement in the spectral reproducibility and the accuracy and precision of wavelength discrimination [28].

The objective of this research was to develop a robust, rapid and inexpensive FT-NIR method for the analysis of DON-contaminated ground durum wheat samples and to verify the compliance with European regulations. Both quantitative and classification models were developed and validated to establish the most suitable approach to estimate DON in unprocessed durum wheat. The proposed FT-NIR classification method as a first screening step for DON detection in samples could provide a high-throughput analysis platform that could improve food and feed safety at mills and could be used for monitoring programs.

2. Results and Discussion

DON levels in the 464 durum wheat samples varied from <50 µg/kg (quantification limit of the HPLC method) to 16,000 µg/kg, with mean and median values of 2,390 and 1,100 µg/kg, respectively. This concentration range covered the majority of DON concentrations found in the routine surveillance samples in wheat supply chains and was appropriate for the scope of the study to develop calibration and classification models. Although the positive skewness and kurtosis values of DON concentrations indicated that the number of highly-contaminated wheat samples was less than that of wheat samples with no or low values of DON contamination, there was an approximately fifty-fifty distribution of DON levels around the EU ML for unprocessed durum wheat (1,750 µg/kg); in particular, 57% of the tested samples contained DON levels less than ML, whereas the remaining 43% of samples exceed this threshold.

Wheat samples were analyzed by FT-NIR spectroscopy, and spectra were recorded as absorbance between 10,000–4,000 cm^{-1}. Figure 1 shows the FT-NIR raw spectra of five different unprocessed ground wheat samples contaminated in the range of <50 µg/kg to about 10,000 µg/kg DON. From the comparison of these spectra, it appears that wheat samples contaminated with low DON levels have FT-NIR bands in common with those containing high DON levels, thus indicating that the major functional groups and chemical constituents co-exist in both types of samples.

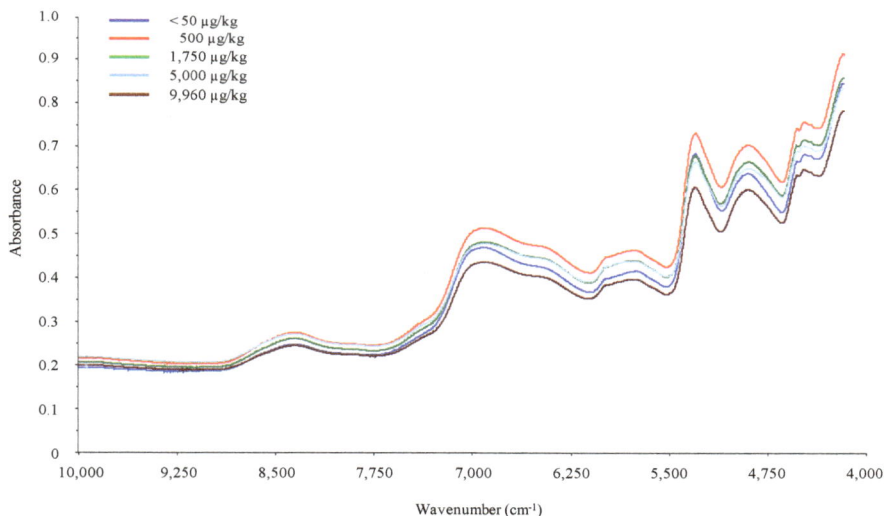

Figure 1. FT-NIR spectra of five different ground unprocessed durum wheat samples contaminated with increasing levels (from <50 to 9,960 µg/kg) of deoxynivalenol (DON) (as measured by the reference HPLC method).

2.1. Quantification of DON in Contaminated Wheat Sample

In our previous work, we evaluated the ability of using FT-NIR spectroscopy for the determination of DON in durum wheat at levels between <50 µg/kg and 2,700 µg/kg by using the statistical approach of partial least squares (PLS) regression [19]. The performance results provided good evidence of the feasibility of using FT-NIR spectroscopy as a screening tool for the determination of DON in wheat samples. However, the inhomogeneous distribution of DON levels around the EU ML for unprocessed durum wheat (*i.e.*, 94% of samples with DON levels <1,750 µg/kg) made the model poorly robust [19]. Based on these findings, we further implemented the PLS model in a larger study involving more calibration and validation samples with a larger concentration range of DON. The resulting PLS model

(PLS I) covered the range of DON concentration from <50 to 16,000 µg/kg and included 232 samples for the calibration set and 232 for the validation one. The model was developed with eight PLS factors, explaining 80% of the total variance of the entire set of calibration data. The slope and the coefficient of determination (r^2) values (both 0.802) of the calibration regression curve indicated that the PLS model can be used for screening and approximate calibrations with a root mean squares error of calibration (RMSEC) value of 1,473 µg/kg DON. However, the r^2 value of the validation regression curve was 0.630, indicating that the PLS model was only usable for a rough screening of wheat samples and showed a root mean squared error of prediction (RMSEP) of 1,977 µg/kg DON that was considered very large with respect to the EU ML. Furthermore, based on the residual predictive deviation (RPD) (1.72) and range error ratio (RER) (6.89) values, the model had a very poor classification ability and was not recommended for any purpose. Figure 2 shows the PLS validation plot of the measured data (by HPLC) in relation to the estimated data (by FT-NIR).

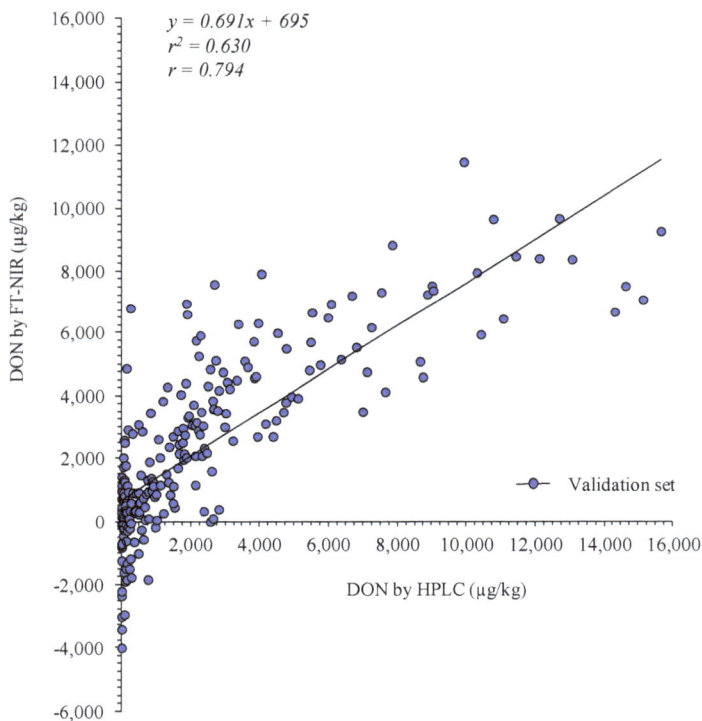

Figure 2. Partial least squares (PLS) regression plot of measured (by HPLC) and estimated (by FT-NIR) DON concentrations in the validation set (model PLS I).

It is well known that fungal infections of kernels cause multiple changes of kernel composition and pigmentation, thus causing spectral variability [38]. To evaluate if high contamination levels of DON were responsible for the poor classification ability of PLS I, samples containing DON levels between 6,000 and 16,000 µg/kg DON were excluded from the dataset, and another PLS model was developed (PLS II). The model included 204 samples for the calibration set and 204 for the validation one with 65% of wheat samples with DON levels less than 1,750 µg/kg. With respect to PLS I, similar results were obtained for PLS II in terms of slope (0.749) and intercept (427 µg/kg), whereas the values of RMSEC and RMSEP were 753 µg/kg and 868 µg/kg, respectively. Although the RMSEP value of model PLS II was lower than that obtained with model PLS I, in both cases, it corresponded to approximately 14% of the entire range of DON concentration in the samples used in the models. As for model PLS I, the RPD value (1.66) indicated that model PLS II was not recommended for any purpose.

PLS results obtained in the present study were in agreement with those reported by Dvoracek *et al.* [23], which applied the FT-NIR spectroscopy to the determination of DON in wheat kernels in the range of 0–13,000 µg/kg and 0–5,000 µg/kg DON. Based on these observations we concluded that the PLS approach was unsuitable for the aim of the study; therefore, the classification one was used.

2.2. Classification of DON Contaminated Wheat Samples

The classification LDA models proposed herein have been developed on a huge number of durum wheat samples with a broad range of DON contamination levels. The resulting models covered the range of DON concentration from <50 to 16,000 µg/kg and included 232 samples for the calibration set and 232 for the validation one. Initially, the spectra were treated using principal components analysis (PCA). The first 10 principal components, accounting for more than 99% of the total variance, were selected as input variables for the LDA. Two different approaches were used to develop LDA models in order to find the most suitable one to estimate DON in durum wheat samples. With the first approach (LDA I), wheat samples were classified into three groups based on DON contamination levels: Class A (DON content less than 1,000 µg/kg), Class B (DON content ranging from 1,000 to 2,500 µg/kg) and Class C (DON content more than 2,500 µg/kg). A discriminant model was then developed to classify ground wheat samples into the three DON contamination groups. The overall classification rate was 82% during the calibration process. In particular, 77% of acceptable samples were correctly classified into Class A; 75% of wheat samples with DON levels to be confirmed with a reference method were correctly classified as Class B; and 94% of rejectable samples were correctly classified into Class C. The model was then validated by using an independent dataset. Results are reported in Table 1.

Table 1. Validation results of the linear discriminant classification model (LDA I). The first column indicates the class (A, B and C) assigned by the HPLC reference analysis, whereas the other three columns refer to the class predicted by LDA analysis.

Assigned class [a] (by HPLC reference analysis)	Number of samples classified in the predicted classes (by FT-NIR analysis)		
	A	B	C
A	92	22	4
B	6	18	13
C	2	12	63
Overall classification rate (%)		75	
FC samples (%) [b]		3	
FNC samples (%) [c]		7	

[a] A: DON ≤ 1,000 µg/kg; B: 1,000 µg/kg < DON ≤ 2,500 µg/kg; C: DON > 2,500 µg/kg; [b] FC, false compliant; [c] FNC, false, not compliant. .

The validated model achieved an overall classification rate of 75% with a comparable classification rate for Class A and Class C (approximately 80%) and lower for Class B (49%). In particular, wheat samples belonging to Class B were classified as either Class A (16%) or Class C (35%) (Table 1). From Figure 3, showing the LDA score plot of wheat samples naturally contaminated with DON and classified into three contamination groups, a scattered distribution of samples of Class A and a partial overlapping of Class B to the contiguous Classes A and C is clearly evident. By looking inside these overlapped samples, five of 22 samples of Class A that were incorrectly classified as Class B had DON levels close to the cut-off of 1,000 µg/kg (*i.e.*, from 700 to 1,000 µg/kg). Similarly, nine of 13 samples of Class B that were incorrectly classified as Class C were contaminated with DON levels close to the cut-off of 2,500 µg/kg (*i.e.*, from 1,900 to 2,500 µg/kg). The low discrimination ability of the model around the cut-off limits was probably due to the lower number of samples representing Class B as compared to the other two classes in the calibration set, thus making the model poorly balanced for wheat samples with contamination levels in the range of 1,000–2,500 µg/kg DON. The European official guidelines for analytical methods recommend that only those validated methods that have a false compliant (FC) rate <5% at the level of interest shall be used for screening purposes, whereas suspected non-compliant results shall be confirmed by a confirmatory method [38]. Despite the low discrimination ability of the model LDA I for samples belonging to Class B, the amount of FC samples (eight out of 232) accounted for 3% and fulfilled the requirement of the European official guidelines for screening methods, making the proposed model suitable for screening purposes.

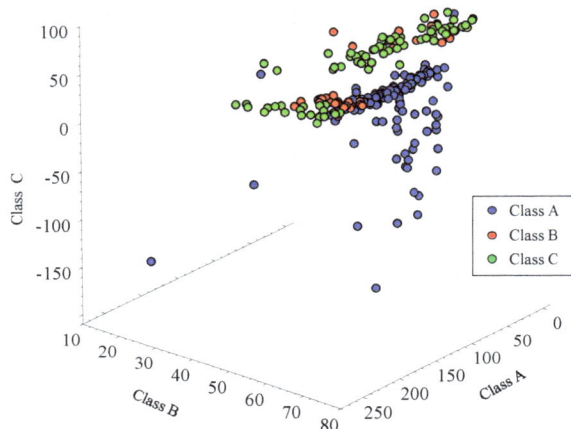

Figure 3. Linear discriminant analysis score plot for FT-NIR spectra of wheat samples naturally contaminated with varying DON content (validation set). Class A: DON ≤ 1,000 µg/kg; Class B: 1,000 µg/kg < DON ≤ 2,500 µg/kg; Class C: DON > 2,500 µg/kg.

With the second approach, wheat samples were classified into two groups based on a cut-off level of DON to distinguish them. Three different LDA models were developed using cut-off levels of 1,000, 1,200 and 1,400 µg/kg DON (LDA II–IV, respectively), in order to find the optimal threshold allowing the highest classification rate of wheat samples. The overall classification rate in the calibration processes was in the range of 91%–92% with FC values of 3%–7%. When the models were validated, the overall classification rates were still good (89%–90%, Table 2) and in agreement with those obtained in the calibration process, indicating the good robustness of the developed LDA models. Moreover, model LDA IV using a cut-off of 1,400 µg/kg fulfilled the requirement of the European official guidelines for analytical methods [38] and was considered the most reliable one for the screening of DON in unprocessed durum wheat. Good results in terms of repeatability (<2%) and within-laboratory reproducibility (<3%) were also obtained for all LDA models, indicating the robustness of the measurements and the instrumental stability over a period of time.

Considering that wheat samples used in the calibration and validation sets belonged to different cultivars or mixtures of cultivars and were obtained in different periods from field crops located in different Italian regions, it was unrealistic to expect 100% correct classification. Although a slight misclassification occurred, advantages in terms of costs and the rapidity of analysis makes classification model LDA IV a useful screening tool for the analysis of a large number of durum wheat samples and to verify the compliance with legislation by reducing the number of samples to be confirmed with a reference method.

Table 2. Validation results of the linear discriminant classification models (LDA II-IV) using different cut-off levels of DON.

Classification results	Discrimination models (cut-off DON, µg/kg)		
	LDA II (1,000)	LDA III (1,200)	LDA IV (1,400)
Overall classification rate (%)	89	90	90
FC samples (%) [a]	7	6	5
FNC samples (%) [b]	4	4	5

[a] FC, false compliant; [b] FNC, false, not compliant.

Classification results obtained in the present paper represent a great improvement in terms of the robustness of the model and DON correct prediction as compared to those reported in our previous work [19]. These findings are also better than those reported by Dvoracek *et al.* [23], which applied the LDA to classify 400 intact wheat grain samples into two classes using four different cut-off levels in the range of 1,250 to 30,000 µg/kg DON. Although performance indexes were 89%–93%, the number of misclassified samples increased with the decreasing of the cut-off level, with the model using a cut-off of 1,250 µg/kg DON giving approximately 40% of misclassified samples. These results were probably related to the use of intact wheat kernels, which made the use of FT-NIR spectroscopy perform poorly at contamination levels close to the EU ML [23]. On the other hand, our findings indicate that the grinding step of unprocessed wheat is a useful step that helps to overcome the inhomogeneity problem and allows one to obtain a representative sample to be analyzed by FT-NIR spectroscopy.

3. Experimental Section

3.1. Chemicals

Acetonitrile of HPLC grade was purchased from Mallinckrodt Baker (Milan, Italy). Ultrapure water was produced by a Millipore Milli-Q system (Millipore, Bedford, MA, USA). Deoxynivalenol (DON) standard was purchased from Sigma-Aldrich (Milan, Italy). DON immunoaffinity columns (DONtest™ HPLC) were obtained from Vicam, a Waters Business (Milford, MA, USA), glass microfiber filters (Whatman GF/A) and paper filters (Whatman No. 4) from Whatman (Maidstone, UK). The PTFE syringe filters with a diameter of 25 mm and a pore size of 0.45 mm were bought from Teknokroma (Barcelona, Spain).

3.2. Durum Wheat Samples

A total of 500 samples (1 kg each) of naturally contaminated durum wheat of several different cultivars or mixtures of cultivars was obtained from field crops located in different Italian regions during the period 2008–2013. After manual homogenization, an aliquot of each wheat sample (about 200 g) was finely ground by the Tecator Cyclotec 1093 (International PBI, Hoganas, Sweden) laboratory mill equipped with a 500-µm sieve and analyzed by FT-NIR spectroscopy and then by HPLC. An additional laboratory quality control (QC) sample of unprocessed durum wheat naturally contaminated with 1,420 ± 39 µg/kg DON was used to evaluate the repeatability and intra-laboratory reproducibility of the best prediction model.

3.3. HPLC Analysis

DON quantitative determination was performed according to the method described by Visconti *et al.* [39], with minor modifications. Briefly, 25 g of ground sample was extracted with 100 mL of phosphate buffer solution (PBS, 10 mM sodium phosphate, 0.85% sodium chloride, pH = 7.4) by blending at high speed for 2 min with a Sorvall Omnimixer (Sorvall Instruments, Norwalk, CN, USA). The extract was filtered through both filter paper (Whatman No. 4) and glass microfiber filter (Whatman GF/A), and 2 mL of the filtered extract were loaded onto the DONtest™ immunoaffinity column. After washing the column by passing 5 mL of water through it, DON was recovered by eluting 1.5 mL of methanol. The eluate was dried under an air stream at 50 °C, re-dissolved into 250 μL of the mobile phase (acetonitrile/water, 8:92; v/v), and 50 μL were injected onto the HPLC system (Agilent 1100 Series, Agilent Technology, Palo Alto, CA, USA) with the diode array detector set at 220 nm.

3.4. FT-NIR Analysis

FT-NIR spectra were recorded using an Antaris II FT-NIR spectrophotometer (Thermo Electron Corporation, Madison, WI, USA) equipped with an interferometer, an integrating sphere working in diffuse reflection and an indium and gallium arsenide (InGaAs) detector. A sample-cup spinner allowed the automatic collection of several sub-scans from each sample that were averaged to obtain representative spectra. The integrating sphere's internal reference was also used to collect the background spectrum. Approximately 30 g of ground wheat samples were placed on the rotary sample-cup spinner, and 64 interferometer sub-scans in ranges from 10,000 to 4,000 cm^{-1}, with a resolution of 8 cm^{-1}, were applied for the collection of each spectrum sample by means of the software, Results Integration v3.0.197 (Thermo Electron Corporation, Madison, WI, USA, 2006).

3.5. Spectral Data Preprocessing and Outlier Identification

The final spectral data were imported as SPA/SPG (Omnic) format into The Unscrambler® X, v10.1 (CAMO Software AS, Oslo, Norway, 2011) software in order to perform multivariate statistical analysis. Some preliminary descriptive analyses were performed by both graphic tools (histogram or spectra visualization) and numerical results (mean, minimum, maximum, standard deviation, number of missing data, *etc.*). Then, to remove the spectral baseline shift, noise and light scatter influence, some spectral preprocessing methods were investigated before linear discriminant analysis (LDA) or partial least squares (PLS) regression analysis. In particular, the spectra were firstly treated by multiplicative scatter correction (MSC), then by de-trending and, finally, smoothed by a 15-point Savitzky–Golay (2nd derivative, 2nd polynomial order) function [40]. Prior to proceeding to LDA or PLS analysis, a principal components analysis (PCA) was also applied to the 500 wheat samples to detect outliers or any clustering of the data. Sample outliers were detected by using the graphical tools of the Unscrambler® X software, *i.e.*, the Hotelling T^2 line plot using a critical limit of *p*-value <5% and the influence plot, displaying samples with high leverage. Samples suggested as outliers were removed from the entire set, because their inclusion would have a detrimental effect on the model. This procedure was repeated several times, until a total of 36 samples were considered as outliers, and the remaining 464 samples were used for LDA and PLS model development.

3.6. Development and Validation of DON Quantitative Models

Quantitative calibration models were developed using partial least squares (PLS) regression algorithms by correlating the DON results from HPLC reference measurements with the FT-NIR spectra results. The PLS is a spectral decomposition technique, which finds the most relevant factors to explain the variance in the dataset and compares the covariance between the spectral data and DON concentration. Two PLS models were developed (PLS I and II). PLS I was developed on 464 samples in the range of concentrations of ≤50 to 16,000 μg/kg DON, whereas PLS II was developed on 408 samples in the range of concentrations of ≤50 to 6,000 μg/kg DON. In both cases, sample spectra were randomly divided into a calibration set (50% of samples) and a validation set (50% samples) for developing and testing the model, respectively. Performance of the PLS models was evaluated by calculating the coefficient of determination (r^2), the slope of the calibration regression curve and the root mean squared error of calibration (RMSEC). The final PLS models were validated using an independent validation dataset, and the performance was evaluated according to the slope and coefficient of determination of the validation regression curve (r^2) and the root mean standard error of prediction (RMSEP). The models' accuracy ability was categorized as follows: r^2 = 0.50–0.64, usable for rough screening; r^2 = 0.66–0.81, usable for screening and some "approximate" calibrations;

r^2 = 0.83–0.90, usable with caution; r^2 = 0.92–0.96, usable in most applications; r^2 > 0.98, usable in any applications [41]. The prediction accuracy of models was also evaluated based on the residual predictive deviation (RPD), defined as the ratio of the standard deviation of the reference DON values in the validation set to the RMSEP [42]. The RPD evaluates how well the developed model can predict the DON in the validation set, and the higher is its value, the greater is the prediction ability of the model. Values of RPD from 3.1 to 4.9 are considered fair, and the model is recommended for very rough screening purposes; values from 5.0 to 6.4 indicate that the model is good for quality control; values from 6.5 to 8 indicate that the model is very good for process control; values higher than 8.1 indicate that the model is excellent and usable for any application [43]. Finally, the range error ratio (RER), calculated by dividing the range of the reference DON values in the validation set by the RMSEP, was used as a useful indicator to assess the practical utility of the calibration as a predictive model. Both the RER and RPD standardize the RMSEP value of the model against the range and standard deviation, respectively, of the reference data in the validation set.

3.7. Development and Validation of DON Classification Models

In order to develop a chemometric calibration model for the classification of DON-contaminated samples, a linear discriminant analysis (LDA) was applied to the preprocessed spectral measurements. The LDA was carried out assuming equal prior probability for each class; moreover, the first 10 principal components (PCs) of the spectra were used as the explanatory variables to evaluate how wheat classes were separated from each other in the spectral space. Two different approaches were followed to develop LDA models. With the first approach (LDA I), three different classes of samples with different DON content as determined by HPLC analysis were defined: samples with DON content less than or equal to 1,000 µg/kg (Class A); samples with DON content ranging from 1,000 to 2,500 µg/kg (Class B); samples with DON content higher than 2,500 µg/kg (Class C) (LDA I). The DON contamination range for Class B was established by adding and subtracting an arbitrary measurement uncertainty of 750 µg/kg with respect to the EC ML for DON in unprocessed durum wheat (*i.e.*, 1,750 µg/kg). These three classes were chosen to represent three different conditions in terms of DON contamination: compliant samples with DON levels lower than EU ML (Class A), samples with DON concentration to be confirmed by HPLC analysis (Class B) and rejectable (or not compliant) samples with high DON levels (Class C). With the second approach, wheat samples were distinguished into two groups using a cut-off limit to classify them. In particular, three classification models (LDA Models II–IV) were developed by using three different cut-off levels of DON, *i.e.*, 1,000 µg/kg, 1,200 µg/kg and 1,400 µg/kg. Compliant samples were considered those with DON concentrations ≤ the cut-off level (Class A), whereas rejectable (or non-compliant) samples to be confirmed by HPLC analysis were considered those with DON concentrations > the cut-off level (Class B). For each classification model, a total of 464 samples spectra were randomly divided into calibration and validation sets by randomly subdividing the available spectral data into two equal sets (*i.e.*, 50% calibration and 50% validation) (Table 3). The two sets covered the same concentration range of DON.

Table 3. Number of durum wheat samples in Classes A, B and C used for Model I and in Classes A and B used for Models II–IV. For each model, the number of samples in the calibration and validation sets was the same.

Model	Number of samples (DON, µg/kg)		
	Class A	Class B	Class C
I	118 (≤1,000)	37 (1,000–2,500)	77 (>2,500)
II	113 (≤1,000)	119 (>1,000)	-
III	118 (≤1,200)	114 (>1,200)	-
IV	123 (≤1,400)	109 (>1,400)	-

The LDA used pooled covariance matrices in calculating the Mahalanobis squared distance [44]. In this algorithm, a sample is assigned into one of the DON concentration groups when the squared distance between the sample and the group is minimized. The Mahalanobis squared distances were then calculated to classify an unknown sample from the validation data set into the DON group with the lowest corresponding distance. The prediction ability of the chemometric models was expressed in terms of overall discrimination rate, false compliant rate and false non-compliant rate, according to the following equations:

$$\text{Overall discrimination rate (\%)} = \frac{TC + TNC}{Ntot} \times 100, \tag{1}$$

$$\text{False compliant rate (\%)} = \frac{FC}{Ntot} \times 100, \tag{2}$$

$$\text{False, not compliant rate (\%)} = \frac{FNC}{Ntot} \times 100 \tag{3}$$

where TC, TNC, FC and FNC denote the number of true compliant, true, not compliant, false compliant and false, not compliant, respectively. N_{tot} denotes the number of samples in the considered class and in the entire set (calibration or validation), respectively. The best classification model was determined as the one with the highest overall discrimination rate and lowest false compliant samples.

The repeatability of the classification models was estimated by analyzing by FT-NIR spectroscopy the laboratory QC sample of ground unprocessed durum wheat. The sample was analyzed ten times under repeatability conditions, and DON levels were predicted by each LDA model (I–IV). For each model, the repeatability was calculated as the coefficient of variation (CV) of the Mahalanobis squared distances in the PCA of the ten measurements. The within-laboratory reproducibility of the classification models was estimated by analyzing the same quality control sample ten times on three different days under within-laboratory reproducibility conditions. For each model the CV was calculated by considering the 30 measurements.

4. Conclusions

Classification and calibration methods using Fourier-transform near-infrared (FT-NIR) spectroscopy have been developed to predict DON levels in unprocessed durum wheat samples. The quantitative PLS models showed a very poor classification ability and were not recommended for any purpose. On the other hand, the classification models showed good predictive performance with high overall classification rates and low misclassification of wheat samples and fulfilled the requirement of the European official guidelines for screening methods when a cut-off value of 1,400 µg/kg of deoxynivalenol was used.

The proposed classification model may have practical applications for screening durum wheat samples for deoxynivalenol content, making FT-NIR spectroscopy a powerful and robust tool for monitoring safety programs. Moreover, the speed of the procedure allows the analysis of a large number of samples and reduces the number of samples to be confirmed with a reference method to verify the compliance with regulation.

Acknowledgments: AGER (Agro-Food and Research) is acknowledged for funding the project "From Seed to Pasta". The valuable technical assistance of Roberto Schena is also acknowledged.

Author Contributions: Annalisa De Girolamo was responsible for the experimental design, FT-NIR analyses, statistical evaluation of data and writing of the manuscript. Salvatore Cervellieri was responsible for HPLC analyses. Angelo Visconti was responsible for the4 experimental design and critical reviewing of the manuscript. Michelangelo Pascale was responsible for the experimental design and writing of the manuscript.

Conflicts of Interest: The authors declare no conflict of interest.

References

1. Canady, R.A.; Coker, R.D.; Egan, S.K.; Krska, R.; Kuiper-Goodman, T.; Olsen, M.; Pestka, J.; Resnik, S.; Schlatter, J. Deoxynivalenol. In *Safety Evaluation of Certain Mycotoxins in Food*; WHO Food Additive Series 47; World Health Organization: Geneva, Switzerland, 2001; pp. 419–555.
2. Shephard, G.S. *Fusarium* mycotoxins and human health. *Plant Breed. Seed Sci.* **2011**, *64*, 113–121.
3. Arunachalam, C.; Doohan, F.M. Trichothecene toxicity in eukaryotes: Cellular and molecular mechanisms in plants and animals. *Toxicol. Lett.* **2013**, *217*, 149–158. [CrossRef]
4. Antonissen, G.; Martel, A.; Pasmans, F.; Ducatelle, R.; Verbrugghe, E.; Vandenbroucke, V.; Li, S.; Haesebrouck, F.; van Immerseel, F.; Croubels, S.; *et al.* The impact of *Fusarium* mycotoxins on human and animal host susceptibility to infectious diseases. *Toxins* **2014**, *6*, 430–452. [CrossRef]
5. Pinton, P.; Oswald, I.P. Effect of deoxynivalenol and other Type B trichothecenes on the intestine: A review. *Toxins* **2014**, *6*, 1615–1643. [CrossRef]
6. Maresca, M. From the gut to the brain: Journey and pathophysiological effects of the food-associated trichothecene mycotoxin deoxynivalenol. *Toxins* **2013**, *5*, 784–820. [CrossRef]
7. European Commission. Commission Regulation (EC) No. 1126/2007 of 28 September 2007 amending Regulation No. 1881/2006 Setting maximum levels for certain contaminants in foodstuffs as regards Fusarium toxins in maize and maize products. *Off. J. Eur. Union* **2007**, *L255*, 14–17.
8. Ran, R.; Wang, W.; Han, Z.; Wu, A.; Zhang, D.; Shi, J. Determination of deoxynivalenol (DON) and its derivatives: Current status of analytical methods. *Food Control* **2013**, *34*, 138–148. [CrossRef]
9. Meneely, J.P.; Ricci, F.; van Egmond, H.P.; Elliott, C.T. Current methods of analysis for thedetermination of trichothecenemycotoxins in food. *Trends Anal. Chem.* **2011**, *30*, 192–203. [CrossRef]
10. Lattanzio, V.M.T.; Pascale, M.; Visconti, A. Current analytical methods for trichothecene mycotoxins in cereals. *Trends Anal. Chem.* **2011**, *28*, 758–768. [CrossRef]
11. Lippolis, V.; Maragos, C. Fluorescence polarization immunoassays for rapid, accurate and sensitive determination of mycotoxins. *World Mycotoxin J.* **2014**, *7*, 479–489. [CrossRef]
12. Li, Y.; Liu, X.; Lin, Z. Recent developments and applications of surface plasmon resonance biosensors for the detection of mycotoxins in foodstuffs. *Food Chem.* **2012**, *132*, 1549–1554. [CrossRef]
13. Lippolis, V.; Pascale, M.; Cervellieri, S.; Damascelli, A.; Visconti, A. Screening of deoxynivalenol contamination in durum wheat by MOS-based electronic nose and identification of the relevant pattern of volatile compounds. *Food Control* **2014**, *37*, 263–271. [CrossRef]
14. McClure, W. Review: 204 years of near infrared technology: 1800–2003. *J. Near Infrared Spectrosc.* **2003**, *11*, 487–518. [CrossRef]
15. Santos, C.; Frafa, M.E.; Kozakiewicz, Z.; Lima, N. Fourier transform infrared as a powerful technique for the identification and characterization of filamentous fungi and yeasts. *Res. Microb.* **2010**, *161*, 168–175. [CrossRef]
16. Pettersson, H.; Aberg, L. Near infrared spectroscopy for determination of mycotoxins in cereals. *Food Control* **2003**, *14*, 229–232. [CrossRef]
17. Kos, G.; Krska, R.; Lohninger, H.; Griffiths, P.R. A comparative study of mid-infrared diffuse reflection (DR) and attenuated total reflection (ATR) spectroscopy for the detection of fungal infection on RWA2-corn. *Anal. Bioanal. Chem.* **2004**, *378*, 159–166. [CrossRef]
18. Abramovic, B.; Jajic, I.; Abramovic, B.; Cosic, J.; Juric, V. Detection of deoxynivalenol in wheat by Fourier transform infrared spectroscopy. *Acta Chim. Slov.* **2007**, *54*, 859–867.
19. De Girolamo, A.; Lippolis, V.; Nordkvist, E.; Visconti, A. Rapid and non-invasive analysis of deoxynivalenol in durum and common wheat by Fourier-Transform Near Infrared (FT-NIR) spectroscopy. *Food Addit. Contam. Part A Chem. Anal. Control Expo. Risk Assess.* **2009**, *26*, 907–917. [CrossRef]
20. Siuda, R.; Balcerowska, G.; Kupcewicz, B.; Lenc, L. A modified approach to evaluation of DON content in scab-damaged ground wheat by use of diffuse reflectance spectroscopy. *Food Anal. Methods* **2008**, *1*, 283–292. [CrossRef]
21. Bolduan, C.; Montes, J.M.; Dhillon, B.S.; Mirdita, V.; Melchinger, A.E. Determination of mycotoxin concentration by ELISA and near-infrared spectroscopy in *Fusarium*-inoculated maize. *Cereal Res. Commun.* **2009**, *37*, 521–529. [CrossRef]

22. Beyer, M.; Pogoda, F.; Ronellenfitsch, F.K.; Hoffmann, L.; Udelhoven, T. Estimating deoxynivalenol contents of wheat samples containing different levels of *Fusarium*-damaged kernels by diffuse reflectance spectrometry and partial least square regression. *Int. J. Food Microb.* **2010**, *142*, 370–374. [CrossRef]

23. Dvořáček, V.; Prohasková, A.; Chrpová, J.; Štočková, L. Near infrared spectroscopy for deoxynivalenol content estimation in intact wheat grain. *Plant Soil Environ.* **2012**, *58*, 196–203.

24. Czechlowski, M.; Laskowska, M. The development and validation of the calibration model for the VIS-NIR spectrometer used for the evaluation of deoxynivalenol content in wheat grain directly during combine harvest. *J. Res. Appl. Agric. Eng.* **2013**, *58*, 27–30.

25. Galvis-Sánchez, A.C.; Barros, A.S.; Delgadillo, I. Method for analysis dried vine fruits contaminated with ochratoxin A. *Anal. Chim. Acta* **2008**, *617*, 59–63. [CrossRef]

26. Bozza, A.; Tralamazza, S.M.; Rodriguez, J.I.; Scholz, M.B.S.; Reynaud, D.T.; Dalzoto, P.R.; Pimentel, I.C. Potential of Fourier Transform infrared Spectroscopy (FT-IR) to detection and quantification of ochratoxin A: A comparison between reflectance and transmittance techniques. *Int. J. Pharm. Chem. Biol. Sci.* **2013**, *3*, 1242–1247.

27. Berardo, N.; Pisacane, V.; Battilani, P.; Scandolara, A.; Pietri, A.; Marocco, A. Rapid detection of kernel rots and mycotoxins in maize by near-infrared reflectance spectroscopy. *J. Agric. Food Chem.* **2005**, *53*, 8128–8134. [CrossRef]

28. Gaspardo, B.; del Zotto, S.; Torelli, E.; Cividino, S.R.; Firrao, G.; Della Riccia, G.; Stefanon, B. A rapid method for detection of fumonisins B1 and B2 in corn meal using Fourier transform near infrared (FT-NIR) spectroscopy implemented with integrating sphere. *Food Chem.* **2012**, *135*, 1608–1612. [CrossRef]

29. Della Riccia, G.; del Zotto, S. A multivariate regression model for detection of fumonisins content in maize from near infrared spectra. *Food Chem.* **2013**, *141*, 4289–4294. [CrossRef]

30. Hernández-Hierro, J.M.; García-Villanova, R.J.; Gonzáles-Martín, I. Potential of near infrared spectroscopy for the analysis of mycotoxins applied to naturally contaminated red paprika found in the Spanish market. *Anal. Chim. Acta* **2008**, *622*, 189–194. [CrossRef]

31. Fernández-Ibañez, V.; Soldado, A.; Martínez-Fernández, A.; de la Roza-Delgado, B. Application of near infrared spectroscopy for rapid detection of aflatoxin B_1 in maize and barley as analytical quality assessment. *Food Chem.* **2009**, *113*, 629–634. [CrossRef]

32. Tekle, S.; Bjørnstad, A.; Skinnes, H.; Dong, Y.; Segtnan, V.H. Estimating deoxynivalenol content of ground oats using VIS-NIR spectroscopy. *Cereal Chem. J.* **2013**, *90*, 181–185. [CrossRef]

33. Ruan, R.; Li, Y.; Lin, X.; Chen, P. Non-destructive determination of deoxynivalenol levels in barley using near-infrared spectroscopy. *Appl. Eng. Agric.* **2002**, *18*, 549–553. [CrossRef]

34. Peiris, K.H.S.; Dong, Y.; Bockus, B.B.; Dowell, F.E. Estimation of bulk deoxynivalenol and moisture content of wheat grain samples by FT-NIR spectroscopy. In Proceedings of the 2013 ASABE Annual International Meeting, St. Joseph, MI, USA, 21–24 July 2013; ASABE Paper N. 131593402. pp. 1–8.

35. Garon, D.; el Kaddoumi, A.; Carayon, A.; Amiel, C. FT-IR spectroscopy for rapid differentiation of *Aspergillus flavus*, *Aspergillus fumigatus*, *Aspergillus parasiticus* and characterization of aflatoxigenic isolates collected from agricultural environments. *Mycopathologia* **2010**, *170*, 131–142. [CrossRef]

36. Dachoupakan Sirisomboon, C.; Putthang, R.; Sirisomboon, P. Application of near infrared spectroscopy to detect aflatoxigenic fungal contamination in rice. *Food Control* **2013**, *33*, 207–214. [CrossRef]

37. Levasseur-Garcia, C.; Kleiber, D.; Surel, O. Infrared spectroscopy used as a decision-making support for the determination of fungal and mycotoxic risk. *Cah. Agric.* **2013**, *22*, 216–227.

38. European Commission. Commission Decision No 657/2002 of 12 August 2002 implementing Council Directive 96/23/EC concerning the performance of analytical methods and the interpretation of results. *Off. J. Eur. Commun.* **2002**, *L221*, 8–36.

39. Visconti, A.; Haidukowski, M.; Pascale, M.; Silvestri, M. Reduction of deoxynivalenol during durum wheat processing and spaghetti cooking. *Toxicol. Lett.* **2004**, *153*, 181–189. [CrossRef]

40. Savitzky, A.; Golay, M.J.E. Smoothing and differentiation of data by simplified least squares procedures. *Anal. Chem.* **1964**, *36*, 1627–1639. [CrossRef]

41. Williams, P.C. Near-infrared technology: Getting the best out of light. In *A Short Course in the Practical Implementation of Near Infrared Spectroscopy for the User*; PDK Projects Inc.: Nanaimo, BC, Canada, 2004; p. 109.

42. Fearn, T. Assessing calibrations: SEP, RPD, RER and R^2. *NIR News* **2002**, *13*, 12–14. [CrossRef]

43. Williams, P.C. Implementation of near-infrared technology. In *Near Infrared Technology in the Agricultural and Food Industries*, 2nd ed.; Williams, P., Norris, K., Eds.; American Association of Cereal Chemists: St. Paul, MN, USA, 2001.

44. Johnson, D.E. Discriminant analysis. In *Applied Multivariate Methods for Data Analysts*; Johnson, D.E., Ed.; Duxbury Press Brookes/Cole Publishing Company: Pacific Grove, CA, USA, 1998; pp. 217–285.

toxins

MDPI

Article

Assessment of Multi-Mycotoxin Exposure in Southern Italy by Urinary Multi-Biomarker Determination

Michele Solfrizzo *, Lucia Gambacorta and Angelo Visconti

Institute of Sciences of Food Production (ISPA), National Research Council (CNR), Bari 70126, Italy; lucia.gambacorta@ispa.cnr.it (L.G.); angelo.visconti@ispa.cnr.it (A.V.)
* Author to whom correspondence should be addressed; michele.solfrizzo@ispa.cnr.it; Tel.: +39-080-592-9367; Fax: +39-080-592-9374.

Received: 13 December 2013; in revised form: 13 January 2014; Accepted: 21 January 2014; Published: 28 January 2014

Abstract: Human exposure assessment to deoxynivalenol (DON), aflatoxin B1 (AFB1), fumonisin B1 (FB1), zearalenone (ZEA) and ochratoxin A (OTA) can be performed by measuring their urinary biomarkers. Suitable biomarkers of exposure for these mycotoxins are DON + de-epoxydeoxynivalenol (DOM-1), aflatoxin M1 (AFM1), FB_1, ZEA + α-zearalenol (α-ZOL) + β-zearalenol (β-ZOL) and OTA, respectively. An UPLC-MS/MS multi-biomarker method was used to detect and measure incidence and levels of these biomarkers in urine samples of 52 volunteers resident in Apulia region in Southern Italy. The presence of ZEA + ZOLs, OTA, DON, FB1 and AFM1 were detected in 100%, 100%, 96%, 56% and 6%, of samples, respectively. All samples contained biomarkers of two or more mycotoxins. The mean concentrations of biomarkers ranged from 0.055 ng/mL (FB1) to 11.89 ng/mL (DON). Urinary biomarker concentrations were used to estimate human exposure to multiple mycotoxin. For OTA and DON, 94% and 40% of volunteers, respectively exceeded the tolerable daily intake (TDI) for these mycotoxins. The estimated human exposure to FB1 and ZEA was largely below the TDI for these mycotoxins for all volunteers.

Keywords: mycotoxins; biomarker; urine; UPLC-MS/MS; immunoaffinity cleanup; exposure

---

1. Introduction

Aflatoxins, deoxynivalenol (DON), zearalenone (ZEA), fumonisins and ochratoxin A (OTA) are recognized as the principal mycotoxins occurring in agricultural products and their levels in food commodities are constantly inspected worldwide. Humans can be daily exposed to mixtures of these mycotoxins through consumption of foods contaminated with several mycotoxins or consumption of different foods contaminated by a single mycotoxin. Data on the co-occurrence of the principal mycotoxins in foods and beverages are increasing due to the availability and use of modern and sensitive LC-MS/MS methodologies suitable for simultaneous determination of mycotoxins and other fungal metabolites [1,2]. In a recent survey on 265 samples of cereal-based products commercialized in Spain, Italy, Marocco and Tunisia, 14% of the analyzed samples were contaminated with at least two mycotoxins and 18% of the analyzed samples were contaminated by more than two mycotoxins simultaneously [3]. The co-occurrence of DON, ZEA and nivalenol (NIV) in winter wheat produced in Sweden has been recently reported [4]. The presence of mixtures of aflatoxin B1 (AFB_1), ZEA and OTA was reported in samples of breakfast cereals commercialized in Spain [5]. The majority of food commodities consumed in Cameroon were found contaminated with mixtures of mycotoxins, 21% contained DON, ZEA and fumonisin B1 (FB1), 11% DON, AFB1, FB1 and ZEA [6]. Recently, the co-occurrence of DON, FB1, fumonisin B2 (FB2), fumonisin B3 (FB3) and ZEA in good and moldy maize and DON, FB1, FB2, ZEA and OTA in samples of maize based foods was reported in the former Transkei region of South Africa [7,8].

Exposure to mycotoxins can also originate from the ingestion of their masked forms (mycotoxins covalently or non-covalently bound to matrix component) that are digested in the gastrointestinal tract and released into the parent mycotoxins that become bioavailable. The occurrence of masked DON and ZEA, either acetylated, conjugated with glucose or sulfate, has been reported in various cereals and food samples [9]. The occurrence of masked fumonisins in processed food has also been reported but the nature of the masking mechanism has not been fully clarified [10]. The formation of β-glucosides of (4R)- and (4S)-5-hydroxy-OTA in germinating cereals and vegetables spiked with OTA has been demonstrated but the occurrence of these compounds in naturally contaminated food has not been reported [11]. The degree of human bioavailability of mycotoxins derived from their masked forms is not known and could vary between individuals depending of the intestinal microbiota composition.

The assessment of human exposure to mycotoxins is usually performed by means of chemical analysis of foods and beverages and results are correlated with the mean intake of analyzed foods/beverages. The heterogeneous distribution of mycotoxins in food samples can affect the accuracy of results obtained with this approach. Duplicate diet studies could avoid sampling issues but requires suitable analytical methods for single or multi-mycotoxin determination and considerable commitment from the participants.

The measurement of specific urinary mycotoxin biomarkers is a valid alternative to measure exposure to mycotoxins providing that the excretion of biomarkers correlate well with mycotoxin intake. Suitable urinary biomarkers for AFB1, FB1, ZEA, DON and OTA are aflatoxin M1 (AFM1), FB1, ZEA + α-zearalenol (α-ZOL) + β-zearalenol (β-ZOL), DON + de-epoxydeoxynivalenol (DOM-1) and OTA, respectively. These biomarkers are excreted as free and conjugated forms therefore urine samples are usually digested with β-glucuronidase/sulfatase in order to deconjugate the conjugated forms and increase the concentration and detectability of free analytes. Human pilot studies and epidemiological studies based on biomarker approach have been performed for DON, AFB1, OTA and FB1. These studies were conducted by using analytical methods tailored for determination of biomarker(s) of a single mycotoxin [12–15]. LC-MS/MS is the ideal approach for simultaneous determination of analytes and its use for multi-biomarker determination in human and animal urine is recently increased [8,16]. *In vivo* experiments demonstrated a good correlation between the amount of mixtures of DON, OTA, ZEA, FB1 and AFB1 administered to piglets and the amount of relevant biomarkers excreted in 24 h post dose urine. Linear dose-response correlation coefficients ranged between 0.68 and 0.78 for the tested couples of mycotoxin/biomarker. Mean percentages of dietary mycotoxins excreted as biomarkers in 24 h post dose urine were 36.8% for ZEA, 28.5% for DON, 2.6% for FB1, 2.6% for OTA and 2.5% for AFB1 [17]. In this paper, we report the results on the occurrence of biomarkers to DON, OTA, ZEA, FB1 and AFB1 in urine samples of 52 volunteers resident in Apulia a region of Southern Italy.

2. Results and Discussion

The urinary concentrations of mycotoxin biomarkers could be very low for several reasons: (a) the levels of mycotoxins in foods and beverages can be very low especially for AFB1, OTA and ZEA; (b) the gastrointestinal absorption can be low as demonstrated for FB1, c) the serum half-life can be very high as demonstrated for OTA. The purification protocol used in our study for simultaneous determination of urinary DON, DOM-1, AFM1, FB1, ZEA, α-ZOL, β-ZOL and OTA was successfully validated in a mini comparison study involving other laboratories that used either another multi-biomarker method or single-biomarker methods for determination of DON and FB1 [18]. To improve the sensitivity of the method we used an UPLC system coupled with a powerful and sensitive mass spectrometer (API 5000 MS/MS system with ESI interface). It was necessary to optimize the chromatographic conditions for optimal biomarker separation on Acquity BEH phenyl column as well as the MS/MS conditions. The optimized MS/MS conditions for each biomarker are reported in Table 1.

The main differences with the MS/MS parameters previously optimized on a QTrap system [19] were: (a) the increase in number of daughter ions from 3 to 4 for FB1, α-ZOL and β-ZOL; (b) minor modification in the clustering potential, entrance potential, collision energy and collision cell exit potential (Table 1). The use of the Acquity BEH phenyl column and the development of a new linear gradient composition of the mobile phase permitted to reduce the run time from 46 min to 15 min. However, each sample extract was analyzed twice in positive and negative mode. AFM1, FB1 and OTA were detected and measured in positive mode whereas DON, DOM-1, ZEA, α-ZOL and β-ZOL were detected and measured in negative mode. The use of an UPLC system permitted to use a 150 mm × 2.1 mm column with a 1.7 μm particles size of stationary phase that produced sharp peak thus increasing peak high with consequent reduction of the limit of detection (LOD) and quantification (LOQ). The use of a powerful mass spectrometer (API 5000 MS/MS system) produced a further increase in method sensitivity. In particular, a marked increase in sensitivity (up to 114 times) was obtained, in descending order, for ZEA, β-ZOL, α-ZOL, FB1, AFM1 and OTA with the new UPLC-MS/MS system as compared to the previous one (Table 2). No increase of sensitivity was obtained for DON whereas for DOM-1 an increase of LOQ value was observed probably due to ion suppression effect.

Table 1. MS/MS conditions for detection of target analytes by MRM method.

Analyte	Precursor ion	Q1 (m/z)	Q3 (m/z)	DP (V)	EP (V)	CE (V)	CXP (V)
DON	$[DON+CH_3COO]^-$	355.0	295.2 265.0 [a] 59.0 [a]	−50	−10	−15 −22 −35	−20
DOM-1	$[DOM-1+CH_3COO]^-$	339.5	279.0 249.2 59.0 [a]	−50 −50 −55	−10	−12 −18 −35	−20
AFM1	$[AFM1+H]^+$	329.5	273.3 [a] 229.0 [a]	80	10	36.5 56	20
FB1	$[FB1+H]^+$	722.4	370.6 [a] 352.6 [a] 334.7 [a] 316.6 [a]	50	10	55 59 65 66	10
α-ZOL	$[α\text{-}ZOL\text{-}H]^-$	319.1	188.2 [a] 174.1 [a] 160.1 [a] 130.0 [a]	−100	−10	−38 −36 −43 −48	−10
β-ZOL	$[β\text{-}ZOL\text{-}H]^-$	319.1	188.2 [a] 174.1 [a] 160.1 [a] 130.0 [a]	−100	−10	−38 −36 −43 −48	−10
ZEA	$[ZEA\text{-}H]^-$	317.2	273.3 175.0 131.0 [a]	−100	−10	−29 −35 −40	−10
OTA	$[OTA+H]^+$	404.2	358.5 257.3 239.2 [a]	90	10	30 38 47	10

Notes: [a] Transitions used for quantitation; Q1: first quadrupole; Q3: third quadrupole; DP: declustering potential; EP: entrance potential; CE. collision energy; CXP. collision cell exit potential

Table 2. LOQ values for DON, AFM1, FB1, β-ZOL, α-ZOL, ZEA and OTA in human urine samples obtained with two different LC-MS/MS systems: LC-QTrap MS/MS system and UPLC-API 5000 MS/MS system.

Biomarker	Apparatus 1: LC-QTrap MS/MS LOQ (ng/mL)	Apparatus 2: UPLC-API 5000 MS/MS LOQ (ng/mL)
ZEA	0.8	0.007
β-ZOL	4.4	0.054
α-ZOL	1.6	0.030
FB1	0.1	0.010
AFM1	0.1	0.020
OTA	0.02	0.006
DON	1.5	1.5
DOM-1	1.5	9.9

In Figure 1 are reported three chromatograms of a naturally contaminated urine sample containing DON, ZEA, α-ZOL, β-ZOL, FB1 and OTA.

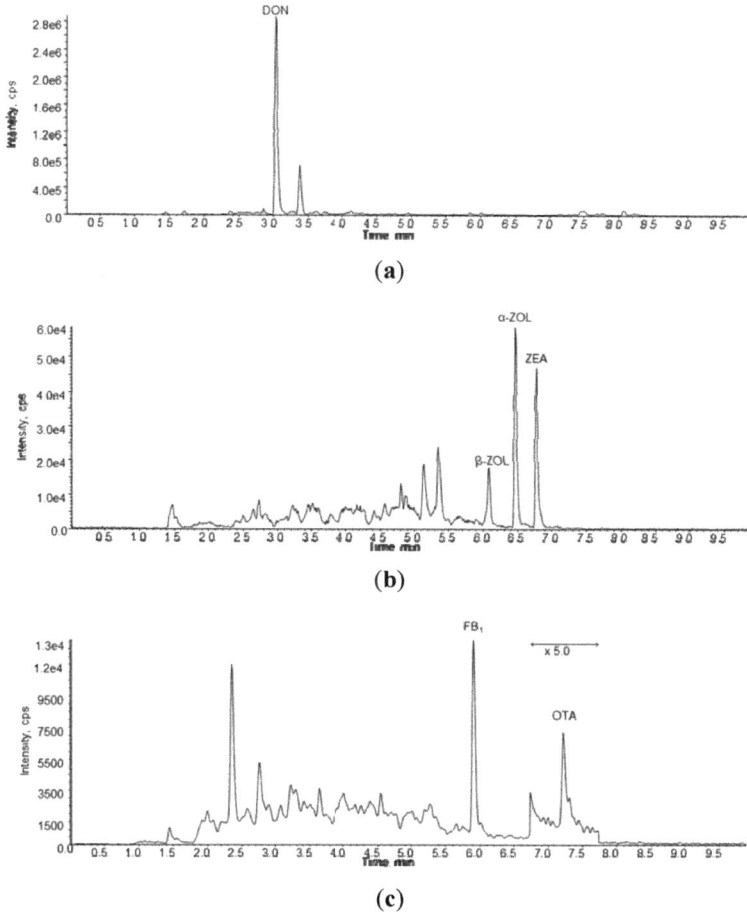

(a)

(b)

(c)

Figure 1. LC-MS/MS chromatograms obtained in negative ion mode (**a**,**b**) and in positive ion mode (**c**) of a naturally urine sample (#4) containing 11.33 ng/mL of DON, 0.108 ng/mL of β-ZOL, 0.123 ng/mL of α-ZOL, 0.082 ng/mL of ZEA, 0.26 ng/mL of FB1 and 0.06 ng/mL of OTA.

A summary of the results of the urine samples collected in this study are reported in Table 3. All urine samples contained ZEA, α-ZOL and OTA whereas β-ZOL was found in 98% of samples. DON, FB1 and AFM1 were found in 96%, 56% and 6% of urine samples, respectively. The highest mean biomarker concentration was found for DON (11.89 ng/mL) followed by OTA (0.144 ng/mL), β-ZOL (0.090 ng/mL), α-ZOL (0.077 ng/mL), AFM1 (0.068 ng/mL), ZEA (0.057 ng/mL) and FB1(0.055 ng/mL). The highest individual biomarker concentration was measured for DON (67.36 ng/mL) followed by OTA (2.129 ng/mL), FB1 (0.352 ng/mL), α-ZOL (0.176 ng/mL), β-ZOL (0.135 ng/mL), AFM1 (0.146 ng/mL) and ZEA (0.120 ng/mL). From this Table it is evident that urinary concentrations of DON are much higher to those of the other biomarkers. Moreover, the highest inhomogeneous distributions of concentrations was observed for OTA as demonstrated by the value of relative standard deviation (RSD) of the mean (217%) followed by FB1 (133%), AFM1 (99%), DON (84%), ZEA (40%), α-ZOL (35%) and β-ZOL (16%).

Table 3. Results of mycotoxin biomarkers in human urine samples collected in Southern Italy and estimated values of PDI for each mycotoxin

Biomarkers	DON	β-ZOL	α-ZOL	ZEA	FB1	OTA	AFM1
N. positive (%)	50 (96)	51 (98)	52 (100)	52 (100)	29 (56)	52 (100)	3 (6)
Mean, ng/mL	11.89	0.090	0.077	0.057	0.055	0.144	0.068
SD, ng/mL	10.05	0.014	0.027	0.023	0.073	0.312	0.067
Median, ng/mL	10.32	0.088	0.074	0.056	0.029	0.061	0.10
Max, ng/mL	67.36	0.135	0.176	0.120	0.352	2.129	0.146
Mycotoxins	DON			ZEA	FB1	OTA	AFB1
Mean PDI[a], μg/kg body weight	1.03			0.015	0.053	0.139	0.068
Max PDI[a], μg/kg body weight	5.90			0.029	0.338	2.047	0.142
% of PDI[a] values exceeding the TDI	40			0	0	94	0
Mean PDI[b], μg/kg body weight	0.59			_[c]	0.274	_[c]	_[c]
Max PDI[b], μg/kg body weight	3.37			_[c]	1.759	_[c]	_[c]
% of PDI[b] values exceeding the TDI	6			_[c]	0	_[c]	_[c]
TDI[d], μg/kg body weight	1.0			0.2	2.0	0.017	_[e]

Notes: [a] calculated based on piglet excretion data; [b] calculated based on human excretion data (50% for DON, 0.5% for FB1) reported in Shephard *et al.* [8]; [c] not estimated due to unavailability of human excretion rate; [d] TDI values are reported in [19] and references therein; [e] there is no TDI value for AFB1 because it is a carcinogenic mycotoxin.

The results of the co-occurrence of multiple biomarkers in the tested urine samples are reported in Table 4. The majority of urine samples (52%) contained biomarkers of DON, ZEA, FB1 and OTA whereas 38% of samples contained biomarkers of DON, ZEA and OTA. The co-occurrence of biomarkers of all mycotoxins was found in two urine samples. Moreover, no individual was found unexposed or exposed to a single mycotoxin since all investigated urine samples contained biomarkers of at least two mycotoxins. No important differences were observed for the results obtained for male and female individuals with the exception that the two urine samples containing biomarkers of all mycotoxins were from females. Mixtures of biomarkers of DON, ZEA, FB1 and OTA were found in human urine samples collected in the former Transkei, South Africa [8]. Co-occurrence of biomarkers of 2 to 5 mycotoxins was reported in human urine samples collected in Cameroon [20,21].

Table 4. Incidence of individuals exposed to mixtures of mycotoxins for a total of 52 volunteers (26 males and 26 females).

Multiple mycotoxins exposure	n. positive samples	% of positive samples
DON, ZEA, FB1, OTA, AFB1	2	4
DON, ZEA, FB1, OTA	27	52
DON, ZEA, OTA	20	38
DON, ZEA, OTA, AFB1	1	2
ZEA, OTA	2	4
TOTAL	52	100

This is the first report on the simultaneous detection of biomarkers of the 5 principal mycotoxins in Italy. The simultaneous presence of DON and OTA in human urine in Italy was previously reported [19] whereas the presence of ZEA, α-ZOL, β-ZOL, FB1 and AFM1 is reported herein for the first time. The presence of DON and OTA in almost all urinary samples is not surprising because these mycotoxins are usually found in cereals and derived products, staple foods of Italian people [22,23]. Interestingly, 56% of urine samples contained FB1, a mycotoxin mainly found in maize and deriving products. Although these products are not staple foods in Italy they are widely consumed as chips, polenta, popcorn, beer, cornflakes, snacks, muesli and mixed cereals [22,24]. The consumption of these products could explain the high percentage of urine samples containing FB1. The low concentrations of biomarkers of ZEA in all tested urine samples is also a new and interesting information and demonstrate that all the 52 volunteers that participated in our study were exposed to low levels of ZEA. On the other hand the results of a large survey conducted in Europe showed that only 32% of 5018 samples of cereals and derived products, the main source for ZEA exposure, were found contaminated with this mycotoxin [22]. A possible explanation of this apparent paradox could be that the high sensitivity of the multi biomarker method used in the present study can detect urinary biomarker deriving from the consumption of foods contaminated with very low levels of ZEA that could not be detected with conventional analytical methods. The presence of AFM1 in only 3 urine samples demonstrates that human exposure to AFB1 is quite limited in this area of Southern Italy. In fact ground nuts and tree nuts, the main source of human AFB1 exposure in Italy, are occasionally consumed in this Country. The absence of AFM1 in almost all urine samples suggests that maize and derived products consumed in Southern Italy are probably not contaminated with AFB1 although it is well known that maize is a major source of AFB1 in some countries. The absence of AFM1 in human urine and AFB1 in maize based food was also reported in South Africa [8].

The urinary biomarker concentrations measured in this study were used to estimate the probable daily intake (PDI) of each mycotoxin by each volunteer according to Equation (1).

$$\text{PDI} = C \times \frac{V}{W} \times \frac{100}{E}$$

(1)

PDI   probable daily intake of mycotoxin (µg/kg body weight);
C human urinary biomarker concentration (µg/L);
V mean 24 h human urine volume (1.5 L);
W mean human body weight (60 kg);
E mean urinary excretion rate of mycotoxin in 24 h post dose in piglets (36.8% for ZEA, 27.9% for DON, 2.6% for FB1, 2.6% for OTA and 2.5% for AFM1 [17]).

Based on mean urinary concentration of each mycotoxin biomarker measured in the 52 human urine samples and the mean urinary excretion rate in piglets for each mycotoxin, the estimated daily mean intake of the 5 investigated mycotoxins were calculated and are reported in Table 3. Due to the unavailability of human excretion rate for all the 5 mycotoxins considered in this study we used the 24 h excretion rate measured in piglets [17] to estimate the PDI in human. However, since sufficient data of human excretion rate of DON and FB1 have been reported [8], additional values of PDI, calculated with human data, were added in Table 3. In this table, the values of max PDI for each mycotoxin and the percentage of individuals that exceeded the tolerable daily intake (TDI) for each mycotoxin are also reported. The estimated mean values of PDI were below or equal to the TDI or provisional maximum TDI (PMTDI) for DON, FB1 and ZEA. Individual analysis of PDI values obtained for DON revealed that 40% of volunteers exceeded the value of TDI of 1 µg/kg body weight established for this mycotoxin with a maximum value of PDI of 5.90 µg/kg body weight. Previous studies conducted in UK, France and Sweden using the urinary DON concentration to estimate DON exposure reported mean values of PDI of 0.12–0.73, 0.61 and 0.16 µg/kg body weight, respectively [25–30]. Human exposure to DON in our study seems to be higher to that estimated in UK since both mean values of PDI and % of individuals that exceed TDI are higher as compared to UK where some 5% of the adult population may exceed the TDI for DON intake [25]. The mean PDI values of DON derived from food analyses in UK (0.14–0.23 µg/kg body weight) seems to be slightly lower as compared to PDI values estimated from urinary DON in that Country [25–28,31]. The mean PDI values of DON derived from food analysis in Europe and reported by SCOOP report [22] and FAO/WHO [32] were 0.34 and 1.4 µg/kg body weight, respectively. The results of our study obtained with the biomarker approach fall within this interval and is more close to the PDI reported by FAO/WHO [32]. A lower mean value of PDI (0.59 µg/kg body weight) was obtained by using the human excretion rate of 50% reported by Shephard *et al.* [8] as well as the percentage of individuals (6%) that exceeded the value of TDI for this mycotoxin.

The mean value of PDI for FB1 calculated in our study with biomarker approach (0.053 µg/kg body weight) is far below the TDI established for this mycotoxin (2 µg/kg body weight) and all individual values are below the TDI for this mycotoxin with a maximum value of PDI of 0.338 µg/kg body weight (Table 3). These results can only be compared with results obtained in Guatemala, South Africa and Mexico because urinary FB1 in European population has not been performed yet. The mean values of PDI estimated with biomarker approach in Guatemala, Mexico and South Africa were 0.45, 0.37 and 0.22 µg/kg body weight, respectively [15,33,34]. Comparison with these data shows that in our study human exposure to FB1 is about 10 times lower to those estimated in Guatemala, Mexico and South Africa. The mean PDI values of FB1 obtained in our study with biomarker approach (0.053 µg/kg body weight) is quite similar to the PDI of 0.056 µg/kg body weight estimated with diet approach by Brera *et al.* in Italy [24]. The estimated mean PDI value reported by FAO/WHO [32] for Europe with the diet approach was 0.2 µg/kg body weight which is 3.8 times higher than the intake estimated in our study. A much higher mean value of PDI (0.274 µg/kg body weight) was obtained by using the human excretion rate of 0.5% reported by Shephard *et al.*, [8]. However even with this estimate all individuals resulted below the value of TDI for this mycotoxin.

The mean value of PDI of ZEA estimated with biomarker approach in our study (0.015 µg/kg body weight) is about ten times below the TDI established for this mycotoxin (0.2 µg/kg body weight). The maximum estimated value of PDI is 0.029 µg/kg body weight (Table 3). In Italy, the mean value of PDI estimated with the diet approach was 0.0008 µg/kg body weight [22] whereas the PDI reported by EFSA for Europe was 0.03–0.06 µg/kg body weight [35]. The mean value of PDI estimated in our study is higher than the Italian value reported in the SCOOP report and lower than the lower limit of the range reported by EFSA [35].

The occurrence of ZEA biomarkers in human urine was recently reported in USA and Cameroon. Mixtures of ZEA, α-ZOL and β-ZOL were detected and measured in 55% of urine samples collected from girls resident in New Jersey with mean concentrations ranging from 0.35 ng/mL (β-ZOL) to 1.82 ng/mL (ZEA) [36]. These concentrations are 4–32 times higher the mean concentrations found in our study for these ZEA biomarkers which means that in New Jersey human exposure to ZEA is higher as compared to Italy. Mixtures of ZEA, α-ZOL and ZEA-glucuronide were found in a low percentage (5%) of human urine samples collected in Cameroon with a total mean concentration of 0.74 ng/mL [20].

As reported in Table 3 the mean value of PDI of OTA estimated with biomarker approach in our study (0.139 µg/kg body weight) is about 8 times higher than the TDI (0.017 µg/kg body weight) established for this mycotoxin [37]. The maximum value of PDI estimated in our study (2.047 µg/kg body weight) is 147 times higher than the TDI and 94% of individuals participating in our study exceeded the value of TDI (Table 3). Several studies reported the occurrence of OTA in human urine with a high percentage of positive samples but the relevant values of PDI from these results were not calculated nor reported. We used the human urinary concentrations of OTA reported in literature to estimate values of PDI of OTA by using the Equation (1) reported above. Urinary concentrations of OTA reported by Fazekas *et al.* [38] and Duarte *et al.* [14,39] produced values of PDI equal or below the TDI, whereas from the results of Pena *et al.* [40], Manique *et al.* [41], Coronel *et al.* [42], Gilbert *et al.* [43] and Domijan *et al.* [44] the estimated values of PDI were higher the TDI. Our result of estimated mean PDI, though higher than the TDI, is lower than the mean PDI values estimated from the urinary concentrations reported by Domijan *et al.* in Croatia [44] and Coronel *et al.* in Spain [42]. When compared to the PDI values estimated with urinary concentrations reported by Fazekas *et al.* [38], Duarte *et al.* [14,39], Pena *et al.* [40], Manique *et al.* [41] and Gilbert *et al.* [43] our estimated PDI values were higher.

In Europe, the mean values of PDI of OTA estimated with the diet approach range from 0.0011 to 0.024 µg/kg body weight [23]. The mean value of PDI estimated with our biomarker study (0.139 µg/kg body weight) is 5.8–127 times higher than the European PDI values estimated with diet approach. All together, these data clearly show that the estimated human exposure to OTA is higher when using the biomarker approach as compared to the diet approach. Possible explanations of these differences could be: a) the urinary excretion rate of OTA in piglets, used in this study to estimate PDI values, is completely different from that in human, b) the food diet approach did not cover all sources of OTA exposure due to the occurrence of this mycotoxin in a very high number of different foods and beverages. It is well known that OTA is the mycotoxin more widespread in several different types of foods and beverages [23,32].

The results shown in Table 3 indicate that AFM1 was detected and measured in only 3 urine samples (6% of total samples) which confirm that human exposure to AFB1 was sporadic. The relevant mean value of PDI of AFB1, calculated for the 3 positive subjects, is 0.068 µg/kg body weight. As reported above, there is no TDI value for AFB1 because this is a carcinogenic mycotoxin. The incidence of positive samples and mean urinary AFM1 concentration measured in positive samples in our study are comparable to those reported by Hatem *et al.* from Marasmus, Egypt [45]. Polychronaki *et al.* reported, for Guinea, a mean urinary concentration of AFM1 in positive samples similar to those found in our study but a higher percentage of positive samples (64%) [13]. The mean value of PDI of AFB1 estimated with biomarker approach for the three volunteers participating in our study (0.068 µg/kg body weight) is largely higher than the PDI of aflatoxin intake measured with the diet approach in Europe (0.00016–0.00055 µg/kg body weight) [23].

3. Experimental Section

3.1. Chemicals and Reagents

Standard solutions were purchased from Romer Labs Diagnostic (Tulln, Austria). In particular, solutions of DON (100 µg/mL), DOM-1 (50 µg/mL), AFM1 (0.5 µg/mL), ZEA (100 µg/mL), α-ZOL (10 µg/mL), β-ZOL (10 µg/mL) and OTA (10 µg/mL) were prepared in acetonitrile (ACN) whereas FB1 solution (50 µg/mL) were prepared in acetonitrile-water (50:50). β-glucuronidase/sulfatase type H-2 from *Helix pomatia* (specific activity 130,200 units/mL β-glucuronidase, 709 units/mL sulfatase). Chromatography-grade methanol (MeOH) and glacial acetic acid were obtained from Carlo Erba (Milan, Italy). Ultrapure water was produced by use of a Milli-Q system (Millipore, Bedford, MA, USA). Myco6in1® immunoaffinity columns were purchased from Vicam L.P (Watertown, MA, USA). OASIS® HLB columns, 60 mg, 3 mL were purchased from Waters (Milford, MA, USA) and regenerated cellulose filters (0.45 µm) were purchased from Sartorius Stedim Biotech (Goettingen, Germany).

3.2. Equipment and Conditions

LC-MS/MS analyses were performed on a triple quadrupole API 5000 system (Applied Biosystems, Foster City, CA, USA), equipped with a ESI interface and an Acquity UPLC system comprising a binary pump and a microautosampler from Waters (Milford, MA, USA). The analytical column was an Acquity UPLC BEH phenyl column (2.1 mm × 150 mm, 1.7 µm particles; Waters). The column oven was set at 40 °C. The flow rate of the mobile phase was 250 µL/min and the injection volume was 10 µL. For multi-biomarker separation a multiple linear binary linear gradient of acidic MeOH (containing 0.5% acetic acid) in water (containing 0.5% acetic acid) was developed and used as mobile phase as follows: from 20% to 80% MeOH in 5 min, then maintained at 80% MeOH for 5 min, then brought to 20% MeOH in 0.5 min and left to equilibrate for 4.5 min before the next run. For LC-MS/MS analyses, the ESI interface was used in positive ion mode for AFM1, FB1 and OTA and in negative ion mode for DON, DOM-1, ZEA, α-ZOL and β-ZOL. The mass spectrometer operated in MRM (multiple reaction monitoring) mode and the optimized MS/MS conditions for each analyte are listed in Table 1. In particular, 4 transitions were monitored for confirmation of FB1, α-ZOL and β-ZOL whereas the sum of the 4 transitions were used for quantification; for DON 3 transitions for confirmation and the sum of 2 transition for quantification; for DOM-1, ZEA and OTA 3 transitions for confirmation and 1 transition for quantification; for AFM1 2 transitions for confirmation and the sum of 2 transition for quantification. Interface conditions were: TEM, 450 °C; CUR, nitrogen, 20 psi;

GS1, air, 60 psi; GS2, air, 40 psi; ionspray voltage +5500 V or −4500 V. The signal of each compound was preliminary optimized with each proposed ionization condition. The tuning procedure included the optimization of source parameters during infusion of 1 μg/mL standard solution (0.1 μg/mL for AFM1) of the individual toxins in MeOH-water (20:80) containing 0.5% acetic acid at 10 μL/min, using an Harvard 11 plus infusion pump, into the UPLC mobile phase (50/50 water/methanol at 250 μL/min) by means of a minimum dead volume T-piece connected after the analytical column.

3.3. Participants and Urine Collection

For the evaluation of human exposure to DON, AFB1, FB1, ZEA and OTA 100 individuals residing in 10 municipalities of Puglia region (Southern Italy) were invited to participate in the urine sampling. Each individual was asked to provide a first morning urine sample and to fill out a questionnaire concerning age, gender and health status. The majority of individuals gave oral informed consent and donated a sample of first morning urine and the filled in questionnaire. Fifty-two individuals (participation rate 52%), 26 males and 26 females, (mean age 41 years, range 3–85 years) were recruited from municipalities of Bari, Triggiano, Mola di Bari, Monopoli, Adelfia, Conversano, Polignano a Mare, Bitonto, Martina Franca and Statte. The remaining 48 individuals failed to donate their morning urine samples. All participants collected their urine samples on 26 April 2011. Urine samples were stored at −18 °C until analysis for identification and determination of DON, DOM-1, AFM1, FB1, ZEA, α-ZOL, β-ZOL and OTA. Eleven volunteers declared to have health problems, in particular three individuals with hypertension, three with allergies, three with diabetes, one with hyperthyroidism and one at risk of thrombosis.

3.4. Calibration Curves

A mixed standard solution with a final concentration of 150 ng/mL DON, 20 ng/mL of DOM-1, 1.8 ng/mL AFM1, 28 ng/mL FB1, 35 ng/mL β-ZOL, 23 ng/mL α-ZOL, 12 ng/mL ZEA and 2 ng/mL OTA was prepared by mixing appropriate volumes of commercially available standard solutions and appropriate volume of ACN. Five standard calibration solutions covering appropriate range of analyte concentrations were prepared by portioning adequate volumes of mixed standard solution that were dried and reconstituted with 200 μL of initial LC-MS/MS mobile phase. In particular, the 5 standard calibration solutions were prepared by drying 25, 125, 250, 375 and 1000 μL of mixed standard solution that were reconstituted with 200 μL of initial LC-MS/MS mobile phase. Matrix-matched calibration solutions were prepared in 5 purified urinary extracts. In particular, urine from 6 individuals were pooled and mixed then 5 aliquots of 6 mL each were purified according to the protocol reported above. The 5 eluates collected from OASIS® HLB and Myco6in1® columns were spiked with 5 aliquots of the mixed standard solution (25, 125, 250, 375 and 1,000 μL), dried and reconstituted with 200 μL of initial LC-MS/MS mobile phase. The ranges of mycotoxin concentrations in the calibration solutions were, therefore: 18.75–750.00 ng/mL for DON, 2.50–100.00 ng/mL for DOM-1, 0.22–8.76 ng/mL for AFM1, 3.50–140.00 ng/mL for FB1, 4.38–175.00 ng/mL for β-ZOL, 2.88–115.00 ng/mL for α-ZOL, 1.50–60.00 ng/mL for ZEA and 0.25–10.00 ng/mL for OTA.

3.5. Analysis of Urinary Biomarkers

Urine samples were hydrolyzed with β-glucuronidase/sulfatase enzyme to hydrolyse glucuronide and/or sulfate conjugates of DON, DOM-1, ZEA, α-ZOL, β-ZOL and then purified with a multi-antibody and OASIS® HLB columns according to the protocol reported elsewhere [19]. In brief, a 6 mL urine sample was hydrolyzed at 37 °C for 18 h with 300 μL of β-glucuronidase/sulfatase type H-2 from *Helix pomatia*. Hydrolyzed sample was diluted with 6 mL of water and purified on a Myco6in1® and OASIS® HLB columns connected in tandem. The OASIS® HLB column was previously conditioned by passing 2 mL MeOH and 2 mL ultrapure water. After sample application and elution, the two columns were separated, the Myco6in1® was washed with water (4 mL) and eluted with methanol (3 mL) and water (2 mL) that were collected in a vial. The OASIS® HLB column was washed

Toxins **2014**, 6, 523–538

with methanol/water (20:80, 1 mL) and DON that had passed through the Myco6in1® and retained on the OASIS® HLB was eluted with methanol/water (40:60, 1 mL). The separate eluates from the two columns were combined, dried down and reconstituted in methanol/water (20:80, 200 μL) with 0.5% acetic acid. Purified extract was filtered through a regenerated cellulose filter and a volume of 10 μL (equivalent to 0.3 mL urine) was analyzed by UPLC-MS/MS.

4. Conclusions

The improved UPLC-MS/MS method for simultaneous determination of urinary biomarker for DON, FB1, OTA, AFB1 and ZEA was suitable to detect and accurately measure the low mycotoxin biomarker concentrations naturally occurring in the human urine samples collected in this study.

A multiple mycotoxin exposure was found for all tested volunteers participating in the study.

This is the first report on the occurrence of urinary AFM1, FB1, ZEA and ZOLs in Italy.

The estimated PDI values of OTA largely exceeded the TDI value for this mycotoxin in 94% of volunteers.

The mean estimated PDI of DON is similar to the TDI value for this mycotoxin but in 40% of volunteers it exceeded the value of TDI.

The values of PDI estimated herein with urinary biomarker approach matched quite well with the intake estimated with the diet approach reported in the literature for DON, FB1 and ZEA whereas for OTA and AFB1 the intake estimated with the biomarker approach was much higher than the intake estimated with the diet approach reported in the literature.

Acknowledgments: This work was supported by the EU-FP7 MYCORED project (grant agreement no. 222690). We thank Water Research Institute, IRSA-CNR, Bari that made available the API 5000 UPLC-MS/MS system and Vito Locaputo for valuable assistance during the LC-MS/MS analyses.

Conflicts of Interest: The authors declare no conflict of interest.

References

1. Shephard, G.S.; Berthiller, F.; Burdaspal, P.; Crews, C.; Jonker, M.A.; Krska, R.; Lattanzio, V.M.T.; MacDonald, S.; Malone, R.J.; Maragos, C.; et al. Developments in mycotoxin analysis: An update for 2011–2012. *World Mycotoxin J.* **2013**, 6, 3–30. [CrossRef]
2. Shephard, G.S.; Berthiller, F.; Burdaspal, P.; Crews, C.; Jonker, M.A.; Krska, R.; MacDonald, S.; Malone, R.J.; Maragos, C.; Sabino, M.; et al. Developments in mycotoxin analysis: An update for 2010–2011. *World Mycotoxin J.* **2012**, 5, 3–30. [CrossRef]
3. Serrano, A.B.; Font, G.; Ruiz, M.J.; Ferrer, E. Co-occurrence and risk assessment of mycotoxins in food and diet from Mediterranean area. *Food Chem.* **2012**, 135, 423–429. [CrossRef]
4. Lindblad, M.; Gidlund, A.; Sulyok, M.; Börjesson, T.; Krska, R.; Olsen, M.; Fredlund, E. Deoxynivalenol and other selected *Fusarium* toxins in Swedish wheat- Occurrence and correlation to specific *Fusarium*. species. *Int. J. Food Microbiol.* **2013**, 167, 284–291. [CrossRef]
5. Ibáñez-Vea, M.; Martínez, R.; Gonzáles-Peñas, E.; Lizarraga, E.; López de Cerain, A. Co-occurrence of aflatoxins, ochratoxin A and zearalenone in breakfast cereals from Spanish market. *Food Control* **2011**, 22, 1949–1955. [CrossRef]
6. Njobeh, P.B.; Dutton, M.F.; Koch, S.H.; Chuturgoon, A.A.; Stoev, S.D.; Mosonik, J.S. Simultaneous occurrence of mycotoxins in human food commodities from Cameroon. *Mycotoxin Res.* **2010**, 26, 47–57. [CrossRef]
7. Shephard, G.S.; Burger, H.M.; Gambacorta, L.; Krska, R.; Powers, S.P.; Rheeder, J.P.; Solfrizzo, M.; Sulyok, M.; Visconti, A.; Warth, B.; et al. Mycological analysis and multimycotoxins in maize from rural subsistence farmers in the former Transkei, South Africa. *J. Agric. Food Chem.* **2013**, 61, 8232–8240. [CrossRef]
8. Shephard, G.S.; Burger, H.M.; Gambacorta, L.; Gong, Y.Y.; Krska, R.; Rheeder, J.P.; Solfrizzo, M.; Srey, C.; Sulyok, M.; Visconti, A.; et al. Multiple mycotoxin exposure determined by urinary biomarkers in rural subsistence farmers in the former Transkei, South Africa. *Food Chem. Toxicol.* **2013**, 62, 217–225. [CrossRef]
9. Vendl, O.; Crws, C.; MacDonald, S.; Krska, R.; Berthiller, F. Occurrence of free and conjugated *Fusarium* toxins in cereal-based food. *Food Add. Contam. Part A* **2010**, 27, 1148–1152. [CrossRef]

10. Falavigna, C.; Cirlini, M.; Galaverna, G.; Dall'Asta, C. Masked fumonisins in processed food: Co-occurrence of hidden and bound forms and their stability under digestive conditions. *World Mycotoxin J.* **2012**, *5*, 325–334. [CrossRef]

11. Berthiller, F.; Crews, C.; Dall'Asta, C.; De Saeger, S.; Haesaert, G.; Karlovsky, P.; Oswald, I.P.; Seefelder, W.; Speijers, G.; Stroka, J. Masked mycotoxins: A review. *Mol. Nutr. Food Res.* **2013**, *57*, 165–186. [CrossRef]

12. Turner, P.C.; Flannery, B.; Isitt, C.; Ali, M.; Pestka, J. The role of biomarkers in evaluating human health concerns from fungal contaminants in food. *Nutr. Res. Rev.* **2012**, *25*, 162–179. [CrossRef]

13. Polychronaki, N.; Wild, C.P.; Mykkänen, H.; Amra, H.; Abdel-Wahhab, M.; Sylla, A.; Diallo, M.; El-Nezami, H.; Turner, P.C. Urinary biomarkers of aflatoxin exposure in young children from Egypt and Guinea. *Food Chem. Toxicol.* **2008**, *46*, 519–526. [CrossRef]

14. Duarte, S.C.; Alves, M.R.; Pena, A.; Lino, C.M. Determinants of ochratoxin A exposure—A one year follow-up study of urine levels. *Int. J. Hyg. Environ. Health* **2012**, *215*, 360–367. [CrossRef]

15. Riley, R.T.; Torres, O.; Showker, J.L.; Zitomer, N.C.; Matute, J.; Voss, K.A.; Gelineau-van Waes, J.; Maddox, J.R.; Gregory, S.G.; Ashley-Koch, A.E. The kinetics of urinary fumonisin B1 excretion in humans consuming maize-based diets. *Mol. Nutr. Food Res.* **2012**, *56*, 1445–1455. [CrossRef]

16. Warth, B.; Sulyok, M.; Krska, R. LC-MS/MS-based multibiomarker approaches for the assessment of human exposure to mycotoxins. *Anal. Bioanal. Chem.* **2013**, *405*, 5687–5695. [CrossRef]

17. Gambacorta, L.; Solfrizzo, M.; Visconti, A.; Powers, S.; Cossalter, A.M.; Pinton, P.; Oswald, I.P. Validation study on urinary biomarkers of exposure for aflatoxin B1, ochratoxin A, fumonisin B1, deoxynivalenol and zearalenone in piglet. *World Mycotoxin J.* **2013**, *6*, 299–308. [CrossRef]

18. Solfrizzo, M.; Gambacorta, L.; Warth, B.; White, K.; Srey, C.; Sulyok, M.; Krska, R.; Gong, Y.Y. Comparison of single and multi-analyte methods based on LC-MS/MS for mycotoxin biomarker determination in human urine. *World Mycotoxin J.* **2013**, *6*, 355–366. [CrossRef]

19. Solfrizzo, M.; Gambacorta, L.; Lattanzio, V.M.T.; Powers, S.; Visconti, A. Simultaneous LC-MS/MS determination of aflatoxin M1, ochratoxin A, deoxynivalenol, de-epoxydeoxynivalenol, α and β-zearalenols and fumonisin B1 in urine as a multi-biomarker method to assess exposure to mycotoxins. *Anal. Bioanal. Chem.* **2011**, *401*, 2831–2841. [CrossRef]

20. Abia, A.W.; Warth, B.; Sulyok, M.; Krska, R.; Tchana, A.; Njobeh, P.B.; Turner, P.C.; Kouanfack, C.; Eyongetah, M.; Dutton, M.; *et al.* Bio-monitoring of mycotoxin exposure in Cameroon using a urinary multi-biomarker approach. *Food Chem. Toxicol.* **2013**, *62*, 927–934. [CrossRef]

21. Ediage, E.N.; Di Mavungu, J.D.; Song, S.; Sioen, I.; De Saeger, S. Multimycotoxin analysis in urines to assess infant exposure: A case study in Cameroon. *Environ. Int.* **2013**, *57–58*, 50–59.

22. Scientific Co-operation on Question relating to Food (SCOOP, Directive 93/5/EEC). Scoop Task 3.2.10: Collection of occurrence data of *Fusarium* toxin in food and assessment of dietary intake by the population of EU member states. Available online: http://ec.europa.eu/food/fs/scoop/task3210.pdf (accessed on 22 January 2014).

23. Food and Agriculture Organization; World Health Organization (FAO/WHO). *Safety Evaluation of Certain Food Additives and Contaminants*; Food Additives Series 59; FAO/WHO: Geneva, Switzerland, 2008.

24. Brera, C.; Angelini, S.; Debegnach, F.; De Santis, B.; Turrini, A.; Miraglia, M. Valutazione analitica dell'esposizione del consumatore alla fumonisina B1. In Proceedings of I Congresso Nazionale—Le micotossine nella filiera agro-alimentare, Superiore di Sanità, Roma, Italy, 29–30 November 2004; pp. 44–52.

25. Turner, P.C.; Rothwell, J.A.; White, K.L.M.; Gong, Y.Y.; Cade, J.E.; Wild, C.P. Urinary deoxynivalenol is correlated with cereal intake in individuals from the United Kingdom. *Environ. Health Perspect* **2008**, *116*, 21–25.

26. Turner, P.C.; Burley, V.J.; Rothwell, J.A.; White, K.L.M.; Cade, J.E.; Wild, C.P. Dietary wheat reduction decreases the level of urinary deoxynivalenol in UK adults. *J. Expo. Sci. Environ. Epidemiol.* **2008**, *18*, 392–399. [CrossRef]

27. Turner, P.C.; White, K.L.; Burley, V.J.; Hopton, R.P.; Rajendram, A.; Fisher, J.; Cade, J.E.; Wild, C.P. A comparison of deoxynivalenol intake and urinary deoxynivalenol in UK adults. *Biomarkers* **2010**, *15*, 553–562. [CrossRef]

28. Turner, P.C.; Hopton, R.P.; Lecluse, Y.; White, K.L.M.; Fisher, J.; Lebailly, P. Determinants of urinary deoxynivalenol and de-epoxy deoxynivalenol in male farmers from Normandy, France. *J. Agric. Food Chem.* **2010**, *58*, 5206–5212.

29. Turner, P.C.; Gong, Y.Y.; Pourshams, A.; Jafari, E.; Routledge, M.N.; Malekzadek, R.; Wild, C.P.; Boffetta, P.; Islami, F. A pilot survey study for *Fusarium* mycotoxin biomarkers in women from Golestan, Northern Iran. *World Mycotoxin J.* **2012**, *5*, 195–199. [CrossRef]

30. Wallin, S.; Hardie, L.J.; Kotova, N.; Warensjö Lemming, E.; Nälsén, C.; Ridefelt, P.; Turner, P.C.; White, K.L.M.; Olsen, M. Biomonitoring study of deoxynivalenol exposure and association with typical cereal consumption in Swedish adults. *World Mycotoxin J.* **2013**, *6*, 439–448. [CrossRef]

31. Hepworth, S.J.; Hardie, L.J.; Fraser, L.K.; Burley, V.J.; Mijal, R.S.; Wild, C.P.; Azad, R.; McKinney, P.A.; Turner, P.C. Deoxynivalenol exposure assessment in a cohort of pregnant women from Bradford, UK. *Food Addit. Contam. Part A* **2012**, *29*, 269–276. [CrossRef]

32. Food and Agriculture Organization; World Health Organization (FAO/WHO). Ochratoxin A. In *Safety Evaluation of Certain Mycotoxins in Food*; World Health Organisation: Geneva, Switzerland, 2001; pp. 281–387.

33. Gong, Y.Y.; Torres-Sanchez, L.; Lopez-Carrillo, L.; Peng, J.H.; Sutcliffe, A.E.; White, K.L.; Humpf, H.U.; Turner, P.C.; Wild, C.P. Association between tortilla consumption and human urinary fumonisin B1 levels in a Mexican population. *Cancer Epidem. Biomar.* **2008**, *17*, 688–694. [CrossRef]

34. Van der Westhuizen, L.; Shephard, G.S.; Rheeder, J.P.; Somdyala, N.I.M.; Marasas, W.F.O. Sphingoid base levels in humans consuming fumonisin-contaminated maize in rural areas of the former Transkei, South Africa: A cross-sectional study. *Food Addit. Contam. Part A* **2008**, *25*, 1385–1391. [CrossRef]

35. European Food Safety Authority (EFSA). Opinion of the Scientific Panel on Contaminants in the Food Chain on a request from the Commission related to Zearalenone as undesirable substance in animal feed. *EFSA J.* **2004**, *89*, 1–35.

36. Bandera, E.V.; Chandran, U.; Buckley, B.; Lin, Y.; Isukapalli, S.; Marshall, I.; King, M.; Zarbl, H. Urinary mycoestrogens, body size and breast development in New Jersey girls. *Sci. Total Environ.* **2011**, *409*, 5221–5227. [CrossRef]

37. European Food Safety Authority (EFSA). Opinion of the Scientific Panel on contaminants in the Food Chain of the EFSA on a request from the Commission related to ochratoxin A in food. Available online: http://www.efsa.europa.eu/de/scdocs/doc/365.pdf (accessed on 22 January 2014).

38. Fazekas, B.; Tar, A.; Kovács, M. Ochratoxin A content of urine samples of healthy humans in Hungary. *Acta. Vet. Hung.* **2005**, *53*, 35–44. [CrossRef]

39. Duarte, S.C.; Pena, A.; Lino, C.M. Ochratoxin A in Portugal: A review to assess human exposure. *Toxins* **2010**, *2*, 1225–1249. [CrossRef]

40. Pena, A.; Seifrtová, M.; Lino, C.; Silveira, I.; Solich, P. Estimation of ochratoxin A in portuguese population: New data on the occurrence in human urine by high performance liquid chromatography with fluorescence detection. *Food Chem. Toxicol.* **2006**, *44*, 1449–1454. [CrossRef]

41. Manique, R.; Pena, A.; Lino, C.M.; Moltó, J.C.; Mañes, J. Ochratoxin A in the morning and afternoon portions of urine from Coimbra and Valencian populations. *Toxicon* **2008**, *51*, 1281–1287. [CrossRef]

42. Coronel, M.B.; Marin, S.; Tarragó, M.; Cano-Sancho, G.; Ramos, A.J.; Sanchis, V. Ochratoxin A and its metabolite ochratoxin alpha in urine and assessment of the exposure of inhabitants of Lleida, Spain. *Food Chem. Toxicol.* **2011**, *49*, 1436–1442. [CrossRef]

43. Gilbert, J.; Brereton, P.; MacDonald, S. Assessment of dietary exposure to ochratoxin A in the UK using a duplicate diet approach and analysis of urine and plasma samples. *Food Addit. Contam.* **2001**, *18*, 1088–1093. [CrossRef]

44. Domijan, A.M.; Peraica, M.; Miletić-Medved, M.; Lucić, A.; Fuchs, R. Two different clean-up procedures for liquid chromatographic determination of ochratoxin A in urine. *J. Chromatogr. B* **2003**, *798*, 317–321. [CrossRef]

45. Hatem, N.L.; Hassab, H.M.; Abd Al-Rahman, E.M.; El-Deeb, S.A.; El-Sayed Ahmed, R.L. Prevalence of aflatoxins in blood and urine of Egyptian infants with protein-energy malnutrition. *Food Nutr. Bull.* **2005**, *26*, 49–56.

Article

Occurrence of Deoxynivalenol and Deoxynivalenol-3-glucoside in Hard Red Spring Wheat Grown in the USA

Senay Simsek [1,*], Maribel Ovando-Martínez [1], Bahri Ozsisli [2], Kristin Whitney [1] and Jae-Bom Ohm [3]

[1] Department of Plant Sciences, North Dakota State University, PO Box 6050, Fargo, ND 58108, USA; maribel.ovando@ndsu.edu (M.O.-M.); kristin.whitney@ndsu.edu (K.W.)

[2] Department of Food Engineering, College of Agriculture, Kahramanmaras Sutcu Imam University, Kahramanmaras 46060, Turkey; bozsisli@ksu.edu.tr

[3] USDA-ARS, Cereal Crops Research Unit, Harris Hall, Hard Red Spring and Durum Wheat Quality Laboratory, North Dakota State University, P.O. Box 6050, Fargo, ND 58108, USA; jae.ohm@ars.usda.gov

* Author to whom correspondence should be addressed; senay.simsek@ndsu.edu; Tel.: +1-701-231-7737; Fax: +1-701-231-8474.

Received: 31 October 2013; in revised form: 12 December 2013; Accepted: 13 December 2013; Published: 18 December 2013

Abstract: Deoxynivalenol (DON) is a mycotoxin found in wheat that is infected with Fusarium fungus. DON may also be converted to a type of "masked mycotoxin", named deoxynivalenol-3-glucoside (D3G), as a result of detoxification of the plant. In this study, DON and D3G were measured using gas chromatographic (GC) and liquid chromatography-mass spectrometry (LC-MS) in wheat samples collected during 2011 and 2012 in the USA. Results indicate that the growing region had a significant effect on the DON and D3G ($p < 0.0001$). There was a positive correlation between both methods (GC and LC-MS) used for determination of DON content. DON showed a significant and positive correlation with D3G during 2011. Overall, DON production had an effect on D3G content and kernel damage, and was dependent on environmental conditions during Fusarium infection.

Keywords: deoxynivalenol; deoxynivalenol-3-glucoside; wheat; USA

---

1. Introduction

One of the most predominant and economically important mycotoxins affecting small-grain cereals, such as wheat, is deoxynivalenol (DON). DON is formed due to the presence of plant pathogenic fungi *Fusarium graminearum* and *F. culmorum*; which are responsible for the disease known as Fusarium head blight (FHB) [1]. The geographical distribution of these fungi is affected by climate. *F. culmorum* occurs more commonly in Europe while *F. graminearum* is common in Europe and North America [1,2]. The association between the FHB intensity level with DON accumulation in spring wheat using a meta-analysis has been reported in multiple studies [3]. The most important environmental factors causing FHB growth and biosynthesis of DON are water availability and temperature [4]. Minimum water activity values for growth and DON production appears to be limited at 0.93 under optimum temperature conditions (25 °C) [4,5]. Also, higher disease severity occurs if the wheat has been exposed to extended periods of wetness [1], which may result in higher DON production. On the other hand, *Fusarium* growth and DON production depends on the growth stages of the plant; in wheat higher damage occurs during flowering (anthesis) and shortly after flowering, the infection can also continue during grain maturation stage [1,6].

The presence of DON in wheat decreases grain quality by rendering the crop unsuitable and unsafe for food, feed and malting process. It is worthwhile noting that DON can have several side effects, such as feed refusal, vomiting, reduced weight gain, diarrhea, hemorrhage, skin lesions, growth depression and immunosuppression [7–9], which have a negative impact on human and animal health [6,10]. Also, DON affects the plant metabolism in wheat because it leads to the inhibition of germination and decreases plant growth. The plant starts to develop a detoxification mechanism in which DON is glycosylated into 3-β-D-glucopyranosil-4-deoxynivalenol (D3G) and stored inside the vacuole or cell wall to combat this situation [11,12]. This product is known as a "masked" mycotoxin, because one or more glucose molecules bind to the DON which reduces the toxicity in the plant and makes it unable to be detected by traditional methods for DON detection. D3G is less active as a protein biosynthesis inhibitor than DON [11]. There is a lack of information about the correlation between the DON and D3G production in wheat. Rasmussen *et al.* [13] found that the concentration of D3G is positively correlated with the increasing of DON content, which is similar to that observed by Lemmens *et al.* [14].

The toxicity of D3G in mammals is currently unknown. Berthiller *et al.* [15] reported that this "masked" mycotoxin resisted the *in vitro* acid conditions, indicating D3G cannot be hydrolyzed to DON in the stomach of mammals; nevertheless, it can be hydrolyzed during fermentation process by bacterial β-glucosidases in the colon. Later, it was demonstrated that D3G is partially absorbed in the gastrointestinal system of rats, but the majority of the D3G ingested was hydrolyzed in the digestion process and excreted in feces, which means that D3G is partially bioavailable in the gastrointestinal system of rats [16]. In Europe, the DON content of food, feed and unprocessed grains of undefined end-use collected in 21 European countries between 2007 and 2012 was recently published. It was found that the levels of DON in wheat, maize and oat may exceed the maximum limits for food or guidance values for feed. Due to the lack of data about D3G, this was not considered in the food safety assessment [17]. However, it may be necessary to analyze the DON and D3G content in wheat to have a clear idea about the total content of DON after the wheat is processed to different products. After determination of DON and D3G levels the assessment must be made to determine if the levels DON are within the permitted levels of DON approved by the Food and Drug Administration (FDA) and also to assess the safety of wheat-based products consumed in USA. The objective of this research was to analyze the DON and D3G content of hard red spring (HRS) wheat between 2011 and 2012 Crop Survey using gas chromatography (GC) and liquid chromatography-quadrupole time of flight mass spectrometry (LC-QTOF-MS) and find the correlation between the DON and D3G production. DON will be measured by both GC and LC-QTOF-MS to evaluate correlation between methods and determine the feasibility of the LC-QTOF-MS methodology for simultaneous measurement of DON and D3G.

2. Results and Discussions

2.1. Wheat Kernel Quality

The crop survey of HRS wheat was conducted in North Dakota (ND), South Dakota (SD), Montana (MT) and Minnesota (MN), which make up the primary HRS wheat growing region in the United States (US) (Figure 1). The Federal Grain Inspection Service (FGIS) use grading and non-grading factors to evaluate the conditions and quality of wheat. The HRS wheat kernel quality of the 2011 and 2012 crop surveys is presented in Table 1. The mean dockage percentage was the same in both survey crops. This factor does not affect the numerical grade but it is one important step in the grading process to eliminate all the material that is not wheat prior to determining the rest of grading and non-grading factors. Other than dockage, there were no differences observed in the grading factors percent shrunken and broken kernels, percent dark hard and vitreous (DHV) and test weight. However, the values of percent damage and percent total defects in the 2011 crop survey were higher than in the 2012 crop survey. According to the HRS wheat 2012 Regional Quality Report, there were differences in the environmental factors during planting, growing and harvest of crops of both years [18]. Higher temperatures and lower rainfall levels were observed in 2012, which decreased the disease pressure and percent damaged kernels, compared to 2011. These environmental factors in turn affected the wheat quality [19,20]. Among the non-grading factors (Table 1), the protein content is very important to determine the suitability of wheat in different final products. The protein percentage of both crop surveys was between the values established by the FGIS (13%–14%). HRS wheat is considered to have high and excellent protein quality for use in bread-baking. The samples from 2011 crop survey presented a lower falling number (386.3 s) compared to 2012 samples (429.3 s). Falling number is an indirect measurement of the α-amylase activity to detect sprout damage in wheat. This is accomplished by the measurement of changes in the physical properties of the starch portion of the wheat kernel caused by this α-amylase. A high enzyme activity indicates faster liquefaction and lower falling number. The falling number may be related to the higher percent damaged kernels and percent total defects presented in samples from 2011. Nevertheless, the falling number of both years were at least 350 s, which is the level at which grain is considered to be sound by the FGIS. Also, the 1000 kernel weight of 2011 crop survey samples was lower than 2012 crop survey samples. This means that the environmental factors (temperature, rainfall and moisture) could increase the percent damaged kernels and reduce the quality of wheat. Kernel quality was also affected by the production of mycotoxins, such as DON and its "masked" mycotoxin D3G, related to the occurrence of the FHB.

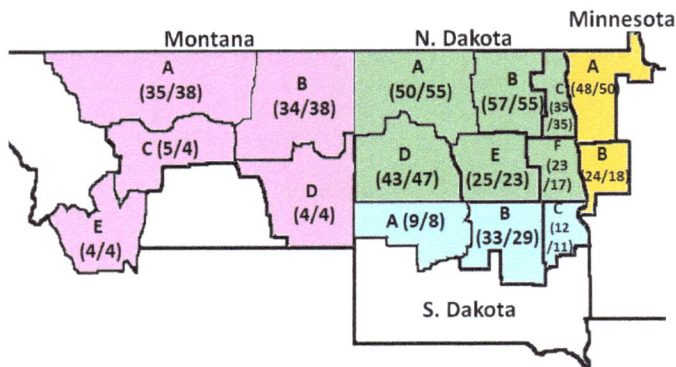

Figure 1. Distribution of hard red spring (HRS) wheat samples from the 2011 and 2012 Crop Surveys from Montana, North Dakota, Minnesota and South Dakota. **A, B, C, D, E** and **F**: regions in which the samples were collected from each state. The numbers inside the parenthesis represent the number of samples taken, from left to right: 2011 Crop Survey and 2012 Crop Survey.

Table 1. Mean, standard deviation (SD), minimum (MIN) and maximum (MAX) values for wheat kernel quality characteristics for 2011 and 2012 crop survey samples.

Factors	2011				2012			
	Mean	SD	MIN	MAX	Mean	SD	MIN	MAX
Dockage (%) [a]	1.1	1.4	0.0	13.4	1.1	1.2	0.0	13.3
Shrunken & broken [a]	1.6	1.4	0.0	10.0	1.2	1.1	0.0	10.1
Damage (%) [a]	0.5	0.8	0.0	10.6	0.1	0.2	0.0	2.8
Hard and vitreous kernel (%) [a]	75.8	21.9	5.0	99.0	74.5	28.4	2.0	99.0
Test weight (lbs/bu) [a]	60.0	2.4	52.5	65.7	60.9	1.9	53.9	65.1
Total defects (%) [a]	2.1	1.7	0.1	13.1	1.3	1.1	0.0	10.1
Protein (12% mb) [b]	14.8	1.3	10.2	18.6	14.6	1.4	10.3	20.1
Falling number (s) [b]	386.3	44.1	226.0	621.0	429.3	48.8	238.0	654.0
1000 kernel weight (g) [b]	26.7	4.1	16.8	40.0	29.1	4.1	16.6	44.1

Notes: [a] Grading factors used to evaluate the kernel quality to ensure general standards of acceptance in flour or semolina production; [b] Non-grade factors used to evaluate the kernel quality; mb = moisture basis.

2.2. Effect of the State and Region on DON, D3G and Damage Kernel

Least square means values of DON analyzed with GC and Liquid chromatography—mass spectrometry (LC-MS), D3G determined with LC-MS, and percent damage kernels are given in Table 2. These variables showed significant differences between state and region mean values, indicating that HRS wheat samples collected from different state or regions might have different levels of DON and D3G. Samples collected from ND presented higher values of DON (both GC and LC-MS methods), D3G, and percent damaged kernels than other states in both years. Samples from MT presented the opposite trend. The regional data showed more specifically that samples collected from ND contained higher levels of DON, D3G and percent damaged kernels than the other states. DON, D3G and percent damaged kernels presented higher values for samples collected from ND regions A, B, C, E and F in 2011. For ND samples collected in 2012, DON, D3G and Kernel damaged showed higher values only for samples from regions A and B. Among MN regions, samples collected in 2011 from region B showed high DON and percent damaged kernels. Samples from SD region B in 2011 also showed high levels of DON, D3G and percent kernel damage. For the 2012 samples, those collected from MN region A and MT region B showed high D3G values. Specifically, samples from MT region B showed high D3G content despite low DON content in 2012. The rest of the regions presented similar content of DON, D3G and percent damaged kernels and the majority did not show significant differences. The 2011 survey samples had high DON content, independent of the method used. Due to the high levels of DON during this year, the percent damaged kernels were also higher compared to 2012 survey samples.

Table 2. Least square mean values for deoxynivalenol (DON), deoxynivalenol-3-glucoside (D3G) and damage kernel in 2011 and 2012 survey.

Growing Area	2011				2012			
State	GC-DON	LC-DON	D3G	Damage (%)	GC-DON	LC-DON	D3G	Damage (%)
MN	1.35 **	1.74 **	0.04	0.35 **	0.89	0.78	0.128 **	0.05
MT	0.03	0.03	0.00	0.03	0.18	0.18	0.090	0.00
ND	2.80***	3.15 ***	0.24 ***	0.63 ***	1.93 ***	1.71 ***	0.142 ***	0.06 ***
SD	1.35 **	1.72 *	0.12	0.36 *	0.37	0.33	0.085	0.03
Region								
MN-A	1.07	1.21	0.01	0.23	0.82	0.72	0.124 *	0.08 *
MN-B	1.63 *	2.27 **	0.06	0.47 *	0.96	0.84	0.133	0.02
MT-A	0.00	0.00	0.00	0.01	0.09	0.12	0.067	0.01
MT-B	0.00	0.00	0.00	0.05	0.04	0.05	0.176 **	0.00
MT-C	0.13	0.13	0.00	0.03	0.11	0.20	0.105	0.00
MT-D	0.00	0.00	0.00	0.08	0.00	0.08	0.015	0.00
MT-E	0.04	0.00	0.00	0.00	0.65	0.47	0.088	0.00
ND-A	3.89 ***	4.26 ***	0.31 ***	0.75 ***	6.91 ***	5.97 ***	0.295 ***	0.15 ***
ND-B	3.77 ***	4.32 ***	0.30 ***	0.69 ***	2.48 ***	2.22 ***	0.174 ***	0.12 ***
ND-C	2.91 ***	3.36 ***	0.30 ***	0.61 ***	0.66	0.60	0.060	0.06
ND-D	1.08	1.10	0.02	0.61 ***	0.34	0.35	0.094	0.01
ND-E	2.19 **	2.54 **	0.09	0.66 ***	0.71	0.72	0.096	0.02
ND-F	3.00 **	3.34 ***	0.39 ***	0.49 **	0.47	0.38	0.133	0.01
SD-A	0.21	0.15	0.00	0.13	0.49	0.42	0.094	0.00
SD-B	2.41 ***	3.20 ***	0.29 **	0.57 **	0.22	0.23	0.097	0.00
SD-C	1.42	1.82	0.06	0.38	0.41	0.34	0.066	0.09

Notes: *, **, and *** means significant differences (H0: least square mean = 0) at $p < 0.05$, 0.01, and 0.001, respectively. Values of DON and D3G are in mg/kg; gas chromatographic (GC).

D3G content in 2011 was not different among states except for ND, but in 2012, MN also showed differences in D3G content. Also, the differences in DON content and D3G among regions and states can be attributed to the environmental conditions presented in each year of survey. The analysis of variance (ANOVA) indicates that state and region had significant effects on variation in DON and D3G content and percent damaged kernels in 2011 survey samples (Table 3). The ANOVA for 2012 survey samples showed slightly different results (Table 3). During 2012 the percent damaged kernels was not affected by the state, region or their interaction. The ANOVA on DON and D3G for both years indicates that the wheat growing environment greatly affects the variations in DON and D3G content. Schmidt-Heydt *et al.* reported that the key genes in the biosynthetic pathway of mycotoxin production could be influenced by the environmental factors and water activity [4]. Therefore, the gene expression resulted in different effects, depending on their interaction with the abiotic conditions, on the DON production levels. On the other hand, D3G is produced by the plant as a detoxification process in response to the DON production [11]. So, it could be possible that the D3G content among states may depends of the infection level of DON in the wheat.

Table 3. Mean square values of state (ST), region (Rga) and county (CTY) on DON, D3G and damaged kernel in 2011 and 2012 survey.

Year	Source	Degrees of freedom	Mean square			
			GC-DON	LC-DON	D3G	Damage
2011	ST	3	221.3 ***	282.8 ***	2.15 ***	10.19 ***
	Rga (ST)	12	28.0 **	40.4 ***	0.43 *	0.45
	CTY (Rga × ST)	100	8.9	13.2	0.19	0.44
	Residual	320	9.7	14.0	0.24	0.70
2012	ST	3	151.2 ***	111.2 ***	0.15 ***	0.22 **
	Rga (ST)	12	104.3 ***	72.8 ***	0.12	0.12 ***
	CTY (Rga × ST)	100	14.5	11.3	0.07	0.04
	Residual	320	12.0	10.9	0.07	0.04

Notes: *, **, and *** means F values are significant at $p < 0.05$, 0.01, and 0.001, respectively.

2.3. Correlation among DON, D3G with Percent Damaged Kernels

The correlation between GC and LC-MS methods used to analyze DON during 2011 and 2012 is shown in Figure 2a,b. DON was measured by both GC and LC-QTOF-MS to evaluate the correlation between methods and determine the feasibility if the LC-QTOF-MS methodology for simultaneous measurement of DON and D3G. The high coefficient of determination (R^2) indicates strong, positive and significant correlation between both methods in both years (Figure 2). When the GC and LC-MS methods for both survey 2011 and 2012 were compared (Figure 2c), an R^2 of 0.947 and mean square error (MSE) of 0.90 were found. To identify relationships between mycotoxin contents and percent damaged kernels, linear and rank correlation coefficients were estimated and given in Table 4. The DON values determined GC and LC methods also showed very high and positive correlation for 2011 and 2012 data, individually (Table 4). These results mean that the LC method is as precise as or better in evaluating DON concentration of HRS wheat lines when compared to the GC method. The linear and rank correlations were significant ($p < 0.01$) and positive between DON concentration data determined for 2011 and 2012 samples. This result indicates that year by region interaction might not be strong, and the region that had higher DON concentration in 2011 samples also had higher DON concentration than other regions in 2012. Specifically, ND regions A and B showed higher DON concentration than other regions for both 2011 and 2012 samples. Further research may be needed to verify this since current results are based on data collected from only 2 years. The correlation between DON and D3G (Figure 3) in survey samples between 2011 and 2012 was significant with a moderate R^2 = 0.521. This means that the D3G production is related positively to the DON content and increasing DON levels also increase the D3G level in wheat. This is in agreement with the results from other researchers [10,13]. However, the moderate R^2 value indicates that DON concentration was partially responsible for D3G variation in this sample set. This is due to the low correlation between DON and D3G for 2012 samples (Table 4). To be more specific, 2012 samples from MN region A and MT region B had high D3G concentrations despite low DON concentrations (Table 2). These results indicate that, for precise evaluation of mycotoxin in HRS wheat samples, the LC-MS method is better to use for determination of DON and D3G content rather than the GC method which can only determine DON.

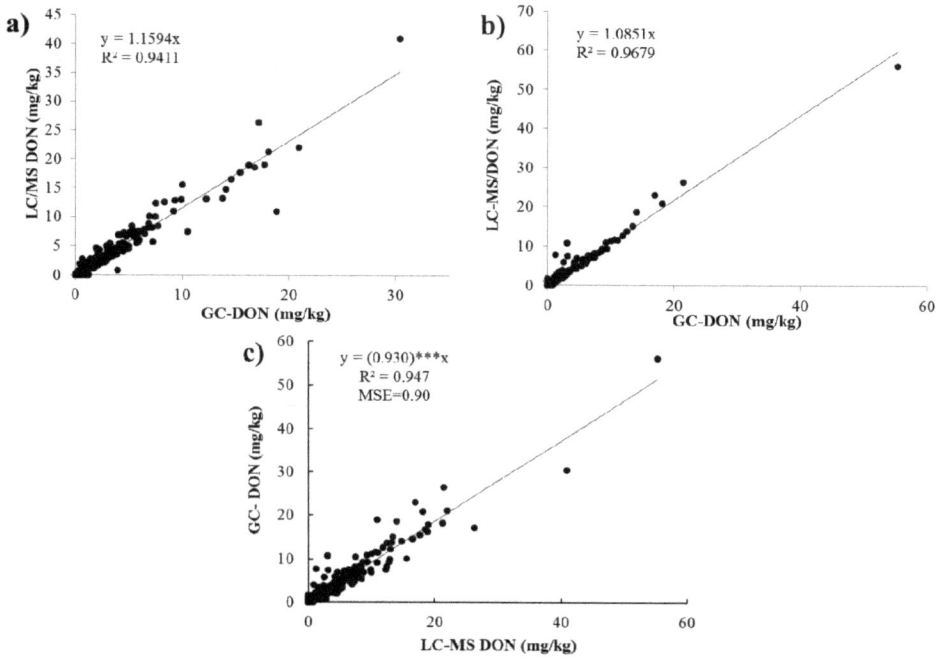

Figure 2. Correlation GC-DON and LC-MS DON content values. (a) 2011 survey samples; (b) 2012 survey samples and (c) 2011 and 2012 survey samples combined. *** Significantly different from 1 at $p < 0.001$.

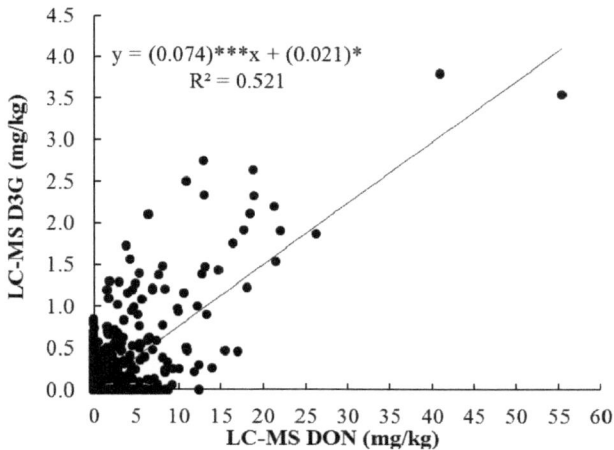

Figure 3. Correlation between DON and D3G levels in survey samples between 2011 and 2012; where ***, and * indicate that regression coefficients are significant at $p < 0.001$ and $p < 0.05$, respectively.

Table 4. Pearson linear and Spearman rank correlation coefficients between DON, D3G and damage kernel for regions.

Year	Variables	GC-DON	LC-DON	D3G	Damage
			Linear correlation		
2011	GC-DON	-	0.99 ***	0.91 ***	0.91 ***
	LC-DON	0.99 ***	-	0.91 ***	0.91 ***
	D3G	0.94 ***	0.92 ***	-	0.74 **
	Damage	0.89 ***	0.89 ***	0.87 ***	-
			Rank correlation		
			Linear correlation		
2012	GC-DON	-	1.00 ***	0.83 ***	0.79 ***
	LC-DON	0.99 ***	-	0.83 ***	0.79 ***
	D3G	0.41^{NS}	0.37^{NS}	-	0.60 *
	Damage	0.72 **	0.69 **	0.28^{NS}	-
			Rank correlation		

Notes: *, **, and *** means correlation coefficient is significant at $p < 0.05$, 0.01, and 0.001, respectively. NS: Not significant ($p \geq 0.05$).

The correlation among DON and D3G with damage kernel during 2011 and 2012 is given in Table 4 and Figure 4. The percent damaged kernels had very highly significant correlation with GC-DON and LC-DON ($p < 0.001$) in 2011; damage also had a very highly significant correlation at $p < 0.01$ with DON (GC and LC) in 2012. The positive correlations indicate that samples which were rated to have higher percent damaged kernels had higher levels of DON in the sample. This was also shown in Figure 4a,b, where the scatter plot between GC-DON and damage was depicted. D3G also had a significant ($p < 0.05$) correlation with percent damaged kernels in 2011 and 2012. This indicates that as the *Fusarium* infection progresses more kernel damage occurs. While DON levels rise which leads to increased production of D3G as the plants detoxification mechanism.

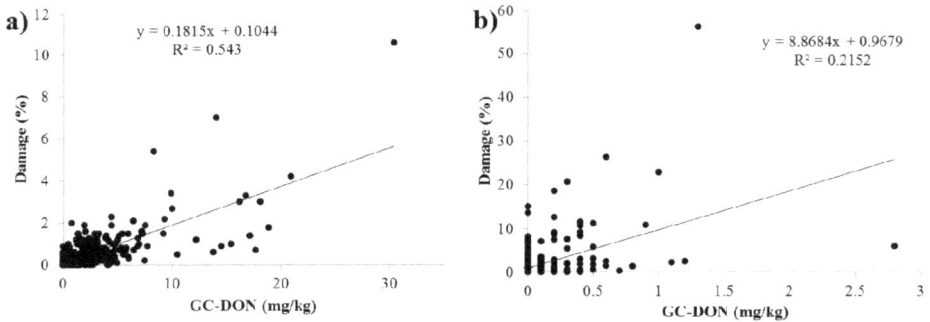

Figure 4. Correlation between GC-DON and damage levels in survey samples from (**a**) 2011 and (**b**) 2012.

3. Experimental Section

3.1. Standards and Chemicals

DON (100.2 μg/mL) and D3G (50.2 μg/mL) both in acetonitrile were purchased from Biopure (Tulln, Austria). For the gas chromatography–electron capture detection (GC–ECD) method, 5 mg/L DON working standard solution was used to make a standard curve prepared in a clean wheat extract. In the LC-MS method, stock solutions were dissolved in acetonitrile and stored in the refrigerator. Composite working standard solutions were prepared by dilution of the stock solutions in a DON-free wheat matrix to prepare matrix-matched calibration standards in concentration of 0 to 20 mg/L for DON and 0 to 10 mg/L for D3G. Acetonitrile was purchased from J. Baker. TMSI (1-(trimethylsilyl)imidazole), TMCS (Chlorotrimethylsilane) and 2,2,4-trimethylpentane (ACS reagent) were obtained from Sigma Aldrich.

3.2. Samples

The hard red spring (HRS) wheat between 2011 and 2012 Crop Survey samples were used as raw material. A total of 441 and 436 samples were selected as wheat grader samples from the 2011 and 2012 HRS wheat crop surveys, respectively, and used in this study. The samples were collected based on production data obtained from the National Agricultural Statistics Service for the 16 regions in the 4 state HRS wheat growing region (Figure 1). The Montana (MT), North Dakota (ND), South Dakota (SD) and Minnesota (MN) state office of the National Agricultural Statistics Service obtained wheat samples during harvest directly from growers either in the fields or farm bins and local elevators. Samples from the 2011 Crop Survey represented a high FHB infection and incidence of DON, while 2012 Crop Survey samples represented wheat with low FHB infection.

3.3. Wheat Kernel Quality

The kernel quality based on non-grading factors consisted of the determination of the protein content (expressed in 12% moisture basis, Method 39.10.01), falling number expressed in seconds (Method 56.81.03), both approved methods of the AACC [21] and thousand kernel weight determined on a 10 g sample of cleaned wheat (free of foreign material and broken kernels) counted by electronic seed counter.

The wheat grade and class of the samples was determined by a licensed grain inspector for the Official United States Standards for Grain. North Dakota Grain Inspection Service, Fargo, ND, provided grades for composite wheat samples. The final grade of the samples was based on dockage (elimination of all material other than wheat), shrunken and broken kernels and percent damaged kernels, test weight measured as pounds per bushel (l b/bu) (Method 55-10, AACC) and percent vitreous kernels (percentage of kernels having vitreous endosperm), as well as the summation of these defects referred to as total defects using an official procedure of USDA (United Stated Department of Agriculture).

3.4. Sample Preparation for GC–ECD and LC-MS

3.4.1. Free DON Analysis with GC–ECD

Free DON was determined using the methodology described by Tacke *et al.* [22] with some modifications [2]. One g of sample was dispersed in 8 mL of acetonitrile:water mixture (84:16, v/v) and shake for 1 h at 180 rpm. The extract (4 mL) was passed through cleanup column (Extract clean™ C18-Al, GRACE, Illinois, USA) and 2 mL of the filtrate were evaporated until dryness with nitrogen at 55 °C. Then, 100 μL of TMSI-TMCS (100:1, v/v) was added to the solution to residue. The samples were vortexed for 10 s and allowed to react for 5 min at room temperature. The internal standard solution (1 mL, 0.5 mg/L Mirex) was added and swirled. Immediately after which, 1 mL of water was added and the tube was shaken for 5 min. The samples were left at room temperature until two layers were observed. The top layer was transferred to vials for analysis by gas chromatography–electron capture detection (GC–ECD) with a series of standards. The GC–ECD equipment was Agilent 6890 GC with dual injector, cool on-column inlet, HP-5 column (30 m, 0.25 mm and 0.25 μm) and ECD detectors (Agilent Technologies, Wilmington, USA). Helium and argon-methane were used as carrier and makeup gas, respectively. The DON recoveries as determined by Tacke *et al.* [22], were 100%, 94% and 94% for 0.5, 4.0 and 20.0 mg/kg DON, respectively. The limit of detection (LOD) was determined to be 0.05 mg/kg and the limit of quantitation (LOQ) was determined to be 0.2 mg/kg [22]. Figure 5a shows a reference chromatogram of 2.0 mg/kg DON determined by GC–ECD.

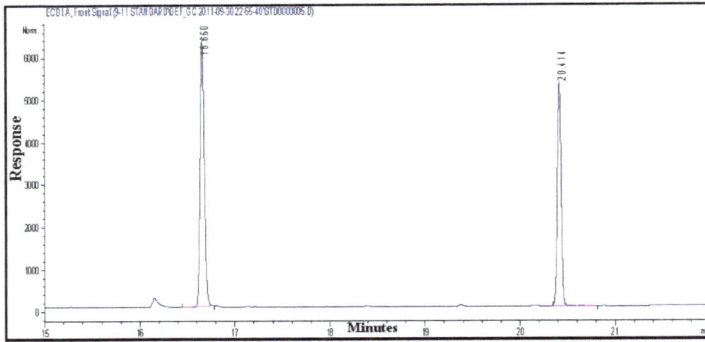

(a)

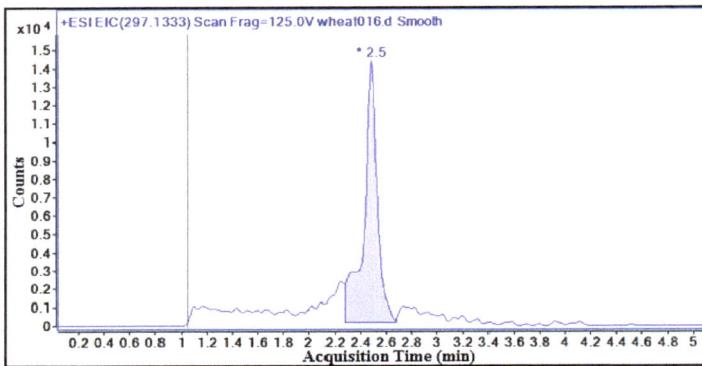

(b)

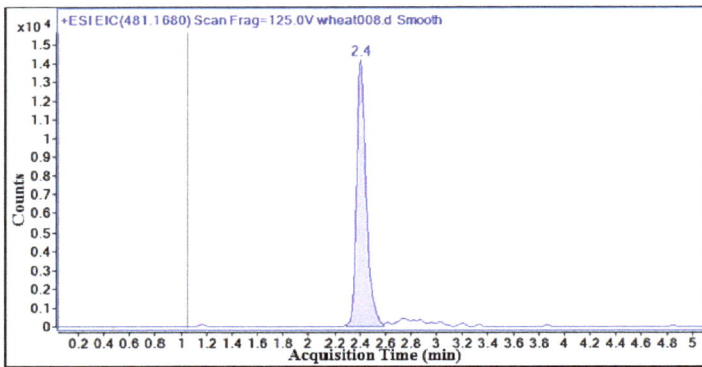

(c)

Figure 5. Representative chromatograms of a wheat sample containing approximately 2.0 mg/kg DON and 1.0 mg/kg D3G. (**a**) gas chromatography–electron capture detection (GC-ECD) chromatogram of DON * and Mirex ** (internal standard); (**b**) LC-MS extracted ion (m/z of (M + H) + ion 297.1333) chromatogram of DON; and (**c**) LC-MS extracted ion (m/z of the (M + Na) + ion 481.1680) chromatogram of D3G.

3.4.2. Free DON and D3G Analysis with LC-MS

The sample preparation was carried out according to Tacke *et al.* [22] with a few modifications. The sample (2.5 g) was extracted with 20 mL of acetonitrile/water mixture (84:16; v/v) for 1 h on an orbital shaker at 180 rpm. The samples were left 20–30 min to settle. One mL of the crude extract was filtered with 0.2 μm nylon syringe filter into a glass vial. The sample was analyzed with a liquid chromatographic coupled with a quadrupole time of flight system (LC-QTOF). A series of DON and D3G standards were also prepared and analyzed with every set of samples run on the LC-QTOF. Matrix matched calibration standards were prepared at several concentrations for DON and D3G to take into consideration any matrix effects from compounds extracted from the wheat. The concentrations of DON were 20.0, 10.0, 6.0, 4.0, 2.0, 1.0, 0.7, 0.5, 0.2 and 0.1 mg/kg and the D3G concentrations were 5.0, 3.0, 2.0, 1.0, 0.75, 0.5, 0.3 and 0.2 mg/kg. Figure 5b,c shows reference chromatograms of DON (2.0 mg/kg) and D3G (1.0 mg/kg) determined by LC-MS.

3.4.3. LC-MS Instrumentation and Methodology

The analysis of mycotoxins by liquid chromatography mass spectrometry (LC-MS) was performed according to the methods of Vendl *et al.* [23] and Simsek *et al.* [2] with some modifications. A 1290 UPLC System (Agilent Technologies, Wilmington, DE, USA) was used for separation of analytes. The DON and D3G separation was carried out with an Eclipse Plus C18 column (Zorbax Rapid Resolution High Definition (RRHD), 2.1 × 100 mm, 1.8-Micron, Agilent Technologies, Wilmington, DE, USA). The column temperature was set to 40 °C. The solvent system consisted of 0.1% formic/water (solvent A) and 0.1% formic/acetonitrile (solvent B). Formic acid was used instead of acetic acid to improve formation of the M+H ion rather than formation of M+Na ion found frequently when using acetic acid. The purge was done with 100% A with a purge flow rate of 4 mL/min during 15 s and the isocratic pump flow was 0.6 mL/min with 100% A. The solvent gradient was modified slightly to reduce run time and improve chromatographic separation of DON and D3G. The gradient program started with 97% A and 3% B with a binary pump flow rate of 0.4 mL/min and was kept until 0.75 min. Afterwards, the proportion of B was increased linearly to 100% within 4 min, followed by a hold time of 6 min at 100% B and 10 min re-equilibration at 97% A, followed by isocratic washout step for 2 min with 97% A. Five μL was the volume injection used.

The analyte detection was determined with a mass spectrometer quadrupole time of flight (Agilent 6540 time-of-flight LC-MS, Agilent Technologies, Wilmington, DE, USA). The ESI interface was used in positive-ionization mode at 300 °C with the following settings: 7 L/min gas flow, 30 psig nebulizer gas, 225 sheath gas temperature and 12 of sheath gas flow. Acquisition mode MS1 parameters were minimal range (m/z) 100, maximum range (m/z) 1700 and scan range of 2 spectra/s. The data analysis was performed using a MassHunter Qualitative Analysis B.05.00 program (Agilent Technologies, Wilmington, USA). Integration and calculation of DON and D3G were prepared by extracting the ion chromatograms for DON and D3G from the total ion chromatogram (TIC) using a mass window of 10 ppm. The DON extracted ion chromatogram used the m/z of the (M+H)$^+$ ion (297.1333) and the D3G EIC used the *m/z* of the (M+Na)$^+$ ion (481.1680). Since the detection method was TOF-MS without fragmentation the accurate mass and retention time of standards were used for identification of compounds (DON and D3G) as determined by Vendl *et al.* [23]. The DON recovery was calculated to be 107% whole the D3G recovery was 51% [23].

3.5. Statistics Analysis

Analysis of variance (ANOVA) was performed for individual years using the "MIXED" procedure in SAS (V 9.2, SAS Institute Inc., Cary, NC, USA). The model for ANOVA was a nested fixed model in which region was nested in state and city was nested in region. Least square mean values were estimated using the "LSMEAN" option. Correlation and regression was performed using "CORR" and "GLM" procedures in SAS, respectively.

4. Conclusions

The results indicated that DON levels varied with the survey crop year and they have a relationship with the kernel quality and D3G detected in wheat. Also, it was found that the growing state cause a larger effect on DON and D3G, but not on percent damaged kernels. The D3G levels were significantly correlated with the percent damaged kernels, but at lower levels than the DON content. DON infection in wheat caused more effect on the kernel quality between years analyzed. Otherwise, the ANOVA and correlation coefficient indicate that both GC and LC-MS can be used to determine DON in HRS. However, due to the ease of the method (sample can be extracted and analyzed without derivatization) and simultaneous determination of the D3G, LC-MS is more advantageous.

Acknowledgments: This work was supported by North Dakota State University Agricultural Experiment Station, Minnesota Wheat Research and Promotion Council and North Dakota Wheat Commission. We would like to thank DeLane Olsen for her help during the analysis of the wheat samples.

Conflicts of Interest: The authors declared no conflict of interest.

References

1. Wegulo, S. Factors influencing deoxinivalenol accumulation in small grain cereals. *Toxins* **2012**, *4*, 1157–1180. [CrossRef]

2. Simsek, S.; Burgess, K.; Whitney, K.L.; Gu, Y.; Qian, S.Y. Analysis of Deoxynivalenol and deoxynivalenol-3-glucoside in wheat. *Food Control* **2012**, *26*, 287–292. [CrossRef]

3. Paul, P.A.; Lipps, P.E.; Madden, L.V. Relationship between visual estimates of fusarium head blight intensity and deoxynivalenol accumulation in harvested wheat grain: A meta-analysis. *Phytopathology* **2005**, *95*, 1225–1236. [CrossRef]

4. Schmidt-Heydt, M.; Parra, R.; Geisen, R.; Magan, N. Modelling the relationship between environmental factors, transcriptional genes and deoxynivalenol mycotoxin production by strains of two Fusarium species. *J. R. Soc. Interface* **2011**, *8*, 117–126. [CrossRef]

5. Hope, R.; Aldred, D.; Magan, N. Comparison of environmental profiles for growth and deoxynivalenol production by Fusarium culmorum and F-graminearum on wheat grain. *Lett. Appl. Microbiol.* **2005**, *40*, 295–300. [CrossRef]

6. Muhovski, Y.; Batoko, H.; Jacquemin, J.M. Identification, characterization and mapping of differentially expressed genes in a winter wheat cultivar (Centenaire) resistant to Fusarium graminearum infection. *Mol. Biol. Reports* **2012**, *39*, 9583–9600. [CrossRef]

7. Pestka, J.J. Deoxynivalenol: Mechanisms of action, human exposure, and toxicological relevance. *Arch. Toxicol.* **2010**, *84*, 663–679. [CrossRef]

8. Arunachalam, C.; Doohan, F.M. Trichothecene toxicity in eukaryotes: Cellular and molecular mechanisms in plants and animals. *Toxicol. Lett.* **2013**, *217*, 149–158. [CrossRef]

9. Maresca, M. From the gut to the brain: Journey and pathophysiological effects of the food-associated trichothecene mycotoxin deoxynivalenol. *Toxins* **2013**, *5*, 784–820. [CrossRef]

10. Desmarchelier, A.; Seefelder, W. Survey of deoxynivalenol and deoxynivalenol-3-glucoside in cereal-based products by liquid chromatography electrospray ionization tandem mass spectrometry. *World Mycotoxin J.* **2011**, *4*, 29–35. [CrossRef]

11. Poppenberger, B.; Berthiller, F.; Lucyshyn, D.; Sieberer, T.; Schuhmacher, R.; Krska, R.; Kuchler, K.; Glossl, J.; Luschnig, C.; Adam, G. Detoxification of the Fusarium mycotoxin deoxynivalenol by a UDP-glucosyltransferase from Arabidopsis thaliana. *J. Biol. Chem.* **2003**, *278*, 47905–47914. [CrossRef]

12. Berthiller, F.; Schuhmacher, R.; Adam, G.; Krska, R. Formation, determination and significance of masked and other conjugated mycotoxins. *Anal. Bioanal. Chem.* **2009**, *395*, 1243–1252. [CrossRef]

13. Rasmussen, P.H.; Nielsen, K.F.; Ghorbani, F.; Spliid, N.H.; Nielsen, G.C.; Jørgensen, L.N. Occurrence of different trichothecenes and deoxynivalenol-3-b-d-glucoside in naturally and artificially contaminated Danish cereal grains and whole maize plants. *Mycotoxin Res.* **2012**, *28*, 181–190. [CrossRef]

14. Lemmens, M.; Scholz, U.; Berthiller, F.; Dall'Asta, C.; Koutnik, A.; Schuhmacher, R.; Adam, G.; Buerstmayr, H.; Mesterhazy, A.; Krska, R.; *et al.* The ability to detoxify the mycotoxin deoxynivalenol colocalizes with a

Toxins **2013**, *5*, 2656–2670

major quantitative trait locus for fusarium head blight resistance in wheat. *Mol. Plant-Microbe Interact.* **2005**, *18*, 1318–1324. [CrossRef]

15. Berthiller, F.; Krska, R.; Domig, K.J.; Kneifel, W.; Juge, N.; Schuhmacher, R.; Adam, G. Hydrolytic fate of deoxynivalenol-3-glucoside during digestion. *Toxicol. Lett.* **2011**, *206*, 264–267. [CrossRef]

16. Nagl, V.; Schwartz, H.; Krska, R.; Moll, W.D.; Knasmuller, S.; Ritzmann, M.; Adam, G.; Berthiller, F. Metabolism of the masked mycotoxin deoxynivalenol-3-glucoside in rats. *Toxicol. Lett.* **2012**, *213*, 367–373. [CrossRef]

17. European Food Safety Authority. Deoxynivalenol in food and feed: Occurrence and exposure. *EFSA J.* **2013**, *11*, 3379–3435.

18. U.S. Hard Red Spring Wheat 2012 Regional Quality Report. Available online: http://www.uswheat. org/cropQuality/doc/1B247A99F59BC22785257C15006659F3/\protect\T1\textdollarFile/HRS2012.pdf? OpenElement# (accessed on 16 December 2013).

19. Serranti, S.; Cesare, D.; Bonifazi, G. The development of a hyperspectral imaging method for the detection of Fusarium-damaged, yellow berry and vitreous Italian durum wheat kernels. *Biosyst. Eng.* **2013**, *115*, 20–30. [CrossRef]

20. Garrido-Lestache, E.; Lopez-Bellido, R.J.; Lopez-Bellido, L. Effect of N rate, timing and splitting and N type on bread-making quality in hard red spring wheat under rainfed Mediterranean conditions. *Field Crops Res.* **2004**, *85*, 213–236. [CrossRef]

21. American Association of Cereal Chemists International (AACCI). *Approved Methods of the American Association of Cereal Chemists*; AACC International: St. Paul, MN, USA, 2010.

22. Tacke, B.K.; Casper, H.H. Determination of deoxynivalenol in wheat, barley, and malt by column cleanup and gas chromatography with electron capture detection. *J. AOAC Int.* **1996**, *79*, 472–475.

23. Vendl, O.; Berthiller, F.; Crews, C.; Krska, R. Simultaneous determination of deoxynivalenol, zearalenone, and their major masked metabolites in cereal-based food by LC-MS-MS. *Anal. Bioanal. Chem.* **2009**, *395*, 1347–1354. [CrossRef]

Article

Analysis of Deoxynivalenol and Deoxynivalenol-3-glucoside in Hard Red Spring Wheat Inoculated with *Fusarium Graminearum*

Maribel Ovando-Martínez [1], Bahri Ozsisli [2], James Anderson [3], Kristin Whitney [1], Jae-Bom Ohm [4] and Senay Simsek [1,*]

[1] Department of Plant Sciences, North Dakota State University, PO Box 6050, Fargo, ND 58108, USA; maribel.ovando@ndsu.edu (M.O.-M.); kristin.whitney@ndsu.edu (K.W.)

[2] Department of Food Engineering, College of Agriculture, Kahramanmaras Sutcu Imam University, Kahramanmaras 46060, Turkey; bozsisli@ksu.edu.tr

[3] University of Minnesota, Agronomy/Plant Genetics, St. Paul, MN 55108, USA; ander319@umn.edu

[4] USDA-ARS, Cereal Crops Research Unit, Hard Red Spring and Durum Wheat Quality Laboratory, Harris Hall, North Dakota State University, P.O. Box 6050, Fargo, ND 58108, USA; jae.ohm@ars.usda.gov

* Author to whom correspondence should be addressed; senay.simsek@ndsu.edu; Tel.: +1-701-231-7737; Fax: +1-701-231-8474.

Received: 31 October 2013; in revised form: 10 December 2013; Accepted: 11 December 2013; Published: 17 December 2013

Abstract: Deoxynivalenol (DON) is a mycotoxin affecting wheat quality. The formation of the "masked" mycotoxin deoxinyvalenol-3-glucoside (D3G) results from a defense mechanism the plant uses for detoxification. Both mycotoxins are important from a food safety point of view. The aim of this work was to analyze DON and D3G content in inoculated near-isogenic wheat lines grown at two locations in Minnesota, USA during three different years. Regression analysis showed positive correlation between DON content measured with LC and GC among wheat lines, locality and year. The relationship between DON and D3G showed a linear increase until a certain point, after which the DON content and the D3G increased. Wheat lines having higher susceptibility to Fusarium showed the opposite trend. ANOVA demonstrated that the line and location have a greater effect on variation of DON and D3G than do their interaction among years. The most important factor affecting DON and D3G was the growing location. In conclusion, the year, environmental conditions and location have an effect on the D3G/DON ratio in response to Fusarium infection.

Keywords: fusarium; wheat; deoxinyvalenol; deoxynivalenol-3-glucoside

---

1. Introduction

Molds can infect almost every agricultural crop—including wheat—worldwide during plant growth and/or after harvest. A great variety of these fungi can produce mycotoxins, which are poisonous for humans and animals and can be found in a great variety of food and feed commodities [1,2]. Deoxynivalenol (DON) or vomitoxin is a trichothecene mycotoxin produced by fungal plant pathogens *Fusarium graminearum* and *F. culmorum*. Both pathogens cause a disease known as Fusarium head blight (FHB) [3]. The severity of FHB during individual seasons depends on precipitation during flowering, and increased levels of DON are often observed in harvest years with frequent rainfall and high humidity during flowering [4]. In determination of the tolerance to FHB in wheat, it has been reported that quantitative trait loci (QTL)-Fhb1 governs resistance towards FHB [5]. In general, the most important factors that influence germination of wheat, growth of *Fusarium* and biosynthesis of DON are temperature and water activity (moisture amount and duration) [3,6]. DON is usually the most prevalent of the trichothecenes found in small grains grown in temperate regions all over the world; the reason for which is extensively studied [4]. The European and Food Safety Authority [2] reported that DON was found in 44.6%, 43.5% and 75.2% of unprocessed grains of undefined end-use, food and feed samples, respectively, with maize, wheat and oat having the highest levels. Also, DON levels were significantly higher in wheat bran than other wheat milling products, while DON levels in processed cereals were significantly lower than the DON levels found in unprocessed grains. So, in the interest of food health and safety, mycotoxin analysis represents a major challenge in the control and inspection of foodstuffs, as a high proportion of cereal based foods are affected [7].

Another emerging food safety concern related to the topic of mycotoxins is their "masked" forms. Deoxynivalenol-3-glucoside (D3G) which has been discovered relatively recently is one of the most common forms of masked DON. D3G is formed as part of a detoxification process in the plant through the glycosylation of DON and is stored in the plant vacuoles [4,8,9]. It has been reported that D3G formation is connected with glycosyltransferases [7]. Poppenberger *et al.* found that the UDP-glucosyltransferase *AtUGT73C5* transferred glucose to the hydroxyl group at carbon 3 of DON forming D3G in *Arabidopsis* [8]. Also, D3G has been found in wheat lines with low FHB susceptibility. In this case, the DON is converted to D3G due to the presence of the quantitative trait locus (QTL)-Fhb1, which encodes a glucosyltransferase or regulates the expression of such enzyme [10]. There was very little data available for D3G content in samples of food, feed and unprocessed grains of undefined end-use collected by 21 European countries between 2007 and 2012. D3G was found in approximately 5% of the samples, almost always together with DON, representing 5.6% of the lower bound sum of DON and D3G. Because of this, the "masked" mycotoxin D3G was not taken into account in the exposure assessment [2]. However, D3G should be measured because it is unknown how this mycotoxin can be reactivated in humans and animals. Berthiller *et al.* have shown that D3G is not affected by stomach conditions, but when D3G is exposed to human lactic acid bacteria, the glucose is cleaved and DON is released, reactivating its toxicity [11]. Therefore, knowledge about the natural occurrence and impact on processing/manufacturing practices on toxin levels in the final products has become important for assessment of potential health risks associated with mycotoxin contamination [12]. The aim of this research was to analyze the DON and D3G content and determine if there is correlation between the DON and D3G production in samples with variation in susceptibility to FHB grown in Minnesota, USA at two locations and three years of study, using liquid chromatography-quadrupole time of flight mass spectrometry (LC-MS).

2. Results and Discussions

2.1. Relationship between DON and D3G

The data obtained showed correlation between DON and D3G. LC-MS was chosen for mycotoxin determination because it is more convenient due to the lack of derivitization step in sample preparation. Another advantage of the LC-MS is that it was possible to determine the D3G content in wheat simultaneously with the DON determination. The correlation between DON and D3G determined by LC-MS is shown in Figure 1. The coefficient of determination was moderate and significant ($R^2 = 0.872$). The equation model obtained with this R^2 value was a second-order curve. The D3G content rose as the DON content increased in samples with DON content between 0 and 30 ppm. However, at higher DON concentration, a decrease in the D3G content was seen. Sasanya *et al.* did not find correlation between DON and D3G content in randomly selected hard red wheat samples, and they also observed that in the samples with the highest DON levels D3G was not detected [13]. Lemmens *et al.* found that samples without the presence of the quantitative trait loci (QTL)-Fhb1 (FHB resistance gene) showed high DON and D3G content [5]. However, in the presence of Fhb1, the conjugation of DON to glucose occurred to a larger extent as compared to the lines without the QTL-Fhb1 [5]. Our results showed the same trend reported by these authors. Possibly, the type of inoculum used between locations to inoculate the samples, and inoculation at slightly different growth stages could be causing the behavior observed in the correlation of DON and D3G in this research. It has been reported that D3G is a "masked" mycotoxin, product of the detoxification of the plant due to DON production after *Fusarium* infection [8]. Such transformation is catalyzed by plant enzymes [1,7], such as glycosyltransferases, which transfer sugars to a wide range of plant receptors. In *Arabidopsis* cloned with HvUGT13248 (UDP-glucosyltranferase (UGT) from barley, capable of detoxifying DON), FHB susceptibility decreased and the capacity to convert DON into D3G increased [9]. Therefore, the results obtained in this study lead us to think that the samples which presented lower FHB susceptibility (lower DON), will produce high levels of D3G; whilst the samples with higher FHB susceptibility will have lower levels of this "masked" mycotoxin. This means the DON and D3G formation exerted by the less FHB susceptible wheat lines is a response towards *Fusarium* infection [14].

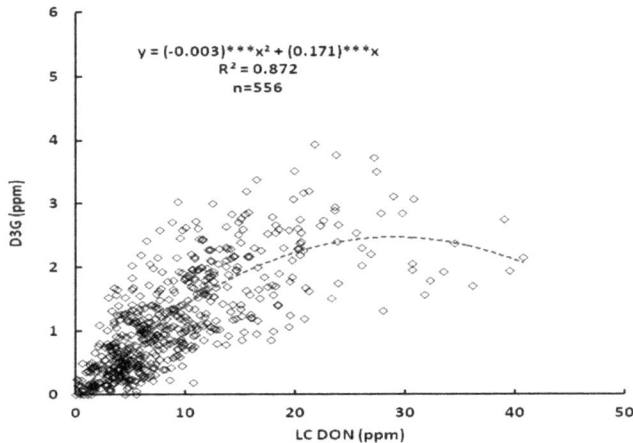

Figure 1. Correlation between liquid chromatography deoxynivalenol (LC-DON) and deoxynivalenol-3-glucoside (D3G) values (combined 2008, 2009 and 2010). *** Significantly different from 1 at $p < 0.001$.

2.2. Effect of the Line, Location and Their Interactions on DON and D3G Content

Table 1 shows the means of DON and D3G content of wheat lines grown in Minnesota collected during 2008, 2009 and 2010. The values for 2008 ranged from 0.1 to 33.9 ppm and 0.1 to 1.9 ppm for LC-DON and D3G, respectively. Overall, the mycotoxin contents for 2009 were lower and ranged from 0.0 to 23.6 ppm and 0.0 to 3.0 ppm, for LC-DON and D3G, respectively. The lowest mycotoxin contents were observed in 2010 and ranged from 0.2 to 17.7 ppm and 0.0 to 2.2 ppm, for LC-DON and D3G, respectively. The analysis of variance (ANOVA) for DON and D3G in the samples for individual years is shown in Table 2. During 2008, DON and D3G contents were not statistically related to the main effects [Line and Location (Loc)] or their interaction (Line × Loc). In 2009, it was observed that the effects of Line and Loc on DON were statistically significant. This means that growing location is the main influence of DON content among samples. Concerning the D3G, during 2009 only the Loc had a significant effect. On the other hand, during 2010, the main effects were significantly related to the DON and D3G content in the samples, whereas the interaction between factors was not significant. These results indicated that genetic and environmental conditions play an important role in the DON and D3G production in 2010. So, it can be concluded that the DON content, dependent on the growing conditions in a particular season, is affected by the wheat line. However, the main influence on DON production is the location where the wheat line is planted, which also showed the highest influence on the D3G content. It has been reported that the environmental conditions affect the gene expression and so affect the mycotoxin occurrence, in this case DON production [6]. Although, factors such as the stage of plant development and concentration and delivery of the toxin may be important in determining when and where DON is relevant in the disease process [10]. As was found for the DON, the D3G production will depend on the tolerance/susceptibility level of the wheat line to FHB and its response to environmental conditions. Lemmens *et al.* found D3G in naturally infected wheat samples, and also that wheat containing the Qfhs.ndsu-3BS QTL have the ability to convert DON to D3G, resulting in a high D3G/DON ratio and a low FHB susceptibility level [10]. It seems that conjugation of DON to glucose is the primary biochemical mechanism for resistance towards DON [10].

The results of this study on D3G could help to increase the data about its occurrence in hard red spring wheat in the USA. For example, in Europe, D3G content is not taken into account in the food safety assessment [2] and in the USA, there is not a statement about the maximum levels of D3G in food and feed, due to the lack of data about this "masked" mycotoxin. Increased D3G content, due to the detoxification process of the plant, should be taken into account in terms of food safety because this "masked" mycotoxin might be converted back to the parental toxin in the end products made from the wheat [9].

Table 1. Means of GC-DON, LC-DON and D3G of wheat samples collected during 2008–2010 in Crookston, Saint Paul and Minnesota (MN).

Year	Location	Range	DON [a]	D3G [a]
2008	Crookston	Min (n = 22)	0.1	0.3
		Max (n = 22)	24.2	1.8
		Average (n = 22)	5.7	1.1
	St. Paul	Min (n = 22)	0.7	0.1
		Max (n = 22)	39.5	1.9
		Average (n = 22)	11.1	0.9
	MN	Min (n = 44)	0.1	0.1
		Max (n = 44)	39.5	1.9
		Average (n = 44)	8.4	1.0
2009	Crookston	Min (n = 35)	0.0	0.0
		Max (n = 35)	25.7	3.8
		Average (n = 35)	11.9	2.1
	St. Paul	Min (n = 35)	0.2	0.0
		Max (n = 35)	21.0	1.5
		Average (n = 35)	4.8	0.5
	MN	Min (n = 70)	0.0	0.0
		Max (n = 70)	25.7	3.8
		Average (n = 70)	8.3	1.3
2010	Crookston	Min (n = 88)	1.7	0.4
		Max (n = 88)	20.2	2.6
		Average (n = 88)	7.9	1.3
	St. Paul	Min (n = 90)	0.2	0.0
		Max (n = 90)	11.5	1.5
		Average (n = 90)	4.2	0.5
	MN	Min (n = 178)	0.2	0.0
		Max (n = 178)	20.2	2.6
		Average (n = 178)	6.1	0.9

Notes: [a] in ppm (parts per million); n = number of lines in each set.

Table 2. ANOVA table for DON and D3G of wheat samples for 2008–2010.

Year	Traits	Source	DF	Sum of squares	Mean square	F Value	Pr > F
2008	DON	Line	21	3369.6	160.5	2.9	0.0095
		Loc	1	418.7	418.7	7.5	0.0122
		Line × Loc	21	1169.0	55.7	3.5	0.015
		Error	12	191.7	16.0		
	D3G	Line	21	12.7	0.60	3.8	0.0016
		Loc	1	0.9	0.88	5.6	0.0275
		Line × Loc	21	3.3	0.16	4.7	0.004
		Error	12	0.4	0.03		
2009	DON	Line	34	5314.4	156.3	5.3	<0.001
		Loc	1	1013.4	1013.4	34.3	<0.001
		Line × Loc	34	1004.5	29.5	1.3	0.177
		Error	72	1639.7	22.8		
	D3G	Line	34	27.9	0.82	2.5	0.0048
		Loc	1	49.3	49.28	149.0	<0.0001
		Line × Loc	34	11.2	0.33	2.6	0.000
		Error	72	9.0	0.13		
2010	DON	Line	89	3479.0	39.1	8.0	<0.0001
		Loc	1	650.1	650.1	132.5	<0.0001
		Line × Loc	87	426.8	4.9	1.0	0.497
		Error	88	419.0	4.8		
	D3G	Line	89	48.1	0.54	4.0	<0.0001
		Loc	1	33.1	33.09	247.4	<0.0001
		Line × Loc	87	11.6	0.13	1.0	0.497
		Error	88	11.8	0.13		

2.3. Correlation of DON and D3G Content between Locations

The Pearson and Spearman's correlations were used to determine the correlation between DON and D3G in wheat grown at two localities of Minnesota, and are shown in Table 3. During 2008, the ANOVA did not show any significant effect of the Loc between these two parameters. However, the Spearman's correlation showed positive correlation coefficients with significant levels ($p < 0.05$, 0.01 and 0.001) among DON and D3G from Crookston and St. Paul (Table 3). With respect to D3G from Saint Paul, the Pearson correlation indicated that there was significant ($p < 0.001$) correlation with DON from Saint Paul for 2008, 2009 and 2010. However, the Spearman correlation determined a correlation coefficient of 0.44 ($p < 0.01$) for DON. This may be related to the trend (second order curve) observed among the DON and D3G content among localities and years of study obtained in Figure 1. The low significance level could be due to the different kind of inoculum used to infect the wheat lines, differences in the growth stage development of the plant when the inoculum was applied, and the differences in the weather conditions between Crookston and St. Paul during the three years of study. In the case of 2010, the correlation among the parameters between both localities showed high and significant correlation ($p < 0.001$). This indicated that the year of study also influenced the DON and D3G content in the wheat lines, probably because of differences in the rainfall or moisture, relative humidity, temperature, factors that have a notable effect on *Fusarium* infection [3]. Incidentally, the significant ($p < 0.05$) correlations that occurred between mycotoxin contents of samples collected from two different locations supports the notion that the interaction of wheat line and location might not have a strong effect on variation of mycotoxin contents as already suggested by ANOVA. These results also indicate that selection of wheat lines that have resistance to mycotoxin production might be possible in one location. Cowger *et al.* found that genetic differences among cultivars may reflect its ability to resist DON under increasing moisture conditions [15], meaning that the higher the disease with varying post-anthesis moisture durations (in this case a nursery), the greater the differential effects. While it is well known that agronomic and climatic conditions play an important role in mycotoxin formation in wheat, cultivar selection and breeding strategies are very important for identification of wheat with low susceptibility to *Fusarium* on the basis of masked mycotoxin formation.

Table 3. Pearson and Spearman's correlation coefficients between DON and D3G of two localities of Minnesota for 2008–2010.

Year	Crk DON	Stp DON	Crk D3G	Stp D3G
		Pearson Correlation		
	-	0.59 **	0.56 **	0.56 **
2008	0.51 *	-	0.47 *	0.90 ***
	0.68 ***	0.46 *	-	0.59 **
	0.58 **	0.87 ***	0.51 *	-
		Spearman correlation		
		Pearson Correlation		
	-	0.52 **	0.66 ***	0.50 **
2009	0.45 **	-	0.24 NS	0.75 ***
	0.56 ***	0.44 **	-	0.41 *
	0.49 **	0.69 ***	0.56 ***	-
		Spearman correlation		
		Pearson Correlation		
	-	0.63 ***	0.69 ***	0.41 ***
2010	0.57 ***	-	0.51 ***	0.67 ***
	0.55 ***	0.45 ***	-	0.48 ***
	0.34 **	0.61 ***	0.44 ***	-
		Spearman correlation		

Notes: Crk: Crookston, Stp: Saint Paul, DON: deoxynivalenol, D3G: deoxynivalenol-3-glucoside; NS: No significant. *, **, and *** means correlation coefficient is significant at $p < 0.05$, 0.01, and 0.001, respectively.

3. Experimental Section

3.1. Standards and Chemicals

DON (100.2 µg/mL) and D3G (50.2 µg/mL) both in acetonitrile were purchased from Biopure (Tulln, Austria). The standard curve for both GC-ECD and LC-MS methods were prepared using clean wheat extract (DON-free wheat matrix). Acetonitrile was purchased from J. Baker. TMSI (1-(trimethylsilyl)imidazole), TMCS (Chlorotrimethylsilane) and 2,2,4-trimethylpentane (ACS reagent) were obtained from Sigma Aldrich.

3.2. Samples

Different wheat lines ranging from moderately susceptible to susceptible to Fusarium head blight (FHB) were analyzed. The samples were collected when the latest maturing lines were at harvest ripeness (14% or less grain moisture content). The sample set is comprised of experimental spring wheat lines from the University of Minnesota wheat breeding program, ranging from first year to third year yield trial lines. The checks Alsen, BacUp, Roblin, Wheaton, and MN00269 are included for each nursery and are represented 53–55 times each in the data set. Therefore, the checks represent a total of 272 samples. The experimental lines, each represented 1–4 times, totaled 287 samples.

All lines were grown under two field screening during 2008, 2009 and 2010 in two locations of Minnesota, USA. The growing locations were St. Paul, MN (44.9441° N, 93.0852° W) and Crookson, MN (47.7742° N, 96.6081° W). For both locations, the weather conditions in 2008 were cool and wet during planting. The growing conditions were hot and slightly dry, but with adequate soil moisture at both locations in 2008. The growing conditions in 2009 at both locations were cooler than average with adequate precipitation. Both growing locations in 2010 had cooler growing temperatures and adequate precipitation.

At the St. Paul location (StP), *F. graminearum* macronidia was applied by backpack sprayer at the rate of 60 mL of a 100,000 conidia/mL per 2.4 m row at anthesis and 3–4 days later. At the Crookston location (Crk), grain spawn inoculum was spread at the rate of 56 kg/ha at the jointing stage and with a second application one week later. Both nurseries were misted periodically overnight to maintain high humidity environments. In Crookston, the grain-spawn inoculum method used in Crookston mimics more closely what happens in nature, and a constant supply of inoculum was possible. In St. Paul, conidia were spray applied to control timing and inoculum dose. The conidia spray method is not subject to as great of a possibility of escapes, unless the climate conditions are particularly non-conducive during inoculation and the 48 hours post inoculation (for example, very windy and dry conditions).

The samples were ground using a UDY mill with a 0.8 mm screen and conserved under refrigeration until their analysis.

3.3. Sample Preparation

The sample preparation was carried out according to Tacke *et al.*, with some modifications [16,17]. The sample (2.5 g) was extracted with 20 mL of acetonitrile/water mixture (84:16; v/v) for 1 h on an orbital shaker at 180 rpm. The samples were left 20–30 min to settle. The crude extract (1 mL) was filtered with 0.2 μm nylon syringe filter into glass vial. The sample was analyzed with a liquid chromatography system coupled with a quadrupole time of flight system (LC-QTOF).

3.4. LC-MS Instrumentation and Methodology

A 1200 Series HPLC System (Agilent Technologies, Wilmington, DE, USA) was used for separation of analytes. The DON and D3G separation was carried out with an Eclipse Plus C18 column (Zorbax Rapid Resolution High Definition (RRHD), 2.1 × 100 mm, 1.8-Micron, Agilent Technologies, Wilmington, DE, USA). The column temperature was set to 40 °C. The solvent system consisted of 0.1% formic acid/water (solvent A) and 0.1% formic acid/acetonitrile (solvent B). The purge was done with 100% A with a purge flow rate of 4 mL/min during 15 s and the isocratic pump flow was 0.6 mL/min with 100% A. The gradient program started with 97% A and 3% B with a binary pump flow rate of 0.4 mL/min and was kept until 0.75 min. Afterwards, the proportion of B was increased linearly to 100% within 4 min, followed by a hold time of 6 min at 100% B and 10 min re-equilibration at 97% A, followed by isocratic washout step for 2 min with 100% A. The volume of injection used was 5 μL.

The analytes' detection was determined with a mass spectrometer quadrupole time of flight (Agilent 6500 series time-of-flight LC/MS, Agilent Technologies, Wilmington, DE, USA). The ESI interface was used in positive-ionization mode at 300 °C with the following settings: 7 L/min gas flow, 30 psig nebulizer gas, 225 sheath gas temperature and 12 of sheath gas flow. Acquisition mode MS1 parameters were minimal range (m/z) 100, maximum range (m/z) 1700 and a scan range of 2 spectra/s. The data analysis was performed using a MassHunter Qualitative Analysis B.05.00 program (Agilent Technologies, Wilmington, DE, USA).

3.5. Statistics Analysis

Analysis of variance (ANOVA) was performed individually for three year data. The "GLM" procedure in SAS (V 9.2, SAS Institute Inc., Cary, NC, USA) was used for ANOVA in which wheat line and location were considered as fixed effects. The main effects of wheat line and location and their interaction were tested for significance using the residual error terms. Correlation and regression was performed using "CORR" and "GLM" procedures in SAS, respectively.

4. Conclusions

In conclusion, the relationship between DON and D3G fit a second order curve, indicating that the tolerance of the wheat lines to the Fusarium infection is related to the ability of the wheat line to convert the DON to D3G during the detoxification process. Also, the most important factor affecting the DON and D3G formation is locality, which may be due to differences in gene expression of the wheat line in different environmental conditions and its response to different inoculum and development stages of the wheat during the inoculation process.

Acknowledgments: This work was supported by North Dakota State University Agricultural Experiment Station and Minnesota Wheat Research and Promotion Council. We would like to thank DeLane Olsen for her help during the analysis of the wheat samples.

Conflicts of Interest: The authors declared no conflict of interest.

References

1. Berthiller, F.; Schuhmacher, R.; Adam, G.; Krska, R. Formation, determination and significance of masked and other conjugated mycotoxins. *Anal. Bioanal. Chem.* **2009**, *395*, 1243–1252.
2. European Food Safety Authority. European Food Safety Authority Deoxynivalenol in food and feed: Occurrence and exposure. *EFSA J.* **2013**, *11*, 3379–3435.
3. Wegulo, S. Factors influencing deoxinivalenol accumulation in small grain cereals. *Toxins* **2012**, *4*, 1157–1180. [CrossRef]
4. Rasmussen, P.H.; Nielsen, K.F.; Ghorbani, F.; Spliid, N.H.; Nielsen, G.C.; Jørgensen, L.N. Occurrence of different trichothecenes and deoxynivalenol-3-β-d-glucoside in naturally and artificially contaminated Danish cereal grains and whole maize plants. *Mycotoxin. Res.* **2012**, *28*, 181–190. [CrossRef]
5. Lemmens, M.; Koutnik, A.; Steiner, B.; Buerstmayr, H.; Berthiller, F.; Schuhmacher, R.; Maier, F.; Schäfer, W. Investigations on the ability of Fhb1 to protect wheat against nivalenol and deoxynivalenol. *Cereal Res. Commun.* **2008**, *36*, 429–435.
6. Schmidt-Heydt, M.; Parra, R.; Geisen, R.; Magan, N. Modelling the relationship between environmental factors, transcriptional genes and deoxynivalenol mycotoxin production by strains of two Fusarium species. *J. Royal Soc. Interface* **2011**, *8*, 117–126.
7. Maul, R.; Müller, C.; Rieß, S.; Koch, M.; Methner, F.J.; Irene, N. Germination induces the glucosylation of the *Fusarium* mycotoxin deoxynivalenol in various grains. *Food Chem.* **2012**, *131*, 274–279. [CrossRef]
8. Poppenberger, B.; Berthiller, F.; Lucyshyn, D.; Sieberer, T.; Schuhmacher, R.; Krska, R.; Kuchler, K.; Glossl, J.; Luschnig, C.; Adam, G. Detoxification of the Fusarium mycotoxin deoxynivalenol by a UDP-glucosyltransferase from Arabidopsis thaliana. *J. Biol. Chem.* **2003**, *278*, 47905–47914. [CrossRef]
9. Shin, S.; Torres-Acosta, J.A.; Heinen, S.J.; McCormick, S.; Lemmens, M.; Paris, M.P.K.; Berthiller, F.; Adam, G.; Muehlbauer, G.J. Transgenic Arabidopsis thaliana expressing a barley UDP-glucosyltransferase exhibit resistance to the mycotoxin deoxynivalenol. *J. Exp. Botany* **2012**, *63*, 4731–4740. [CrossRef]
10. Lemmens, M.; Scholz, U.; Berthiller, F.; Dall'Asta, C.; Koutnik, A.; Schuhmacher, R.; Adam, G.; Buerstmayr, H.; Mesterházy, Á.; Krska, R. The ability to detoxify the mycotoxin deoxynivalenol colocalizes with a major quantitative trait locus for Fusarium head blight resistance in wheat. *Mol. Plant-Mic. Interact.* **2005**, *18*, 1318–1324. [CrossRef]
11. Berthiller, F.; Krska, R.; Domig, K.J.; Kneifel, W.; Juge, N.; Schuhmacher, R.; Adam, G. Hydrolytic fate of deoxynivalenol-3-glucoside during digestion. *Toxicol. Lett.* **2011**, *206*, 264–267.
12. Malachova, A.; Dzuman, Z.; Veprikova, Z.; Vaclavikova, M.; Zachariasova, M.; Hajslova, J. Deoxynivalenol, deoxynivalenol-3-glucoside, and enniatins: The major mycotoxins found in cereal-based products on the Czech market. *J. Agricul. Food Chem.* **2011**, *59*, 12990–12997. [CrossRef]
13. Sasanya, J.J.; Hall, C.; Wolf-Hall, C. Analysis of deoxynivalenol, masked deoxynivalenol, and Fusarium graminearum pigment in wheat samples, using liquid chromatographyuvmass spectrometry. *J. Food Prot.* **2008**, *71*, 1205–1213.
14. Dall'Asta, C.; Dall'Erta, A.; Mantovani, P.; Massi, A.; Galaverna, G. Occurrence of deoxynivalenol and deoxynivalenol-3-glucoside in durum wheat. *World Mycotoxin J.* **2013**, *6*, 83–91. [CrossRef]
15. Cowger, C.; Patton-Ozkurt, J.; Brown-Guedira, G.; Perugini, L. Post-anthesis moisture increased Fusarium head blight and deoxynivalenol levels in North Carolina winter wheat. *Phytopathology* **2009**, *99*, 320–327. [CrossRef]
16. Tacke, B.K.; Casper, H.H. Determination of deoxynivalenol in wheat, barley, and malt by column cleanup and gas chromatography with electron capture detection. *J. AOAC Intern.* **1996**, *79*, 472–475.
17. Simsek, S.; Burgess, K.; Whitney, K.L.; Gu, Y.; Qian, S.Y. Analysis of deoxynivalenol and deoxynivalenol-3-glucoside in wheat. *Food Control* **2012**, *26*, 287–292. [CrossRef]

toxins

MDPI

Article

Exposure Assessment for Italian Population Groups to Deoxynivalenol Deriving from Pasta Consumption

Carlo Brera *, Valentina Bertazzoni, Francesca Debegnach, Emanuela Gregori, Elisabetta Prantera and Barbara De Santis

Istituto Superiore di Sanità, Dipartimento di Sanità Pubblica Veterinaria e Sicurezza Alimentare, Reparto OGM e Xenobiotici di origine fungina, Viale Regina Elena, Rome 299-00161, Italy; bertazzonivale@gmail.com (V.B.); francesca.debegnach@iss.it (F.D.); emanuela.gregori@iss.it (E.G.); elisabetta.prantera@gmail.com (E.P.); barbara.desantis@iss.it (B.D.S.)

* Author to whom correspondence should be addressed; carlo.brera@iss.it;
  Tel.: +39-06-4990-2377; Fax: +39-06-4990-2363.

Received: 17 October 2013; in revised form: 18 November 2013; Accepted: 19 November 2013; Published: 26 November 2013

Abstract: Four hundred and seventy-two pasta samples were collected from long retail distribution chain sales points located in North, Central and South Italy. Representative criteria in the sample collection were followed in terms of number of samples collected, market share, and types of pasta. Samples were analysed by an accredited HPLC-UV method of analysis. The mean contamination level (64.8 µg/kg) of deoxynivalenol (DON) was in the 95th percentile (239 µg/kg) and 99th percentile (337 µg/kg), far below the legal limit (750 µg/kg) set by Regulation EC/1126/2007, accounting for about one tenth, one third and half the legal limit, respectively. Ninety-nine percent of samples fell below half the legal limit. On the basis of the obtained occurrence levels and considering the consumption rates reported by the Italian official database, no health concern was assessed for all consumer groups, being that exposure was far below the Tolerable Daily Intake (TDI) of 1000 ng/kg b.w./day. Nevertheless, despite this, particular attention should be devoted to the exposure to DON by high consumers, such as children aged 3–5 years, who could reach the TDI even with very low levels of DON contamination.

Keywords: deoxynivalenol; pasta; exposure assessment; risk assessment; consumer groups; children; cereals

1. Introduction

Deoxynivalenol (DON) is a natural-occurring mycotoxin produced at pre-harvest stage by several *Fusarium* species, mainly *F.graminearum* and *F. culmorum* [1], and belongs to a wide family of mycotoxins known as trichothecenes. It is also known as vomitoxin due to its strong emetic effects after consumption, because it is transported into the brain where it runs dopaminergic receptors.

Chemically, DON (Figure 1) is a sesquiterpenoid polar organic compound, which belongs to the type B trichothecenes since it contains carbonyl group in C-8. Its empirical formula is $C_{15}H_{20}O_6$. DON is highly hydrosoluble and stable at cooking temperatures (120 °C), and in storage conditions and milling processes [2,3].

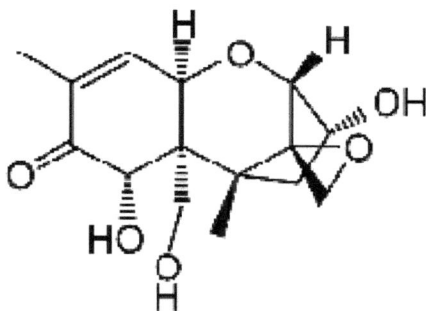

Figure 1. Chemical structure of deoxynivalenol (DON).

DON is one of the most pervasive mycotoxins that predominantly colonize wheat ears and also corn leaves. Fungi attack mainly occurs in the field before harvest [4].

The optimal range of temperatures for DON production is 21–29 °C at moisture levels >20%. DON is considered as a marker for the presence of other mycotoxins such as zearalenone [5].

The fungus has two distinct growth cycles corresponding to mould growth during warm daytime temperatures and toxin production during cool night-time temperatures [6].

Red ear rot caused by *F. graminearum* is favoured by warm wet weather after silking. This plant disease tends to be more risky in conditions of no rotation between subsequent cultivations, and in general when corn or wheat precedes wheat and corn crops respectively as a consequence of the permanence of contamination in the debris. Reduced tillage situations, provides additional elements for an increase of the probability of fungi attack.

To date, all animal species are susceptible to DON in the following order: pigs > mice > rats > poultry ≈ ruminants [7]. Differences in metabolism, absorption, distribution, and elimination of DON among animal species might account for this differential sensitivity.

Specifically in swine, DON intake reduces weight gain and hinders animal feeding. At high concentrations (more than 10 ppm) typical signs are emesis and total feed refusal [8]. In terms of DON bioavailability, sheep and cows show very low rates (10% for single-doses administered) [9] in contrast with swine where approximately 95% of the administered dose was recovered as deoxynivalenol [10].

Acute exposure of pigs to DON causes abdominal distress, increased salivation, malaise, diarrhea, and emesis [11,12].

DON is detectable also in blood and serum in high amounts immediately after ingestion, but is rapidly cleared from the blood stream.

Although JECFA established that DON is a probable factor for acute pathologies in humans, there is not enough data yet to set an Acute Reference Dose (RfD) [13].

In 1993, IARC classified DON in Group 3, corresponding to not classifiable for its carcinogenicity to humans; in 2002 the Scientific Committee for Food set a tolerable daily intake of 1 μg/kg bw/d.

In humans, deoxynivalenol causes gastro-intestinal problems, immunosuppression and interferences with reproduction and development [14–18]. DON effects in humans have not yet been widely registered but new evidence that has led to a report of immunotoxicity in humans even at low doses of contamination that could create proteomic changes in human B (RPMI1788) and T (JurkatE6.1) lymphocyte cell lines is currently under consideration. These potential effects are to be considered much more alarming if transferred to the fetus where, according to a Norwegian study, 21% of the toxin is transferred, with no activation of DON required and a very poor detoxification route. More generally, a correlation between low sanitary quality of cereals and increase of abortions during pregnancy was also noted [19–21].

In vitro studies using human intestinal Caco-2 cells suggested that DON crosses the intestinal mucosa through a para-cellular pathway, though contribution by passive trans-cellular diffusion could not be ruled out [17,22].

From the above, this study provides an evaluation of the exposure to deoxynivalenol deriving from the consumption of pasta by different groups of Italian consumers, with a focus to a very sensitive subgroup such as children.

1.1. Legislation on DON

Almost 40 countries have established regulatory limits or guidelines for DON in wheat and other cereal-based products. The guideline levels for cereals and finished cereal products for humans range from 100 to 2000 µg/kg, depending on consumer age and stage of processing of the grain [23]; levels in diets for swine, poultry, and cattle range from 500 to10.000 µg/kg [24].

The European Commission set maximum tolerable limits for DON in food products. EC Regulation 1126/2007 applies to the unprocessed durum wheat and oat (1750 µg/kg), soft wheat (1250 µg/kg) and to milled intermediate products, e.g finished products (500 µg/kg), dry pasta, cereals destined for direct human consumption, such as cereal flour, bran and germ (750 µg/kg) and processed cereal-based baby foods and foods for young children (200 µg/kg) [25]. Currently, no legal limits but only guidance values have been set by the EU Recommendation 576/2006 for complementary and complete feeding stuffs at various levels depending on the animal species susceptibility such as pigs, calves (<4 months), lambs and kids [26].

However, it should be considered that DON contribution to animal origin food products as carried over from feeds is generally negligible since no residues in eggs, milk and edible tissues were found in the literature. It was shown, in fact, that DON is rapidly metabolized by de-epoxydation and glucuronization leading to the formation of reduced toxicity metabolites [27].

1.2. Exposure Assessment by DON in Humans

Generally, human exposure assessment derived from mycotoxin-contaminated diet is a more and more challenging issue in the worldwide scenario. What is still pending is the real aetiological role of these hazards in the development of pathologies such as cancer or mycotoxin-induced immunodepression diseases or other pathologies not yet related, like autism or celiac disease [28,29].

From the data available in the literature, in humans, the emetic effects of this mycotoxin were firstly described in Japanese men consuming mouldy barley containing *Fusarium* fungi in 1972 [30,31].

In China, between 1961 and 1985, about 35 outbreaks of acute human illness were reported. The symptoms of nausea, vomiting, diarrhea, abdominal pain, headache, dizziness, and fever were attributed to DON and other trichothecenes contaminated cereals, with at least 7818 victims.

In an outbreak in 1984 in Xingtai County, 94% persons who ate moldy maize became ill. The range of DON levels was from 3.8 to 93 mg/kg [13].

DON was detected in all 15 urine samples of female inhabitants of Linxian County and Gejiu, two Chinese high and low, respectively, risk exposure regions for DON and oesophageal cancer, with mean levels of 37 and 12 ng/mL, respectively [32].

In one-year-old Dutch children exposed to DON levels above the Provisional Maximum Tolerable Daily Intake (PMTDI), reductions in body weight and relative liver weight were estimated at 2.2% and 2.7% (confidence interval: 0.2%–25%), respectively [33].

In a study performed in UK, DON was detected in the urine of 296 out of 300 healthy subjects showing a strong association between the cereal intake and urinary DON concentrations ($p < 0.0005$). From a multivariable analysis, wholemeal and white bread as well as other cereal-based food products including pasta were consistently related to urinary DON excretion.

The geometric mean concentrations were 6.55, 9.63, and 13.24 µg DON/day for low-, medium-, and high-cereal intake groups, respectively. Consumption of other grain-based foods such as cereal products and pasta was also significantly associated with urinary DON concentrations [34].

In another study by Hepworth *et al.*, DON exposure assessment was evaluated in a group of pregnant women aged 16–44 from Bradford, UK. The urinary DON was detected in all samples in a range from 0.5 to 116.7 ng/mg creatinine. From a food questionnaire, bread, particularly chapattis in South Asian women, was the major contributor to DON exposure [35].

In 2003, the study performed by the European Commission within the SCOOP task 3.2.10 revealed that the total intake of DON calculated from data coming from 12 member States ranged from 14.45%–46.1% to 11.3%–95.9% of TDI for adults and children, respectively. The contribution deriving from wheat and derived products accounted for 76%–90%. In this study, among foodstuffs, pasta showed a lower rate in the overall contribution to DON intake (25%) compared to wheat flour and bread [36].

Analogous results were obtained by Larsen *et al.* [37] where higher intakes were calculated considering the 95th percentile of consumption data multiplied by the mean DON concentration resulted in an intake very close to or even higher than TDI, with specific emphasis for children and infants.

It should be noted that in another study performed on baby foods in Italy by Pietri *et al.* in 2004, DON intakes higher than the TDI (121%) were observed. In this study, a consumption of 100 grams of cereal-based products was considered [38].

Occurrence of DON in Wheat Products

So far, in order to estimate the real amount of DON ingested with human diet, a reliable assessment should be made, taking into account (i) the metabolic pathways of DON leading to the formation of various DON-metabolites, DOM-1, glucuronic-DON, 3-AcOH-DON and 15-AcOH-DON; (ii) the lifecycle of DON from the raw kernel to the finished products such as flour or bread and pasta; and (iii) the frequency and degree of the occurrence of DON contamination levels in the wheat products ingested by the final consumer.

The results obtained in a recent study by Brera *et al.* [39] showed a significant DON reduction from the caryopsis to cooked pasta, accounting for a mean DON contamination decrease of 78%. Moreover, the overall DON reduction observed from wheat grain to dry pasta was 66%.

Another study conducted by Visconti *et al.* concluded that the retention level of DON from grains on the market to cooked pasta on the plate can be conservatively assessed at 25% or less [40].

As far as DON bio-accessibility, a recent *in vitro* study has demonstrated differences in levels of DON during the child digestion processes, attributable to different typologies of pasta and initial contamination levels [41].

L. González-Osnaya *et al.* evaluated the occurrence of DON at a rate of 28% in bread whereas in pasta the occurrence was higher, varying from 9.3% to 62.7%. The mean content of deoxynivalenol in bread was 42.5 µg/kg while in pasta the content of deoxynivalenol was higher (137.1 µg/kg). The estimated daily intake of deoxynivalenol from the consumption of the mentioned products represented 8.4% of the tolerable daily intake [42].

Bockhorn *et al.* analysed 29 pasta samples purchased from retail shops in Berlin in April and May 2001 and were analysed for their content of deoxynivalenol. Ninety percent of the raw samples contained less than 0.5 mg DON/kg, but three out of 29 samples had contamination of up to 0.84 mg/kg. The amount of DON decreased after cooking, resulting in 60%–80% lower DON levels in the ready to eat products [43].

2. Results and Discussion

2.1. Exposure Assessment

Exposure of different population groups was calculated by a deterministic approach using the following equation:

$$EXPOSURE = \frac{mean\ contamination\ \left(\frac{ng}{g}\right)\ \times\ mean\ consumption\ (g)}{weight\ (kg)} \tag{1}$$

As far as the calculation of DON intake derived from only pasta, three different parameters were taken into account: the mean contamination value, the mean consumption rate expressed in grams, and the body weight expressed in kg. The exposure was calculated as ng/kg body weight/day. This unit was chosen to compare the resulting values with the TDI of DON that was set by the Scientific Committee for Food at 1000 ng/kg bw/day [44].

2.2. Occurrence Values

The statistical description analysis of the obtained results is shown in Tables 1 and 2. The mean (64.8 µg/kg) and median (35 µg/kg) DON values of all samples were far from the legal limit of 750 µg/kg.

Ninety-nine percent of samples did not exceed the threshold of 50% of the legal limit. A percentage of 78.6% of samples was lower than the limit of quantification (70 µg/kg). Even the DON contamination levels corresponding to the 95th percentile and 99th percentile were around 50% of the legal limit.

The contamination profile of a subgroup of pasta samples (N = 43), namely short shape typology, more commonly consumed by elderly and children but not corresponding to baby foods, showed slight different values for the mean (101.5 µg/kg) that were higher than the overall mean. All the other values, *i.e.*, median, 95th percentile, 99th percentile and the maximum contamination value, overlapped with the overall scenario.

No cluster contamination was observed for a specific brand and generally the contamination was equally distributed among the different brands.

The obtained contamination levels generally confirm previous findings cited before.

Table 1. Descriptive statistics of DON contamination in pasta samples.

Parameter	Numerical value
Number of samples	472
Samples <LOQ	371 (78.6%)
Samples ≥LOQ	101 (21.4%)
Mean contamination (µg/kg)	64.8*
Median contamination (µg/kg)	35*
95th percentile of contamination (µg/kg)	239.4
99th percentile of contamination (µg/kg)	337.0
MAX contamination (µg/kg)	385.7

Note: *Mean and median values have been computed assigning to <LOQ results, the value of LOQ/2 = 35 µg/kg.

Table 2. Descriptive statistics of DON contamination in small size pasta samples.

Parameter	Numerical value
Number of samples	43
Samples <LOQ	26 (60.5%)
Samples ≥LOQ	17 (39.5%)
Mean contamination (µg/kg)	101.5*
Median contamination (µg/kg)	35*
95th percentile of contamination (µg/kg)	279.6
99th percentile of contamination (µg/kg)	320.9
MAX contamination (µg/kg)	336.4

Note: *Mean and median values have been computed assigning to <LOQ results, the value of LOQ/2 = 35 µg/kg.

2.3. Consumption Rate

Mean consumption rates of pasta related to specific subgroups of population were taken by the Italian official reference database published by Leclercq in 2009 [45]. The study was conducted randomly selecting households after geographical stratification of the national territory. Food consumption was assessed on three consecutive days through individual estimated dietary records. The study sample encompassed 3323 subjects (1501 males and 1822 females) aged 0.1 to 97.7 years belonging to 1329 households.

In Table 3, mean, 95th percentile and 99th percentile consumption rates for total population, consumers only, children, adolescents, adults and elderly, are reported. For children, the data are reported combining males and females. *Vice versa*, for adolescents, adults and elderly, a distinction of gender is provided.

The range of consumption rates is between 54.2 g/d and 161.7 g/d being this latter value the worst in absolute terms since it corresponds to children 3–9.9 years whose body weight is to be considered unfavorable for the intake.

For this reason, the exposure assessment was calculated considering in more detail the status only for children.

Table 3. Mean, 95th percentile* and 99th percentile** of individual daily consumption of pasta in the total population (TP), in consumers only (C) and in males (M) and females (F) of different ages (g/d).

Category		Gender	Consumption (g/day)		
			Mean	95th percentile	99th percentile
Total population			54.2	108.7	140.1
Consumers only			59.5	110.7	141.9
Children (3–9.9 years)	Total population		58.2	104.9	161.7
	Consumers only		59.8	104.9	161.7
Adolescent (10–17.9 years)	Total population	M	63.6	128.0	133.3
		F	56.6	105.3	133.3
	Consumers only	M	66.7	128.0	133.3
		F	61.0	105.3	133.3
Adult (18–64.9 years)	Total population	M	60.3	118.4	156.1
		F	47.7	100.0	134.8
	Consumers only	M	66.0	121.6	156.9
		F	53.8	102.2	137.8
Elderly (≥65 years)	Total population	M	61.1	109.6	129.8
		F	50.7	100.6	117.4
	Consumers only	M	64.3	116.5	131.2
		F	54.5	110.9	121.5

2.4. Body Weights

Body weights of children between 3 and 5 years were taken from the WHO official database [46], and between 6–14 years, 15–18 years and over 18 years values as reported from EFSA [47] were considered. In Table 4, the corresponding values are reported.

Table 4. Mean weight (kg) in all groups of population between 3 and >18 years.

AGE (years)	Boys	Girls
3	14.3	13.9
4	16.3	16.1
5	18.3	18.2
9.9	31.2	31.9
10–14		45
15–18		60
>18		70

2.5. DON Exposure in Adolescents and Adults

In all cases, considering mean DON occurrence levels as a fixed parameter and mean, 95th percentile and 99th percentile consumption rates, the overall exposure of the Italian population sub-groups has to be considered not at risk. For instance, for adolescents, even in the worst case, *i.e.*, 133.3 g/d corresponding to the 99th percentile consumption rate—a body weight of 45 kg corresponding to the subgroup of adolescents aged 10–14 years and a mean DON contamination level of 64.8 µg/kg—the exposure would account for 192 ng/kg bw/day corresponding to almost one fifth of the TDI. By taking as a reference situation the exposure of adolescents, all the other population subgroups—adults and elderly—accounted for an even lower exposure rate considering approximately similar consumption rates and higher body weights. For instance, for adults with a 99th percentile consumption rate of 156.1 g/d, a body weight of 70 kg and considering the DON mean contamination level, the exposure resulted in 144.5 ng/kg bw/day.

2.6. DON Intakes in Children

As far as children, different scenarios for their exposure assessment were taken into consideration consistently with different consumption rates (mean, 95th percentile and 99th percentile) and contamination levels (mean level, threshold level, legal limit) (Table 5). By accounting for the daily mean consumption rate of pasta corresponding to 59.8 g for children aged from 3 to 9.9 years, the exposure, only related to the consumption of pasta, did not reveal any alarming situation, being quite far below the TDI.

More specifically, the obtained exposure rates decreased from 3 year old children to 10 year old children accounting for a mean value, combining male and female data, of 275 ng/kg bw/day (27% of the TDI) to 123 ng/kg/bw/day (12% of TDI), respectively.

A second scenario was evaluated taking into consideration double the mean contamination level, *i.e.*, 130 µg/kg; also in this condition, the exposure equalled 50% of the TDI in the worst case (children aged 3 years). No meaningful difference between male (543 ng/kg bw/day) and female (559 ng/kg bw/day) groups was noted.

A third scenario considering the calculated DON contamination level leading to an exposure corresponding to the TDI for every subgroup of children was also considered. The resulting levels ranged from one third to two third of the legal limit, as set by the Regulation 1126/2007 [25].

Nevertheless, it must be considered that 99% of the results obtained from this study fell below these contamination levels.

Table 5. DON exposure in children (males (M) and females (F)) assuming a mean consumption of pasta of 59.8 g for children aged from 3 to 9.9 years.

DON contamination (µg/kg)	Weight (kg) /age (years) males	Exposure of males (ng/kg bw/day)	Weight (kg) /age (years) females	Exposure of females (ng/kg bw/day)
64.8*	14.3/3	271	13.9/3	279
	16.3/4	237	16.1/4	240
	18.3/5	211	18.2/5	213
	31.2/10	124	31.9/10	121
130	14.3/3	543	13.9/3	559
	16.3/4	476	16.1/4	482
	18.3/5	424	18.2/5	427
	31.2/10	249	31.9/10	243
239 (M)–233 (F)	14.3/3	999	13.9/3	1002
273 (M)–269 (F)	16.3/4	1001	16.1/4	999
306 (M)–305 (F)	18.3/5	999	18.2/5	1002
522 (M)–533 (F)	31.2/10	1000	31.9/10	999

Note: *: DON contamination level (N = 472).

In Table 6, the same scenarios were calculated but for consumption rates corresponding to the 99th percentile, *i.e.*, 161.7 grams. Even in this case, the exposure corresponding to the mean occurrence level obtained in the present study always fell below the TDI with a maximum value for the exposure accounting for the 75% of the toxicological threshold in the worst case (females aged 3 years).

Table 6. DON exposure in children (males (M) and females (F)) assuming a 99th percentile consumption of pasta of 161.7 g for children aged from 3 to 9.9 years.

DON contamination (μg/kg)	Weight (kg) /age (years) males	Exposure of males (ng/kg bw/day)	Weight (kg) /age (years) females	Exposure of females (ng/kg bw/day)
64.8*	14.3/3	732	13.9/3	753
	16.3/4	643	16.1/4	650
	18.3/5	572	18.2/5	575
	31.2/10	335	31.9/10	328
89 (M)–86 (F)	14.3/3	1006	13.9/3	1000
101 (M)–100 (F)	16.3/4	1002	16.1/4	1004
114 (M)–113 (F)	18.3/5	1007	18.2/5	1003
193 (M)–198 (F)	31.2/10	1000	31.9/10	1003

Note: *: DON contamination level (N = 472).

Analogously to the previous description of different scenarios, the DON concentration levels leading to an exposure equal to the TDI were calculated resulting in all children subgroups as a higher value than the DON mean contamination level obtained in this study.

A last scenario, shown in Figures 2 and 3, leads to a very challenging issue since if the legal limit of 750 μg/kg is considered, for any consumption rate, both mean and high values (99th percentile), and for all children subgroups, an increase of 1.5–8 times the TDI would be reached, respectively.

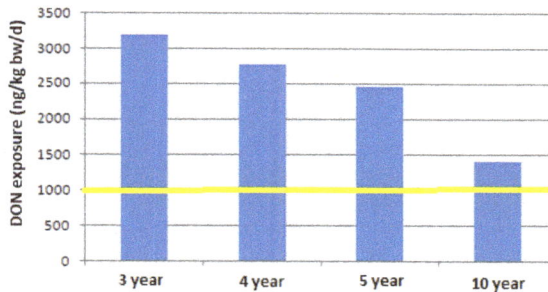

Figure 2. DON exposure for a consumption of pasta of 59.8 g (mean consumption of pasta by children between 3 and 9.9 years old, [45]) at a contamination level of 750 μg/kg.

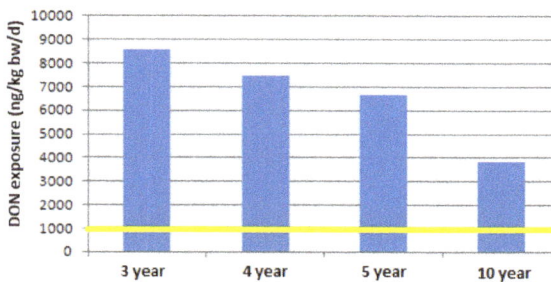

Figure 3. DON exposure for a consumption of pasta of 161.7 g (99th percentile of consumption of children between 3 and 9.9 years old, [45]) at a contamination level of 750 μg/kg.

3. Experimental Section

3.1. Sampling

In order to assess the level of DON contamination in pasta, 472 packages of commercially produced pasta taken from four large-scale retail traders distributed in different geographical areas in Italy were collected over a period between November 2010 and May 2011. Pasta samples were chosen with the criterion to guarantee the maximum representativeness of the most diffused marketed and consumed products. The selection was chosen on the basis of the market share and consumption rates in the north, south and centre of Italy. Forty-one brands were sampled.

The collected samples had to meet the following characteristics:

- Typology of pasta products (dry, fresh (10%), whole (10%), not egg-pasta, not addressed to baby food chain);
- Random selection among different shape of pasta products (spaghetti, medium size like rigatoni, short size like penne);
- Only one 500 g package per lot, typology, shape and brand.

3.2. Principle of the Method

The analytical method used for the analysis of pasta samples got accreditation, number 0779, by the national accreditation body ACCREDIA, for DON determination in wheat and derived products.

DON is extracted from the matrix by homogenizing the sample with water. After centrifugation, the extract is cleaned-up step by immunoaffinity columns and the toxin is eluted with methanol, dried under nitrogen and re-dissolved in an aqueous solution of methanol (9.5%). DON is quantified by reversed-phase HPLC by UV detection.

According to the obtained linearity, the used analytical method for DON determination in pasta samples is to be considered applicable in the range 70 µg/kg ÷ 2000 µg/kg.

3.3. Sample Preparation

The first step was to choose between a wet (slurry) and a dry homogenization of pasta samples. The preparation of slurried samples by using a lab-scale 4-lt Waring Blender and adding to the sample tap water in a ratio 1:1, led to very sticky products. For this reason, this approach was abandoned and a processing of samples in dry conditions was preferred.

All pasta samples were ground with a laboratory mill (Retsch ZM 200, Retsch GmbH, Haan, Germany) to obtain a particle size of at least 0.75 mm. After grinding and before drawing the test aliquot, a thorough homogenization of the test portion was done.

3.3.1. Extraction

DON is extracted by weighing 20.0 g (±0.1 g) of the dry ground test portion directly in the blender. One hundred and sixty milliliters of deionized water as extraction solvent was added. After 3 min of blending, the extract is centrifuged at 8000 rpm per 10 min.

The extract is then transferred to the IAC.

3.3.2. Immunoaffinity Clean-Up

To clean samples, manufacturers' instructions of IAC (DONPREP, R-Biopharm-Rhone, Glasgow, UK) were followed. IAC columns were kept at room temperature before the conditioning.

Under slight vacuum conditions, 3 mL (equivalent to 0.375 grams of pasta) of the centrifuged extract was pipetted into the syringe connected to the IAC. During the clean-up, the flow was maintained at a constant flow of not more than 3 mL/min, (in any case, the flow speed must not exceed 5 mL/min).

After the complete passage of the filtered sample, IAC was washed with 5 mL of de-ionized water. The column was successively dried by flowing at least 2 volumes of air.

DON was eluted from the IAC in two steps: firstly, 750 µL of methanol were flown by gravity and then, after 1 min, a second amount of 750 µL of methanol was applied to the IAC.

To collect the entire residual methanol, the elution is completed by pushing one volume of air inside the IAC by a 10-mL syringe.

The eluate is taken to dryness with a nitrogen flow at a temperature of 40 °C. The residue is then redissolved in 750 µL of the injection mixture MeOH:H$_2$O (9.5%:90.5%, v:v), and mixed thoroughly by vortex. The sample to be injected was stored at 4 °C until the HPLC analysis.

3.4. HPLC Analysis

One-hundred and fifty microliters (equivalent to 0.075 grams of pasta sample) out of 750 µL of the reconstituted analytical sample were injected onto HPLC, by partial loop.

HPLC Operating Conditions

The following operating conditions were used:

- Chromatographic column: Symmetry® C18 (Waters, Milford, Massachusetts, UK) reversed phase, 5 µm, 4.6 mm × 150 mm, kept at constant temperature of 40 °C;
- Mobile phase: deionized water: methanol 85:15 v:v;
- Flow rate: 1.0 mL/min;
- UV spectrophotometer regulated at a wavelength of 220 nm;
- Injection Volume: 150 µL.

3.5. Spiking Procedure

To assess the recovery factors to be used for correcting the analytical results, six replicates of fortified samples were prepared by weighing directly in the homogenizer 20.0 g ± 0.1 g of blank pasta samples and pipetting proper volumes of working solution of DON reference standard to obtain spiking concentration levels of 200 µg/Kg and 750 µg/Kg. The fortified samples were left under hood for two hours before the successive processing.

3.6. Validation Study

With the aim to test the reliability of the analytical results, a single-laboratory validation study was performed on the used method according to the IUPAC protocol [48].

All method performance characteristics, including precision and accuracy, were compliant with the criteria set by the European Commission Regulation (CE) 401/2006 [49].

Limit of Quantification

The limit of quantification (LoQ) of the used method was calculated over replicate analyses of blank samples and calculating mean values and standard deviation of the areas of the background. The standard deviation was then multiplied by 10.

From this approach a LoQ of 70 µg/kg was calculated. In these analyses, a value of repeatability ≤10% was verified.

3.7. Calibration Curve

Fit for purpose calibration curve was built up by preparing six working solutions in 5-mL flasks. Each level was injected in triple.

The stock solution of the primary certified standard material (Trilogy®, R-Biopharm: 100 µg/mL) was diluted with a mixture water:methanol 90.5:9.5 (v:v) to 10 µg/mL, by pipetting 1000µL of it in

10-mL flask. To prepare working solutions necessary for the building up of the calibration curve, proper amounts of DON were drawn from the diluted solution.

4. Conclusions

Estimates of dietary exposure to DON based on the obtained occurrence data in pasta samples are far below the TDI of 1000 ng/kg b.w. for populations of all age groups, and therefore do not represent a health concern.

Just in some cases regarding population subgroups' exposure to DON, some borderline situations were noted. In fact, considering a DON contamination level slightly higher than the mean value obtained in this study—like for instance 90–100 µg/kg, for children from 3 to 5 years with high consumption rates (99th percentile)—the exposure levels would reach the TDI, considering pasta as the only daily contribution to exposure.

However, it should be noted that the abovementioned concentration value would correspond to almost one seventh of the legal limit; therefore, creating a quite challenging situation where even for very low levels of DON contamination in a food product highly consumed such as pasta, a health warning for some sensitive consumer groups could be raised.

It should be also noted that, due to the high DON hydro-solubility, the overall exposure assessment should, however, be re-examined considering a loss of up to 20% of DON after cooking pasta, except for when the cooking water remains in the meal as in the case of soups, consumed especially by the elderly and young children.

The scientific information derived from this study can provide a basis for more strict consideration of children's exposure through diet, since even for contamination levels far lower than the legal limit, borderline situations can arise.

This consideration is more pertinent if applied to DON that is present in the wide spectrum of food products such as bread, breakfast cereals, pizza, biscuits and other cereal-based products.

For the above, reconsideration both of the maximum tolerable legal limit and of the toxicological threshold could represent interesting and challenging topics to be examined with the aim to guarantee an even higher level of safety for those population subgroups, such as children and adolescents, for whom in some cases issues could arise.

Conflicts of Interest: The authors declare no conflict of interest.

References

1. Kushiro, M. Effects of milling and cooking processes on the Deoxynivalenol content in wheat. *Int. J. Mol. Sci.* **2008**, *9*, 2127–2145. [CrossRef]
2. Bretz, M.; Beyer, M.; Cramer, B.; Knecht, A.; Humpf, H.U. Thermal degradation of the Fusarium mycotoxin deoxynivalenol. *J. Agric. Food Chem.* **2006**, *54*, 6445–6451. [CrossRef]
3. Hazel, C.M.; Scudamore, K.A.; Patel, S.; Scriven, F. Deoxynivalenol and other Fusarium mycotoxins in bread, cake, and biscuits produced from UK-grown wheat under commercial and pilot scale conditions. *Food Addit. Contam.* **2009**, *26*, 1191–1198. [CrossRef]
4. Payne, G.A. Ear and kernel rots. In *Compendium of Corn Diseases*; White, D.G., Ed.; The American Phytopathology Society (APS Press): St. Paul, MN, USA, 1999; pp. 44–47.
5. Lawlor, P.G.; Lynch, P.B. Mycotoxins in pig feeds. 2: Clinical aspects. *Irish Vet. J.* **2001**, *54*, 172–176.
6. Cheeke, P.R.; Shull, L.R. *Natural Toxicants in Feeds and Poisonous Plants*; Avi Publishing Company Inc.: Westport, CT, USA, 1985; p. 492.
7. Prelusky, D.; Rotter, B.; Rotter, R. Toxicology of mycotoxins. In *Mycotoxins in Grain: Compounds Other than Aflatoxins*; Miller, J., Trenholm, H., Eds.; Eagan Press: St. Paul, MN, USA, 1994; pp. 359–403.
8. Pestka, J.J.; Smolinski, A.T. Deoxynivalenol: Toxicology and potential effects on humans. *J. Toxicol. Environm. Health Part B* **2005**, *8*, 39–69. [CrossRef]
9. Prelusky, D.B.; Veira, D.M.; Trenholm, H.L. Plasma pharmacokinetics of the mycotoxin Deoxynivalenol following oral and intravenous administration to sheep. *J. Environ. Sci. Health B* **1985**, *20*, 603–624.

10. Prelusky, D.B.; Hartin, K.E.; Trenholm, H.L.; Miller, J.D. Pharmacokinetic fate of carbon-14-labeled Deoxynivalenol in swine. *Fundam. Appl. Toxicol.* **1988**, *10*, 276–286. [CrossRef]

11. Pestka, J.J.; Lin, W.S.; Miller, E.R. Emetic activity of the trichothecene 15 acetyldeoxynivalenol in pigs. *Food Chem. Toxicol.* **1987**, *25*, 855–858. [CrossRef]

12. Prelusky, D.B.; Trenholm, H.L. The efficacy of various classes of anti-emetics in preventing Deoxynivalenol-induced vomiting in swine. *Nat. Toxins* **1993**, *1*, 296–302. [CrossRef]

13. Canady, R.A.; Coker, R.D.; Egan, S.K.; Krska, R.; Kuiper-Goodman, T.; Olsen, M.; Pestka, J.; Resnik, S.; Schlatter, J. Deoxynivalenol. In *Safety Evaluation of Certain Mycotoxins in Food*; WHO Food Additive Series 47; World Health Organization: Geneva, Switzerland, 2001; pp. 419–555.

14. Berthiller, F.; Crews, C.; Dall'Asta, C.; Saeger, S.D.; Haesaert, G.; Karlovsky, P.; Oswald, I.P.; Seefelder, W.; Speijers, G.; Stroka, J. Masked mycotoxins: A review. *Mol. Nutr. Food Res.* **2013**, *57*, 165–186. [CrossRef]

15. Maresca, M. From the gut to the brain: Journey and pathophysiological effects of the food-associated trichothecene mycotoxin Deoxynivalenol. *Toxins* **2013**, *23*, 784–820. [CrossRef]

16. Visconti, A. Problems associated with Fusarium mycotoxins in cereals. *Bull. Inst. Compr. Agric. Sci.* **2001**, *9*, 39–55.

17. Pestka, J.J. Deoxynivalenol: Mechanisms of action, human exposure, and toxicological relevance. *Arch. Toxicol.* **2010**, *84*, 663–679. [CrossRef]

18. Pestka, J.J. Deoxynivalenol-induced proinflammatory gene expression: Mechanisms and pathological sequelae. *Toxins* **2010**, *2*, 1300–1317. [CrossRef]

19. Rocha, O.; Ansari, K.; Doohan, F.M. Effects of trichothecene mycotoxins on eukaryotic cells: A review. *Food Addit. Contam.* **2005**, *22*, 369–378. [CrossRef]

20. Sobrova, P.; Adam, V.; Vasatkova, A.; Beklova, M.; Zeman, L. Deoxynivalenol and its toxicity. *Interdiscip. Toxicol.* **2010**, *3*, 94–99.

21. Rotter, B.A.; Prelusky, D.B.; Pestka, J.J. Toxicology of deoxynivalenol (vomitoxin). *J. Toxicol. Environm. Health* **1996**, *48*, 1–34.

22. Sergent, T.; Parys, M.; Garsou, S.; Pussemier, L.; Schneider, Y.J.; Larondelle, Y. Deoxynivalenol transport across human intestinal Caco-2 cells and its effects on cellular metabolism at realistic intestinal concentrations. *Toxicol. Lett.* **2006**, *164*, 167–176. [CrossRef]

23. FAO. *Worldwide Regulations for Mycotoxins in Food and Feed in 2003*; Food and Nutrition Paper 81; Food and Agriculture Organization of the United Nations: Rome, Italy, 2003.

24. Proposed draft maximum levels for Deoxynivalenol in cereals and cereal-based products and associated sampling plans (CX/CF 13/7/7). In Proceedings of Codex Committee on Contaminants in Food 7th Session, Moscow, Russian Federation, 8–12 April 2013.

25. European Commission. Commission Regulation (EC) No 1126/2007 of 28 September 2007 amending Regulation (EC) No 1881/2006 setting maximum levels for certain contaminants in foodstuffs as regards Fusarium toxins in maize and maize products. *Off. J. Eur. Union* **2007**, L 255/14.

26. European Commission. Commission Recommendation of 17 August 2006 on the presence of deoxynivalenol, zearalenone, ochratoxin A, T-2 and HT-2 and fumonisins in products intended for animal feeding. *Off. J. Eur. Union* **2006**, L 229/7.

27. Swanson, S.P.; Rood, H.D.; Behrens, J.C.; Sanders, P.E. Preparation and characterization of the deepoxy trichothecenes: deepoxy HT-2, deepoxy T-2 triol, deepoxy T-2 tetraol, deepoxy 15-monoacetoxyscirpenol, and deepoxy scirpentriol. *Appl. Environm. Microb.* **1987**, *53*, 2821–2826.

28. Maresca, M.; Fantini, J. Some food-associated mycotoxins as potential risk factors in humans predisposed to chronic intestinal inflammatory diseases. *Toxicon* **2010**, *56*, 282–294. [CrossRef]

29. Mezzelani, A.; Landini, M.; Facchiano, F.; Raggi, M.E.; Villa, L.; Molteni, M.; De Santis, B.; Brera, C.; Caroli, A.M.; Milanesi, L.; *et al.* Environment, dysbiosis, immunity, and sex-specific susceptibility: An evidence-based translational hypothesis for regressive autism pathogenesis. *Nutr. Neurosci.* **2013**. submitted for publication.

30. Ueno, Y. The toxicology of mycotoxins. *CRC Crit. Rev. Toxicol.* **1985**, *14*, 99–132. [CrossRef]

31. Ueno, Y. Toxicology of trichothecene mycotoxins. *ISI Atlas Sci. Pharm.* **1988**, *2*, 121–124.

32. Meky, F.A.; Turner, P.C.; Ashcroft, A.E.; Miller, J.D.; Qiao, Y.L.; Roth, M.J.; Wild, C.P. Development of a urinary biomarker of human exposure to Deoxynivalenol. *Food Chem. Toxicol.* **2003**, *41*, 265–273.

33. Pieters, M.N.; Freijer, J.L.; Baars, A.J.; Fiolet, D.C.M.; Van Klaveren, J.; Slob, W. Risk assessment of Deoxynivalenol in food. Concentration limits, exposure and effects. *Adv. Exp. Med. Biol.* **2002**, *504*, 235–248. [CrossRef]

34. Turner, P.C.; Rothwell, J.A.; White, K.L.M.; Cade, J.E.; Wild, C.P. Urinary Deoxynivalenol is correlated with cereal intake in individuals from the United Kingdom. *Environ. Health Persp.* **2008**, *116*, 21–25.

35. Hepworth, S.J.; Hardie, L.J.; Fraser, L.K.; Burley, V.J.; Mijal, R.S.; Wild, C.P.; Azad, R.; McKinney, P.A.; Turner, P.C. Deoxynivalenol exposure assessment in a cohort of pregnant women from Bradford, UK. *Food Addit. Contam. Part A* **2012**, *29*, 269–276.

36. *Collection of OCCURRENCE DATA of Fusarium TOXIns in FOod and ASSESSment of DIETARY INTake by the POpulation of EU Member States*; Scientific Cooperation (SCOOP) 3.2.10.; Directorate General Health and Consumer Protection: Brussels, Belgium, 2003.

37. Larsen, J.C.; Hunt, J.; Perrin, I.; Ruckenbauer, P. Workshop on trichothecenes with a focus on DON: Summary report. *Toxicol. Lett.* **2004**, *153*, 1–22. [CrossRef]

38. Pietri, A.; Bertuzzi, T.; Zanetti, M.; Rastelli, S. Presenza di tricoteceni e di ocratossina A in baby-foods e prodotti dietetici ricchi di crusca. In *Rapporti ISTISAN 05/42*; Miraglia, M., Carlo, B., Eds.; Istituto Superiore di Sanità: Rome, Italy, 2005; pp. 39–42.

39. Brera, C.; Peduto, A.; Debegnach, F.; Pannunzi, E.; Prantera, E.; Gregori, E.; De Giacomo, M.; De Santis, B. Study of the influence of the milling process on the distribution of Deoxynivalenol content from the caryopsis to cooked pasta. *Food Control* **2012**, *32*, 309–312.

40. Visconti , A.; Haidukowski, M.; Pascale, M.; Silvestri, M. Reduction of Deoxynivalenol during durum wheat processing and spaghetti cooking. *Toxicol. Lett.* **2004**, *153*, 181–189. [CrossRef]

41. Raiola, A.; Meca, G.; Mañes, J.; Ritieni, A. Bioaccessibility of Deoxynivalenol and its natural co-occurrence with Ochratoxin A and Aflatoxin B1 in Italian commercial pasta. *Food Chem. Toxicol.* **2012**, *50*, 280–287. [CrossRef]

42. González-Osnaya, L.; Cortés, C.; Soriano, J.M.; Moltó, J.C.; Mañes, J. Occurrence of Deoxynivalenol and T-2 toxin in bread and pasta commercialised in Spain. *Food Chem.* **2011**, *124*, 156–161. [CrossRef]

43. Bockhorn, I.; Bockhorn, A.; Pohler, S. Deoxynivalenol (DON) in raw and cooked pasta. *Mycotoxin Res.* **2001**, *17* (Suppl. 1), 67–70. [CrossRef]

44. SCF (Scientific Committee on Food). Opinion of the Scientific Committee on Food on Fusarium toxins. Part 6: Group Evaluation of T-2 Toxin, HT-2 Toxin, Nivalenol and Deoxynivalenol: SCF/CS/CNTM/MYC/27 Final. 2002. Available online: http://europa.eu.int/comm/food/fs/sc/scf/out123_en.pdf (accessed on 15 June 2007).

45. Leclercq, C.; Arcella, D.; Piccinelli, R.; Sette, S.; Le Donne, C.; Turrini, A. The Italian National Food Consumption Survey INRAN-SCAI 2005–06: Main results in terms of food consumption. *Public Health Nutr.* **2009**, *12*, 2504–2532. [CrossRef]

46. World Health Organization (WHO). Child Growth Standards. Available online: http://www.who.int/childgrowth/standards/en/ (accessed on 21 November 2013).

47. European Food Safety Authority (EFSA). Scientific opinion—Guidance on selected default values to be used by the EFSA Scientific Committee, Scientific Panels and Units in the absence of actual measured data. *EFSA J.* **2012**, *10*, 2579–2611.

48. Thompson, M.; Ellison, S.L.R.; Wood, R. Harmonized guidelines for single-laboratory validation of methods of analysis (IUPAC Technical Report). *Pure Appl. Chem.* **2002**, *74*, 835–855. [CrossRef]

49. European Commission. Commission Regulation (EC) No 401/2006 of 23 February 2006 laying down the methods of sampling and analysis for the official control of the levels of mycotoxins in foodstuffs. *Off. J. Eur. Union* **2006**, L70/1.

Section 2:
Impacts of DON on Plants

toxins

MDPI

Article

Light Influences How the Fungal Toxin Deoxynivalenol Affects Plant Cell Death and Defense Responses

Khairul I. Ansari [1,†], Siamsa M. Doyle [2,†], Joanna Kacprzyk [3], Mojibur R. Khan [4], Stephanie Walter [5], Josephine M. Brennan [6], Chanemouga Soundharam Arunachalam [3], Paul F. McCabe [3] and Fiona M. Doohan [3,*]

[1] Neurosurgery, Brigham and Women's Hospital, Harvard Medical School, 4 Blackfan Circle, Boston, MA 02115, USA; kansari@partners.org

[2] Department of Forest Genetics and Plant Physiology, Umeå Plant Science Centre, Swedish University of Agricultural Sciences (SLU), Umeå 90 183, Sweden; Siamsa.Doyle@slu.se

[3] UCD Earth Institute and School of Biology and Environmental Science, College of Science, University College Dublin, Belfield, Dublin 4, Ireland; jkacprzyk@gmail.com (J.K.); chansundar@yahoo.com (C.S.A.); paul.mccabe@ucd.ie (P.F.M.)

[4] Institute of Advanced Study in Science and Technology, Guwahati-35, India; mrk6@rediffmail.com

[5] Department of Integrated Pest Management, Research Centre Flakkebjerg, Forsøgsvej 1, Slagelse DK-4200, Denmark; stephanie-walter@gmx.net

[6] Plant Health Laboratory, Department of Agriculture and Food, Backweston, Co. Kildare, Ireland; josephine.brennan@agriculture.gov.ie

* Author to whom correspondence should be addressed; Fiona.doohan@ucd.ie; Tel.: 00353-1-7162248; Fax: 00353-1-7161102.

† These authors contributed equally to this work.

Received: 3 December 2013; in revised form: 6 February 2014; Accepted: 8 February 2014; Published: 20 February 2014

Abstract: The *Fusarium* mycotoxin deoxynivalenol (DON) can cause cell death in wheat (*Triticum aestivum*), but can also reduce the level of cell death caused by heat shock in Arabidopsis (*Arabidopsis thaliana*) cell cultures. We show that 10 µg mL^{-1} DON does not cause cell death in Arabidopsis cell cultures, and its ability to retard heat-induced cell death is light dependent. Under dark conditions, it actually promoted heat-induced cell death. Wheat cultivars differ in their ability to resist this toxin, and we investigated if the ability of wheat to mount defense responses was light dependent. We found no evidence that light affected the transcription of defense genes in DON-treated roots of seedlings of two wheat cultivars, namely cultivar CM82036 that is resistant to DON-induced bleaching of spikelet tissue and cultivar Remus that is not. However, DON treatment of roots led to genotype-dependent and light-enhanced defense transcript accumulation in coleoptiles. Wheat transcripts encoding a phenylalanine ammonia lyase (*PAL*) gene (previously associated with *Fusarium* resistance), non-expressor of pathogenesis-related genes-1 (*NPR1*) and a class III plant peroxidase (*POX*) were DON-upregulated in coleoptiles of wheat cultivar CM82036 but not of cultivar Remus, and DON-upregulation of these transcripts in cultivar CM82036 was light enhanced. Light and genotype-dependent differences in the DON/DON derivative content of coleoptiles were also observed. These results, coupled with previous findings regarding the effect of DON on plants, show that light either directly or indirectly influences the plant defense responses to DON.

Keywords: Arabidopsis; β-1,3-glucanase; cell death; *Fusarium*; light; non-expressor of pathogenesis-related genes-1 (NPR1); peroxidase; phenylalanine ammonia lyase; wheat

1. Introduction

Fusarium graminearum Schwabe [teleomorph *Gibberella zeae* (Schweinitz) Petch] and *F. culmorum* (W.G. Smith) Saccardo cause diseases on the roots, stems and heads of cereal plants [1]. *Fusarium* head blight (FHB) receives significant attention because of both the yield losses and mycotoxin contamination of grain associated with this disease. *F. graminearum* and *F. culmorum* commonly produce the trichothecene mycotoxin deoxynivalenol (DON) in infected plant tissue and this toxin acts as an aggressiveness factor for the pathogen during the development of root rot and FHB disease [2,3]. DON inhibits protein synthesis and its effect on wheat (*Triticum aestivum* L.) head tissue is similar to that of FHB disease, in that it bleaches the tissue [4,5]. Wheat genotypes differ in their response to DON; resistance to DON-induced bleaching is associated with resistance to the spread of FHB disease (type II resistance to FHB), but not with resistance to *Fusarium* infection (type I resistance to FHB) [4].

Studies have shown that DON treatment induces defense gene transcription in wheat [5–7], the production of reactive oxygen species (ROS) and, thereafter, an increase in programmed cell death (PCD) [6]. Diamond *et al.* [8], however, showed that lower levels of DON (10 *vs.* 100–200 µg mL^{-1}) and a DON-producing strain of *F. graminearum* did not cause cell death in *Arabidopsis thaliana* cell cultures, but they did reduce the level of cell death caused by heat shock. The opposing effects of DON on cell viability and death in Arabidopsis and wheat may be due to many factors, including the differences in DON concentrations used. We postulated that it might in part be due to light-dependent signaling. The Arabidopsis experiments were conducted using light-grown cell cultures (that contained mature chloroplasts), while the wheat experiments were conducted using seedlings. Therefore, the light exposure of cells was quite different, and it is known that DON-induced bleaching of barley tissue is light dependent [9]. The opposing effects of DON on cell death in Arabidopsis *vs.* wheat may also reflect host-dependent responses to the toxin, or a specific type of resistance to DON that is inherited in a genotype-dependent manner. DON resistance inherent to some wheat genotypes is associated with the capacity to convert DON to the less toxic DON-3-glucoside and co-segregated with the QTL *Fhb1* [4]. This may be in and of itself a light-dependent phenomenon, because, in Arabidopsis, a UDP glucosyltransferase (UGT) catalyzes the glucosylation of DON [10] and an analysis of Arabidopsis microarray experiments available in public repositories shows that the transcription of the encoding gene is light regulated.

The first objective of this work was to try and determine if light plays a role in how DON influences plant cell death. We show the ability of DON and DON-producing *F. graminearum* to retard cell death caused by heat shock in Arabidopsis cell cultures is light dependent, and that DON actually enhanced heat-induced cell death in dark-grown cells. The fact that the effect of DON and DON-producing *Fusarium* on cell viability was light dependent, coupled with the previously reported light-dependent bleaching of barley leaves by DON [9], indicated that light might be an important determinant of the plant response to this toxin and its producer fungi. The second objective was to determine if light influences the ability of wheat seedlings to mount defense in response to DON treatment. Using seedlings whose roots were treated with DON, we show that light does enhance defense transcript accumulation in coleoptiles and increases the DON metabolite content of coleoptiles. The extent to which light influences defense transcript accumulation and DON metabolite translocation in DON-treated seedlings, however, is wheat genotype-dependent. The implications of these results are discussed.

2. Results

2.1. The Effect of DON on the Viability of Heat-Shocked Arabidopsis Cell Cultures Is Light Dependent

We used heat as an abiotic cell death inducer and determined whether light was necessary for DON-mediated inhibition of heat-induced cell death (a phenomenon previously discovered by Diamond *et al.*, [8]). We compared light- and dark-incubated *Arabidopsis thaliana* ecotype Landsberg erecta cell cultures with respect to the effect of 10 µg mL^{-1} DON on heat-induced cell death. Light-grown cultures contained mature chloroplasts, while dark-grown cultures contained plastids, as determined by electron microscopy [11]. Cell cultures were treated with 10 µg mL^{-1} DON or water (controls) 24 h prior to heat treatment (55 °C for 10 min). Cells were treated with fluorescein diacetate and cell fluorescence and morphology (epifluorescent microscopy) were used to distinguish between viable and non-viable cells and to determine whether non-viable cells displayed apoptotic-like cell death morphology or necrotic morphology (Figure S1) [8]. Results showed that the effect of DON on cell death was light-dependent. When cells were incubated in the dark (Figure 1A, B), DON pre-treatment did not enhance the viability of heat-shocked cells. Indeed, when comparing DON and water-pretreated cells examined 5 h post-heat treatment, toxin treatment resulted in 2.6-fold higher numbers of cells exhibiting apoptotic-like cell death morphology and reduced cell viability by 5.94-fold ($P \leq 0.01$). However, in light-grown cells harvested 5 h post-heat treatment, DON, as compared to water, reduced cell mortality (6.2 and 2.8-fold reductions, respectively in cells exhibiting apoptotic-like cell death and necrotic morphology) and enhanced the level of cell viability (by 5.0-fold; $P \leq 0.02$; Figure 1C). Another interesting observation was that the ability of DON to inhibit heat-induced cell death in light-grown cultures was temporal. At 24 h post-treatment, the DON (relative to water) pretreatment did not influence the viability of heat-shocked cells ($P \geq 0.31$) (Figure 1D).

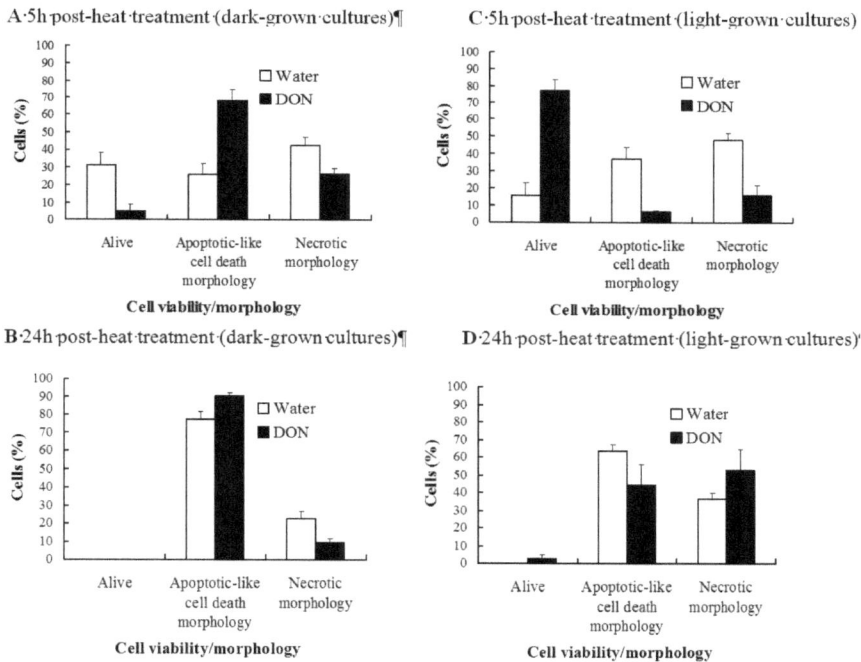

Figure 1. The effect of DON on the viability of heat-stressed Arabidopsis cells. Cells were cultured under dark conditions (**A** and **B**) or light conditions (**C** and **D**) and were treated with DON (10 µg mL^{-1}) or water (controls) 24 h pre-heat treatment (55 °C, 10 min) and cells were examined at either (**A** and **C**) 5 h, or (**B** and **D**) 24 h post-heat treatment. Cells were treated with fluorescein diacetate and examined under phase contrast microscopy with or without UV fluorescence (490 nm) in order to determine if cells were viable, or non-viable and exhibiting either programmed cell death (PCD) or necrotic morphology. Results represent the mean percentage (+/− standard error) of cells in a given state, based on five independent experiments, and in each experiment 200 cells were scored per treatment per time point. In control, water-treated, non-heat shocked cells, ≥84% were viable and ≤14 and 3% respectively displayed PCD or necrotic morphology in dark-/light-grown cultures at the time points analyzed.

Diamond *et al.* [8] showed that, like DON, pre-treatment with DON-producing *F. graminearum* reduced the level of PCD caused by subsequent heat treatment in Arabidopsis cell cultures. We conducted similar experiments, where light- and dark-incubated Arabidopsis cell cultures were pre-treated with conidia of wild-type DON producing *F. graminearum* (strain GZ3639) or its DON-minus mutant derivative (strain GZT40), 20 h prior to heat treatment (55 °C, 10 min). Cell viability was assessed at 5 h post-heat treatment, as described above (Figure 2). For dark-grown cultures, neither the wild type nor mutant *Fusarium* significantly affected either cell viability or the morphology of dead cells ($P \geq 0.30$) (Figure 2A). This contrasted with the DON which enhanced death under dark conditions (Figure 1). In light grown cultures, the wild-type, but not the mutant, significantly enhanced cell viability (by 2.5-fold) and reduced cell necrosis (by 1.6-fold; $P \leq 0.05$; Figure 2B). Neither fungal strain significantly influenced the level of apoptotic-like cell death morphology observed in heat-shocked cells at this time point ($P \geq 0.27$). Therefore we concluded that the effect of toxigenic *F. graminearum* on plant cell death is light dependent, possibly dependent on chloroplasts, and under light conditions, the ability of DON to prevent cell death caused by abiotic stress is temporal.

2.2. DON Induction of Defense Gene Expression in Wheat Is Light Enhanced and Genotype Dependent

The second objective was to determine if light influenced the wheat response to DON. In this study, we used two wheat cultivars which differ with respect to the ability of their spikelets to resist DON; cv. CM82036 is DON-resistant, while cv. Remus is susceptible [4,5]. The response analyzed was defense gene expression, namely genes encoding *POX*, *PAL*, *NPR1* and *GLC1*. These were analyzedbecause a previous study showed that these were all DON-upregulated in wheat spikelets, and that *NPR1* transcription was more highly DON-upregulated in the toxin-treated spikelets of cv. CM82036, as compared to cv. Remus ([12]). Furthermore, the upregulation of the *PAL* gene is associated with two quantitative trait loci (QTL) that confers spikelets of cv. CM82036 with enhanced resistance to both FHB and DON [13]. In this study, the roots of seedlings grown under light or dark conditions were treated with DON and defense gene transcription in both roots and coleoptile was analyzed at 4 and 24 h post-treatment. The localized effect of DON on defense gene expression in roots of the two cultivars was not light enhanced (results not shown). In coleoptiles of cv. Remus, *GLC1* was the only transcript that was significantly DON-upregulated; this phenomenon was light-enhanced (3.1-fold higher in DON *vs.* water treated, light-grown cells observed 24 h post-treatment; $P = 0.02$) (Figure 3). In coleoptiles of dark-grown cv. CM82036 seedlings, *POX* was the only transcript significantly upregulated in response to DON treatment (2.3-fold at 24 h, $P = 0.05$). But, under light conditions, and by 24 h post-root treatment, DON had transcriptionally upregulated all four defense genes in coleoptiles of this cultivar (2.0–4.3-fold upregulation; $P \leq 0.02$) (Figure 3). Moreover, the highest transcript levels were detected at 24 as compared to 4 h post-treatment.

2.3. Both Genotype and Light Affect the Movement of DON Metabolites within Wheat Seedlings

Using an ELISA test, we determined the level of DON and DON derivatives in coleoptiles 24 h post-root treatment with toxin. The test used did not discriminate between DON and DON 3-glucoside (and thus compounds detected are described as DON metabolites). Both the incubation conditions (light/dark) and wheat genotype influenced the level of DON metabolites detected within coleoptiles (Figure 4). Coleoptiles of cv. CM82036 contained more DON metabolites than those of cv. Remus, irrespective of incubation conditions (≥ 1.7-fold more; $P = 0.003$) and coleoptiles of cv. CM82036, but not of cv. Remus, accumulated significantly more DON metabolites by this time when incubated under light as compared to dark conditions (1.4-fold more; $P = 0.046$). By 24 h post-DON treatment, neither light nor cultivar-dependent differences in coleoptile dry weight were observed (mean = 32–35 mg; $P \geq 0.20$). However, coleoptiles of cv. CM82036 were more elongated than those of cv. Remus (results not shown).

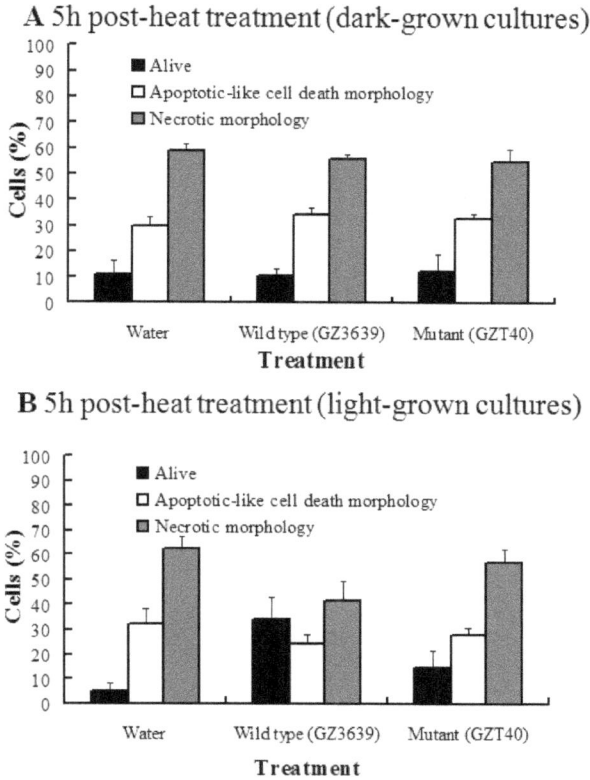

Figure 2. The effect of DON production by *Fusarium graminearum* on the viability of heat-stressed Arabidopsis cells. Cells were cultured under dark conditions (**A**) or light conditions (**B**) and were treated with either water, or conidia of either wild type, DON-producing *F. graminearum* (strain GZ3639) or its mutant, non-DON-producing derivative (strain GZT40). After 20 h, cells were incubated at either 23 or 55 °C for 10 min, and thereafter at 23 °C. Cells were examined at 5 h post-heat treatment. Cells were treated with fluorescein diacetate and examined under phase contrast microscopy with or without UV fluorescence (490 nm) in order to determine if cells were viable, or non-viable and exhibiting either programmed cell death (PCD) or necrotic morphology. Results represent the mean percentage (+/− standard error) of cells in a given state, based on five independent experiments, and in each experiment, a minimum of 200 cells were scored per treatment. In control, water-treated, non-heat shocked cells, ≥94% were viable and ≤4.6 and 1.4% respectively displayed PCD or necrotic morphology in dark-/light-grown cultures.

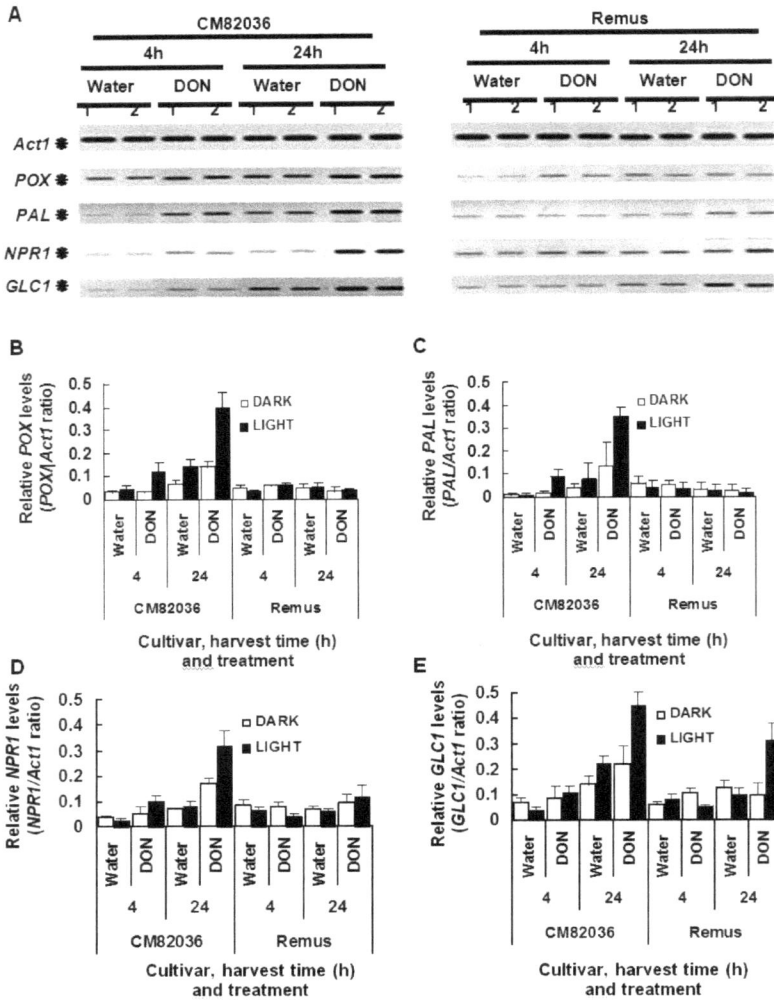

Figure 3. Influence of light on deoxynivalenol (DON)-induced accumulation of defense transcripts in coleoptiles of wheat cultivars CM82036 and Remus. (**A**) Visualization and (**B–E**) quantification of the respective *POX*, *PAL*, *NPR1* and *GLC1* transcript (relative to *Act1* transcript) levels. Results represent the mean (+/− standard error), based on two independent experiments, each including two replicates per treatment. The roots of germinated seedlings (48 h) were incubated in DON (20 mgmL^{-1}) or water under light or dark conditions at 20 °C. RNA extracted from coleoptiles at either 4 or 24 h post-treatment was used for RT-PCR analyses. Gene codes: *Act1* = actin; *POX* = class III plant peroxidase; *PAL* = phenylalanine ammonia lyase; *NPR1* = a non-expressor of pathogenesis-related genes-1; GLC = β-1,3-glucanase. Arrows indicate *Act1*, *POX*, *PAL*, *NPR1* and *GLC1* RT-PCR products (270, 272, 241, 219 and 201 bp, respectively).

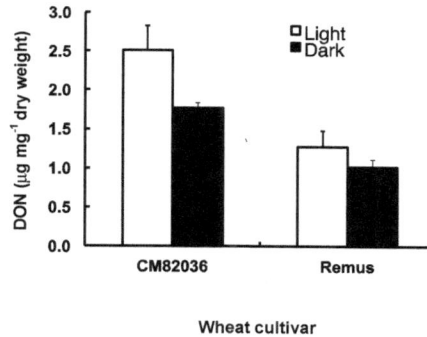

Figure 4. The translocation of DON metabolites within seedlings of wheat cultivars CM82036 and Remus. The roots of germinated seedlings (48 h) were incubated in DON (20 mg mL^{-1}). Coleoptiles were harvested 24 h post-treatment and DON was extracted and quantified by ELISA analysis. Results represent the mean (+/− standard error), based on two independent experiments, each including two replicates per treatment.

3. Discussion

This research has established that light is an important determinant of how plants cells respond to the *Fusarium* mycotoxin DON and that wheat genotypes differ with respect to their ability to mount light-dependent defense responses to DON. Plant defense responses against pathogens, including the activation of PAL, the accumulation of SA, expression of PR proteins and the hypersensitive response, are often light dependent [14–18]. Light can influence defense responses via its effects on chloroplast metabolism, ROS generation and phytochrome signaling [19]. Light influences chloroplast function and several lines of evidence point to the possible role of the chloroplast as an important determinant of the plant response to DON. The light-grown Arabidopsis cells used in this study contained chloroplasts, while the dark-grown did not. DON damage of chloroplasts is a light-dependent phenomenon [9]. This could result in the accumulation of photosensitive pigments, and these can directly generate ROS in the light [20]. DON has been shown to induce ROS production in plants [6]. Excess energy produced via the photosynthetic electron transport chain [19] may contribute to the DON-induced ROS accumulation.

It is difficult to screen for the inhibition of cell death; such a phenomenon may be interpreted as a null effect and would only be obvious in situations where cell death is induced by other factors, such as heat stress. We thus used heat to induce PCD as it provided a means to activate this conserved pathway in plants. The fact that DON and DON-production inhibits heat-induced PCD supports the previous deduction from gene expression studies. Based on microarray studies, it was deduced that ROS scavenging and the promotion of cell survival are key early defense strategies that are more effectively employed by DON-resistant as compared to susceptible wheat genotypes [7]. It would be logical for *Fusarium*-resistant plants to promote the survival, and hence the potential for defense, of cells that are being attacked by necrotrophic *Fusarium* fungi that can colonize dead plant tissue. Babaeizad *et al.* [21] showed that overexpression of a gene encoding a cell death suppressor, BAX inhibitor-1, retarded *F. graminearum* colonisation of barley seedlings. While traditionally regarded as a necrotroph, there is increasing evidence and belief that *F. graminearum* is actually a hemibiotrophic pathogen, with a short biotrophic phase preceding the necrotrophic phase of disease spread. DON suppression of death suggests that the role of DON may be to disable PCD during the initial biotrophic infection stages in plant cells, with the accumulation of higher concentrations causing PCD and facilitating nectrotrophism and disease spread. In the dark, DON enhanced the rate of apoptotic cell death. However, the fungal DON-producing strain did not do so relative to the mutant strain. There are many possible reasons for these contradictory results, including the variation in DON concentrations in toxin *versus* fungal studies and the fact that fungal factors other than DON are also very likely to affect cell death.

This study showed that defense genes were light regulated and that *NPR1, POX and PAL* were more DON responsive in seedlings of the DON-resistant cultivar CM82036, as compared to the susceptible cultivar Remus. *NPR1* and *PAL* genes have been associated with enhanced FHB resistance, but not with enhanced DON resistance. NPR1 is a central regulator of plant defense responses, including systemic acquired resistance (SAR), induced systemic resistance (ISR) and salicylic acid (SA)/jasmonic acid (JA) cross-talk [22]. An Arabidopsis *npr1-1* mutant was more susceptible to *F. graminearum* and accumulated higher concentrations of DON in buds and flowers than did wild type plants [23]. Overexpression of Arabidopsis *NPR1* (At*NPR1*) transcript in wheat conferred heritable resistance to FHB disease spread, but not initial infection [24]. These data, together with the facts that DON is associated with disease spread and that it induces the accumulation of *NPR1* transcript in wheat, suggest that the NPR1 protein is linked to Type II resistance to FHB (resistance to disease spread). In rice, overexpression of a *NPR1* homolog led to the constitutive expression of defense genes, including *POX* and *PAL* [25].

Steiner *et al.* [13] showed that a *PAL* transcript was *Fusarium* responsive in wheat spikelets and its responsiveness was associated with the presence of two quantitative trait loci that confer enhanced resistance to FHB, namely *Fhb1* and *Qfhs-ifa-5A*. This transcript is 100% homologous to that used herein for PCR primer design. The results of Steiner *et al.* [13] and the higher accumulation of cinnamic acid, benzoic acid, and glutamine in *F. graminearum*-infected spikelets of the FHB resistant wheat cv. Sumai-3 (a parent of cv. CM82036 that carries QTL *Fhb1*) than in spikelets of the susceptible cv. Roblin [26] provide evidence that the phenylpropanoid pathway plays a role in host defense against DON-producing *Fusaria*. QTL *Fhb1* also confers wheat with enhanced resistance to DON-induced bleaching of spikelets [4]. It is therefore likely that components of phenylpropanoid pathway play a light-dependent role in wheat resistance to DON. This is independent of DON conversion to DON-3-glucoside as Gunnaiah *et al.* [27] recently showed that *Fhb1* derived from the wheat genotype Nyubai is mainly associated with cell wall thickening due to deposition of hydroxycinnamic acid amides, phenolic glucosides and flavonoids, but not with the conversion of DON to less toxic DON-3-glucoside.

The cv. CM82036 differs from cv. Remus in its enhanced ability to convert DON to DON-3-glucoside [4]. This derivative is detected by the ELISA test [28] and, based on spikelet studies [4], it is likely to be the predominant DON metabolite in cv. CM82036, though not in cv. Remus coleoptiles. It is possible that light affects the conversion of DON to DON-3-glucoside in a genotype-dependent manner. The light-enhanced translocation in cv. CM82036 may be an indirect consequence of enhanced sugar availability for the formation of the DON-3-glucoside. Whether or not translocated DON metabolites contributed to the light-enhanced defense transcript accumulation in coleoptiles is unknown.

4. Experimental Section

4.1. Maintenance, Growth and Treatment of Arabidopsis Cells

Arabidopsis thaliana (ecotype Landsberg erecta) cells were grown in liquid Murashige and Skoog (MS) media ([29,30]. Cells were sub-cultured by pipetting 10 mL of culture into 100 mL of fresh media every 7 days, and were grown on a rotary shaker at 100 rpm (5 cm rotation), a constant temperature of 23 °C, and either in darkness or at a continuous light intensity of approximately 4 μmol photons $m^{-2} s^{-1}$. DON (Sigma, UK) was dissolved in water at a concentration of 2000 μg mL^{-1} and stored at 4 °C. Conidia of *Fusarium graminearum* strain GZ3639 and its trichothecene-minus mutant derivative (strain GZT40) [31] were produced as described previously [32] and adjusted to a concentration of 5×10^4 mL^{-1} H_2O (fresh conidia were prepared for each experiment). Arabidopsis cell samples (10 mL) were transferred to sterile 100 mL conical flasks and were treated with 0.5 mL DON 24 h prior to heat treatment or with 2 mL of conidial inoculum 20 h prior to heat treatment. Controls were treated with equivalent volumes of water. For heat treatment, cell culture flasks were placed in a shaking water bath (80 rpm) that was pre-equilibrated to either 23 or 55 °C, for 10 min. Between DON/fungal and heat treatment, and subsequent to heat treatment, samples were returned to their prior growth conditions (as above, either light or dark). Samples were morphologically analyzed at either 5 or 24 h after heat treatment. Five independent experiments compared the effect of DON and water on heat/non-heat-treated cells, and another five compared the effect of *F. graminearum* wild type, mutant and water on heat/non-heat-treated cells. Each experiment included one flask per treatment per harvest time point and 200 cells were morphologically analyzed per treatment per time point per experiment.

4.2. Morphological Analysis of Arabidopsis Cells

Cells were examined under a Leica DM LB microscope with an attached fluorescence lamp and camera (Leica Microsystems GmBH, Wetzlar, Germany). Cells were scored as being viable or non-viable and exhibiting either necrotic or apoptotic-like cell death morphology. The vital stain fluorescein diacetate (FDA) was used to assay for live cells [30]. When FDA is excited by light at a wavelength of 490 nm, a bright green fluorescence is observed in viable cells whose plasma membrane is intact. Cells that die by necrosis do not display the protoplast retraction associated with apoptotic-like cell death and do not fluoresce, while cells that have undergone apoptotic-like cell death show a characteristic retraction of the protoplast away from the cell wall and cannot cleave FDA [8] (Figure S1).

4.3. Growth and Treatment of Wheat Seedlings

Wheat (*Triticum aestivum*) cvs. CM82036 and Remus were used in this study. Cultivar CM82036 carries a major quantitative trait locus (QTL) on the short arm of chromosome 3B that is associated with resistance to both FHB disease and DON-induced bleaching of spikelets (*Fhb1*; syn. *Qfhs.ndsu-3BS*) and it carries another QTL on chromosome 5A that is associated with FHB resistance but not with DON resistance [4,33]. Cultivar Remus is susceptible to FHB and DON-induced bleaching [4,33]. Seeds were pre-germinated in the dark at 20 °C for 48 h in Petri dishes containing filter paper moistened with 7 mL of sterile water (12 seeds per plate). Germinated seeds were then air-dried for 10 min on filter paper and carefully placed in Petri dishes containing 7 mL of either water or DON (20 mgmL^{-1} water) (12 seeds per plate) such that coleoptiles were not in contact with the treatment solution. Wheat seedlings are less sensitive to DON than Arabidopsis cell cultures ([34]), hence the reason for the higher concentration as compared to the Arabidopsis studies. Plates were incubated at 20 °C under either constant darkness or constant light (~110 mmol m^{-2} s^{-1}). The roots and coleoptile were harvested at 4 or 24 h post-treatment, flash frozen in liquid N$_2$ and stored at −70 °C prior to either RNA or DON extraction. Seedling experiments conducted for RNA and DON analysis each included two replica plates per treatment and each experiment was conducted twice.

4.4. Quantification of DON

Freeze-dried coleoptile tissue was homogenized as previously described [5]. DON/DON derivatives was extracted from coleoptiles and quantified using the Ridascreen® DON Fast immunoassay (R-Biopharm AG, Darmstadt, Germany) according to the manufacturer's instructions. The antibodies used in this assay detect DON and DON derivatives including the less phytotoxic DON-3-glucoside [27]. Values were based on the average obtained for two replicates per sample.

4.5. RNA Extraction and Gene-Specific RT-PCR Analyses

Freeze-dried root or coleoptile samples were homogenized and total RNA was extracted and DNase1-treated as described by Ansari *et al.* [5]. Reverse transcription of total RNA was conducted as described by Ansari *et al.* [5], except that the primer used was oligo dT$_{12-18}$ (Life Technologies, Paisley, UK). Both the genes of interest and wheat actin (*Act1*) were PCR-amplified (separately) using gene-specific primers (Supplemental Table S1 lists GenBank accession numbers, gene-specific primer sequences and expected product sizes); *Act1* served as a control gene that was constitutively expressed in roots, coleoptile and head tissue. RT products were diluted to 100 mL and 3 mL was PCR-amplified in a 10 mL reaction containing 1 unit of Taq DNA polymerase and $1\times$ PCR buffer (Life Technologies, Paisley, UK), 1.5 mM $MgCl_2$, 150 mM each of dATP, dGTP, dCTP and dTTP, and 100 nM each of forward and reverse transcript-specific primers. PCR reactions were conducted in a Peltier thermal cycler DNA engine (MJ Research, St. Bruno, Canada) and the programme constituted 30 cycles of 94 °C for 30 s, 60 °C for 20 s and 72 °C for 45 s, with a final extension at 72 °C for 5 min. PCR products were electrophoresed through 2% (w v^{-1}) agarose gels containing 0.5 mg mL^{-1} ethidium bromide and visualized using Imagemaster VDS and Liscap software (GE Healthcare Life Sciences, Buckinghamshire, UK).

4.6. Data Analysis

All data analyses were conducted using Minitab (Minitab release $13^{©}$, 1994 Minitab Ltd, Coventry, UK). No data set followed a normal distribution, as determined using the Normality test and none could be transformed to fit a normal distribution using the Johnson Transformation tool. Non-normally distributed data (cell viability and morphology data, gene expression data (transcript/*Act1* levels in DON relative to water-treated samples) and DON data) were analyzed using the Mann-Whitney Rank sum test (confidence level 95%, alternatives of greater than, less than or equal chosen as appropriate).

5. Conclusions

We have shown that the effect of DON on the viability of abiotically stressed cells and on defense gene expression in wheat is light enhanced. Future studies should investigate the role of cell death suppression, cell survival pathways and light-regulated pathways in the resistance of wheat to *Fusarium* fungi. The combination of DON and heat stress offers a valuable means by which to unravel some of the complexity of plant programmed cell death.

Acknowledgments: This research was funded by Science Foundation Ireland Principal Investigator project (10-IN1-B3028) and EU FP5 project FUCOMYR (QLRT-2000-02044). We thank Hermann Buerstmayr (IFA-Tulln, Austria) for providing wheat seed and Robert Proctor (USDA Agricultural Research Service, Peoria, IL, USA) for providing the *Fusarium* strains used in this work.

Conflicts of Interest: The authors declare no conflict of interest.

References

1. Parry, D.W.; Jenkinson, P.; McLeod, L. *Fusarium* ear blight (scab) in small-grain cereals—A review. *Plant Pathol.* **1995**, *44*, 207–238. [CrossRef]
2. Langevin, F.; Eudes, F.; Comeau, A. Effect of trichothecenes produced by *Fusarium graminearum* during *Fusarium* head blight development in six cereal species. *Eur. J. Plant Pathol.* **2004**, *110*, 735–746. [CrossRef]
3. Wang, H.; Hwang, S.F.; Eudes, F.; Chang, K.F.; Howard, R.J.; Turnbull, G.D. Trichothecenes and aggressiveness of *Fusarium graminearum* causing seedling blight and root rot in cereals. *Plant Pathol.* **2006**, *55*, 224–230. [CrossRef]
4. Lemmens, M.; Scholz, U.; Berthiller, F.; Dall'Asta, C.; Koutnik, A.; Schuhmacher, R.; Adam, G.; Buerstmayr, H.; Mesterhazy, A.; Krska, R.; *et al.* The ability to detoxify the mycotoxin deoxynivalenol colocalizes with a major quantitative trait locus for *Fusarium* head blight resistance in wheat. *Mol. Plant. Microbe Interact.* **2005**, *18*, 1318–1324. [CrossRef]
5. Ansari, K.I.; Walter, S.; Brennan, J.M.; Lemmens, M.; Kessans, S.; McGahern, A.; Egan, D.; Doohan, F.M. Retrotransposon and gene activation in wheat in response to mycotoxigenic and non-mycotoxigenic-associated *Fusarium* stress. *Theor. Appl. Genet.* **2007**, *114*, 927–937. [CrossRef]
6. Desmond, O.J.; Manners, J.M.; Stephens, A.E.; MaClean, D.J.; Schenk, P.M.; Gardiner, D.M.; Munn, A.L.; Kazan, K. The *Fusarium* mycotoxin deoxynivalenol elicits hydrogen peroxide production, programmed cell death and defence responses in wheat. *Mol. Plant Pathol.* **2008**, *9*, 435–445. [CrossRef]

7. Walter, S.; Brennan, J.M.; Arunachalam, C.; Ansari, K.I.; Hu, X.; Khan, M.R.; Trognit, Z.F.; Trognitz, B.; Leonard, G.; Egan, D.; *et al.* Components of the gene network associated with genotype-dependent response of wheat to the *Fusarium* mycotoxin deoxynivalenol. *Funct. Integr. Genomics* **2008**, *8*, 421–427. [CrossRef]

8. Diamond, M.; Reape, T.J.; Rocha, O.; Doyle, S.M.; Doohan, F.M.; McCabe, P.F. The mycotoxin Deoxynivalenol produced by necrotrophic *Fusarium* species can inhibit plant apoptotic-like programmed cell death. *PLoS ONE* **2013**. [CrossRef]

9. Bushnell, W.R.; Seeland, T.M.; Perkins-Veazie, P.M.; Krueger, D.E.; Collins, J.K.; Russo, V.M. The effects of deoxynivalenol on barley leaf tissues. In *Genomic and Genetic Analysis of Plant Parasitism and Defence*; Tsuyumu, S., Leach, J.E., Shiraishi, T., Wolpert, T., Eds.; APS Press: St. Paul, MN, USA, 2004; pp. 270–284.

10. Poppenberger, B.; Berthiller, F.; Lucyshyn, D.; Sieberer, T.; Schuhmacher, R.; Krska, R.; Kuchler, K.; Glössl, J.; Luschnig, C.; Adam, G. Detoxification of the *Fusarium* mycotoxin Deoxynivalenol by a UDP-glucosyltransferase from *Arabidopsis thaliana*. *J. Biol. Chem.* **2003**, *278*, 47905–47914. [CrossRef]

11. Doyle, S.M.; Diamond, M.; McCabe, P.F. Chloroplast and reactive oxygen species involvement in apoptotic-like programmed cell death in Arabidopsis suspension cultures. *J. Exp. Bot.* **2010**, *61*, 473–482. [CrossRef]

12. Doohan, F.M. University College Dublin: Dublin, Ireland, Unpublished data. 2008.

13. Steiner, B.; Kurz, H.; Lemmens, M.; Buerstmayr, H. Differential gene expression of related wheat lines with contrasting levels of head blight resistance after *Fusarium graminearum* inoculation. *Theor. Appl. Genet.* **2009**. [CrossRef]

14. Chamnongpol, S.; Willekens, H.; Langebartels, C.; VanMontagu, M.; Inze, D.; VanCamp, W. Transgenic tobacco with a reduced catalase activity develops necrotic lesions and induces pathogenesis-related expression under high light. *Plant J.* **1996**, *10*, 491–503.

15. Genoud, T.; Buchala, A.J.; Chua, N.H.; Metraux, J.P. Phytochrome signalling modulates the SA-perceptive pathway in Arabidopsis. *Plant J.* **2002**, *31*, 87–95. [CrossRef]

16. Metraux, J.P.; Nawrath, C.; Genoud, T. Systemic acquired resistance. *Euphytica* **2002**, *124*, 237–243. [CrossRef]

17. Zeier, J.; Pink, B.; Mueller, M.J.; Berger, S. Light conditions influence specific defence responses in incompatible plant-pathogen interactions: Uncoupling systemic resistance from salicylic acid and PR-1 accumulation. *Planta* **2004**, *219*, 673–683.

18. Griebel, T.; Zeier, J. Light regulation and daytime dependency of inducible plant defenses in arabidopsis: Phytochrome signaling controls systemic acquired resistance rather than local defense. *Plant Physiol.* **2008**, *147*, 790–801. [CrossRef]

19. Roberts, M.R.; Paul, N.D. Seduced by the dark side: Integrating molecular and ecological perspectives on the influence of light on plant defence against pests and pathogens. *New Phytol.* **2006**, *170*, 677–699. [CrossRef]

20. Apel, K.; Hirt, H. Reactive oxygen species: Metabolism, oxidative stress, and signal transduction. *Annu. Rev. Plant Biol.* **2004**, *55*, 373–399. [CrossRef]

21. Babaeizad, V.; Imani, J.; Kogel, K.H.; Eichmann, R.; Hückelhoven, R. Over-expression of the cell death regulator BAX inhibitor-1 in barley confers reduced or enhanced susceptibility to distinct fungal pathogens. *Theor. Appl. Genet.* **2008**, *118*, 455–463.

22. Durrant, W.E.; Dong, X. Systemic acquired resistance. *Annu. Rev. Phytopathol.* **2004**, *42*, 185–209. [CrossRef]

23. Cuzick, A.; Lee, S.; Gezan, S.; Hammond-Kosack, K.E. NPR1 and EDS11 contribute to host resistance against *Fusarium. culmorum* in Arabidopsis buds and flowers. *Mol. Plant Pathol.* **2008**, *9*, 697–704. [CrossRef]

24. Makandar, R.; Essig, J.S.; Schapaugh, M.A.; Trick, H.N.; Shah, J. Genetically engineered resistance to *Fusarium* head blight in wheat by expression of Arabidopsis NPR1. *Mol. Plant Microbe Interact.* **2006**, *19*, 123–129. [CrossRef]

25. Chern, M.; Fitzgerald, H.A.; Canlas, P.E.; Navarre, D.A.; Ronald, P.C. Overexpression of a rice NPR1 homolog leads to constitutive activation of defense response and hypersensitivity to light. *Mol. Plant-Microbe Interact.* **2005**, *18*, 511–520. [CrossRef]

26. Hamzehzarghani, H.; Kushalappa, A.C.; Dion, Y.; Rioux, S.; Comeau, A.; Yaylayan, V.; Marshall, W.D.; Mather, D.E. Metabolic profiling and factor analysis to discriminate quantitative resistance in wheat cultivars against *Fusarium* head blight. *Physiol. Mol. Plant Pathol.* **2005**, *66*, 119–133. [CrossRef]

27. Gunnaiah, R.; Kushalappa, A.C.; Duggavathi, R.; Fox, S.; Somers, D.J. Integrated metabolo-proteomic approach to decipher the mechanisms by which wheat QTL (Fhb1) contributes to resistance against *Fusarium graminearum*. *PLoS One* **2012**, *7*, e40695.

28. Ruprich, J.; Ostry, J. Immunochemical methods in health risk assessment: Cross reactivity of antibodies against mycotoxin deoxynivalenol with deoxynivalenol-3-glucoside. *Cent. Eur. J. Public Health* **2008**, *16*, 34–37.

29. May, M.J.; Leaver, C.J. Oxidative stimulation of glutathione synthesis in *Arabidopsis thaliana* suspension cultures. *Plant Physiol.* **1993**, *103*, 621–627.

30. McCabe, P.F.; Leaver, C.J. Programmed cell death in cell cultures. *Plant Mol. Biol.* **2000**, *44*, 359–368. [CrossRef]

31. Proctor, R.H.; Hohn, T.M.; McCormick, S.P. Reduced virulence of *Gibberella zeae* caused by disruption of a trichothecene toxin biosynthetic gene. *Mol. Plant-Microbe Interact.* **1995**, *8*, 593–601. [CrossRef]

32. Li, W.L.; Faris, J.D.; Muthukrishnan, S.; Liu, D.J.; Chen, P.D.; Gill, B.S. Isolation and characterization of novel cDNA clones of acidic chitinases and beta-1,3-glucanases from wheat spikes infected by *Fusarium graminearum*. *Theor. Appl. Genet.* **2001**, *102*, 353–362. [CrossRef]

33. Buerstmayr, H.; Steiner, B.; Hartl, L.; Griesser, M.; Angerer, N.; Lengauer, D.; Miedaner, T.; Schneider, B.; Lemmens, M. Molecular mapping of QTLs for *Fusarium* head blight resistance in spring wheat. II. Resistance to fungal penetration and spread. *Theor. Appl. Genet.* **2003**, *107*, 503–508. [CrossRef]

34. Doohan, F.M. University College Dublin: Dublin, Ireland, Unpublished work. 2010.

toxins

MDPI

Article

Deoxynivalenol and Oxidative Stress Indicators in Winter Wheat Inoculated with *Fusarium graminearum*

Agnieszka Waśkiewicz [1], Iwona Morkunas [2], Waldemar Bednarski [3], Van Chung Mai [2,4], Magda Formela [2], Monika Beszterda [1], Halina Wiśniewska [5] and Piotr Goliński [1,*]

[1] Department of Chemistry, Poznań University of Life Sciences, Wojska Polskiego 75, Poznań 60-625, Poland; agat@up.poznan.pl (A.W.); monika.beszterda@up.poznan.pl (M.B.)

[2] Department of Plant Physiology, Poznań University of Life Sciences, Wołyńska 35, Poznań 60-637, Poland; morkunas@jay.up.poznan.pl (I.M.); chungmai@up.poznan.pl (V.C.M.); formelamagda@o2.pl (M.F.)

[3] Institute of Molecular Physics, Polish Academy of Sciences, Smoluchowskiego 17, Poznań 60-179, Poland; waldemar.bednarski@ifmpan.poznan.pl

[4] Department of Plant Physiology, Vinh University, Le Duan 182, Vinh City, Vietnam; chungmv@vinhuni.edu.vn

[5] Department of Genomics, Institute of Plant Genetics, Polish Academy of Sciences, Strzeszyńska 34, Poznań 60-479, Poland; hwis@igr.poznan.pl

* Author to whom correspondence should be addressed; piotrg@up.poznan.pl; Tel.: +48-61-848-78-37; Fax: +48-61-848-78-24.

Received: 5 November 2013; in revised form: 14 January 2014; Accepted: 20 January 2013; Published: 7 February 2014

Abstract: This study comprises analyses of contents of mycotoxins, such as deoxynivalenol and zearalenone, as well as the level of oxidative stress in ears of a susceptible wheat cultivar Hanseat and cv. Arina, resistant to a pathogenic fungus *Fusarium graminearum*. Starting from 48 h after inoculation, a marked increase was observed in the contents of these mycotoxins in ears of wheat; however, the greatest accumulation was recorded in the late period after inoculation, *i.e.*, during development of disease. Up to 120 h after inoculation, in ears of both wheat cultivars, the level of deoxynivalenol was higher than that of zearalenone. The susceptible cultivar was characterized by a much greater accumulation of deoxynivalenol than the resistant cultivar. At the same time, in this cultivar, in the time from 0 to 72 h after inoculation, a marked post-infection increase was observed in the generation of the superoxide radical ($O_2^{\bullet-}$). Additionally, its level, at all the time points after inoculation, was higher than in the control. In wheat cv. Arina, a markedly higher level of $O_2^{\bullet-}$ generation in relation to the control was found up to two hours after inoculation and, next, at a later time after inoculation. In turn, the level of semiquinone radicals detected by electron paramagnetic resonance (EPR) increased at later culture times, both in cv. Hanseat and Arina; however, in infested ears of wheat, it was generally lower than in the control. Analysis of disease symptoms revealed the presence of more extensive lesions in ears of a susceptible wheat cv. Hanseat than resistant cv. Arina. Additionally, ergosterol level as a fungal growth indicator was higher in ears of susceptible wheat than in the resistant cultivar.

Keywords: deoxynivalenol; semiquinone radicals; *Fusarium graminearum*; oxidative stress; winter wheat; zearalenone; ergosterol

1. Introduction

Interactions of plants and their pathogenic fungi now constitute an interesting and rapidly developing field in plant science, with a significant impact on new strategies for plant protection. The plant response to infection is determined by the genetic background of the host, as well as the pathogen [1]. The type of induced response that is effective against a given pathogen varies, depending on the lifestyle of the pathogen [2]. Pathogens have devised different strategies to invade a plant, as well as to feed on and reproduce in the plant. Biotrophic pathogens need living tissue for growth and reproduction; in many interactions the tissue will die in the late stages of the infection (hemi-biotrophic pathogens). By contrast, necrotrophic pathogens kill the host tissue at the beginning of the infection and feed on the dead tissue [3].

As plants are confined to the place where they grow, they have to develop a broad range of defense responses to cope with pathogenic infections. Oxidative burst, a rapid, transient production of huge amounts of reactive oxygen species (ROS), is one of the earliest observable manifestations of a plant's defense strategy [4–6]. Various aspects, mechanisms, and functions of the oxidative burst with generation of superoxide anions ($O_2^{\bullet-}$) in plant cells, which is stimulated by active defense-inducing fungal infection or elicitor treatment, were reviewed mainly on the basis of experimental evidence obtained in different pathosystems [7–9]. Free radicals, including ROS, may function in defense through their direct toxicity to pathogens, or may activate various metabolic pathways. Enhanced generation of free radicals, such as ROS, plays a significant role especially at the early plant-pathogen interaction, whereas, at a later stage of the disease development—when not coordinated with an effective system of their removal—it may enhance destructive changes in plants and facilitate the spread of a pathogen [6,10,11]. Recently, concluding evidence suggests that the ROS network is essential to induce disease resistance [12]. On the other hand, investigations show also that necrotrophic pathogens can use oxidative processes during their attack and invasion of plant tissues [13]. Therefore, host cell death can occur through the action of fungal toxins and an oxidative burst generated by both the pathogen and the host [14].

This study, next to oxidative stress indexes indicating early defense responses of plants, also investigated the accumulation of mycotoxins formed by a pathogenic fungus *Fusarium graminearum*. Reverberi and co-workers reported that several secondary metabolites are synthesized by fungi during morphological and metabolic transitions when the accumulation of ROS occurs [15]. Plant compounds involved in plant-fungi interactions are able to interfere with mycotoxin biosynthesis in host tissues [16]. Mycotoxins are harmful and often carcinogenic secondary metabolites produced by a range of widespread fungi, including *Fusarium*. In general, they are low-molecular-weight compounds synthesized by filamentous fungi and are capable of causing disease and death in plants, animals and humans [17]. While in the literature there are many reports indicating high toxicity of mycotoxins, little is known about their role in plant-pathogen interactions. The relationship between the decrease in cell proliferation, the presence of oxidative stress generated by the enhancement of intracellular ROS production, and ROS-induced lipid peroxidation by mycotoxins is a priority direction of research [18]. *Fusarium* mycotoxins, currently considered of importance from the toxicological point of view, include zearalenone, trichothecenes and fumonisins, and their occurrence is now regulated by legal limits in all developed countries [19,20]. Among trichothecenes, deoxynivalenol (DON) is the most popular mycotoxin formed mainly by *Fusarium graminearum* and *F. culmorum* [19]. *Fusarium graminearum* is most common in moist and warm continental climates, such as Central and South-Eastern Europe, whereas *F. culmorum* is found more often in maritime and cooler European countries [21–23]. The primary sources of DON are cereals, including wheat, barley, maize, and oat [24,25]. Toxicity is associated with the presence of both 9, 10 double bond, 12, 13 epoxide group and varied substituent groups in the deoxynivalenol structure [26]. DON is responsible for the inhibition of protein biosynthesis, reduction of enzymatic activity, disturbance in cytoplasmic membrane permeability, and cell division disorders [26]. Another important mycotoxin, similar to DON, produced mainly by the same fungi, is zearalenone (ZON) [27].

The aim of the present study was to examine the interdependence between the level of oxidative stress and mycotoxin contents in ears of two winter wheat cultivars, *i.e.*, the susceptible cv. Hanseat and cv. Arina, resistant to a pathogenic fungus *Fusarium graminearum*. Therefore, the level of superoxide anion radical generation and concentrations of free radicals, such as semiquinones, were estimated in non-inoculated (control) and *F. graminearum* - inoculated ears of winter wheat. The semiquinone radicals analyzed in this study using electron paramagnetic resonance (EPR) spectrometry are among the relatively stable radicals that readily donate electrons to molecular oxygen (O_2), forming $O_2^{\bullet-}$. Moreover, changes in mycotoxin contents, such as deoxynivalenol and zearalenone, were determined in ears of the above-mentioned wheat cultivars. Additionally, disease symptoms were analyzed and ergosterol level as a fungal growth indicator was estimated in both wheat cultivars.

106

2. Results

2.1. Mycotoxin Contents

Starting from 48 h after inoculation with a pathogenic fungus *F. graminearum* a marked increase was observed in the contents of mycotoxins, such as deoxynivalenol and zearalenone, in ears of wheat—both the susceptible cv. Hanseat and the resistant Arina (Figure 1). The highest accumulation of mycotoxins was recorded at late time points after inoculation (at 168 h and in week two), *i.e.*, during development of disease. Up to 120 h after inoculation in ears of both wheat cultivars, a higher level of deoxynivalenol (DON) was found in comparison to zearalenone (ZON). Analysis of variance (ANOVA) results showed that the differences in concentrations of DON and ZON in inoculated ears of wheat cultivars were highly statistically significant. It needs to be stressed that the susceptible cultivar (Figure 1A,B) was characterized by a much greater accumulation of deoxynivalenol than the resistant cultivar (Figure 1C,D). In the susceptible cultivar, the level of deoxynivalenol ranged from 1.1 to 109.88 ng g^{-1} FW, while in the resistant cultivar it was from 1.76 to 62.41 ng g^{-1} FW. Only at 168 h and in week two after inoculation in ears of the resistant wheat cv. Arina, zearalenone level was higher than that of deoxynivalenol (Figure 1C,D). ANOVA results showed that the differences in DON concentration in infected tissue of cultivars Hanseat/Arina and the control plants at 72, 120, 168 h, and two weeks were highly statistically significant (e.g., $p = 0.00006/0.00005$, $p = 0.00016/0.00033$, $p = 0.00047/0.0069$, $p = 0.00005/0.00011$, respectively). Moreover, ANOVA results showed that the differences in ZON concentration in infected tissue of cultivars Hanseat/Arina and the control plants at 168 h and two weeks were highly statistically significant (e.g., $p = 0.00027/0.00028$ and $p = 0.00006/0.00001$, respectively).

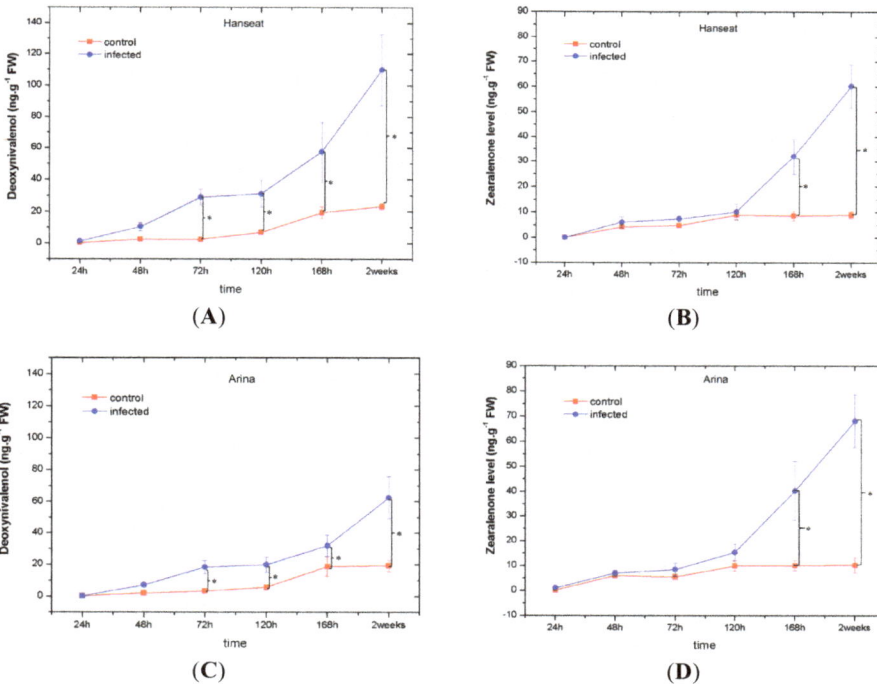

Note: * significant differences on figures using asterisks are shown.

Figure 1. The effect of pathogenic fungus *Fusarium graminearum* on the content of mycotoxins, such as deoxynivalenol and zearalenone, in ears of a susceptible wheat cv. Hanseat (**A**,**B**) and resistant cv. Arina (**C**,**D**). Significant differences ($p < 0.05$) were observed between control and infected ears.

2.2. Generation of Superoxide Anion

In the period from 0 to 72 h after inoculation in ears of wheat cv. Hanseat, a marked post-infection increase was observed in the generation of superoxide anion ($O_2^{\bullet-}$), while starting from 120 h after inoculation it fluctuated (Figure 2A). Moreover, in the susceptible cv. Hanseat at all time points after inoculation a higher post-infection level of $O_2^{\bullet-}$ generation was found in comparison to the control. In turn, in the wheat resistant cv. Arina up to 2 h after inoculation a higher level of $O_2^{\bullet-}$ was recorded than in the control and next at later time points after inoculation with *F. graminearum*, *i.e.*, at 120, 168 h, and in week two after inoculation, $O_2^{\bullet-}$ generation level was markedly higher than in the control (Figure 2B). It is of interest that at 24 and 48 h after inoculation a strong increase was found in the generation of $O_2^{\bullet-}$, both in the control and in inoculated ears of wheat cv. Arina, whereas at 72 h a marked reduction of $O_2^{\bullet-}$ was recorded in these tissues, while in inoculated ears the concentration of $O_2^{\bullet-}$ was lower than in the control. Starting from 120 h after inoculation, the post-infection level of $O_2^{\bullet-}$ was much higher than in the control. The significant differences in the level of superoxide anion were observed among the experimental variants as analyzed by ANOVA. ANOVA results showed that the differences in the concentration of $O_2^{\bullet-}$ in infected tissue of cultivar Hanseat and the control plants at 0.5, 4, 72, and 168 h were highly statistically significant (e.g., $p = 0.0001$, $p = 0.0011$, $p = 0.0007$ and $p = 0.0005$), while in infected tissue of cultivar Arina and the control plants at 72, 120, 168 h, and two weeks they were highly statistically significant (e.g., $p = 0.005$, $p = 0.00016$, $p = 0.00006$ and $p = 0.00009$, respectively).

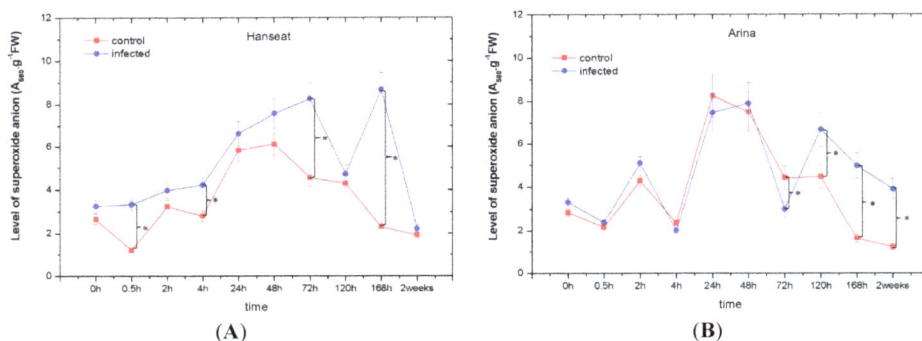

Note: * significant differences on figures using asterisks are shown.

Figure 2. The effect of pathogenic fungus *Fusarium graminearum* on the generation of superoxide anion radical in ears of a susceptible wheat cv. Hanseat (**A**) and resistant cv. Arina (**B**). Significant differences ($p < 0.005$) were observed between control and infected ears.

2.3. Generation of Semiquinone Radicals

Levels of semiquinone radicals detected by electron paramagnetic resonance (EPR) increased at later time points of culture both in cv. Hanseat and Arina; however, in inoculated ears of wheat it was lower than in the control (Figure 3), except for 168 h after inoculation in ears of cv. Arina. Moreover, in the period from 0 to 120 h of culture in both wheat cultivars, Hanseat and Arina, slight fluctuations were observed in the concentration of semiquinone radicals both in the control and in inoculated tissues. However, the range of generation of these radicals in 168-h and two-week-old control tissues in the susceptible cv. Hanseat was two-fold greater than in the resistant cv. Arina. ANOVA results showed that the differences in concentrations of semiquinone radicals both in the control and in inoculated tissues were highly statistically significant. ANOVA results showed that the differences in semiquinone radical concentrations in infected tissue of the susceptible cv. Hanseat and the control plants at 168 h and two weeks were highly statistically significant (e.g., $p = 0.0027$ and

p = 0.00064, respectively), while in the infected tissue of the resistant cv. Arina and the control plants at two weeks they were highly statistically significant (e.g., p = 0.00378).

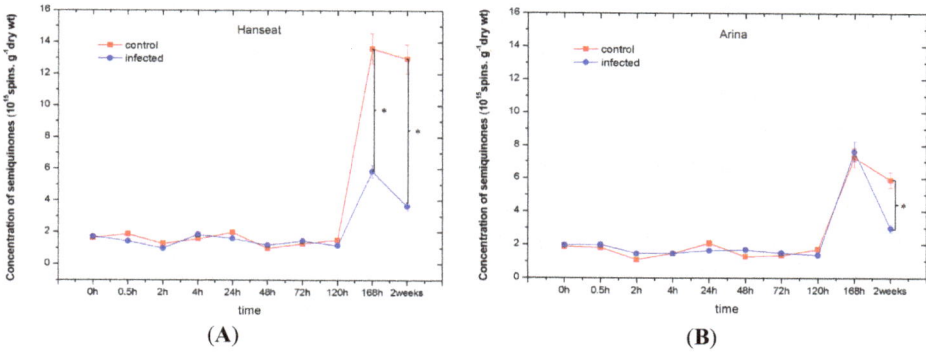

Note: * significant differences on figures using asterisks are shown.

Figure 3. The effect of pathogenic fungus *Fusarium graminearum* on the generation of semiquinone radicals in ears of a susceptible wheat cv. Hanseat (**A**) and resistant cv. Arina (**B**). Significant differences (p < 0.005) were observed between control and infected ears.

2.4. Analysis of Disease Symptoms and the Level of Ergosterol

Table 1 showed disease development in ears of the susceptible wheat cv. Hanseat and resistant cv. Arina after inoculation with *Fusarium graminearum*. From 120 h after inoculation a pronounced severity of disease development was observed in ears of the susceptible wheat cv. Hanseat, it was stronger than in resistant cv. Arina. Symptoms first began as water-soaked brownish spots at the base of the glumes and ultimately turned into bigger brown discolorations. Moreover, masses of black spores occurred along the base of the glumes or over the infected head. In addition, from 120 h after inoculation the level of ergosterol in infected tissue of the susceptible wheat cv. Hanseat was higher than the resistant cv. Arina (Figure 4). ANOVA results showed that the differences in ergosterol concentration in infected tissue of cultivar Arina/Hanseat and the control plants at 48, 72, 120, 168 h, and two weeks were highly statistically significant (e.g., p = 0.03293/0.14943, p = 0.02159/0.06088, p = 0.01926/0.00921, p = 0.00777/0.00048 and p = 0.0001/0.00091, respectively).

Table 1. Disease development in ears of a susceptible wheat cv. Hanseat and resistant cv. Arina after inoculation with *Fusarium graminearum* (− lack of disease symptoms, + severity of disease symptoms, *i.e.*, strong discolorations and browning of ears where +++++ bigger brown discolorations in over 50% ears, sometimes black spores were found along ears and + single, light brown discolorations).

Time after inoculation	Disease development			
	Arina-Resistant		Hanseat-Susceptible	
	Control	Infected	Control	Infected
24 h	−	−	−	−
48 h	−	−	−	+
72 h	−	++	−	++
120 h	−	++	−	++++
168 h	−	+++	−	++++
2 weeks	−	++++	−	+++++

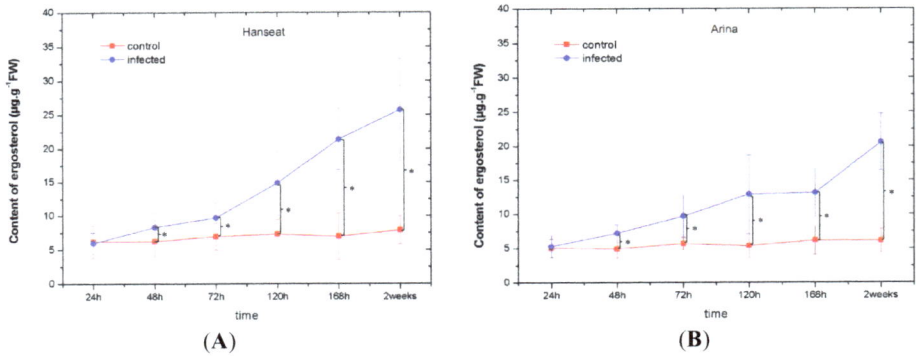

Note: * significant differences on figures using asterisks are shown.

Figure 4. The level of ergosterol as a fungal growth indicator in ears of a susceptible wheat cv. Hanseat (**A**) and resistant cv. Arina (**B**). Significant differences ($p < 0.05$) were observed between control and infected ears.

3. Discussion

This study investigated the interdependence between the level of oxidative stress and contents of mycotoxins, such as deoxynivalenol and zearalenone, in ears of two wheat cultivars, *i.e.*, susceptible Hanseat and Arina, resistant to a pathogenic fungus *Fusarium graminearum*. Our objective was to understand the difference in the invasion process of the host plant by the pathogen and the differential defense response in the resistant and susceptible cultivars. We detected changes in the redox status of cells in wheat ears, associated with the generation of superoxide anion and semiquinone radicals accompanying the accumulation of mycotoxins produced by the pathogenic fungus *F. graminearum*. The pathogenic fungus *F. graminearum* produces a range of sesquiterpenoid mycotoxins, including several types of B trichothecenes, such as DON and its acetylated derivatives 15Ac-DON and 3Ac-DON, which are required for full virulence on wheat ears [28–31]. Bin-Umer and co-workers reported that trichothecene toxins can inhibit mitochondrial translation independent of their effects on cytosolic translation and mitochondrial membrane integrity [32].

As a consequence of contact between the pathogen and the plant cell, biochemical reactions are initiated, limiting development of infection and disease. The first step of defense is the rapid generation of free radicals, including reactive oxygen species (ROS) and the activation of pre-existing components, such as the liberation of toxic compounds (e.g., phenolics and subsequent oxidative reactions) [33]. Thus, in the present study, in wheat ears of the susceptible cultivar we observed a marked post-infection increase in the generation of the superoxide anion radical ($O_2^{\bullet-}$) in the period from 0 to 72 h after inoculation (Figure 2A). Additionally, its level at all time points after inoculation was higher than in the control. In turn, in wheat cv. Arina a markedly greater level of $O_2^{\bullet-}$ generation in relation to the control was recorded at 2 h after inoculation and next at later time points after inoculation, *i.e.*, at the phase of disease development (Figure 2B). The difference in the level of $O_2^{\bullet-}$ generation between the resistant wheat cultivar and the susceptible cultivar was connected with the earlier reduction in the generation of $O_2^{\bullet-}$ in the resistant cultivar. In the resistant cultivar Arina fluctuations were observed in its generation *versus* time after inoculation. Perhaps this is related to the capture of electrons originating from superoxide anion ($O_2^{\bullet-}$) by semiquinones.

Although superoxide anion ($O_2^{\bullet-}$) is the proximal product generated, the more stable hydrogen peroxide (H_2O_2) species is detected in many studies [34]. Oxidative burst could have a direct effect on the pathogen or the defenses because of its reactivity. ROS could directly kill the pathogen, especially in the case of the more reactive species such as hydroxyl radicals [35]. ROS could also contribute to the establishment of physical barriers at the large papillae that are formed at the site of interaction of many pathogens by cross linking of cell wall glycoproteins [36] or via oxidative cross-linking of precursors during the localized biosynthesis of lignin and suberin polymers [37]. Gupta and co-workers reported that ROS are known to play pivotal roles in pathogen perception, recognition, and downstream defense signaling [38]. However, how these redox alarms coordinate, *in planta*, into a defensive network is still intangible.

Published literature sources comprise reports concerning modulation of the cellular redox status by mycotoxins produced by pathogenic fungi, but it is mainly in cells of animals and the human body, while there are scarce studies showing the above dependencies in plant cells. For example, recent studies showed modulation of the cellular redox status in human body cells by toxins of a pathogenic fungus, such as the *Alternaria* [39]. Therefore, the mycotoxins alternariol (AOH) and alternariol monomethyl ether (AME) were found to modulate the redox balance of HT29 cells from the human body, but without an apparent negative effect on DNA integrity. Additionally, Arunachalam and Doohan reported that trichothecene mycotoxins inhibit eukaryotic protein synthesis and are toxic to plants, humans and farm animals [40]. At the cellular level, they induce oxidative stress cell-cycle arrest and apoptosis, and affect membrane integrity. In animals, trichothecenes can be either immunostimulatory or immunosuppressive and induce apoptosis *via* mitochondria-mediated or -independent pathways. In turn, in plants trichothecenes induce programmed cell death via production of reactive oxygen species and they can induce genes involved in oxidative stress, cell death, and plant defense signaling. Studies by Gilchrist revealed a connection of mycotoxins between plants and animals in apoptosis and ceramide signaling [41]. Dobosz and co-workers showed also a relationship between the increase of the free form of salicylic acid (SA), free radical (FR) concentration, and propagation of *F. proliferatum* and *F. oxysporum* as a consequence mycotoxin formation, such as moniliformin and fumonisin B_1 in infected plants of *Asparagus officinalis* [42]. In plants, the use of the *Arabidopsis* model system to understand molecular events in trichothecene-induced phytotoxicity has identified the involvement of MAPK signaling pathways and downstream transcription factors that manifest the toxicity effects [43,44]. Additionally, Desmond and co-workers demonstrated that infusion of wheat leaves with DON induced hydrogen peroxide production within 6 h, followed by cell death within 24 h that was accompanied by DNA laddering, a hallmark of programmed cell death [45].

In this study with an enhanced post-infection generation of $O_2^{\bullet-}$ (Figure 2) a marked increase was observed in the contents of mycotoxins, such as DON and ZON; however, the highest accumulation of these toxins was recorded at the late period after inoculation in ears of both wheat cultivars, *i.e.*, at the disease development phase (Figure 1, Table 1). Up to 120 h after inoculation in ears of both wheat cultivars, the level of DON was higher than that of ZON, while at later time points (168 h and two weeks after inoculation) the resistant cultivar was characterized by a lower accumulation of DON than the sensitive cultivar. In parallel, in the resistant cultivar the development of disease symptoms (necrotic changes, discoloration of tissue) was limited (Table 1) and the level of ergosterol was lower than in the sensitive cultivar (Figure 4). Ergosterol (ERG) is a specific component of the fungal cell membrane [46]. It is also present in membranes in the cell walls and mitochondria in some yeasts, but is not produced in significant quantities by higher plants, rust fungi, or phycomycetes, hence, it can be used as a tool to estimate fungal biomass from any kind of mixtures [47,48]. A good positive correlation has been established between ergosterol content and fungal growth in other studies [49–53].

At the same time in this study, next to the increased generation of $O_2^{\bullet-}$, which may be one of the lines of defense against *F. graminearum*, the concentration of free radicals, such as semiquinones was also determined (Figure 3). These free radicals detected in ears of wheat give signals characterized by *g*-value of $2.0037 - 2.0039 \pm 0.0005$, similarly as in previous reports [6,54–57], indicating that they are semiquinone-derived radicals.

It should also be mentioned that quinones, which represent the largest group of redox cycling compounds, are particularly active in ROS generation. Semiquinone radicals exhibit high reactivity and cytotoxicity and are formed during the oxidation of phenols by phenolases, peroxidases, and also by polyphenol oxidase activity. Moreover, the semiquinone radicals analyzed in this study using EPR spectrometry are among the relatively stable radicals. These oxidized phenolic species have an enhanced antimicrobial activity and thus may be directly involved in stopping pathogen development. During the pathogen-plant interaction, oxidation processes are stimulated, which enhances the effectiveness of defense mechanisms [6,10,11,56].

Measurements of semiquinone radicals using electron paramagnetic resonance (EPR) showed that the level of these radicals, in the period from 0 to 120 h, both in the control and in the infested ears of the sensitive and resistant cultivars showed fluctuations and ranged from 0.9 to 2×10^{15} spins g^{-1} dry weight. In turn, at the time points the concentration of these radicals increased rapidly in tissues both in cv. Hanseat and Arina, although, in infested ears of wheat, it was generally lower than in the control. We assume that the lower level of these radicals in relation to the control may indicate their involvement in the stimulation of defense mechanisms connected with strengthening of cell walls. It is also possible that these radicals in plant cells may be incorporated into polymers, such as lignins and by combining with reactive free radicals that propagate depolymerization through the lignin matrix, these protective free radicals could prevent the breakdown of associated cell walls [11,58]. Additionally, in the resistant cultivar the concentration of semiquinone radicals was lower than in the susceptible cultivar (Figure 3).

Summing up, recorded results indicate that the accumulation of mycotoxins produced by *F. graminearum* in ears of winter wheat was accompanied by a markedly enhanced generation of superoxide anion as an indicator of oxidative stress. A lower level of semiquinone radicals at later time points after inoculation may probably indicate their incorporation into polymers, e.g., such as lignins, by bonding with reactive oxygen species especially superoxide anion ($O_2^{\bullet-}$) and, thus, strengthen the cell wall. The resistant cultivar was characterized by a lower level of semiquinone radicals than the sensitive cultivar especially at the late phase after inoculation. Development of disease was inhibited in the resistant cultivar and ergosterol content was lower than in the sensitive cultivar. Additionally, in the resistant cultivar production of the mycotoxin DON and the level of generation of superoxide anion ($O_2^{\bullet-}$) at the late phase after inoculation was lower than in the sensitive cultivar.

4. Experimental Section

4.1. Plant Material and Growth Conditions

Plant material comprised two popular winter wheat cultivars with different susceptibility to *Fusarium*, *i.e.*, —a susceptible cv. Hanseat and a resistant cv. Arina. The experiment was performed in Cerekwica (Central West Poland, 30 km northwest of Poznań), in the randomized complete block design in triplicate, with plot size of 1 m × 1 m. Seeds of both winter wheat cultivars were sown in three independent plots both in the control and infected *F. graminarum*. Both cultivars, *i.e.*, Hanseat (susceptible) and the resistant Arina, originate from the Plant Breeding Company in Poznań, Poland.

4.2. Fusarium Strain and Inoculum Preparation

Fusarium graminearum strain KF 2870 (elsewhere referred to as *F. graminearum*) was obtained from the Collection of Plant Pathogenic Fungi held by the Institute of Plant Genetics Polish Academy of Sciences, Poznan. The pathogen was incubated in the dark at 25 °C in Petri dishes (+9 cm diameter) on potato dextrose agar (PDA) medium (Difco; pH 5.5). After three weeks of growth the *F. graminearum* spore suspension was prepared. The spore suspension was obtained by washing the mycelium with sterile water and shaking with glass pearls. At mid-anthesis (Zadoks scale 65), 30 winter wheat heads of each replication were inoculated individually (by brushing) with the conidial suspension (2×10^6 spores) isolate of *Fusarium graminearum* (KF 2870). Non-inoculated plots of the same genotypes were used as the control. Inoculated and control samples (heads) for the determination of superoxide anion and semiquinone radicals were collected at 0, 0.5, 2, 4, 24, 48, 72, 120, 168 h, and two weeks after inoculation. In turn, for the determination of mycotoxins and ergosterol content, and analyses of disease development plant samples were collected at 24, 48, 72, 120, 168 h and two weeks after inoculation. To evaluate the disease, 50 ears of control plants and plants infected with *F. graminearum* were collected for both varieties, *i.e.*, resistant and sensitive.

4.3. Standards, Chemicals, and Reagents

Deoxynivalenol, zearalenone, and ergosterol standards and organic solvents (HPLC grade) were purchased with a standard grade certificate from Sigma-Aldrich (Steinheim, Germany). All chemicals used for extraction and purification of mycotoxins were purchased from POCh (Gliwice, Poland). Water for the HPLC mobile phase was purified using a Milli-Q system (Millipore, Bedford, MA, USA).

4.4. Extraction and Purification Procedure for Mycotoxins

Samples of 10 g homogenized winter wheat ears were prepared for analyses. Both mycotoxins (ZON and DON) were extracted and purified according to the detailed procedure described by Wiśniewska *et al.* [19]. The eluate was evaporated to dryness at 40 °C under a stream of nitrogen. Dry residue was stored at −20 °C until HPLC analyses.

4.5. HPLC Analysis of Mycotoxins

The chromatographic system consisted of a Waters 2695 high-performance liquid chromatograph (Waters, Milford, PA, USA) with detectors:

- Waters 2996 Photodiode Array Detector with a Nova Pak C-18 column (300 mm × 3.9 mm) for DON (λ_{max} = 224 nm) analysis,
- Waters 2475 Multi λ Fluorescence Detector (λ_{ex} = 274 nm, λ_{em} = 440 nm) and a Waters 2996 Photodiode Array Detector with a Nova Pak C-18 column (150 mm × 3.9 mm) for ZON analysis.

Quantification of mycotoxins was performed by measuring the peak areas at retention time according to the relevant calibration curve. The presence of mycotoxins was confirmed by a comparison of retention times with the external standard and by co-injection of every tenth sample with mycotoxin standards. Limits of detection were 0.001 μg g^{-1} for ZON and 0.01 μg g^{-1} for DON.

4.6. Ergosterol Extraction, Purification, and HPLC Analysis

Plant samples (100 mg) were suspended in 2 mL methanol in a culture tube, treated with 0.5 mL of 2 M aqueous sodium hydroxide, and sealed tightly. Samples were irradiated twice in a microwave oven (370 W) for 20 s. After 15 min contents of cultures tubes were neutralized with 1 M aqueous hydrochloric acid, then 2 mL methanol were added and samples were extracted with n-pentane (3 × 4 mL). The combined pentane extracts were evaporated to dryness in a stream of nitrogen, before analysis dissolved in 1 mL of methanol and 20 μL of thus prepared mixture were analyzed by HPLC. The ergosterol separation was performed on a 3.9 mm Nova Pak C-18, 4 mm column with methanol:acetonitrile (90:10, v/v) as the mobile phase at a flow rate of 1.0 mL min^{-1}. EGR was detected with a Waters 2996 Photodiode Array Detector (Waters Division of Millipore, Milford, MA, USA) set at 282 nm. The presence of ergosterol was confirmed by a comparison of retention times with the external standard and by co-injection of every tenth sample with an ERG standard. The detection limit was 0.01 μg g^{-1} and standard deviation was below 7%.

4.7. Determination of Superoxide Anion Radical Content

Determination of superoxide anion radical ($O_2^{\bullet -}$) content in biological samples was based on its ability to reduce nitro blue tetrazolium (NBT) [59]. The superoxide anion was detected according to Mai and co-workers [57], ears of wheat (0.30 g fr. wt) were cut into fragments (3 mm × 3 mm) and immersed in 10 mM potassium phosphate buffer (pH 7.8) containing 0.05% NBT and 10 mM NaN$_3$ in a final volume of 3 mL and incubated for 1 h at room temperature. After incubation, 2 mL of the reaction solution were heated at 85 °C for 15 min and rapidly cooled. The levels of $O_2^{•-}$ in ears of wheat were expressed as absorbance at 580 nm per 1 g of fresh materials ($A_{580}.g^{-1}$ fr. wt). The measurement was carried out in the Perkin Elmer Lambda 15 UV-Vis spectrophotometer (Norwalk, CT, USA).

4.8. Determination of Semiquinone Radicals

Samples of several g fresh weight of wheat ears were frozen in liquid nitrogen and lyophilized in a Jouan LP3 freeze dryer. The lyophilized material was transferred to EPR-type quartz tubes of 6 mm in diameter. Electron paramagnetic resonance measurements were performed with a Bruker ELEXSYS spectrometer operating at the X-band. The EPR spectra were recorded at room temperature as derivatives of microwave absorption. A magnetic field modulation of about 2 Gs and a microwave power of 5 mW were typically used for all experiments to avoid line saturation and deformation. EPR spectra of free radicals were recorded in the magnetic field range of 3000–3650 Gs and with 4096 data points. In order to determine the number of paramagnetic centers (free radicals) in the samples, the spectra were double-integrated and compared with the intensity of the monocrystal standard chromium-doped corundum (Al_2O_3:Cr^{3+}) with a known spin concentration [6,11,55,57,60,61]. Before and after the first integration of the spectra, small background corrections were made to obtain a reliable absorption signal before the second integration. Double integration of the free radicals was performed separately and this value was subtracted from the value obtained for the full 3000–3650 Gs scan range integration. As samples placed in quartz tubes were of equal volume, but of different weights, EPR intensity data were recalculated per 1 g of dry sample.

4.9. Statistical Analysis

All determinations were performed in three independent experiments. Analysis of variance (ANOVA) was applied to verify whether means from independent experiments within a given experimental variant were significant. Data shown in the figures are means of triplicates for each variant and standard errors of mean (SE). In individual figures significant differences are shown using asterisks.

5. Conclusions

A marked increase was found for the contents of mycotoxins, such as deoxynivalenol and zearalenone, in ears of both wheat cultivars in relation to the time after inoculation.

1. The susceptible cultivar was characterized by a much greater accumulation of deoxynivalenol than the resistant cultivar.
2. In the susceptible cultivar a marked post-infection increase in $O_2^{•-}$ level was found up to 120 h after inoculation.
3. The level of $O_2^{•-}$ generation in infested ears of both wheat cultivars was generally greater than in the control.
4. An earlier reduction in the level of $O_2^{•-}$ generation with the time after inoculation was observed in the resistant rather than in the susceptible cultivar.
5. The concentration of semiquinone radicals, detected by EPR, increased at later culture times; however, in infested ears of wheat it was generally lower than in the control.
6. The resistant cultivar was characterized by a lower level of semiquinone radicals than the sensitive cultivar especially at the late phase after inoculation. It may probably indicate their incorporation into polymers, such as lignins, and strengthening of the cell wall.
7. Development of disease was inhibited in the resistant cultivar and ergosterol content was lower than in the sensitive cultivar.
8. Production of the mycotoxin DON and the level of generation of superoxide anion ($O_2^{•-}$) in the resistant cultivar at the late phase after inoculation was lower than in the sensitive cultivar.

Acknowledgments: The study was supported by the Polish Ministry of Science and Higher Education (PMSHE) Project NN 3103019 34 and NN 3107203 40.

Conflicts of Interest: The authors declare no conflict of interest.

References

1. Molodchenkova, O.O.; Adamovskaya, V.G.; Yu, A.L.; Gontarenko, O.V.; Sokolov, V.M. Maize response to salicylic acid and *Fusarium moniliforme*. *Appl. Biochem. Microbiol.* **2002**, *38*, 381–385. [CrossRef]

2. Thatcher, L.F.; Manners, J.M.; Kazan, K. *Fusarium oxysporum* hijacks COI1-mediated jasmonate signaling to promote disease development in *Arabidopsis*. *Plant J.* **2009**, *588*, 927–939. [CrossRef]

3. Berger, S.; Benediktyova, Z.; Matous, K.; Bonfig, K.B.; Mueller, M.J.; Nedbal, L.; Roitsch, T. Visualiztion of dynamics of plant–pathogen interaction by novel combination of chlorophyll fluorescence imaging and statistical analysis: Differential effects of virulent and avirulent strains of *P. syringae* and of oxylipins on *A. thaliana*. *J. Exp. Bot.* **2007**, *58*, 797–806.

4. Wojtaszek, P. Mechanisms for the generation of reactive oxygen species in plant defence response. *Acta Physiol. Plant* **1997**, *19*, 581–589. [CrossRef]

5. Hückelhoven, R.; Kogel, K.H. Reactive oxygen intermediates in plant–microbe interactions: Who is who in powdery mildew resistance? *Planta* **2003**, *216*, 891–902.

6. Morkunas, I.; Bednarski, W. *Fusarium oxysporum*—Induced oxidative stress and antioxidative defenses of yellow lupine embryo axes with different sugar levels. *J. Plant Physiol.* **2008**, *165*, 262–277. [CrossRef]

7. Doke, N.; Miura, Y.; Sanchez, L.M.; Park, H.-J.; Noritake, T.; Yoshioka, H.; Kawakita, K. The oxidative burst protects plants against pathogen attack: Mechanism and role as an emergency signal for plant bio-defense— A review. *Gene* **1996**, *179*, 45–51. [CrossRef]

8. Govrin, E.M.; Levine, A. The hypersensitive response facilitates plant infection by the necrotrophic pathogen *Botrytis cinerea*. *Curr. Biol.* **2000**, *10*, 751–757. [CrossRef]

9. Apell, K.; Hirt, H. Reactive oxygen species: Metabolism, oxidative stress, and signal transduction. *Annu. Rev. Plant Biol.* **2004**, *55*, 373–399. [CrossRef]

10. Hammerschmidt, R. Phenols and plant–pathogen interactions: The saga continues. *Physiol. Mol. Plant Pathol.* **2005**, *66*, 77–78. [CrossRef]

11. Morkunas, I.; Bednarski, W.; Kopyra, M. Defense strategies of pea embryo axes with different levels of sucrose to *Fusarium oxysporum* and Ascochyta pisi. *Physiol. Mol. Plant Pathol.* **2008**, *72*, 167–178. [CrossRef]

12. Kotchoni, S.O.; Gachomo, E.W. The reactive oxygen species network pathways: An essential prerequisite for perception of pathogen attack and the acquired disease resistance in plants. *J. Biosci.* **2006**, *31*, 389–404. [CrossRef]

13. Mayer, A.M.; Staples, R.C.; Gil-ad, N.L. Mechanisms of survival necrotrophic fungal plant pathogens in hosts expressing the hypersensitive response. *Phytochemistry* **2001**, *58*, 33–41. [CrossRef]

14. Choquer, M.; Fournier, E.; Kunz, C.; Levis, C.; Pradier, J.-M.; Simon, A.; Viaud, M. Botrytis cinerea virulence factors: New insights into a necrotrophic and polyphageous pathogen. *FEMS Microbiol. Lett.* **2007**, *277*, 1–10. [CrossRef]

15. Reverberi, M.; Zjalic, S.; Ricelli, A.; Punelli, F.; Camera, E.; Fabbri, C.; Picardo, M.; Fanelli, C.; Fabbri, A. Modulation of antioxidant defense in *Aspergillus parasiticus* is involved in aflatoxin biosynthesis: A role for the ApyapA gene. *Eukaryot. Cell* **2008**, *7*, 988–1000. [CrossRef]

16. Boutigny, A.L.; Richard-Forget, F.; Barreau, C. Natural mechanisms for cereal resistance to *Fusarium* mycotoxins accumulation. *Rev. Eur. J. Plant Pathol.* **2008**, *121*, 411–423. [CrossRef]

17. Bennett, J.W.; Klich, M. Mycotoxins. *Clin. Microbiol. Rev.* **2003**, *16*, 497–516. [CrossRef]

18. Ferrer, E.; Juan-Gracia, A.; Font, G.; Ruiz, M.J. Reactive oxygen species induced by beauvericin, patulin and zearalenone in CHO-K1 cells. *Toxicol. In Vitro* **2009**, *23*, 1504–1509.

19. Wiśniewska, H.; Stępień, Ł.; Waśkiewicz, A.; Beszterda, M.; Góral, T.; Belter, J. Toxigenic *Fusarium* species infecting wheat heads in Poland. *Cent. Eur. J. Biol.* **2014**, *9*, 163–172. [CrossRef]

20. Goliński, P.; Waśkiewicz, A.; Wiśniewska, H.; Kiecana, I.; Mielniczuk, E.; Gromadzka, K.; Kostecki, M.; Bocianowski, J.; Rymaniak, E. Reaction of winter wheat (*Triticum aestivum* L.) cultivars to infection with *Fusarium* spp.: Mycotoxin contamination in grain and chaff. *Food Addit. Contam.* **2010**, *27*, 1015–1024. [CrossRef]

21. Audenaert, K.; van Broeck, R.; Bekaert, B.; de Witte, F.; Heremans, B.; Messens, K.; Höfte, M.; Haesaert, G. Fusarium head blight (FHB) in Flanders: Population diversity, inter-species associations and DON contamination in commercial winter wheat varieties. *Eur. J. Plant Pathol.* **2009**, *125*, 445–458. [CrossRef]

22. Spanic, V.; Lemmens, M.; Drezner, G. Morphological and molecular identification of *Fusarium* species associated with head blight on wheat in East Croatia. *Eur. J. Plant Pathol.* **2010**, *128*, 511–516. [CrossRef]

23. Jestoi, M.; Paavanen-Huhtala, S.; Parikka, S.; Yli-Mattila, T. *In vitro* and *in vivo* mycotoxin production of *Fusarium* species isolated from Finnish grains. *Arch. Phytopathol. Plant Protect.* **2008**, *41*, 545–558. [CrossRef]

24. Mishra, S.; Ansari, K.M.; Dwivedi, P.D.; Pandey, H.P.; Das, M. Occurrence of deoxynivalenol in cereals and exposure risk assessment in Indian population. *Food Control* **2013**, *30*, 549–555. [CrossRef]

25. Wegulo, S.N. Factors influencing deoxynivalenol accumulation in small grain cereals. *Toxins* **2012**, *4*, 1157–1180. [CrossRef]

26. Pestka, J.J. Deoxynivalenol: Mechanisms of action, human exposure and toxicological relevance. *Arch. Toxicol.* **2010**, *84*, 663–679. [CrossRef]

27. Zinedine, A.; Soriano, J.M.; Molto, J.C.; Manes, J. Review on the toxicity, occurrence, metabolism, detoxification, regulations and intake of zearalenone: An oestrogenic mycotoxin. *Food Chem. Toxicol.* **2007**, *45*, 1–18. [CrossRef]

28. Brown, N.A.; Antoniw, J.; Hammond-Kosack, K.E. The predicted secretome of the plant pathogenic fungus *Fusarium graminearum*: A refined comparative analysis. *PLoS ONE* **2012**, *7*, e33731.

29. Proctor, R.H.; Hohn, T.M.; McCormick, S.P.; Desjardins, A.E. Tri6 encodes an unusual zinc finger protein involved in regulation of trichothecene biosynthesis in *Fusarium sporotrichioides*. *Appl. Environ. Microbiol.* **1995**, *61*, 1923–1930.

30. Cuzick, A.; Urban, M.; Hammond-Kosack, K. *Fusarium graminearum* gene deletion mutants map1 and tri5 reveal similarities and differences in the pathogenicity requirements to cause disease on *Arabidopsis* and wheat floral tissue. *New Phytol.* **2008**, *177*, 990–1000. [CrossRef]

31. Kimura, M.; Kaneko, I.; Komiyama, M.; Takatsuki, A.; Koshino, H.; Yoneyama, K.; Yamaguchi, I. Trichothecene 3-O-acetyltransferase protects both the producing organism and transformed yeast from related mycotoxins. Cloning and characterization of Tri101. *J. Biol. Chem.* **1998**, *273*, 1654–1661.

32. Bin-Umer, M.A.; McLaughlin, J.E.; Basu, D.; McCormick, S.; Tumer, N.E. Trichothecene mycotoxins inhibit mitochondrial translation—Implication for the mechanism of toxicity. *Toxins* **2011**, *3*, 1484–1501. [CrossRef]

33. Torres, M.A.; Jones, J.D.G.; Dangl, J.L. Reactive oxygen species signaling in response to pathogens. *Plant Physiol.* **2006**, *141*, 373–437. [CrossRef]

34. Torres, M.A. ROS in biotic interactions. *Physiol. Plant* **2010**, *138*, 414–429. [CrossRef]

35. Chen, S.-X.; Schopfer, P. Hydroxyl-radical production in physiological reactions: A novel function of peroxidase. *Eur. J. Biochem.* **1999**, *260*, 726–735. [CrossRef]

36. Bradley, D.J.; Kjellbom, P.; Lamb, C.J. Elicitor- and wound-induced oxidative cross-linking of a proline-rich plant cell wall protein: A novel, rapid defense response. *Cell* **1992**, *70*, 21–30. [CrossRef]

37. Hückelhoven, R. Cell wall-associated mechanisms of disease resistance and susceptibility. *Annu. Rev. Phytopathol.* **2007**, *45*, 101–127. [CrossRef]

38. Gupta, R.K.; Banerjee, A.; Pathak, S.; Sharma, C.; Singh, N. Induction of mitochondrial-mediated apoptosis by *Morinda citrifolia* (noni) in human cervical cancer cells. *Asian Pac. J. Cancer Prev.* **2013**, *14*, 237–242. [CrossRef]

39. Tiessen, C.; Fehr, M.; Schwarz, C.; Baechler, S.; Domnanich, K.; Böttler, U.; Pahlke, G.; Marko, D. Modulation of the cellular redox status by the *Alternaria* toxins alternariol and alternariol monomethyl ether. *Toxicol. Lett.* **2013**, *216*, 23–30. [CrossRef]

40. Arunachalam, C.; Doohan, F.M. Trichothecene toxicity in eukaryotes: Cellular and molecular mechanisms in plants and animals. *Toxicol. Lett.* **2013**, *21*, 149–158. [CrossRef]

41. Gilchrist, D.G. Mycotoxins reveal connections between plants and animals in apoptosis and ceramide signaling. *Cell Death Differ.* **1997**, *4*, 689–698.

42. Dobosz, B.; Drzewiecka, K.; Waskiewicz, A.; Irzykowska, L.; Bocianowski, J.; Karolewski, Z.; Kostecki, M.; Kruczynski, Z.; Krzyminiewski, R.; Weber, Z.; *et al.* Free radicals, salicylic acid and mycotoxins in Asparagus after inoculation with *Fusarium proliferatum* and *F. oxysporum*. *Appl. Magn. Reson.* **2011**, *41*, 19–30. [CrossRef]

43. Nishiuchi, T.; Masuda, D.; Nakashita, H.; Ichimura, K.; Shinozaki, K.; Yoshida, S.; Kimura, M.; Yamaguchi, I.; Yamaguchi, K. *Fusarium* phytotoxin trichothecenes have an elicitor-like activity in *Arabidopsis thaliana*, but the activity differed significantly among their molecular species. *Mol. Plant Microbe Interact.* **2006**, *198*, 512–520.

44. Masuda, D.; Ishida, M.; Yamaguchi, K.; Yamaguchi, I.; Kimura, M.; Nishiuchi, T. Phytotoxic effects of trichothecenes on the growth and morphology of *Arabidopsis thaliana*. *J. Exp. Bot.* **2007**, *588*, 1617–1626.

45. Desmond, O.J.; Manners, J.M.; Stephens, A.E.; Maclean, D.J.; Schenk, P.M.; Gardiner, D.M.; Munn, A.L.; Kazan, K. The *Fusarium* mycotoxin deoxynivalenol elicits hydrogen peroxide production, programmed cell death and defence responses in wheat. *Mol. Plant Pathol.* **2008**, *9*, 435–445. [CrossRef]

46. Parsi, Z.; Gorecki, T. Determination of ergosterol as an indicator of fungal biomass in various samples using non-discriminating flash pyrolysis. *J. Chromatogr. A* **2006**, *1130*, 145–150.

47. Hippelein, M.; Rügamer, M. Ergosterol as an indicator of mould growth on building materials. *Int. J. Hyg. Environ. Health* **2004**, *207*, 379–385. [CrossRef]

48. Bhosle, S.R.; Sandhya, G.; Sonawane, H.B.; Vaidya, J.G. Ergosterol content of several wood decaying fungi using a modified method. *Int. J.Pharm. Life Sci.* **2011**, *2*, 916–918.

49. Bankole, S.A.; Adenusi, A.A.; Lawal, O.S.; Adesanya, O.O. Occurrence of aflatoxin B1 in food products derivable from "egusi" melon seeds consumed in southwestern Nigeria. *Food Control* **2010**, *7*, 974–976.

50. Janardhana, G.R.; Raveesha, K.A.; Shetty, H.S. Mycotoxin contamination of maize grains grown in Karnataka (India). *Food Chem. Toxicol.* **1999**, *37*, 863–868. [CrossRef]

51. Waśkiewicz, A.; Goliński, P.; Karolewski, Z.; Irzykowska, L.; Bocianowski, J.; Kostecki, M.; Weber, Z. Formation of fumonisins and other secondary metabolites by *Fusarium oxysporum* and *F. proliferatum*—A comparative study. *Food Addit. Contam.* **2010**, *27*, 608–615. [CrossRef]

52. Waśkiewicz, A.; Wit, M.; Goliński, P.; Chełkowski, J.; Warzecha, R.; Ochodzki, P.; Wakuliński, W. Kinetics of fumonisin B_1 formation in maize ears inoculated with *Fusarium verticillioides*. *Food Addit. Contam.* **2012**, *29*, 1752–1761. [CrossRef]

53. Waśkiewicz, A.; Irzykowska, L.; Bocianowski, J.; Karolewski, Z.; Weber, Z.; Goliński, P. Fusariotoxins in asparagus—their biosynthesis and migration. *Food Addit. Contam.* **2013**, *30*, 1332–1338. [CrossRef]

54. Barbehenn, R.V.; Poopat, U.; Spencer, B. Semiquinone and ascorbyl radicals in the gut fluids of caterpillars measured with EPR spectrometry. *Insect Biochem. Mol. Biol.* **2003**, *33*, 125–130. [CrossRef]

55. Bednarski, W.; Ostrowski, A.; Waplak, S. Low temperature short-range ordering caused by Mn^{2+} doping of $Rb_3H(SO_4)_2$. *J. Phys. Condens. Matter* **2010**, *22*. [CrossRef]

56. Morkunas, I.; Formela, M.; Floryszak-Wieczorek, J.; Marczak, Ł.; Narożna, D.; Nowak, W.; Bednarski, W. Cross-talk interactions of exogenous nitric oxide and sucrose modulates phenylpropanoid metabolism in yellow lupine embryo axes infected with *Fusarium oxysporum*. *Plant Sci.* **2013**, *211*, 102–121. [CrossRef]

57. Mai, V.C.; Bednarski, W.; Borowiak-Sobkowiak, B.; Wilkaniec, B.; Samardakiewicz, S.; Morkunas, I. Oxidative stress in pea seedling leaves in response to *Acyrthosiphon pisum* infestation. *Phytochemistry* **2013**, *93*, 49–62. [CrossRef]

58. Pearce, R.B.; Edwards, P.P.; Green, T.L.; Anderson, P.A.; Fisher, B.J.; Carpenter, T.A.; Hall, L.D. Immobilized long-lived free radicals at the host–pathogen interface in sycamore (*Acer pseudoplatanus* L.). *Physiol. Mol. Plant Pathol.* **1997**, *50*, 371–390. [CrossRef]

59. Doke, N. Involvement of superoxide anion generation in the hypersensitive response of potato tuber tissues to infection with an incompatible race of *Phytophthora infestans* and to the hyphal wall components. *Physiol. Mol. Plant Pathol.* **1983**, *23*, 345–355. [CrossRef]

60. Morkunas, I.; Garnczarska, M.; Bednarski, W.; Ratajczak, W.; Waplak, S. Metabolic and ultrastructural responses of lupine embryo axes to sugar starvation. *J. Plant Physiol.* **2003**, *160*, 311–319. [CrossRef]

61. Morkunas, I.; Bednarski, W.; Kozłowska, M. Response of embryo axes of germinating seeds of yellow lupine to *Fusarium oxysporum*. *Plant Physiol. Biochem.* **2004**, *42*, 493–499. [CrossRef]

toxins

MDPI

Article

Durum Wheat (*Triticum Durum* Desf.) Lines Show Different Abilities to Form Masked Mycotoxins under Greenhouse Conditions

Martina Cirlini [1], Silvia Generotti [2], Andrea Dall'Erta [1], Pietro Lancioni [3], Gianluca Ferrazzano [3], Andrea Massi [3], Gianni Galaverna [1] and Chiara Dall'Asta [1,*]

[1] Department of Food Science, University of Parma, Parco Area delle Scienze 95/A, Parma 43124, Italy; martina.cirlini@unipr.it (M.C.); andrea.dallerta@unipr.it (A.D.); gianni.galaverna@unipr.it (G.G.)

[2] Barilla G. R. F.lli SpA, Food Research Labs, Parma 43124, Italy; silvia.generotti@barilla.com

[3] Società Produttori Sementi Spa, Via Macero 1, Argelato 40050, Italy; pietro.lancioni@sygenta.com (P.L.); g.ferrazzano@prosementi.com (G.F.); a.massi@prosementi.com (A.M.)

* Author to whom correspondence should be addressed; chiara.dallasta@unipr.it; Tel.: +39-0521-905431; Fax: +39-0521-905472.

Received: 4 November 2013; in revised form: 10 December 2013; Accepted: 12 December 2013; Published: 24 December 2013

Abstract: Deoxynivalenol (DON) is the most prevalent trichothecene in Europe and its occurrence is associated with infections of *Fusarium graminearum* and *F. culmorum*, causal agents of Fusarium head blight (FHB) on wheat. Resistance to FHB is a complex character and high variability occurs in the relationship between DON content and FHB incidence. DON conjugation to glucose (DON-3-glucoside, D3G) is the primary plant mechanism for resistance towards DON accumulation. Although this mechanism has been already described in bread wheat and barley, no data are reported so far about durum wheat, a key cereal in the pasta production chain. To address this issue, the ability of durum wheat to detoxify and convert deoxynivalenol into D3G was studied under greenhouse controlled conditions. Four durum wheat varieties (Svevo, Claudio, Kofa and Neodur) were assessed for DON-D3G conversion; Sumai 3, a bread wheat variety carrying a major QTL for FHB resistance (QFhs.ndsu-3B), was used as a positive control. Data reported hereby clearly demonstrate the ability of durum wheat to convert deoxynivalenol into its conjugated form, D3G.

Keywords: masked mycotoxins; fusarium head blight; pasta; deoxynivalenol; virulence factor

1. Introduction

Fusarium head blight (FHB) is one of the most deleterious fungal diseases affecting wheat worldwide: it is related to infection by pathogenic fungi of the *Fusarium* spp. and it is widely diffused, especially in those areas with inductive climatic conditions (hot/warm temperatures and high/medium high humidity) [1–3]. FHB causes severe yield losses (up to 70%), affecting the quality of grains which show low protein content and color defects. Moreover, fungal infection may lead to the accumulation of mycotoxins: depending on the chemotype of the fungus, the type B trichothecenes deoxynivalenol (DON), nivalenol, 3-acetyldeoxynivalenol (3-ADON), and 15-acetyldeoxynivalenol (15-ADON) often accumulate in the developing grain [4].

This contamination is especially critical for durum wheat, which is used primarily for human consumption. Although the best economic and ecological strategy for reducing FHB damage is the utilization of resistant cultivars, the attempts to define durum wheat resistant lines have been unsuccessful so far.

In soft wheat, numerous QTL related to FHB resistance have been described [5]. In particular, several studies performed using the high resistant Chinese Spring wheat line Sumai-3 showed that the two most effective QTLs related to FHB resistance are positioned on the short arm of chromosome 3B (*Fhb1*) and on chromosome 5A (*Qfhs.ifa-5A*) [6–8].

Durum wheat cultivars were generally considered to be susceptible to FHB. Indeed, no variation in resistance to FHB has been found within *T. durum* lines, even among large germplasm collections of several thousand lines [9,10].

This fact may be due to several factors: a narrow genetic base compared to hexaploid wheat might be linked to the fact that durum wheat is tretraploid and that limited breeding efforts have been undertaken on this crop [11]. Attempts to transfer resistance from hexaploid into tetraploid wheat have been met with limited success [12,13].

Several types of resistance to FHB are known in wheat, classified as type I (resistance towards initial infection of spikelets), type II (resistance against spread of pathogen within spike) [14,15] and type III (resistance to DON accumulation in grains) [16,17].

In particular, DON was proven to inhibit protein synthesis in eukaryotic cells and acts as a virulence factor during fungal pathogenesis, therefore resistance to DON is considered an important component of resistance against FHB [18]. As reported by several studies, one mechanism of resistance to DON is the conversion of DON into the less toxic metabolites deoxynivalenol-3-*O*-glucoside (D3G) [19–21]. In particular, in a wheat population segregating for Fhb1, lines containing the Fhb1 resistance allele efficiently conjugate DON to the less toxic D3G [19]. In this work, the authors reported a good correlation between FHB resistance and DON conversion rate, expressed as [D3G]/[DON] ratio. This hypothesis was recently questioned by Gunnaiah *et al.* [22], whose study showed that DON resistance is not a major mechanism of FHB resistance associated with Nyubai Alleles of *Fhb1*.

The co-occurrence of DON and D3G in durum wheat harvested in Northern Italy was recently reported [23], showing a diffuse contamination of most samples with both compounds present at significant levels.

The present work is aimed at the study of the DON-to-D3G conversion ability of wheat lines (four durum wheat genotypes and one soft wheat genotype) under controlled greenhouse conditions.

In particular, the soft wheat line Sumai-3 was chosen as reference standard based on its well-known resistance towards FHB [24–26], while durum wheat lines (Kofa, Svevo, Neodur and Claudio) were chosen on account of their technological performances and because have been widely used in the most relevant durum breeding programs.

In a first trial, plants have been treated with *F. graminearum* and with DON under controlled growing conditions and samples have been analyzed for free and masked mycotoxins content. Then, those lines that had showed strong differences in the first trial were further considered in a second trial, just focusing on the DON-treatment at anthesis.

2. Results and Discussion

2.1. Set up of the First Trial

Two different trials were performed within this study, following the general scheme reported in Figure 1. The first trial involved four genotypes of *Triticum durum* (Kofa, Svevo, Claudio, Neodur) and one genotype of *T. aestivum* (Sumai-3), selected on the basis of their resistance towards Fusarium head blight (FHB) under in-field conditions and their genetic background. In particular, the DON detoxification ability related to FHB resistance was largely studied in soft wheat [27], while very few studies only reported the occurrence of D3G in durum wheat under natural infection conditions, so far [23].

We decided to compare contamination data obtained upon fungal conidia inoculation and DON-treatment, in consideration of the possible different effects pointed out by these treatment. In particular, the inoculation with *F. graminearum* at the flowering stage (Zadoks 60) [28] easily resembles the plant infection occurring in field, causing thus a similar plant-pathogen cross-talk. On the other hand, the direct treatment with DON, although extremely simplified compared to the natural phenomenon, allows to maximize those enzymatic mechanisms carried out by the plant to limit the DON toxic action during fungal infection.

Plants (*n* = 750) were split into three groups and underwent different treatments at the flowering stage, namely inoculation with fungal conidia suspension, treatment with DON solution and mock-treatment with distilled water (control group). Each group was sampled after five days and after 15 days from the treatment (see Figure 1a). Sampling times were chosen in order to highlight the DON detoxification ability in the first phases after the fungal infection, since reports published until now commonly focused on samples collected at harvest (full maturation stage) [19].

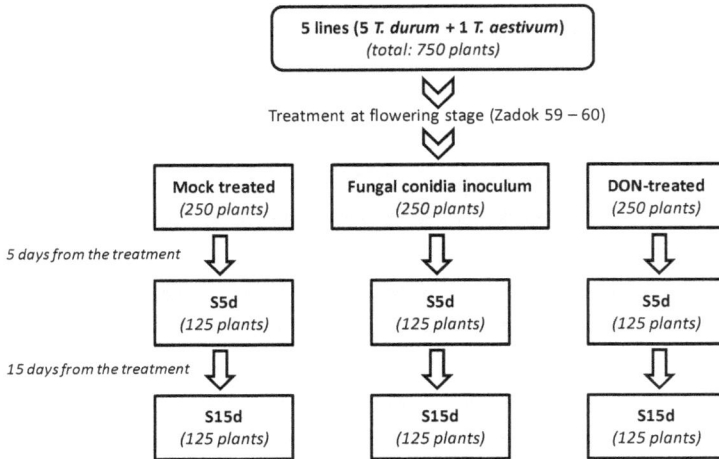

(a)

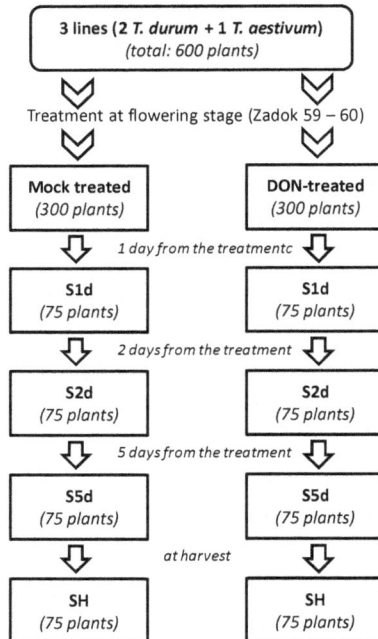

(b)

Figure 1. Experiment set up: (**a**) first trial; (**b**) second trial.

2.1.1. Fusarium graminearum *Treated Samples*

In the first trial, four selected durum wheat lines (Kofa, Svevo, Claudio and Neodur) and a soft wheat line (Sumai-3) were inoculated with *F. graminearum*, following the scheme reported in Figure 1a.

Since also other trichothecenes could be present besides DON and D3G, acetylated-DON derivatives as well as other type B and type A trichothecenes were also screened.

No detectable contamination was found in the control spikes, thus ensuring the reliability of the experiment, while all treated samples were found to be positive for both DON and D3G.

Acetylated-DON were found in most of the samples collected at S15d, while other type B and type A trichothecenes were not detected at any stage. Since the fungal growth was not comparable in the considered samples, the ergosterol content was used for data normalization. Ergosterol can be efficiently used as fungal growth biomarker as the artificially inoculated fungal strain was the only one responsible for the fungal biomass in the considered system.

Total contamination expressed as the sum of DON and analogues at S5d was in the range 4.8–17.4 mg/Kg, thus indicating a significant mycotoxin production already after five days from the fungal inoculation. Besides Claudio that showed a significantly higher toxin amount, the other wheat lines were found to be comparable in terms of total contamination. D3G was found in all the considered samples at levels ranging from 0.2 to 4.4 mg/Kg, with an average [D3G]/[DON] ratio of about 0.3.

Since significant differences between lines were not outlined at this stage, results obtained at S15d were better considered for the discussion.

Total contamination at S15d, expressed as the sum of DON, D3G and ADON, ranged from 49 to 168 mg/Kg, as reported in Table 1, thus indicating that the trend obtained at S5d stage was confirmed also after 15 days from the fungal inoculation.

Table 1. Total contamination (TD) expressed as the sum of deoxynivalenol (DON), deoxynivalenol-3-*O*-glucoside (D3G), acetyldeoxynivalenol (ADON) (mg/Kg) and ergosterol content determined in *F. graminearum* treated wheat lines at S15d stage. Data are referred to the first trial, described in Figure 1a.

Cultivars	TD (mg/Kg)	Ergosterol (mg/Kg)
Claudio	168 ± 34	22.0 ± 7.7
Neodur	90 ± 20	5.2 ± 1.8
Kofa	79 ± 5	4.0 ± 1.3
Svevo	129 ± 6	10.1 ± 0.7
Sumai-3	49 ± 6	2.5 ± 0.2

Among durum wheat varieties, Claudio and Svevo showed the highest contamination at S15d, while concentrations found in Neodur and Kofa were comparable. According to the ergosterol content, Claudio supported also the highest fungal growth compared to other lines.

Data, normalised on the ergosterol content and expressed as relative percentage of DON and D3G found in the five biological replicates considered for both maturation stages, are reported in Figure 2.

At first, from the plot, Sumai-3 showed the lower toxin accumulation compared to the durum wheat lines. This result is supported by a lower fungal growth compared to the other lines, measured as ergosterol content (see Table 1). These data are in agreement with the literature, since Sumai-3 is often reported as a highly resistant line towards FHB and DON accumulation [29,30]. Furthermore, the strongest toxin accumulation after 15 days from the inoculum was recorded for Claudio cv. also in this case in agreement with ergosterol data.

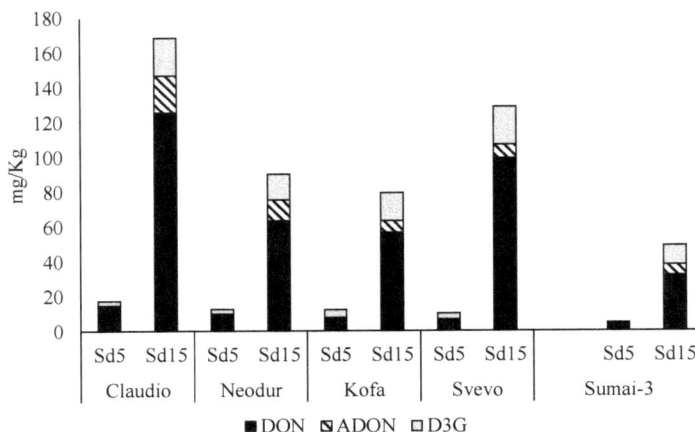

Figure 2. DON, ADON and D3G amount found in *F. graminearum* treated samples collected at S5d and S15d stages. Data referred to the first trial, described in Figure 1a.

Concentration ratio values obtained for durum wheat lines and expressed as the ratio between D3G and DON concentrations, were statistically compared by ANOVA analysis followed by a *post-hoc* Tuckey test; results are reported in Table 2. The sampling time was found significant for all the considered samples ($p = 0.000$), thus indicating that the DON conversion starts after the treatment and progressively increases during the maturation. Kofa and Svevo showed a good conversion rate (0.25 and 0.28, respectively), while [D3G]/[TDON] ratio was the lowest in Claudio (0.14).

Table 2. Conversion rate expressed as [D3G]/[TDON] (where TDON is the sum of DON, D3G and ADON) found in durum wheat lines at S15d stage. Different letters indicate significantly different values. Data are referred to the first trial, described in Figure 1a.

Cultivars	[D3G]/[TDON]	
	DON-Treated Samples	*F. graminearum Inoculated Samples*
Claudio	0.292 b	0.144 b
Neodur	0.371 a	0.182 b
Kofa	0.407 a	0.252 a
Svevo	0.273 b	0.282 a

2.1.2. DON-Treated Samples

Besides fungal inoculation, wheat lines were also treated with DON, as already performed by Lemmens *et al.* [19]. The use of a simplified system, although very far from the complicated cross-talk phenomena occurring upon in field fungal inoculation, allows distinguishing the possible differences in the transformation occurring in plants. Differently from Lemmens *et al.* [19], we decided to apply an amount of DON 10-times lower (total amount: 80 μg) to better resemble the common conditions experimented under in-field infection.

All the DON-treated and control spikes were separately analysed for DON and D3G content; each analysis was performed in duplicate.

As already reported for *F. graminearum* treated plants, no detectable DON contamination was found in the control spikes, while all DON-treated samples were found to be positive for both DON and D3G. Data, expressed as concentration of DON and D3G found in the five biological replicates at both maturation stages, are reported in Figure 3.

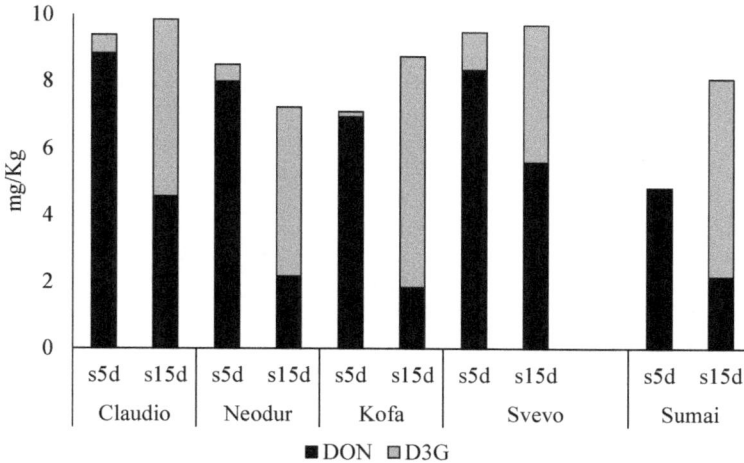

Figure 3. DON and D3G concentrations found in DON-treated samples collected at S5d and S15d stages. Data are referred to the first trial, described in Figure 1a.

As a general comment, all the conversion rates observed in the DON-treated samples were extremely higher than those reported for plants treated with *F. graminearum*. This could be explained considering two points: (a) at first, the toxin concentration levels reached in *F. graminearum* treated samples are definitely higher than those obtained in the DON-treatment; (b) then, the fungal growth under inductive conditions was strong, with severe pathogenesis symptoms. In consideration of these conditions, it is reasonable to deduce that the detoxification occurred at a lower extent.

D3G was found to be higher or comparable to DON at S15d stage in all the considered lines, but an important conversion was already noticed five days from the treatment (S5d).

The *Sumai-3* line confirmed the strong DON-to-D3G conversion ability already 15 days from the toxin treatment, being the ratio between D3G and total DON-related compounds (TDON) 0.731 at S15d.

Among durum wheat lines, Kofa showed the highest conversion ability, as reported in Table 2; on the other hand, Svevo showed the lowest conversion rate. It should be noticed that DON concentration was found higher than D3G only in Svevo samples.

As further observation, the total DON concentration at S15d was higher in Claudio and Svevo than in Neodur and Kofa, thus suggesting that possible other conjugation mechanisms may be active in plant.

In addition, differently from other lines, Svevo and Claudio also presented comparable total DON concentration at S5d and S15d: also in this case, lower bioconversion ability could be considered as a possible explanation.

Interestingly, our results clearly showed that D3G formation *in planta* already started five days from the treatment.

Similarly to the results obtained for *F. graminearum* treated samples, Kofa was found to show the highest conversion rate; on the contrary, Svevo showed a significantly higher conversion rate upon fungal inoculation compared to what was observed upon DON-treatment. This fact could be due to the absence of fungal growth, and should be carefully considered in further studies.

2.2. Set up of the Second Trial

Starting from the results obtained in the first trial, a second experiment was set up. According to the data obtained, Kofa and Svevo durum wheat lines showed the most different detoxification ability and thus were selected for further investigation, in comparison to Sumai-3.

Plants ($n = 600$) were split into two groups and alternatively underwent DON-treatment and mock-inoculation (control group) at the flowering stage (see Figure 1b).

DON-treated plants were compared to control group by collecting samples one day (Sd1), two days (Sd2), and five days (Sd5) from the treatment, and at harvest. A recent study reported that UDP-glucosyltransferase proteins in Sumai-3, involved in the DON detoxification process, are over-expressed already 32–48 h from the fungal inoculation [29]; thus, short times after treatment were considered to better understand the D3G formation at the early-stage. In addition, spikes at full maturation were harvested and analysed to evaluate the final content of DON and D3G.

DON-Treated Samples

The general trend observed in the first trial was confirmed, although this time the conversion rate at harvest was lower for the considered lines, as reported in Table 3. This fact may be due to the inter-individual differences between plants and to the slightly different environmental conditions not directly dependent from the controlled agronomical parameters (e.g., external climatic factors), experimented in the greenhouse during the trials.

Table 3. Conversion rate expressed as [D3G]/[TDON] (where TDON is the sum of DON, D3G and ADON) obtained at each sampling stage for the DON-treated lines considered in the second trial.

Sampling dates	Kofa	Svevo	Sumai-3
Sd1	0.046	0.056	0.246
Sd2	0.391	0.154	0.655
Sd5	0.421	0.115	0.619
SH	0.530	0.312	0.843

Conversion rate values were statistically compared by ANOVA followed by *post-hoc* Tuckey test. The sampling stage was found to be significant for all the considered lines ($p = 0.000$). D3G was found in all the analysed samples already after 24 hours from the treatment, thus indicating that the detoxification pathway is promptly active towards DON in plant. This fact is in agreement with the expression data recently reported by Gottwald *et al.* [29] for FHB resistant soft wheat lines, among those Sumai-3. In particular, the authors stated that the expression profiles of genes related to detoxification processes in resistant genotypes are inducted in the early stage, while susceptible lines typically show late inductions.

As reported in Table 3, Sumai-3 showed the highest conversion rate already after 24 h from the treatment and its detoxification ability is confirmed as very high at harvest. Concerning durum wheat, also in this second trial, Kofa showed a higher detoxification ability compared to Svevo already after two days from the treatment ($p = 0.003$ at Sd2) and this difference became particularly evident at harvest. The data here presented are clearly in agreement with Gottwald *et al.* [29], since Svevo, which is usually assumed to be less resistant than Kofa towards FHB, showed a delayed conversion of DON to D3G.

3. Experimental Section

3.1. Chemicals

Sodium acetate, formic acid, methanol, hexane, ethyl acetate, bis-silyltrifluoroacetamide (BSTFA) and acetonitrile were purchased from Sigma-Aldrich (Taufkirchen, Germany). All solvents were HPLC grade. Ultra-pure water was in-house produced by using a Milli-Q System (Millipore, Bedford, MA, USA). The following mycotoxin standard solutions were purchased from RomerLabs (Tulln, Austria): ^{13}C-Deoxynivalenol (internal standard, 25 mg/L in acetonitrile), D3G (50 mg/L in acetonitrile), mix A + B trichothecenes containing DON, NIV, T-2 and HT-2 (10 mg/L each in acetonitrile). All standard solutions were stored at $-20\ ^\circ$C and brought to room temperature before use. A working solution (2.5 μg/mL) containing all the target mycotoxins was prepared in water/methanol (80/20, v/v) by combining and properly diluting suitable aliquots of each standard. Ergosterol and cholestanol were purchased from Sigma-Aldrich (Taufkirchen, Germany). *F. graminearum* strain was purchased from Department of Agronomy (DiPSA) University of Bologna.

3.2. Wheat Lines Background Information

For this study, four durum wheat lines (Kofa, Svevo, Neodur and Claudio) and one soft wheat genotype (Sumai-3) have been chosen.

Kofa, is a Southwestern United States cv. released by Western Plant Breeders (Arizona) obtained from a population based on multiple parents (dicoccum alpha pop-85 S-1) mainly related to the American and CIMMYT germplasm, with the inclusion of emmer accessions. Svevo, an Italian cv. released by Società Produttori Sementi, has been obtained from the cross between a CIMMYT line (pedigree rok/fg//stil/3/dur1/4/sapi/teal//hui) related to the cv. Yavaros, genetic background (Jori/Anhinga//Flamingo), and the cv. Zenit originating from a cross between Italian and American accessions (Valriccardo/Vic). Both Kofa and Svevo are well adapted to the Mediterranean climate and can be classified as early-flowering genotypes in such conditions, and are susceptible to FHB. Neodur (pedigree 184-7/Valdur//Edmore) is a late flowering variety cultivated in north Italy carrying a major QTL for resistance against the soil-borne cereal mosaic virus. Claudio (pedigree Sel.Cimmyt35/Durango//ISEA1938xGrazia) is a medium late variety widely cultivated in Italy and France well known for its yield stability. Both Neodur and Claudio can be considered moderately tolerant to FHB under Mediterranean growing conditions.

The four elite varieties of durum wheat (Svevo, Claudio, Kofa e Neodur) were selected for a preliminary study on the ability to convert DON into D3G. Lines were initially chosen based on the different susceptibility to DON accumulation and to FHB symptoms, Kofa and Svevo showing medium susceptibility and Claudio and Neodur medium tolerance. Moreover, the selected lines well represent the genetic diversity of the main ameliorated pool of durum wheat adapted to grow in the Mediterranean area. The soft wheat cv. Sumai 3 was used as control for its known resistance ability towards FHB (type II resistance, linked to the QTL *QFhs.ndsu-3B*).

3.3. Fusarium graminearum Strain Background Information

Fusarium graminearum strain F566 was used for plant inoculation; the strain was isolated from wheat samples harvested in Northern Italy (Bologna area, Emilia Romagna region); the chemotype was fully characterized as reported by Prodi *et al.* [30], as well as its aggressiveness [31]. The strain is conserved in the Phytopatological Mycology Collection of the Department of Agricultural Sciences of the University of Bologna.

3.4. Design of the Greenhouse Experiment

The artificial inoculation experiment on durum wheat plants has been carried out according to a RCB (Randomized Complete Block) scheme within a greenhouse plant at Produttori Sementi S.p.A. (Argelato, Bologna, Italy). All the plants were singularly sown and maintained at low temperature (5 °C) for 30 days to mimic vernal conditions (vernalization), then they were transferred into the greenhouse, fixing the ambient conditions in order to mimic first the spring period and then the summer season. For the first experiment, four lines of durum wheat (Svevo, Claudio, Kofa e Neodur) and one line of soft wheat chosen as a reference line for its resistance to FHB (Sumai 3) were selected.

For each line, four groups were considered, each formed by five plants: at the flowering stage (zadok 59–60) the first group was artificially inoculated with a *F. graminearum* inoculum, the second group was contaminated with a DON solution, whereas the third group was considered as the control one. All the contaminated groups, as well as the negative (not contaminated) control, were replicated five times. The entire experiment was performed in triplicate in order to allow two different sampling steps at two different maturation stages. Sampling steps were fixed from the inoculation step as follows: after five days (S5d) and after 15 days (S15d). The total plant number was 750. The treatment was performed on four spikelets in the central part of the spike: each spikelet was inoculated with 10 µL of a fungal conidia solution (1×10^5 conidia/mL) or with 20 µL DON standard solution (exact title: 0.828 mg/mL). The control samples were mock treated with distilled water.

The second trial was focused on two lines of durum wheat (Svevo, and Kofa) and one line of soft wheat (Sumai 3). In this case, four sampling steps at four maturation stages were considered. Sampling steps were fixed from the treatment step as follows: after one day (S1d), after two days (S2d), after five days (S5d) and at harvest (SH). The total number of plants was 600. The treatment with DON solution was performed as above. Control plants were mock-treated with distilled water.

At each sampling stage, spikes were separately collected and milled, before LC-MS analysis. For fungal contaminated samples, data were normalized on the amount of fungal biomass developed, using ergosterol as fungal growth biomarker.

3.5. Free and Masked Mycotoxins Analysis

Samples were prepared according to Berthiller *et al.* [32] with slight modifications. Briefly, the whole ears were finely grounded and mixed 0.5 g of grounded wheat was extracted for 90 min at 200 strokes/min on a shaker with 2 mL of acetonitrile/water (80/20, v/v) acidified with 0.1% of formic acid. The extract was collected and centrifuged for 10 min at 1,4000 rpm at room temperature, then 1 mL was evaporated to dryness under nitrogen. After addition of the internal standard (^{13}C-DON, 20 µL), the residue was dissolved in 1 mL of water/methanol (80/20, v/v) and analyzed by UPLC-ESI/MS.

The UPLC-ESI/MS analyses were carried out according to Dall'Asta *et al.* [23], using an Acquity UPLC separation system (Waters Co., Milford, MA, USA) equipped with an Acquity Single Quadrupole MS detector with an electrospray source. Chromatographic conditions were as follows: column, Acquity UPLC BEH C18 (1.7 µm, 2.1 × 50 mm); flow rate, 0.35 mL/min; column temperature, 40 °C; injection volume, 5 µL; gradient elution using 0.1 mM sodium acetate solution in water (eluent A) and methanol (eluent B), both acidified with 0.2% formic acid. Gradient conditions: initial conditions were set at 2% B for 1 min, then eluent B was increased to 20% in 1 min; after an isocratic step (6 min), eluent B was increased to 90% in 9 min; after a 3 min isocratic step, the system was re-equilibrated to initial conditions for 3 min. The total analysis time was 23 min. The ESI source was operated in positive ionization mode. MS parameters were as follows: capillary voltage, 2.50 kV; cone, 30 V; source block temperature, 120 °C; desolvation temperature, 350 °C; cone gas, 50 L/h; desolvation gas, 850 L/h. Detection was performed using single ion monitoring mode and monitoring the [M + Na]$^+$ ion, as reported by Dall'Asta *et al.* [23].

Matrix-matched calibration curves (calibration range 100–2500 µg/kg) were used for target analyte quantification. A good linearity was obtained for all the considered mycotoxins ($r^2 > 0.99$). For all the target compounds, limit of quantification (LOQ) and limit of detection (LOD) were lower than 30 µg/kg and 10 µg/kg, respectively.

3.6. Ergosterol Extraction and Analysis

Lipid fraction was extracted from wheat by stirring 1 g of milled sample in 20 mL of n-hexane for 1 h at room temperature. After that, the extract was filtered on paper filter and dried under vacuum.

Lipid fraction, containing fatty acids and sterols, were dissolved in 1 mL of hexane, added of 0.3 mL of an internal standard solution (cholestanol, 100 µg/mL) and transesterified by adding 1 mL of 5% KOH in methanol and shaking vigorously. The organic phase containing ergosterol was then purified on a silica gel cartridge eluting the fraction of interest with 3 mL of ethyl acetate, in accordance with Annaratone *et al.* [33]. The residue was dissolved in 0.5 mL of hexane, with 50 μL of BSTFA added, and reacted at 60 °C for 1 h. Analysis and quantification of ergosterol was performed by GC-MS, injecting 1 μL of the obtained solution.

GC-MS analysis was performed on a 6890N gas chromatograph coupled to a 5973N mass selective detector (Agilent technologies, Santa Clara, CA, USA), using a SLB-5MS capillary column (Supelco, Bellefonte, PA, USA). The programmed temperature gradient was as follows: 240 °C for 3 min, then increasing to 280 °C in 2 min and holding this final condition for 20 min. Injection was performed in split mode. Data acquisition mode was SIM, monitoring 460.4 and 445.4 m/z for cholestanol (internal standard) and 468.4 and 363.3 m/z for ergosterol detection.

4. Conclusions

The study presented herein showed for the first time the DON-to-D3G conversion ability of several durum wheat lines in comparison to Sumai-3 soft wheat under greenhouse conditions. In addition, for the first time, the D3G formation in plants already 24 h from the DON treatment and five days from the fungal infection was reported. Among the considered lines, Claudio cv. supported the highest fungal growth and the highest DON accumulation upon *F. graminearum* inoculation, showing also the lowest [D3G]/[TDON] ratio. This result was further confirmed upon DON-treatment, thus suggesting that low glycosylation ability is also related to a high susceptibility towards fungal infection and toxin accumulation, at least under greenhouse conditions. Furthermore, Kofa cv. showed a good glucosylation activity towards DON in both the assays and is a promising candidate for further studies to better define the DON detoxification activity in durum wheat.

Data here represented support thus the hypothesis reported by Lemmens *et al.* [19]: the ability to convert DON to D3G in these lines seems actually to be related to their resistance towards FHB. The role played by D3G as DON detoxification products in durum wheat was demonstrated, thus pinpointing as well the necessity for careful monitoring of this masked mycotoxin in durum wheat.

Acknowledgments: This work was supported by AGER-From Seed to Pasta (2009–2012). The authors kindly acknowledge Antonio Prodi and Andrea Tonti from Department of Agronomy (DiPSA), University of Bologna for the preparation of *Fusarium* strains. The authors are also grateful to Alessandro Tonelli, Department of Food Science, University of Parma, for the technical assistance and to Michele Suman, Barilla G. R. F.lli SpA, Food Research Labs for the fruitfull discussion.

Conflicts of Interest: The authors declare no conflict of interest.

References

1. McMullen, M.; Jones, R.; Gallenberg, D. Scab of wheat and barley: A re-emerging disease of devastating impact. *Plant Dis.* **1997**, *81*, 1340–1348. [CrossRef]
2. Champeil, A.; Doré, T.; Fourbet, J.F. Fusarium head blight: Epidemiological origin of the effects of cultural practices on head blight attacks and the production of mycotoxins by Fusarium in wheat grains. *Plant Sci.* **2004**, *166*, 1389–1415.
3. Leonard, K.J.; Bushnell, W.R. *Fusarium Head Blight of Wheat and Barley*; APS Press: Saint Paul, MN, USA, 2003.
4. Desjardins, A.E.; Hohn, T.M.; McCormick, S.P. Trichothecene biosynthesis in Fusarium species: Chemistry, genetics, and significance. *Microbiol. Rev.* **1993**, *57*, 595–604.
5. Buerstmayr, H.; Ban, T.; Anderson, J.A. QTL mapping and marker-assisted selection for Fusarium head blight resistance in wheat: A review. *Plant Breed.* **2009**, *128*, 1–26. [CrossRef]
6. Anderson, J.A.; Stack, R.W.; Liu, S.; Waldron, B.L.; Fjeld, A.D.; Coyne, C.; Moreno-Sevilla, B.; Fetch, J.M.; Song, Q.J.; Cregan, P.B.; *et al.* DNA markers for Fusarium head blight resistance QTLs in two wheat populations. *Theor. Appl. Genet.* **2001**, *102*, 1164–1168. [CrossRef]
7. Buerstmayr, H.; Lemmens, M.; Hartl, L.; Doldi, L.; Steiner, B.; Stierschneider, M.; Ruckenbauer, P. Molecular mapping of QTLs for Fusarium head blight resistance in spring wheat. I. Resistance to fungal spread (type II resistance). *Theor. Appl. Genet.* **2002**, *104*, 84–91. [CrossRef]

8. Buerstmayr, H.; Stierschneider, M.; Steiner, B.; Lemmens, M.; Griesser, M.; Nevo, E.; Fahima, T. Variation for resistance to head blight caused by Fusarium graminearum in wild emmer (Triticum dicoccoides) originating from Israel. *Euphytica* **2003**, *130*, 17–23. [CrossRef]

9. Stack, R.W.; Elias, E.M.; Mitchell Fetch, J.; Miller, J.D.; Joppa, L.R. Fusarium head blight reaction of Langdon durum-Triticum dicoccoides chromosome substitution lines. *Crop Sci.* **2002**, *42*, 637–642. [CrossRef]

10. Somers, D.J.; Fedak, G.; Clarke, J.; Cao, W. Mapping of FHB resistance QTLs in tetraploid wheat. *Genome* **2006**, *49*, 1586–1593. [CrossRef]

11. Oliver, R.E.; Cai, X.; Friesen, T.L.; Halley, S.; Stack, R.W.; Xu, S.S. Evaluation of Fusarium head blight resistance in tetraploid wheat (Triticum turgidum L.). *Crop Sci.* **2008**, *48*, 213–222. [CrossRef]

12. Gilbert, J.; Procunier, J.D.; Aung, T. Influence of the D genome in conferring resistance to fusarium head blight in spring wheat. *Euphytica* **2000**, *114*, 181–186. [CrossRef]

13. Oliver, R.E.; Stack, R.W.; Miller, J.D.; Cai, X. Reaction of wild emmer wheat accessions to fusarium head blight. *Crop Sci.* **2007**, *47*, 893–899. [CrossRef]

14. Lu, Q.; Lillemo, M.; Skinnes, H.; He, X.; Shi, J.; Ji, F.; Dong, Y.; Bjørnstad, A. Anther extrusion and plant height are associated with Type I resistance to Fusarium head blight in bread wheat line "Shanghai-3/Catbird". *Theor. Appl. Genet.* **2013**, *126*, 317–334. [CrossRef]

15. Lin, F.; Xue, S.L.; Zhang, Z.Z.; Zhang, C.Q.; Kong, Z.X.; Yao, G.Q.; Tian, D.G.; Zhu, H.L.; Li, C.J.; Cao, Y.; et al. Mapping QTL associated with resistance to Fusarium head blight in the Nanda2419 × Wangshuibai population. II: Type I resistance. *Theor. Appl. Genet.* **2006**, *112*, 528–535. [CrossRef]

16. Buerstmayr, M.; Alimari, A.; Steiner, B.; Buerstmayr, H. Genetic mapping of QTL for resistance to Fusarium head blight spread (type 2 resistance) in a Triticum dicoccoides × Triticum durum backcross-derived population. *Theor. Appl. Genet.* **2013**, *126*, 2825–2834. [CrossRef]

17. Lin, F.; Kong, Z.X.; Zhu, H.L.; Xue, S.L.; Wu, J.Z.; Tian, D.G.; Wei, J.B.; Zhang, C.Q.; Ma, Z.Q. Mapping QTL associated with resistance to Fusarium head blight in the Nanda2419 × Wangshuibai population. I. Type II resistance. *Theor. Appl. Genet.* **2004**, *109*, 1504–1511. [CrossRef]

18. Arunachalam, C.; Doohan, F.M. Trichothecene toxicity in eukaryotes: Cellular and molecular mechanisms in plants and animals. *Toxicol. Lett.* **2013**, *217*, 149–158. [CrossRef]

19. Lemmens, M.; Scholz, U.; Berthiller, F.; Dall'Asta, C.; Koutnik, A.; Schumacher, R.; Adam, G.; Buerstmayr, H.; Mesterhazy, A.; Krska, R.; et al. The ability to detoxify the mycotoxin deoxynivalenol colocalizes with a major quantitative trait locus for Fusarium Head Blight resistance in wheat. *Mol. Plant Microbe. Interact.* **2005**, *18*, 1318–1324. [CrossRef]

20. Schweiger, W.; Boddu, J.; Shin, S.; Poppenberger, B.; Berthiller, F.; Lemmens, M.; Muehlbauer, G.J.; Adam, G. Validation of a candidate deoxynivalenol-inactivating UDP-glucosyltransferase from barley by heterologous expression in yeast. *Mol. Plant Microbe Interact.* **2010**, *23*, 977–986. [CrossRef]

21. Poppenberger, B.; Berthiller, F.; Lucyshyn, D.; Sieberer, T.; Schuhmacher, R.; Krska, R.; Kuchler, K.; Glössl, J.; Luschnig, C.; Adam, G. Detoxification of the Fusarium mycotoxin deoxynivalenol by a UDP-glucosyltransferase from Arabidopsis thaliana. *J. Biol. Chem.* **2003**, *278*, 47905–47914. [CrossRef]

22. Gunnaiah, R.; Kushalappa, A.C.; Duggavathi, R.; Fox, S.; Somers, D.J. Integrated metabolo-proteomic approach to decipher the mechanisms by which wheat QTL (Fhb1) contributes to resistance against *Fusarium graminearum*. *PLoS One* **2012**, *7*. [CrossRef]

23. Dall'Asta, C.; Dall'Erta, A.; Mantovani, P.; Massi, A.; Galaverna, G. Occurrence of deoxynivalenol and deoxynivalenol-3-glucoside in durum wheat. *World Mycotoxin J.* **2013**, *6*, 83–91. [CrossRef]

24. Zhuang, Y.; Gala, A.; Yen, Y. Identification of functional genic components of major fusarium head blight resistance quantitative trait loci in wheat cultivar Sumai 3. *Mol. Plant Microbe Interact.* **2013**, *26*, 442–450. [CrossRef]

25. Zhou, M.P.; Hayden, M.J.; Zhang, Z.Y.; Lu, W.Z.; Ma, H.X. Saturation and mapping of a major Fusarium head blight resistance QTL on chromosome 3BS of Sumai 3 wheat. *J. Appl. Genet.* **2010**, *51*, 19–25. [CrossRef]

26. Golkari, S.; Gilbert, J.; Ban, T.; Procunier, J.D. QTL-specific microarray gene expression analysis of wheat resistance to Fusarium head blight in Sumai-3 and two susceptible NILs. *Genome* **2009**, *52*, 409–418. [CrossRef]

27. Karlovsky, P. Biological detoxification of the mycotoxin deoxynivalenol and its use in genetically engineered crops and feed additives. *Appl. Microbiol. Biotechnol.* **2011**, *91*, 491–504. [CrossRef]

28. Zadoks, J.C.; Chang, T.T.; Konzak, C.F. A decimal code for the growth stages of cereals. *Weed Res.* **1974**, *14*, 415–421. [CrossRef]

29. Gottwald, S.; Samans, B.; Lück, S.; Friedt, W. Jasmonate and ethylene dependent defence gene expression and suppression of fungal virulence factors: Two essential mechanisms of Fusarium head blight resistance in wheat? *BMC Genomics* **2012**, *13*, 369. [CrossRef]
30. Prodi, A.; Tonti, S.; Nipoti, P.; Pancaldi, D.; Pisi, A. Identification of deoxynivalenol and nivalenol producing chemotypes of Fusarium graminearum isolates from durum wheat in a restricted area of northern Italy. *J. Plant Pathol.* **2009**, *91*, 727–731.
31. Purahong, W.; Nipoti, P.; Pisi, A.; Lemmens, M.; Prodi, A. Aggressiveness of different Fusarium graminearum chemotypes within a population from northern-central Italy. *Mycoscience* **2013**. [CrossRef]
32. Berthiller, F.; Schuhmacher, R.; Buttinger, G.; Krska, R. Rapid simultaneous determination of major type A- and B-trichothecenes as well as zearalenone in maize by high performance liquid chromatography–tandem mass spectrometry. *J. Chrom. A* **2005**, *1062*, 209–216.
33. Annaratone, C.; Caligiani, A.; Cirlini, M.; Toffanin, L.; Palla, G. Sterols, sterol oxides and CLA in typical meat products from pigs fed with different diets. *Czech J. Food Sci.* **2009**, *27*, 219–222.

toxins

MDPI

Article

Correlation of ATP Citrate Lyase and Acetyl CoA Levels with Trichothecene Production in *Fusarium graminearum*

Naoko Sakamoto [1], Rie Tsuyuki [1], Tomoya Yoshinari [2], Jermnak Usuma [1], Tomohiro Furukawa [1], Hiromichi Nagasawa [1] and Shohei Sakuda [1,*]

[1] Department of Applied Biological Chemistry, University of Tokyo, 1-1-1 Yayoi, Bunkyo-ku, Tokyo 113-8657, Japan; m.n.s@pony.ocn.ne.jp (N.S.); rie_yoshinari@takasago.com (R.T.); plenamioga@hotmail.com (J.U.); mountainxtc@live.jp (T.F.); anagahi@mail.ecc.u-tokyo.ac.jp (H.N.)

[2] National Institute of Health Sciences, 1-18-1 Kamiyoga, Setagaya-ku, Tokyo 158-8501, Japan; t-yoshinari@nihs.go.jp

* Author to whom correspondence should be addressed; asakuda@mail.ecc.u-tokyo.ac.jp; Tel.: +81-3-5841-5133; Fax: +81-3-5841-8022.

Received: 1 November 2013; in revised form: 18 November 2013; Accepted: 18 November 2013; Published: 20 November 2013

Abstract: Thecorrelation of ATP citrate lyase (ACL) and acetyl CoA levels with trichothecene production in *Fusarium graminearum* was investigated using an inhibitor (precocene II) and an enhancer (cobalt chloride) of trichothecene production by changing carbon sources in liquid medium. When precocene II (30 μM) was added to inhibit trichothecene production in a trichothecene high-production medium containing sucrose, ACL expression was reduced and *ACL* mRNA level as well as acetyl CoA amount in the fungal cells were reduced to the levels observed in a trichothecene trace-production medium containing glucose or fructose. The *ACL* mRNA level was greatly increased by addition of cobalt chloride in the trichothecene high-production medium, but not in the trichothecene trace-production medium. Levels were reduced to those level in the trichothecene trace-production medium by addition of precocene II (300 μM) together with cobalt chloride. These results suggest that ACL expression is activated in the presence of sucrose and that acetyl CoA produced by the increased ALC level may be used for trichothecene production in the fungus. These findings also suggest that sucrose is important for the action of cobalt chloride in activating trichothecene production and that precocene II may affect a step down-stream of the target of cobalt chloride.

Keywords: ATP citrate lyase; precocene II; acetyl CoA; trichothecene production; cobalt chloride; sucrose; *Fusarium graminearum*

1. Introduction

Fusarium graminearum is a worldwide predominant plant pathogen that causes Fusarium head blight of wheat and other grain cereals and produces trichothecene mycotoxins in infected grains. Contamination of important cereal crops by trichothecenes, mainly deoxynivalenol and nivalenol, is a serious human and livestock health concern and also has the potential to cause drastic economic consequences [1]. Presently, the use of fungicide is the most effective method for controlling contamination, but inhibition of fungal growth may lead to the spread of resistant fungal strains [2]. Therefore, it is critical to develop other effective means for controlling trichothecene contamination. To determine the optimal target for developing an effective method, a better understanding of the basic regulatory mechanisms for trichothecene production in the fungus is required.

Trichothecenes are biosynthesized from farnesyl pyrophosphate, which is produced *via* the mevalonate pathway. Mevalonate, the key intermediate in the pathway, is biosynthesized from three acetyl CoA molecules. A number of *Tri* genes are responsible for trichothecene biosynthesis from farnesyl pyrophosphate [3]. In *F. graminearum*, *Tri6* encodes a key regulatory protein for trichothecene biosynthesis [4,5]. The expression of *Tri* genes encoding trichothecene biosynthetic enzymes is under the positive control of TRI6. TRI6 also up-regulates the expression of genes encoding enzymes involved in the mevalonate pathway [6]. Therefore, the overall biosynthetic pathway from acetyl CoA to trichothecenes is activated by TRI6. However, the upstream events that lead to TRI6 expression have not yet been clarified.

Supply of acetyl CoA to the biosynthetic pathway may be necessary for trichothecene production, but the mechanism by which acetyl CoA is supplied is not clear. Three pathways, in which acetyl CoA synthetase, carnitine acetyltransferase, or ATP citrate lyase (ACL) is a key enzyme, are known to produce acetyl CoA in the fungal cytosol [7]. Occurrence of ACL and carnitine acetyltransferase activity was shown to be spread widely in filamentous fungi [8]. It has been shown that each subunit of ACL of *F. graminearum* is encoded by two genes (*ACL1* and *ACL2*) and deletion of *ACL1* and/or *ACL2* results in reduction of trichothecene production in the fungus [9]. It also has been shown that two carnitine acetyltransferases (CAT1 and CAT2) are present in the fungus and deletion of *CAT1* decreases trichothecene production [10]. In our proteome analysis studies on the mode of action of precocene II, a trichothecene production inhibitor, we found that ACL2 expression was reduced by precocene II. In a microarray experiment with the *Tri6* deletion mutant, *ACL* mRNA levels were not affected by the deletion, suggesting that its expression is not under the control of TRI6 [5]. Although acetyl CoA levels are thought to be a key factor for trichothecene production, the relationship of ACL expression levels or acetyl CoA amounts with trichothecene production has not been investigated. Therefore, in this study we investigated the relationship using factors that affect trichothecene production.

Carbon source is an important factor for trichothecene production in *F. graminearum* [11]. Sucrose and some oligosaccharides containing a sucrose moiety are necessary for active production of trichothecene. Other sugars, such as glucose or fructose, do not induce the production. The molecular mechanism of sucrose for induction of trichothecene production is not clear. Precocene II and cobalt chloride are known as a specific inhibitor and enhancer for trichothecene production of *F. graminearum*, respectively [12,13]. Precocene II reduced the mRNA levels of *Tri6*, whereas cobalt chloride enhanced levels. The target molecules of precocene II and cobalt chloride have not yet been elucidated.

In this paper, we describe the results of the proteome analysis and experiments designed to investigate the correlation of the ACL expression level and acetyl CoA amount with trichothecene production in *F. graminearum* using precocene II and/or cobalt chloride under different carbon sources.

2. Results and Discussion

2.1. Effects of Precocene II on ACL Expression

Fusarium graminearum MAFF101551 accumulates 3-acetyldeoxynivalenol (3-ADON) in its culture supernatant in SYEP liquid medium containing sucrose as a carbon source. The production of 3-ADON by the strain begins 2 days after cultivation and reaches a plateau at 4 days of cultivation in the medium [12]. Precocene II, a constituent of essential oils, almost completely inhibited 3-ADON production of strain MAFF101551 at a concentration of 30 μM without affecting fungal growth and ergosterol production. Precocene II may affect a regulatory mechanism leading to expression of TRI6 in the trichothecene production pathway in the fungus. To investigate the mode of action of precocene II, proteome analysis by two-dimensional differential gel electrophoresis (2D-DIGE) was performed [14,15].

Strain MAFF101551 was incubated without or with precocene II (30 μM) for 2 or 4 days and the proteins were extracted from the cells. Control and precocene II-treated samples were labeled with Cy3 (green) and Cy5 (red), respectively, and combined. The mixture was separated by two-dimensional gel electrophoresis followed by fluorescent imaging (Figure 1). Green spots (decreased with precocene II treatment) and red spots (increased with precocene II treatment), showing changes in abundance greater than 1.2-fold, were clearly observed on the gels. Proteins in the spots were identified by

MALDI-TOFMS analysis after in-gel tryptic digestion and by MASCOT research (Table 1). Among the identified proteins, diphosphomevalonate decarboxylase (spot 11 on the sample gel from day 4), an enzyme involved in the mevalonate pathway, was identified as a protein that had decreased expression after precocene II treatment. Since expression of diphosphomevalonate decarboxylase is promoted by TRI6, inhibition of TRI6 expression by precocene II may lead to a decrease of the protein expression level. We also focused on the decrease of ACL (ACL2 encoded by FGSG_06039.3, [8]) expression (spot 5 on sample gels from days 2 and 4) because ACL was a key enzyme for acetyl CoA supply into the cytosol as described above. The relationship of the other identified proteins listed in Table 1 with precocene II's function is currently under investigation.

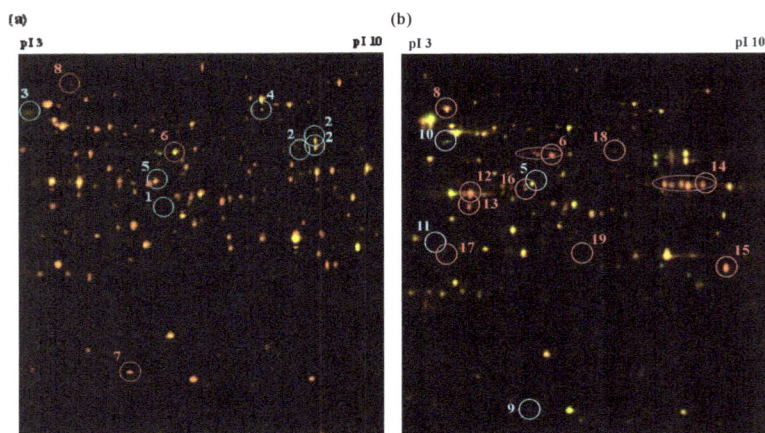

Figure 1. A two-dimensional differential gel electrophoresis (2D-DIGE) gel image of *F. graminearum* proteome after precocene II treatment for 2 days (**a**) or 4 days (**b**). Green, red, and yellow spots indicate proteins with decreased, increased, and unchanged abundance levels, respectively. Selected proteins (>1.2 and <1.2 change in abundance; $p < 0.05$; $n = 3$) are numbered.

Table 1. Analysis of the effect of precocene II by 2D-DIGE.

Spot	Protein	Fold change [a] 2 days	Fold change [a] 4 days	pI [b]	kDa	Expect	Matched peptides	Sequence coverage (%)
1	Formamidase	−1.43	-	5.6	44.2	3.1×10^{-10}	9	29
2	Serine carboxypeptidase	−1.53	-	6.5	59.7	0.0019	5	10
3	Serine carboxypeptidase	−1.54	-	5.1	63.6	0.029	6	13
4	Peroxidase catalase 2	−1.43	-	5.8	81.3	3.1×10^{-11}	11	17
5	ATP citrate lyase	−1.27	−1.59	5.5	53.3	5.0×10^{-7}	8	26
6	Pyruvate decarboxylase	+1.60	+4.29	5.6	63.5	2.0×10^{-13}	12	34
7	Glycolipid transfer protein	+1.33	-	5.7	22.4	6.0×10^{-5}	5	30
8	Cell division cycle protein 48	+1.35	-	4.9	90.5	0.00087	6	11
9	Superoxide dismutase	-	−2.27	5.6	27.7	4.0×10^{-7}	6	29
10	Kinesin heavy chain	-	−2.56	5.7	104.1	19	4	6
11	Diphosphomevalonate decarboxylase	-	−1.77	5.3	41.0	2.2	3	7
12	Enolase	-	+2.32	5.0	47.5	1.0×10^{-10}	10	35
13	Carboxypeptidase S1	-	+2.18	5.5	52.3	6.3×10^{-7}	8	28
14	Acid phosphatase	-	+1.85	6.6	47.8	0.15	4	10
15	Alcohol dehydrogenase 1	-	+2.39	7.6	42.0	0.0061	5	19
16	Aldehyde dehydrogenase	-	+1.60	5.4	53.9	1.0×10^{-12}	11	32
17	40S ribosomal protein 50	-	+2.34	4.8	31.8	3.1×10^{-10}	8	30
18	Pyruvate kinase	-	+2.78	5.8	60.0	4.8×10^{-5}	7	13
19	UDP glucose epimerase	-	+1.69	5.8	42.0	0.0051	5	20

[a] Normalized spot intensity: precocene II treated *vs.* untreated cells (average of triplicate gels from each of the three independent cultures). [b] Theoretical pI (isoelectric point) was calculated from amino acid sequence.

2.2. Effects of Carbon Sources on Acetyl CoA Amount and ACL2 mRNA Level

As reported by Jiao *et al.* [11], strain MAFF101551 produced a very small amount of 3-ADON in GYEP or FYEP liquid medium containing glucose or fructose as a carbon source instead of sucrose in SYEP medium at 3 and 7 days cultivation (Figure 2). We analyzed the levels of acetyl CoA and *ACL2* mRNA in the fungal cells cultured in these three media. Analysis of acetyl CoA amount was performed using the method by Ruijter *et al.* with some modification [16]. The acetyl CoA amount in the fungal cells cultured in SYEP liquid medium was much higher than that in GYEP or FYEP medium at 3 days of cultivation (Figure 3). A decrease of the acetyl CoA amount was observed after 7 days of cultivation in SYEP medium. The fungal cells cultured in SYEP medium showed the highest *ACL2* mRNA level among the three media at 3 days of cultivation (Figure 4). However, after 7 days of cultivation, the *ACL2* mRNA level was nearly the same among all three conditions. These results suggest that the acetyl CoA amount and *ACL2* mRNA level correlate with trichothecene production of *F. graminearum* and that sucrose is a key factor for activating *ACL2* transcription and acetyl CoA production.

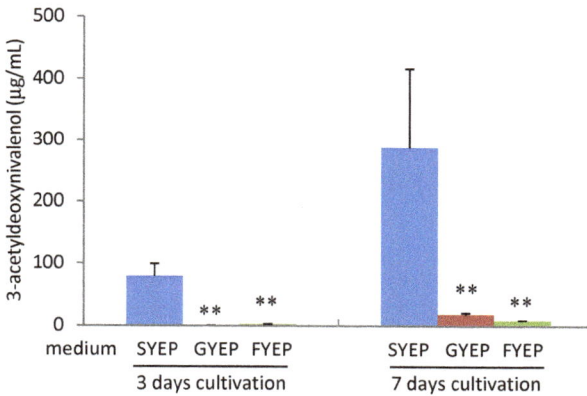

Figure 2. Effects of carbon sources on 3-acetyldeoxynivalenol production by *F. graminearum*. $n = 3$, ** $p < 0.01$, *vs.* control.

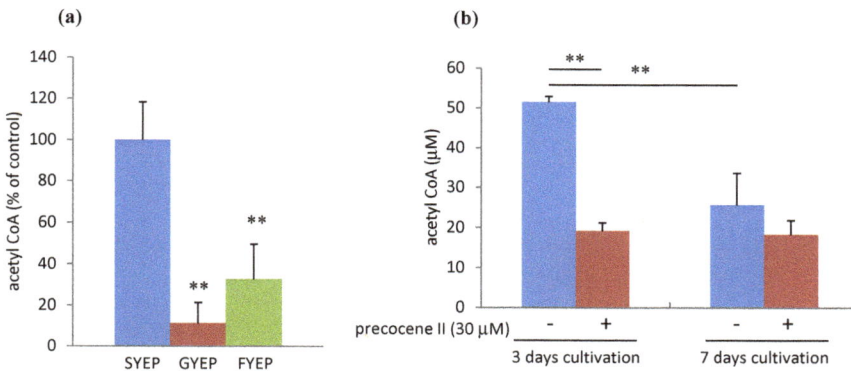

Figure 3. Effects of carbon sources (**a**) and precocene II in SYEP medium (**b**) on acetyl CoA amount. $n = 3$, ** $p < 0.01$, *vs.* control.

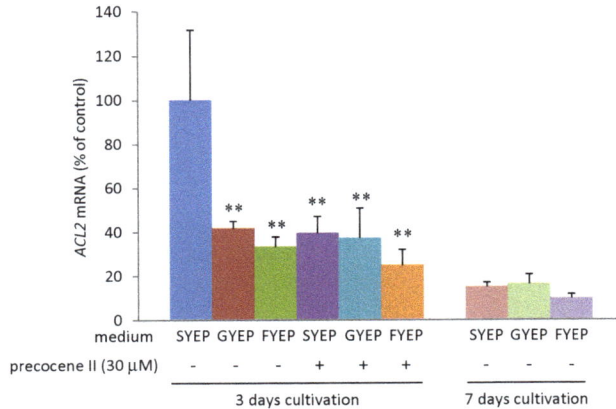

Figure 4. Effects of carbon sources and precocene II on *ACL2* mRNA level. $n = 4$ (3 days cultivation), $n = 3$ (7 days cultivation), ** $p < 0.01$, *vs.* control.

2.3. Effects of Precocene II on Acetyl CoA Amount and ACL2 mRNA Level

The acetyl CoA amount in the fungal cells cultured in SYEP medium was strongly reduced by addition of 30 μM precocene II at 3 days of cultivation (Figure 3), but it was not significantly affected after 7 days of cultivation in SYEP medium. The *ACL2* mRNA level observed in SYEP medium was also reduced to the level detected in GYEP or FYEP by addition of precocene II at 3 days of cultivation (Figure 4). In addition, precocene II did not have a strong effect on the transcription level of *ACL2* in GYEP or FYEP. These results suggest that precocene II may suppress the activation of *ACL2* transcription and acetyl CoA production by sucrose.

2.4. Effects of Cobalt Chloride on Acetyl CoA Amount and ACL2 mRNA Level

The production of 3-ADON in strain MAFF101551 is known to be dramatically enhanced by addition of 30 μM cobalt chloride in SYEP medium without any effect on fungal mycelial weight and ergosterol amount [13]. Cobalt chloride strongly up-regulated transcription of *Tri6* and genes regulated by TRI6 as well as genes encoding ergosterol biosynthetic enzymes. Higher concentrations (300 μM) of precocene II were necessary to almost completely inhibit 3-ADON production in SYEP medium containing 30 μM cobalt chloride. In the presence of 30 μM of cobalt chloride and 300 μM of precocene II, transcription of *Tri6* was suppressed, but the transcription of ergosterol biosynthetic enzyme genes was still up-regulated [13]. After addition of 30 μM cobalt chloride, the *ACL2* mRNA level in the fungal cells was enhanced in SYEP liquid medium, but not significantly affected in GYEP medium (Figure 5). The acetyl CoA amount in the fungal cells was slightly increased by addition of 30 μM cobalt chloride in SYEP medium, but not in GYEP medium (Figure 6). Co-addition of 300 μM of precocene II into SYEP medium containing 30 μM of cobalt chloride reduced the *ACL2* mRNA level and acetyl CoA amount to those observed in GYEP medium (Figures 5 and 6).

Based on our results, we propose a regulatory mechanism for trichothecene production (Figure 7) whereby the expression of ACL and TRI6 is activated in the presence of sucrose, and the acetyl CoA produced by the increased ALC levels may be used for trichothecene production in *F. graminearum*. Sucrose is important for the action of cobalt chloride in activating transcription of *ACL* and *Tri6*, which leads to promotion of trichothecene production. Precocene II may affect a step down-stream of the target of cobalt chloride. *F. graminearum* can produce many secondary metabolites including polyketides and terpens other than trichothecenes [17]. Since acetyl CoA is the key precursor common to biosynthesis of polyketides and terpens produced by the fungus, increasing of acetyl CoA amount may affect their production. Therefore, the results obtained in this study may have a possibility to provide clues for clarifying the regulatory mechanism of production of not only trichothecenes but also other secondary metabolites produced by the fungus.

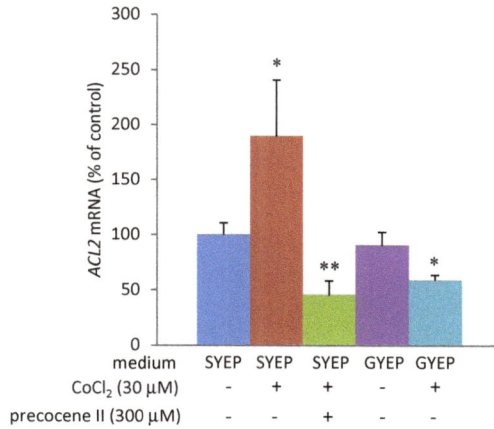

Figure 5. Effects of cobalt chloride on *ACL2* mRNA level. $n = 4$, ** $p < 0.01$, * $p < 0.05$, *vs.* control.

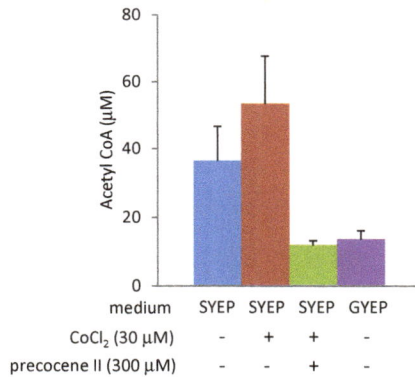

Figure 6. Effects of cobalt chloride on acetyl CoA amount. $n = 3$.

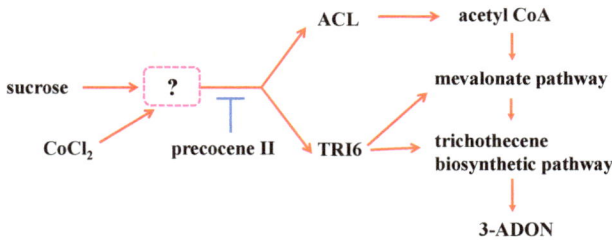

Figure 7. Possible regulatory mechanism of trichothecene production by *F. graminearum*.

3. Experimental Section

3.1. F. graminearum *Culture Conditions and Analysis of 3-Acetyldeoxynivalenol*

A Japanese isolate strain, *Fusarium graminearum* MAFF101551, described previously [13] was used as a 3-acetyldeoxynivalenol producer. A spore suspension of the strain was prepared using carnation leaf agar medium. After cultivation of the strain on the medium at 28 °C for 10 d, the leaf was rinsed in 20% glycerol aqueous solution and the mixture was filtered with miracloth to obtain a filtrate containing spores, which was stored at −80 °C and used as a spore suspension. Liquid medium (5 mL) of SYEP (sucrose 5%, yeast extract 0.1%, polypeptone 0.1%), GYEP (D-glucose 5%, yeast extract 0.1%, polypeptone 0.1%), or FYEP (D-fructose 5%, yeast extract 0.1%, polypeptone 0.1%) was put into test tubes (1.6 cm × 18 cm) and autoclaved. Autoclaved aqueous $CoCl_2$ solution (5 µL) and/or methanolic precocene II solution (15 µL) was added to the medium. Each tube was inoculated with a spore suspension of the strain (1×10^5 spores/tube) and then incubated with continuous shaking (300 rpm) at 26.5 °C for 2–7 d. The resulting culture broth was filtered to obtain the mycelia and filtrate. The filtrate (1 mL) was extracted with 200 µL of ethyl acetate, the ethyl acetate solution was evaporated to dryness, and the obtained residue was dissolved in 200 µL of 10% acetonitrile in water, which was subjected to LC/MS analysis using a 2695 HPLC system (Waters, Milford, MA, USA) equipped with a 150 mm × 2 mm i.d. Capcell-Pak C_{18} column eluted with a gradient of 10%–80% acetonitrile in water containing 10 mM ammonium acetate in 20 min. The flow rate was 0.2 mL/min and the retention time of 3-acetyldeoxynivalenol was 9.4 min. MS analysis was done with a micromassZQ (Waters) by ESI, in positive ion mode; spray chamber parameters: source temperature, 120 °C; desolvation temperature, 350 °C; cone, 30 V; desolvation gas, 600 L/h; cone gas, 50 L/h; capillary voltage, 2800 V. MS ions were monitored in single-ion recording mode using the extracted ion m/z 339 $(M + H)^+$.

3.2. 2D-DIGE Analysis

A spore suspension of the strain MAFF101551was inoculated into SYEP medium (5 mL) in a test tube with or without precocene II (30 µM), and incubated with continuous shaking (300 rpm) at 26.5 °C for 2 d or 4 d. After incubation, mycelia were harvested and lyophilized. The dried mycelia were ground in a mortar with a pestle in liquid nitrogen, and incubated in CelLytic Y Yeast Cell Lysis/Extraction Reagent (600 µL, Sigma-Aldrich, St. Louis, MO, USA) containing 1 M DTT solution (2 µL) and phosphatase inhibitor cocktail (2 µL, Sigma-Aldrich) for 30 min at room temperature to extract total proteins. After centrifugation of the mixture at 15,000 g for 10 min at 4 °C, the supernatant was mixed with acetone (800 µL) and incubated for 1 h at −80 °C. The obtained precipitates were collected by centrifugation, washed with ethanol, and dissolved in 2D-DIGE lysis buffer (2 M thiourea, 7 M urea, 4% (*w*/*v*) CHAPS, 30 mM Tris-HCl, pH 8.5). Total protein contents were determined by the Bradford assay using BSA as a standard.Main text paragraph

Precocene II-treated and control protein samples were quantified and labeled with NHS-Cy2, -Cy3 and -Cy5 (GE Healthcare, Buckinghamshire, UK). Protein (50 µg) taken from precocene II-treated or control samples were minimally labeled with 160 pmol of Cy3 or Cy5 in triplicate. An equal pool of precocene II-treated and control samples were labeled with Cy2 (3.2 pmol/µg protein) and run as a standard on all gels to aid in spot matching and cross-gel quantitative analysis. Protein labeling was performed on ice in the dark for 30 min. Reactions were quenched by 10 min incubation with a 20-fold molar excess of free lysine. The labeled samples were mixed and carrier Pharmalyte (pH 3–10, GE Healthcare) was added to a final concentration of 2%.

The mixture was separated by isoelectric focusing in the first dimension using 24 cm pH 3–10 NL (nonlinear) strip (GE Healthcare) and by SDS-PAGE in the second dimension using 10% SDS-PAGE gel bonded to low-fluorescence glass plates. Gels were run in Ettan 6 gel tanks at 400 mA per gel at 20 °C until the dye front had run off the bottom.

After 2DE, gels were scanned using a Typhoon™ 9400 variable mode imager (GE Healthcare) and ImageQuant software (GE Healthcare). The photomultiplier tube voltage was adjusted for each dye channel for preliminary low-resolution scans to give maximum pixel values within 5%–10% for each channel and below saturation, prior to the acquisition of 100 μm high-resolution images. Images were cropped and analyzed using DeCyder™ V5.0 (GE Healthcare). Ratios of spot intensities detected in precocene II-treated or control samples to those of corresponding spots in standard sample were calibrated and obtained values were averaged across triplicates for each experimental condition. Statistical analysis was performed to pick spots matching across all images, displaying a ≥1.2 average-fold increase or decrease in abundance between drug-treated and control samples and with p values < 0.05 (Student's t-test).

3.3. In-Gel Digestion and Protein Identification by Mass Spectrometry

Visible spots were excised from the CBB R250 (Nacalai tesque, Kyoto, Japan) -stained gels and transferred into microcentrifuge tubes. Excised fragments were washed successively with water, 25 mM NH_4HCO_3, acetonitrile/25 mM NH_4HCO_3 (1:1) and acetonitrile. Destained gel fragments were dried under vacuum with a centrifugal evaporator. Tryptic digestion was performed overnight at 37 °C using 10 μL of 10 μg/mL trypsin (Roche, Basel, Switzerland) in 50 mM NH_4HCO_3, pH 7.8. The resulting tryptic fragments were extracted twice with 100 μL of acetonitrile/water (3:2) containing 0.1% TFA, in an ultrasonic bath for 15 min. The supernatants were concentrated to 10 μL in a centrifugal evaporator and passed through a Zip-Tip™ C18 (Millipore, Bedford, MA, USA). The adsorbed peptides were eluted with 2.0 μL of acetonitrile/water (3:2) containing 0.1% TFA and 1% α-cyano-4-hydroxycinnamic acid. The eluent was loaded onto a mass spectrometer sample plate. Mass spectra were acquired on a Voyager-DE STR MALDI-TOF mass spectrometer (Applied Biosystems, Foster City, CA, USA) operated in the reflectron-delayed extraction mode. Spectra were internally calibrated using trypsin autodigestion products. MASCOT (Matrix Science, Boston, MA, USA) was used to search databases. The expect value, number of matched peptides and percentage of sequence coverage for each identified protein are listed in Table 1.

3.4. Analysis of Acetyl CoA

The strain MAFF101551 was incubated in liquid medium (5 mL), and the mycelia were obtained and frozen by liquid nitrogen. The frozen mycelia were incubated for 30 min at −45 °C with a solution of methanol (6 mL) and chloroform (10 mL) containing 3,4-dimethoxyaniline (1 μM) as the internal standard. After adding an aqueous 5 mM triethanolamine solution (4 mL) to the mixture, the mixture was incubated for 10 min at −45 °C and then shaken for 30 min at room temperature. After centrifuging the mixture at 5000 g for 5 min at −10 °C, the upper layer was transferred to a 50 mL tube. The remaining lower layer was mixed with 5 mM triethanolamine (2 mL) and shaken. After centrifuging the mixture at 5000 g for 5 min at −10 °C, the upper layer was transferred into the 50 mL tube in which the first upper layer has been collected. The pooled upper layers were shaken with diethyl ether (15 mL) and centrifuged at 5000 g for 5 min at −10 °C to obtain the lower layer. After the solution was filtered and evaporated, the obtained residue was dissolved in 200 μL of 25% methanol in water, which was subjected to LC/MS analysis under the same conditions as 3-acetyldeoxynivalenol was analyzed above mentioned except for the following ones. Isocratic elution by solvent A (5 mM hexylamine in water whose pH was adjusted to 6.3 by acetic acid) from 0 to 2 min and then gradient elution by changing the ratio of solvent A and solvent B (90% methanol in water containing 10 mM ammonium acetate whose pH was adjusted to 8.5 by ammonia water) from 100:0 to 0:100 from 2–50 min were used for the HPLC elution. The retention time of acetyl CoA was 33.2 min. MS ions were monitored in single-ion recording mode using the extracted ion m/z 810 $(M + H)^+$.

3.5. RT-PCR Analysis

The strain MAFF101551 was incubated in liquid medium (5 mL) and mycelial cake was harvested by filtration and lyophilized. Total RNA was extracted using a TRIzol plus RNA Purification Kit (Invitrogen, Carlsbad, CA, USA) according to the manufacturer's protocol. First-strand cDNA was prepared using the SuperScript III First Strand Synthesis System (Invitrogen) with random hexamer primers, according to the protocol. The cDNA derived from 0.005 µg of total RNA was used as a template. Real-time quantitative RT-PCR was carried out using the SYBR Green Master Mix (Applied Biosystems, Foster City, CA, USA), in a final volume of 25 µL for each reaction, and an ABI PRISM 7300 thermal cycler (Applied Biosystems). Two-step PCR conditions were as follows: after an initial incubation at 95 °C for 10 min, 40 cycles of 95 °C for 15 s and 60 °C for 1 min were performed. The PCR primers for each gene were as follows: ACL2 5′-CGCCAACTACGGCGAGTAC-3′ and 5′-AGTTCGGGCGTAGTGGTAAGTC-3′: β-tubulin (control gene) 5′-CCTGACCTGCTCTGCCATCT-3′ and 5′-TGGTCCTCAACCTCCTTCATG-3′. The amount of each mRNA was normalized to the amount of β-tubulin mRNA in each sample.

3.6. Data Analysis

Data are presented as the mean ± SD. Differences between groups were assessed with one-way ANOVA followed by Dunnett's test. Values of $p < 0.05$ were considered to be significant.

4. Conclusions

The relationship of ATP citrate lyase and acetyl CoA levels to trichothecene production in F. graminearum was clearly observed in the trichothecene-production medium containing sucrose, which suggested that up-regulation of expression of ATP citrate lyase by sucrose is a key event for inducing the trichothecene biosynthesis.

Acknowledgments: This work was partly supported by the grant from Research project for improving food safety and animal health.

Conflicts of Interest: The authors declare no conflict of interest.

References

1. Pestka, J.J.; Smolinski, A.T. Deoxynivalenol: Toxicology and potential effects on humans. *J. Toxicol. Environ. Health* 2005, 8, 39–69. [CrossRef]
2. Chen, C.; Wang, J.; Luo, Q.; Yuan, S.; Zhou, M. Characterization and fitness of carbendazim-resistant strains of *Fusarium graminearum* (wheat scab). *Pest. Manag. Sci.* 2007, 63, 1201–1207. [CrossRef]
3. Kimura, M.; Tokai, T.; Takahashi-Ando, N.; Ohsato, S.; Fujimura, M. Molecular and genetic studies of *Fusarium* trichothecene biosynthesis: Pathways, genes, and evolution. *Biosci. Biotechnol. Biochem.* 2007, 71, 2105–2123. [CrossRef]
4. Hohn, T.M.; Krishna, R.; Proctor, R.H. Characterization of a transcriptional activator controlling trichothecene toxin biosynthesis. *Fungal Genet. Biol.* 1999, 26, 224–235. [CrossRef]
5. Seong, K.-Y.; Pasquali, M.; Zhou, X.; Song, J.; Hilburn, K.; McCormick, S.; Dong, Y.; Xu, J.-R.; Kistler, H.C. Global gene regulation by *Fusarium* transcription factor Tri6 and Tri10 reveals adaptation for toxin biosynthesis. *Mol. Microbiol.* 2009, 72, 354–367. [CrossRef]
6. Peplow, A.W.; Tag, A.G.; Garifullina, G.F.; Beremand, M.N. Identification of new genes positively regulated by Tri10 and a regulatory network for trichothecene mycotoxin production. *Appl. Environ. Microbiol.* 2003, 69, 2731–2736.
7. Lee, S.; Son, H.; Lee, J.; Min, K.; Choi, G.J.; Kim, J.-C.; Lee, Y.-W. Functional analysis of two acetyl coenzyme A synthases in the ascomycetes *Gibberella zeae*. *Eukaryot. Cell* 2011, 10, 1043–1052. [CrossRef]
8. Wynn, J.P.; Hamid, A.A.; Midgley, M.; Ratledge, C. Widespread occurrence of ATP: citrate lyase and carnitine acetyltransferase in filamentous fungi. *World J. Microbiol. Biotechnol.* 1998, 14, 145–147.
9. Son, H.; Lee, J.; Park, A.R.; Lee, Y.-W. ATP citrate lyase is required for nomal sexual ans asexual development in *Gibberella zeae*. *Fungal Genet. Biol.* 2011, 48, 408–417. [CrossRef]
10. Son, H.; Min, K.; Lee, J.; Choi, G.J.; Kim, J.-C.; Lee, Y.-W. Mitochondrial carnitine-dependant acetyl coenzyme A transport is required for normal sexual and asexual development of the ascomycete *Gibberella zeae*. *Eukaryot. Cell* 2012, 11, 1143–1153. [CrossRef]

11. Jiao, F.; Kawakami, A.; Nakajima, T. Effects of different carbon sources on trichothecene production and *Tri* gene expression by *Fusarium graminearum* in liquid culture. *FEMS Microbiol. Lett.* **2008**, *285*, 212–219. [CrossRef]

12. Yaguchi, A.; Yoshinari, T.; Tsuyuki, R.; Takahashi, H.; Nakajima, T.; Sugita-Konishi, Y.; Nagasawa, H.; Sakuda, S. Isolation and identification of precocenes and piperitone from essential oils as specific inhibitors of trichothecene production by *Fusarium graminearum. J. Agric. Food Chem.* **2009**, *57*, 846–851. [CrossRef]

13. Tsuyuki, R.; Yoshinari, T.; Sakamoto, N.; Nagasawa, H.; Sakuda, S. Enhancement of trichothecene production in *Fusarium graminearum* by cobalt chloride. *J. Agric. Food Chem.* **2011**, *59*, 1760–1766. [CrossRef]

14. Marough, R.; David, S.; Hawkins, E. The development of the DIGE system. 2D fluorescence difference gel analysis technology. *Anal. Bioanal. Chem.* **2005**, *382*, 669–678. [CrossRef]

15. Yoshinari, T.; Noda, Y.; Yoda, K.; Sezaki, H.; Nagasawa, H.; Sakuda, S. Inhibitory activity of blasticidin A, a strong aflatoxin production inhibitor, on protein synthesis of yeast: Selective inhibition of aflatoxin production by protein synthesis inhibitors. *J. Antibiot.* **2010**, *63*, 309–314. [CrossRef]

16. Ruijter, G.J.G.; Visser, J. Determination of intermediary metabolites in *Aspergillus niger. J. Microbiol. Methods* **1996**, *25*, 295–302. [CrossRef]

17. Cuomo, C.A.; Guldener, U.; Xu, J.R.; Trail, F.; Turgeon, B.G.; di Pietro, A.; Walton, J.D.; Ma, L.J.; Baker, S.E.; Rep, M.; *et al.* The *Fusarium graminearum* genome reveals a link between localized polymprphism and pathogen specialization. *Science* **2007**, *317*, 1400–1402. [CrossRef]

Review

Deoxynivalenol: A Major Player in the Multifaceted Response of *Fusarium* to Its Environment

Kris Audenaert [1],*, Adriaan Vanheule [1], Monica Höfte [2] and Geert Haesaert [1]

[1] Department of Applied BioSciences, Faculty Bioscience Engineering, Ghent University, Valentin
     Vaerwyckweg, 1, Ghent 9000, Belgium; adriaan.vanheule@ugent.be (A.V.); geert.haesaert@ugent.be (G.H.)
[2] Department of Crop Protection, Laboratory of Phytopathology, Faculty Bioscience Engineering,
     Ghent University, Coupure links 653, Ghent 9000, Belgium; monica.hofte@ugent.be
*   Author to whom correspondence should be addressed. kris.audenaert@ugent.be; kris.audenaert@ugent.be;
     Tel.: +32-477-97-00-75; Fax: +32-9-242-42-93.

Received: 24 October 2013; in revised form: 16 December 2013; Accepted: 16 December 2013; Published: 19
December 2013

Abstract: The mycotoxin deoxynivalenol (DON), produced by several *Fusarium* spp., acts as a virulence factor and is essential for symptom development after initial wheat infection. Accumulating evidence shows that the production of this secondary metabolite can be triggered by diverse environmental and cellular signals, implying that it might have additional roles during the life cycle of the fungus. Here, we review data that position DON in the saprophytic fitness of *Fusarium*, in defense and in the primary C and N metabolism of the plant and the fungus. We combine the available information in speculative models on the role of DON throughout the interaction with the host, providing working hypotheses that await experimental validation. We also highlight the possible impact of control measures in the field on DON production and summarize the influence of abiotic factors during processing and storage of food and feed matrices. Altogether, we can conclude that DON is a very important compound for *Fusarium* to cope with a changing environment and to assure its growth, survival, and production of toxic metabolites in diverse situations.

Keywords: trichothecene; oxidative stress; virulence factor; fungicides; primary metabolism

1. Introduction

Fusarium head blight (FHB) is an important disease of small-grain cereals that is caused by a diverse set of *Fusarium* species. Although yield reduction is a serious consequence of *Fusarium* infection in the field, the primary interest in FHB research is driven mainly by the ability of *Fusarium* to produce mycotoxins that have toxic effects on plants, animals and humans [1,2]. Deoxynivalenol (DON) is one of the most prevalent mycotoxins encountered in grain fields. Consequently, although it is not the most toxic one, DON is considered to be the most economically important mycotoxin. DON belongs to the structural group of trichothecenes all bearing a common tricyclic 12,13-epoxytrichothec-9-ene core structure. Type A, B, C and D trichothecenes can be distinguished based on substitutions at position C-4, C-7, C-8 and/or C15 [3]. DON belongs to the type B trichothecenes and is mainly produced by *Fusarium graminearum* and *F. culmorum*, two important members of the FHB-causing species complex [4]. Historically, DON, also called vomitoxin, has been notorious because it provokes acute and chronic disease symptoms in humans and animals that consume contaminated grains [5]. Its toxic effects range from diarrhea, vomiting, gastro-intestinal inflammation, necrosis of the intestinal tract, the bone marrow and the lymphoid tissues. It causes inhibition of protein, DNA and RNA synthesis and inhibition of mitochondrial function. In addition, it has effects on cell division and membrane integrity and induces apoptosis [6]. Only after its toxicity for mammals had been established, were dedicated efforts initiated to unravel the conditions under which *Fusarium* species produce DON.

Many environmental factors are reported to affect DON levels during the infection process [7,8]. For instance, humidity and intensive rainfall during and after anthesis result in increased DON production and proliferated FHB symptoms [9–16]. Moreover, the weather conditions during the vegetative growth of wheat are important parameters determining *Fusarium* and DON load, reflecting the importance of survival of the primary inoculum present in soil and on crop debris during winter [14]. Furthermore, FHB and DON are influenced by many agronomic and other anthropogenic factors: no-, minimal-, or non-inversive tillage systems are beneficial for *Fusarium* [17]. Crop rotation, nitrogen fertilization, and weed management shape the structure of the soil biota and influence *Fusarium* survival [14,18,19]. Finally, the germplasm of the host has been shown to influence FHB and DON synthesis for example by the ability of resistant genotypes to metabolize DON [20,21].

Although this information is very valuable, in most studies no mechanistic clues are provided on how these factors affect the toxigenic machinery of the fungus. In addition, there are many other abiotic factors affecting DON of which the physiological relevance is not always clear. Obviously, a thorough insight into the functional rationale of DON production may provide hints towards an adjustment of control measures in order to avoid DON presence in the field. Therefore, we have placed the factors known to induce DON production in a relevant physiological frame, namely the different phases in the life cycle of *Fusarium* during the growing season of wheat (*Triticum* sp.) as a model host. Where possible we combine this information into working models that should be experimentally validated to obtain a holistic view on DON production by *Fusarium*.

2. The Saprophytic Phase

2.1. Survival of the Fittest

During the saprophytic phase, *F. graminearum* can survive on dead organic matter to persist in the absence of a living host, which is an important asset during the active invasion of hosts later on in the season. Therefore, saprophytic fitness is a significant component of the overall pathogen vigor [22]. Strikingly, information on the role of DON during this saprophytic period is scarce, although it covers a major part in the pathogen's life cycle and determines the primary inoculum load. Indeed, recently, DON production during the saprophytic survival on wheat stubble has been shown to be correlated with the aggressiveness of the isolates during their pathogenic phase [22].

The ability of most *F. graminearum* isolates to produce DON provides a dual advantage at the saprophytic state in the competition for niches on crop residues and organic matter. Firstly, DON is an antimicrobial metabolite that is effective against other eukaryotic soil organisms because of its interference with protein biosynthesis [5]. Secondly, DON can affect the metabolite production of other soil-residing fungi, such as *Trichoderma* sp., that are known for their strong outcompeting capacity by mycoparasitism, orchestrated by chitinases and other degrading enzymes [23]. In co-inoculation experiments, DON proved to repress the chitinase activity in *T. atroviride* [24], although a reduction in the *Trichoderma* biomass due to DON production by *F. graminearum* could not be observed [25].

Despite the very limited amount of information on the role of DON during the saprophytic phase, indirect evidence may come from comparative studies on the saprophytic survival of different *Fusarium* species. Apparently, *F. poae* which is considered a weak pathogen, is a better saprophytic survivor that outcompetes *F. graminearum* from soil and crop debris samples [26,27]. Since *F. poae* produces a more toxic blend of mycotoxins than *F. graminearum*, comprised of both type A and type B trichothecenes, it is tempting to speculate that this feature accounts for its better saprophytic survival capacity. The remarkable omnipresence of *F. poae* in the subsequent growth phase on living plant tissue, may thus originate from a "strength in numbers" strategy, originating from an inoculum build-up during the saprophytic phase.

2.2. Linkage between DON Production and Formation of Conidia and Ascospores

As the infection of *F. graminearum* is realized via production of conidia and ascospores, the formation of these reproductive structures is a very important phase in the pathogen's life cycle. Recent research has shown that both DON production and conidia/ascospore formation are under tight regulation by overlapping cellular factors [28], some of which are mentioned below. APSES proteins are a conserved class of transcription factors regulating development, secondary metabolism and pathogenicity [29,30]. Recently, *FgStuA*, a *F. graminearum* gene encoding a protein with high homology to APSES transcription factors has been characterized. Using a knock-out approach, *FgStuA* was shown to influence spore development and DON biosynthesis amongst other processes [31]. Several other regulatory cellular proteins such as the C-type cyclin like protein CID1, the ZIF1 b-zip transcription factor and the Wor-1 like nuclear protein *Fgp1* are all involved in sexual reproduction and influence DON production [32–35]. These results highlight a tight link between reproductive fungal development and secondary metabolite production.

3. DON in the Pathogenic Phase: A Lethal Weapon of a Hemibiotrophic Cereal Killer

3.1. Plant Defense: A Matter of Making the Good Choices at the Right Time

Plants are endowed with a sophisticated set of plant defense mechanisms that can be activated upon pathogen infection. These defense responses can be divided into two main signaling pathways. One pathway involves a prompt induction of reactive oxygen species (ROS) followed by the accumulation of salicylic acid (SA), activating the plant's defense machinery. This type of defense often coincides with a programmed cell death (PCD)-type response and a hypersensitive response (HR) that isolate the pathogen and deprive it from nutrients. This SA-type defense is generally accepted to be efficient against biotrophic pathogens that need viable cells for survival. The other pathway involves jasmonic acid (JA). This type of response is especially activated during the plant defense against necrotrophic pathogens [36,37].

However, some pathogens, such as DON-producing *Fusarium* spp, are hemibiotrophic and have both a biotrophic and a necrotrophic phase during the colonization of their host. Hence, in such interactions, a coordinated and ordered expression of SA- and JA-dependent defense responses in the plant is crucial to halt the fungus [38], but at the same time, it provides multiple opportunities for interference by the pathogen.

3.2. DON and the Plant Defense Response: Hijacking the Plants Oxidative Armor

There is ample evidence suggesting that DON production during infection is a sophisticated strategy of the fungus to circumvent and hijack the plant's defense system. When a rain-splashed conidium or wind-dispersed ascospore lands on the exposed vulnerable parts of a crop plant (glumae, floral cavity, lemma, palea, or anthers) during or just after anthesis, it can germinate and penetrate the plant [39]. An initial superficial and intercellular growth of the fungus is eventually followed by the actual penetration of the plant, which involves the formation of infection cushions and foot-like structures invaginating the host tissue [40,41]. In this first phase, the fungus grows biotrophically into the intercellular spaces and the role of DON is assumed to be unimportant. Still, during this biotrophic phase several reports describe *Tri* gene expression at the hyphal tip [42–44]. Recently, the ability of very low DON concentrations to inhibit PCD has been illustrated [45] which could interfere with PCD, thus disrupting the biotroph-type defense (Figure 1).

Afterward, the fungus switches to a more invasive intracellular growth, including necrosis and cell death [40]. During this second necrotrophic infection phase, the production of the mycotoxin DON becomes apparent and is necessary for the spread of the fungus in the rachis of wheat [46]. Previously, studies have demonstrated that *tri5* knockout mutants, which cannot produce DON because the inactive *Tri5* gene does not convert farnesyl pyrophosphate to trichodiene, are less virulent due to the lack of spread in the rachis, implying that DON is crucial in ear colonization [42,47,48].

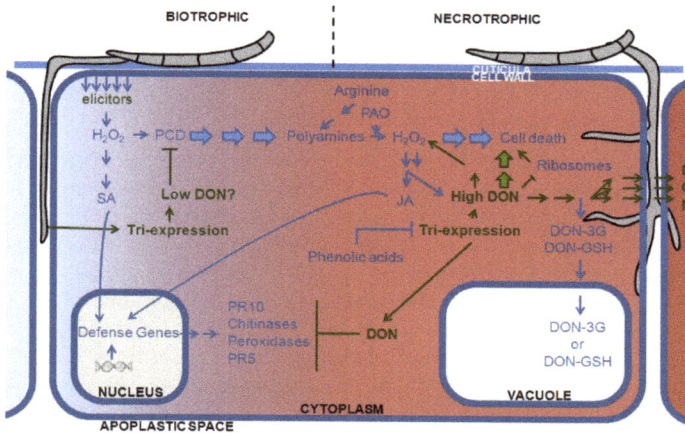

Figure 1. Hypothetical model of the effect of DON during the biotrophic and necrotrophic phases of *F. graminearum* infection of wheat, based on defense-related responses in wheat. The left part depicts the biotrophic phase and the right and red parts indicate the necrotrophic phase of the fungus. Green lines and arrows mark pathways of the fungus, whereas the blue lines reflect pathways of the plant. DON: deoxynivalenol; DON-3G: DON-glucoside; DON-GSH: DON-gluthatione; JA: jasmonic acid; PAO: polyamine oxidases; PCD: programmed cell death; PR: pathogenesis related; SA: salicylic acid; Tri: trichothecenes.

The induction of cell death is a well-known defense strategy of plants against biotrophic but not against necrotrophic fungi [49]. In this context, it is interesting that high DON concentrations were shown to trigger H_2O_2 synthesis and subsequent cell death (Figure 1). Moreover, using an *in vitro* approach, several research groups demonstrated that H_2O_2 is an efficient inducer of DON production, especially when applied at early stages of spore germination [50–52]. Physiologically, these observations indicate that if H_2O_2 is one of the first defense molecules encountered by the invading *Fusarium* hyphae, it also establishes a positive feedback loop leading both to increased DON and H_2O_2 levels. Consequently, DON production by *Fusarium* in the necrotrophic infection phase may interfere with the two-step defense response against hemibiotrophs, because it directs the plant towards an oxidative burst which is not effective against necrotrophs. The eventual activation of H_2O_2-mediated defense responses comprising phenolic acids, chitinases, glucanases and peroxidases [46], might come too late or at the wrong time point for the plant to defend itself against the invasive necrotrophic growth of *F. graminearum*. Indeed, it is generally recognized that both timing and localization of defense or signaling compounds determine the outcome of a plant-pathogen interaction. The importance of H_2O_2 in the induction of DON was confirmed by the effectiveness of anti-oxidative phenolic acids, such as ferulic acid, to inhibit trichothecene accumulation at a transcriptional level *in vitro* [53–55]. In addition, *in planta*, the presence of ferulic acid in wheat cultivars correlated negatively with the accumulation of DON during *F. graminearum* infection [56].

Finally, it seems that DON-producing *Fusarium* species also interfere with the plant defense pathway further downstream of the oxidative burst. Indeed, SA can be used by *F. graminearum* as a carbon source [36], which may result in reduced expression of the typical SA-dependent defense genes such as pathogenesis-related protein 1 (PR1), nonexpressor of PR genes 1 (NPR1), and PR4, possibly impeding the control of symptoms development [36]. Moreover, the production of other defense-related compounds, such as PR10, chitinases, peroxidases, PR5, and PR10, is inhibited by DON at later time points during infection [49].

Nevertheless, DON is not essential in all *F. graminearum* plant interactions. For instance, although eventually a high DON load is measured as well, the infection of barley and rice with *F. graminearum* strains does not involve this mycotoxin [47,57,58].

3.3. Directing DON to the Vacuoles: Deprivation of the Pathogen of Its Virulence Factor

From the above, it is clear that DON is a powerful tool of *F. graminearum* to grow within the wheat host. Nevertheless, the plant is endowed with detoxification mechanisms to dampen the detrimental effects of the mycotoxin (for review [21]). Most important is the covalent binding of DON to hydrophilic molecules, such as glucose and glutathione (γ-glutamyl-cysteinyl-glycine, GSH). Conjugated DON is then transported via membrane-bound transporters to the vacuoles or apoplastic space [59,60]. The detoxifying effect of the conjugation is beyond dispute, but intriguingly, glutathione, a product derived from glyoxylate in the Calvin cycle, also plays an important role in modulating the redox status of the host cell, which determines the outcome of plant-pathogen interactions. Hence, it is tempting to speculate that through conjugation to DON, the fungus sequesters glutathione that affects the antioxidative status and, consequently, the defense machinery of the host cell. Still, it is important to notice that the oxidative status of plant cells is very complex with amongst others catalases, ascorbate peroxidases, superoxide dismutases and NADPH oxidases establishing the oxidative equilibrium.

4. The Plant's Primary Carbohydrate and Nitrogen Metabolism Feed into DON Production and Fungal Growth

Although current research particularly focuses on downstream defense signaling, the energy and carbon skeletons used in the defense reactions activated in wheat upon infection with *Fusarium* require the redistribution of energy from the primary metabolism of the plant. Interestingly, pathogens themselves seemingly also drain energy from the primary metabolism of the host to the advantage of their own pathogenic growth and production of their virulence factors [61,62].

When a plant is attacked by a pathogen, the availability of ready-to-use energy, reducing agents, and carbon skeletons is a prerequisite for optimal activation of defense. In many plant pathosystems, photosynthesis, which generates ATP and NADPH, decreases at the site of infection, establishing novel sink tissues [61]. Carbohydrate partitioning between source and sink tissues is a highly dynamic process during the plant's life cycle and the physiological balance can easily be disrupted. Because of reduced photosynthesis, the plant will mobilize monosaccharides to the infection site by activating membrane-bound invertases that cleave apoplastic sucrose, thus generating energy and carbohydrate skeletons for diverse metabolic processes, including defense. However, sucrose is also an important inducer of the *Tri* gene machinery. Especially *Tri5* and *Tri4*, which are both involved in the initial steps of trichothecene biosynthesis by converting farnesylpyrophosphate to trichodiene and the latter to 15-decalonectrin, respectively, are strongly upregulated by sucrose, resulting in increased DON biosynthesis [63].

In the *F. graminearum*-wheat interaction, several plant invertases are upregulated, indicating that the fungus exploits sucrose not only as a trigger for DON biosynthesis, but also as a monosaccharide source that can be used for its own growth [64]. However, the contributions to the metabolism of the plant and of the fungus are difficult to distinguish. Indeed, pathogenic fungi also produce invertases that can potentially disturb the source-sink balance and the repartitioning of the carbon sources in the plant and, hence, affect the infection process.

The importance of nitrogen in plant defense is mainly situated at three levels. Firstly, nitrogen is indirectly involved as an energy source. Inorganic nitrogen is usually taken up as NH_4 or NO_3 after which it is incorporated into amino acids, such as glutamate, glutamine, asparagine, and aspartate via glutamine synthase. Subsequently, these amino acids are transported or stored in the plant by the glutamine-oxoglutarate aminotransferase (GOGAT) cycle. When the energy demand of the plant cells increases, for example upon pathogen infection, these amino acids are diverted to the energy-generating tricarboxylic acid (TCA) cycle, in part via the γ-aminobutyrate (GABA) shunt, leading to reducing equivalents and ultimately ATP [61,62]. Secondly, nitrogen is a main compound in the regulation of the redox status of plant cells. Reactive nitrogen species, such as nitric oxide (NO), but also polyamines, produced from the precursor L-arginine, can be directly involved in plant defense through HR induction [65]. Moreover, N-containing glutathione is an important antioxidant alleviating oxidative damage during an HR [66]. Thirdly, the plant's nitrogen metabolism has been suggested to be involved in the defense response through a pivotal mechanism of evasion or endurance [62]. During the evasion process, implicated in a successful defense response against biotrophic pathogens, nitrogen is uploaded in the phloem as asparagine or glutamine and transported away from the invaded area to deprive the pathogen from the necessary nitrogen sources. During the endurance process nitrogen is remobilized from noninfected tissues providing infected cells with sufficient nitrogen to keep them alive; a strategy that is very efficient against necrotrophic pathogens [62].

Just as with the carbohydrate metabolism, pathogens, including DON-producing *Fusarium* species, appear to hijack the primary nitrogen metabolism of the plant for their own benefit. For instance, several pathogens can use the plant's amino acids as N-sources. Moreover, upon infection with *F. graminearum*, the primary GOGAT cycle appears to be redirected toward the production of ornithine and arginine, resulting in the formation of polyamines [67] (Figure 2). Indeed, a metabolo-proteomics approach revealed the induction of the agmatin-to-polyamine conversion [68]. As described above, the accumulation of polyamines can lead to ROS through the formation of NO and the action of polyamine oxidases [38,69], which could hypothetically contribute positively to the necrotrophic phase of *F. graminearum*. Finally, in an *in vitro* study, polyamines have been shown to induce DON production as well, further contributing to the fungus pathogenicity [70].

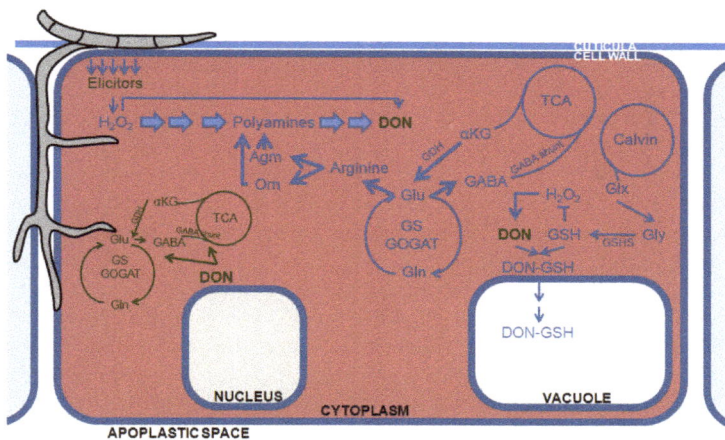

Figure 2. Hypothetical model of the interaction of DON with the primary metabolism of the host and the pathogen. Green lines and arrows indicate pathways of the fungus, blue lines reflect pathways of the plant. Bullet lines represent inhibitory actions. Agm: agmatine; αKG: α-ketoglutarate; DON: deoxynivalenol; GABA: γ-aminobutyric acid; GDH: glutamate dehydrogenase; Gln: glutamine; Glu: glutamate; Glx: glyoxylate; Gly: glycine; GOGAT: glutamine oxoglutarate aminotransferase; Orn: ornithine; TCA: tricarboxylic acid.

Although evidence is scarce, DON may interfere with aspects of the primary metabolism of the fungus itself. Although DON is considered to be a secondary metabolite, knocking out of the *Tri5* gene has a very profound impact on the primary metabolism of the fungus leading to decreased levels of glutamate and GABA and reduced glutamine synthase and GABA transferase activities [71]. Consequently the complete GABA shunt, TCA cycle, and polyamine metabolism are negatively affected (Figure 2). Conversely, upon infection, the GABA shunt becomes activated in DON-producing

F. graminearum strains, suggesting a replenishment of the TCA cycle during the interaction with a host [72]. Moreover, metabolomic studies of wheat ears have revealed that the TCA cycle of the host is disturbed as well upon infection with *F. graminearum*, resulting in an increased activity of glutamate hydrogenase that converts α-ketoglutarate to glutamate although a direct link with DON production was not investigated. Interestingly, in other pathosystems involving necrotrophic and/or toxin-producing plant pathogens, a similar exhaustion of the TCA cycle of the host takes place, suggesting this might be a conserved and effective virulence strategy [73,74].

5. Of Crops and Men: The DON Molecule and Man's Chemical Warfare

Because *Fusarium* infects an important economic crop cultivated within an agro-ecosystem, the plant-fungus interaction is more complex than in a natural ecosystem. Indeed, farmers interfere to minimize the presence of DON and other mycotoxins in the crop. Whereas the effect of chemical fungicides on fungal outgrowth is quite straightforward and generally results in reduced fungal load, reports on the impact of fungicides on the production of fungal secondary metabolites, especially mycotoxins, are rather inconsistent and fragmentary. Still, careful analysis of the information reveals important insights into the function of DON in the reaction of fungi to fungicide applications.

The effect of the strobilurin fungicide azoxystrobin on DON production varies from an increase [75–78] to a reduction [79] depending on environmental factors. Some other fungicides, such as carbendazim and thiram, have been tested for their efficiency to reduce DON in grain samples, but no clear effect was observed [80]. Nevertheless, the mycotoxin chemotype and the sensitivity toward carbendazim fungicides correlated well. As such, most strains producing nivalenol (NIV) or 15-acetyl-deoxynivalenol (15ADON) were susceptible, whereas all carbendazim-resistant isolates were 3-acetyl-deoxynivalenol (3ADON) producers [81].

The most important fungicides currently used to control *Fusarium* are the azoles. A multi-year and multi-location experiment carried out in Belgium illustrated that the effect of azole fungicides with respect to DON depended on the DON concentration in the wheat host. In plants containing low and high DON amounts, fungicide applications often resulted in an increase and a reduction of DON load, respectively. These field trials also demonstrated that it was impossible to decrease the DON levels by more than 75% of the control fields (Figure 3). This observation may imply that highly contaminated fields, in which DON levels exceed the legislative values multiple fold, cannot be rescued by fungicide applications.

Within the group of azole fungicides, field doses of tebuconazole [75,77,82–85], metconazole [79,82,85], and prothioconazole [85] consistently reduced DON biosynthesis or content. In contrast, application of another azole fungicide, propiconazole, either decreased or increased DON levels [76,85]. Intriguingly, DON amounts are increased by application of a sublethal dose of prothioconazole, which is meticulously regulated through the production of H_2O_2 as an oxidative stress response of the fungus. Indeed, oxidative stress as a booster of toxigenic pathways is now considered a trait common to various toxigenic fungi from different genera of the fungal kingdom [86]. Moreover, qRT-PCR analyses have revealed that the expression of *Tri4*, *Tri5*, and *Tri11* was higher in cultures of *F. graminearum* isolates supplemented with sublethal concentrations of tebuconazole and propioconazole than that in nontreated controls, although the fold change in the *Tri* transcript levels differed according to the type of azole used [87].

Typically, azole sensitivity in fungi is modified by either point mutations in the cytochrome P450 monooxygenase-encoding target gene *CYP51* [88,89], overexpression of *CYP51* [90], presence of paralogous CYP51 genes [91], the presence of fungal drug transporters, belonging to the ABC or MDR classes [92], or an altered composition of the sterol content [93]. However, considering the effect of low fungicide levels on DON production, the question arises whether DON interferes with the fungicide effectiveness. Indirect proof comes from *in vitro* fungicide assays with a *tri5* knockout mutant of *F. graminearum*. The overall fitness and fecundity of the mutant was comparable to that of the parent strain; but, when homeopathic levels of azole fungicides were applied, only the mutant fungus promptly stopped growing [94]. Apparently, when a strain cannot produce its toxic secondary metabolite DON, it becomes hypersensitive to azole fungicides. Additionally, when *F. graminearum* strains were allowed to adapt to azole fungicides, they showed an increased production of the B-type trichothecene NIV [95].

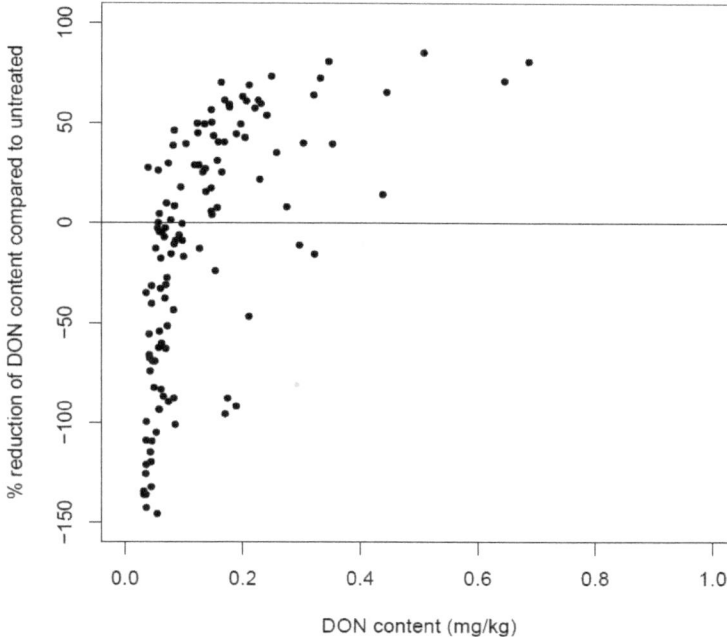

Figure 3. Percentage of reduced DON content after application of triazole fungicides at GS 39 and GS 55 on different wheat varieties in function of the DON content present in the untreated experimental field trials. All data points are the result of four independent replications and experiments were carried out at several locations in a three-year-experiment.

Although direct evidence on the role of DON in fungicide resistance is currently lacking, *Sacharomyces cerevisiae* is known to be very resistant against DON because of the presence of multiple ABC transporters that pump the mycotoxin out of the cell [96]. In addition, expression of ABC transporters of the plant pathogen *Mycosphaerella graminicola* in an ABC transporter-lacking mutant of *S. cerevisiae* clearly indicated a wide functional overlap between the ABC transporters induced by azole fungicides and those by the A-type trichothecene diacetoxyscirpenol [97]. Finally, transcriptional profiling of ABC transporters upon fungicide application points toward a mechanism alleviating the impact of the fungicide [95]. Together, this fragmentary information seems to imply that the mycotoxin production capacity and resistance against fungicides converge at the ABC efflux level or other MDR pumps. It is not unlikely that (some) efflux pumps activated upon mycotoxin biosynthesis are also activated during exposure to fungicides. Interestingly, at least one ABC transporter has been shown to be important in the virulence of *F. graminearum*, but an effect on the mycotoxin efflux from the deletion mutant was not reported [98].

6. Abiotic Factors Influencing DON Biosynthesis in the Field and during Storage

The impact of abiotic factors on mycotoxin production has recently been reviewed [7,8]. Therefore, we highlight only new research findings that deal with environmental effects on DON biosynthesis.

6.1. pH

Although it is currently unknown whether and how the pH fluctuates during a wheat infection with *Fusarium*, it is well established that a low extracellular pH results in an increased trichothecene production [99]. *Tri* gene expression is regulated by a zinc finger transcription factor *Fg*Pac1 at acidic pH values [100], but the regulation at neutral or basic pH remains unclear. As information on the extracellular pH during wheat infection by *F. graminearum* and during grain storage remains scarce, it is very difficult to place the results on the pH effects in a physiological context. Probably, a dynamic window of pH fluxes influences DON production during the infection process.

6.2. a$_w$ and Temperature

The availability of free water (a$_w$) and the incubation temperature will determine whether there will be an outgrowth of *F. graminearum*, especially during storage of wheat grains after harvest. In addition, the toxigenic outcome of fungal growth also depends on the a$_w$ value and the temperature. Indeed, high a$_w$ values increase DON production in contaminated wheat grain batches [101] as well as elevating the incubation temperature from 15 °C to 30 °C [102]. Several reports also describe a clear interaction between temperature and a$_w$ value [103].

6.3. Light

In plants, several important pathways follow a diurnal regulation based on the day/night regime. Although fungi do not depend on photosynthesis for their energy supply, their secondary metabolism is often fine-tuned by light. One of the most important light-regulatory protein complexes is the velvet complex, comprising at least *Fg*Ve1 (VeA) and *Fg*VeB. Although the velvet complex has been elaborately investigated with regard to the switch between asexual and sexual phases of the fungus, recent research highlights its significance in the regulation of the *Tri* gene machinery. By means of a gene replacement strategy, VeA has been demonstrated to regulate trichothecene production at the level of the biosynthetic genes *Tri4* and *Tri5* and the transcriptional regulator genes *Tri6* and *Tri10* [104,105]. Results with knockout mutants have revealed that *Fg*VeB plays a role in the regulation of *Tri5* and *Tri6* as well [106].

6.4. Post-Harvest Anthropogenic Factors Influencing the DON Content

After harvest, grains are often stored for some time in silos before final use as animal feed or human food. Although DON production during storage is, exceptions notwithstanding, rather rare, effects of changed storage conditions on fungal outgrowth and DON production have been reported. Modified storage atmosphere, chemical preservation systems, and biocontrol with lactic acid bacteria have been proposed as antifungal measures [102]. Detailed insights into the effect of these measures on DON production are still lacking. Chemical compounds, such as antioxidants and essential oils applied during storage of wheat grains clearly have a very variable impact on the DON levels. In an experiment in which wheat grains were inoculated with *F. graminearum* and subsequently treated with neutralized electrolyzed water, the ROS present in the electrolyzed water reduced the fungal load in the wheat commodities. However, at sublethal levels, this decrease in biomass coincided with an increase in DON level. The ROS liberated from the electrolyzed water oxidatively stimulated the *Tri* gene machinery to produce DON [94].

7. Conclusions and Challenges for the Future

In the present review, we gathered available data on diverse factors known to affect DON production by *Fusarium*. We combined this information into hypothetical models on the effect of DON on defense-related processes and the primary metabolism of wheat as a model host. Altogether, based on the present literature, we claim that DON is a molecule that is crucial throughout the fungal life

Toxins **2014**, *6*, 1–19

cycle. During saprophytic survival, DON might be involved in competition for niche. Furthermore, DON production and conidia- and/or ascospore formation are tightly linked processes.

During the interaction with its host, it seems that *Fusarium* uses DON to disturb the defense system at several critical time points of infection assuring successful colonization and symptom development (Figure 1). Moreover, DON appears to be deployed to hijack the primary C and N metabolism of the plant to improve fungal growth and production of virulence factors (Figure 2). Although parts of the proposed models are still highly speculative and not supported by direct experimental evidence, we hope they provide valuable working hypotheses for future research.

An additional challenge is to decipher the function of other type A and type B trichothecenes produced by other members of the FHB disease complex. Is the importance of DON in the life cycle of *Fusarium* spp. unique or can the functions be extrapolated to other mycotoxins? More generally, searching for parallels between *Fusarium* and other toxin-producing plant pathogens might reveal conserved infection strategies typical for this type of phytopathogens.

Acknowledgments: We greatly acknowledge Maarten Ameye, Martine De Cock, and Danny Vereecke for critical reading of the manuscript. Kris Audenaert and Adriaan Vanheule are indebted to the Research Fund of University College Ghent for a post-doctoral and PhD grant, respectively.

Conflicts of Interest: The authors declare no conflict of interest.

References

1. Arunachalam, C.; Doohan, F.M. Trichothecene toxicity in eukaryotes: Cellular and molecular mechanisms in plants and animals. *Toxicol. Lett.* **2013**, *27*, 149–158. [CrossRef]
2. Maresca, M. From the gut to the brain: Journey and pathophysiological effects of the food-associated trichothecene mycotoxin deoxynivalenol. *Toxins* **2013**, *23*, 784–820. [CrossRef]
3. McCormick, S.P.; Stanley, A.M.; Stover, N.A.; Alexander, N.J. Trichothecenes: From simple to complex mycotoxins. *Toxins* **2011**, *3*, 802–814. [CrossRef]
4. Goswami, R.S.; Kistler, H.C. Heading for disaster: *Fusarium graminearum* on cereal crops. *Mol. Plant Pathol.* **2004**, *5*, 515–525. [CrossRef]
5. Bennett, J.W.; Klich, M. Mycotoxins. *Clin. Microbiol. Rev.* **2003**, *16*, 497–516. [CrossRef]
6. Pestka, J.J. Toxicological mechanisms and potential health effects of deoxynivalenol and nivalenol. *World Mycotoxin J.* **2010**, *3*, 323–347. [CrossRef]
7. Wegulo, S.N. Factors influencing deoxynivalenol accumulation in small grain cereals. *Toxins* **2012**, *4*, 1157–1180. [CrossRef]
8. Merhej, J.; Richard-Forget, F.; Barreau, C. Regulation of trichothecene biosynthesis in *Fusarium*: Recent advances and new insights. *Appl. Microbiol. Biotechnol.* **2011**, *91*, 519–528. [CrossRef]
9. Hooker, D.C.; Schaafsma, A.W.; Tamburic-Ilincic, L. Using weather variables pre- and post-heading to predict deoxynivalenol content in winter wheat. *Plant Dis.* **2002**, *86*, 611–619. [CrossRef]
10. Schaafsma, A.W.; Tamburic-Ilinic, L.; Miller, J.D.; Hooker, D.C. Agronomic considerations for reducing deoxynivalenol in wheat grain. *Can. J. Plant Pathol. Rev. Can. Phytopathol.* **2001**, *23*, 279–285. [CrossRef]
11. Moschini, R.C.; Fortugno, C. Predicting wheat head blight incidence using models based on meteorological factors in Pergamino, Argentina. *Eur. J. Plant Pathol.* **1996**, *102*, 211–218. [CrossRef]
12. Klem, K.; Vanova, M.; Hajslova, J.; Lancova, K.; Sehnalova, M. A neural network model for prediction of deoxynivalenol content in wheat grain based on weather data and preceding crop. *Plant Soil Environ.* **2007**, *53*, 421–429.
13. Kriss, A.B.; Paul, P.A.; Xu, X.M.; Nicholson, P.; Doohan, F.M.; Hornok, L.; Rietini, A.; Edwards, S.G.; Madden, L.V. Quantification of the relationship between the environment and *Fusarium* head blight, *Fusarium* pathogen density, and mycotoxins in winter wheat in Europe. *Eur. J. Plant Pathol.* **2012**, *133*, 975–993. [CrossRef]
14. Landschoot, S.; Waegeman, W.; Audenaert, K.; Vandepitte, J.; Baetens, J.M.; De Baets, B.; Haesaert, G. An empirical analysis of explanatory variables affecting *Fusarium* head blight infection and deoxynivalenol content in wheat. *J. Plant Pathol.* **2012**, *94*, 135–147.

15. Lindblad, M.; Borjesson, T.; Hietaniemi, V.; Elen, O. Statistical analysis of agronomical factors and weather conditions influencing deoxynivalenol levels in oats in Scandinavia. *Food Add. Contam. Part A Chem.* **2012**, *29*, 1566–1571. [CrossRef]

16. Gourdain, E.; Piraux, F.; Barrier-Guillot, B. A model combining agronomic and weather factors to predict occurrence of deoxynivalenol in durum wheat kernels. *World Mycotoxin J.* **2011**, *4*, 129–139. [CrossRef]

17. Leplat, J.; Friberg, H.; Abid, M.; Steinberg, C. Survival of *Fusarium graminearum*, the causal agent of *Fusarium* head blight. A review. *Agron. Sustain. Dev.* **2013**, *33*, 97–111. [CrossRef]

18. Bernhoft, A.; Torp, M.; Clasen, P.E.; Loes, A.K.; Kristoffersen, A.B. Influence of agronomic and climatic factors on *Fusarium* infestation and mycotoxin contamination of cereals in Norway. *Food Add. Contam. Part A Chem.* **2012**, *29*, 1129–1140. [CrossRef]

19. Lemmens, M.; Haim, K.; Lew, H.; Ruckenbauer, P. The effect of nitrogen fertilization on *Fusarium* head blight development and deoxynivalenol contamination in wheat. *J. Phytopathol.* **2004**, *152*, 1–8. [CrossRef]

20. Miedaner, T.; Korzun, V. Marker-assisted selection for disease resistance in wheat and barley breeding. *Phytopathology* **2012**, *102*, 560–566. [CrossRef]

21. Berthiller, F.; Crews, C.; Dall'Asta, C.; De Saeger, S.; Haesaert, G.; Karlovsky, P.; Oswald, I.P.; Seefelder, W.; Speijers, G.; Stroka, J. Masked mycotoxins: A review. *Mol. Nutr. Food Res.* **2013**, *57*, 165–186. [CrossRef]

22. Tunali, B.; Obanor, F.; Erginbas, G.; Westecott, R.A.; Nicol, J.; Chakraborty, S. Fitness of three *Fusarium* pathogens of wheat. *FEMS Microbiol. Ecol.* **2012**, *81*, 596–609. [CrossRef]

23. Lorito, M.; Farkas, V.; Rebuffat, S.; Bodo, B.; Kubicek, C.P. Cell wall synthesis is a major target of mycoparasitic antagonism by *Trichoderma harzianum*. *J. Bacteriol.* **1996**, *178*, 6382–6385.

24. Lutz, M.P.; Feichtinger, G.; Defago, G.; Duffy, B. Mycotoxigenic *Fusarium* and deoxynivalenol production repress chitinase gene expression in the biocontrol agent *Trichoderma atroviride* P1. *Appl. Environ. Microbiol.* **2003**, *69*, 3077–3084. [CrossRef]

25. Naef, A.; Senatore, M.; Defago, G. A microsatellite based method for quantification of fungi in decomposing plant material elucidates the role of *Fusarium graminearum* DON production in the saprophytic competition with *Trichoderma atroviride* in maize tissue microcosms. *FEMS Microbiol. Ecol.* **2006**, *55*, 211–220. [CrossRef]

26. Pereyra, S.A.; Dill-Macky, R. Colonization of the residues of diverse plant species by *Gibberella zeae* and their contribution to *Fusarium* head blight inoculum. *Plant Dis.* **2008**, *92*, 800–807. [CrossRef]

27. Landschoot, S.; Audenaert, K.; Waegeman, W.; Pycke, B.; Bekaert, B.; De Baets, B.; Haesaert, G. Connection between primary *Fusarium* inoculum on gramineous weeds, crop residues and soil samples and the final population on wheat ears in Flanders, Belgium. *Crop Protect.* **2011**, *30*, 1297–1305. [CrossRef]

28. Calvo, A.M.; Wilson, R.A.; Bok, J.W.; Keller, N.P. Relationship between secondary metabolism and fungal development. *Microbiol. Mol. Biol. Rev.* **2002**, *66*, 447–459. [CrossRef]

29. Twumasi-Boateng, K.; Yu, Y.; Chen, D.; Gravelat, F.N.; Nierman, W.C.; Sheppard, D.C. Transcriptional profiling identifies a role for BrlA in the response to nitrogen depletion and for StuA in the regulation of secondary metabolite clusters in *Aspergillus fumigatus*. *Eukaryot. Cell* **2009**, *8*, 104–115. [CrossRef]

30. Tong, X.Z.; Zhang, X.W.; Plummer, K.M.; Stowell, K.M.; Sullivan, P.A.; Farley, P.C. GcSTUA, an APSES transcription factor, is required for generation of appressorial turgor pressure and full pathogenicity of *Glomerella cingulata*. *Mol. Plant Microbe Interact.* **2007**, *20*, 1102–1111. [CrossRef]

31. Lysoe, E.; Pasquali, M.; Breakspear, A.; Kistler, H.C. The transcription factor FgStuAp influences spore development, pathogenicity, and secondary metabolism in *Fusarium graminearum*. *Mol. Plant Microbe Interact.* **2011**, *24*, 54–67. [CrossRef]

32. Pasquali, M.; Spanu, F.; Scherm, B.; Balmas, V.; Hoffmann, L.; Hammond-Kosack, K.E.; Beyer, M.; Migheli, Q. FcStuA from *Fusarium culmorum* controls wheat foot and root rot in a toxin dispensable manner. *PLoS ONE* **2013**, *8*, 1–15.

33. Zhou, X.Y.; Heyer, C.; Choi, Y.E.; Mehrabi, R.; Xu, J.R. The CID1 cyclin C-like gene is important for plant infection in *Fusarium graminearum*. *Fungal Genet. Biol.* **2010**, *47*, 143–151. [CrossRef]

34. Wang, Y.; Liu, W.D.; Hou, Z.M.; Wang, C.F.; Zhou, X.Y.; Jonkers, W.; Ding, S.L.; Kistler, H.C.; Xu, J.R. A novel transcriptional factor important for pathogenesis and ascosporogenesis in *Fusarium graminearum*. *Mol. Plant Microbe Interact.* **2011**, *24*, 118–128. [CrossRef]

35. Jonkers, W.; Dong, Y.H.; Broz, K.; Kistler, H.C. The Wor1-like protein Fgp1 regulates pathogenicity, toxin synthesis and reproduction in the phytopathogenic fungus *Fusarium graminearum*. *PLoS Pathog.* **2012**, *8*, 1–18.

36. Qi, P.F.; Johnston, A.; Balcerzak, M.; Rocheleau, H.; Harris, L.J.; Long, X.Y.; Wei, Y.M.; Zheng, Y.L.; Ouellet, T. Effect of salicylic acid on *Fusarium graminearum*, the major causal agent of fusarium head blight in wheat. *Fungal Biol.* **2012**, *116*, 413–426. [CrossRef]

37. Robert-Seilaniantz, A.; Grant, M.; Jones, J.D.G. Hormone crosstalk in plant disease and defense: More than just jasmonate-salicylate antagonism. *Annu. Rev. Phytopathol.* **2011**, *49*, 317–343. [CrossRef]

38. Ding, L.N.; Xu, H.B.; Yi, H.Y.; Yang, L.M.; Kong, Z.X.; Zhang, L.X.; Xue, S.L.; Jia, H.Y.; Ma, Z.Q. Resistance to hemi-biotrophic *F-graminearum* infection is associated with coordinated and ordered expression of diverse defense signaling pathways. *PloS ONE* **2011**, *6*, 1–17.

39. Parry, D.W.; Jenkinson, P.; McLeod, L. *Fusarium* ear blight (Scab) in small grain cereals—A review. *Plant Pathol.* **1995**, *44*, 207–238. [CrossRef]

40. Kazan, K.; Gardiner, D.M.; Manners, J.M. On the trail of a cereal killer: Recent advances in *Fusarium graminearum* pathogenomics and host resistance. *Mol. Plant Pathol.* **2012**, *13*, 399–413. [CrossRef]

41. Boenisch, M.J.; Schafer, W. *Fusarium graminearum* forms mycotoxin producing infection structures on wheat. *BMC Plant Biol.* **2011**, *11*, 1–13. [CrossRef]

42. Desjardins, A.E.; Proctor, R.H.; Bai, G.H.; McCormick, S.P.; Shaner, G.; Buechley, G.; Hohn, T.M. Reduced virulence of trichothecene-nonproducing mutants of *Gibberella zeae* in wheat field tests. *Mol. Plant Microbe Interact.* **1996**, *9*, 775–781. [CrossRef]

43. Cowger, C.; Arellano, C. *Fusarium graminearum* infection and deoxynivalenol concentrations during development of wheat spikes. *Phytopathology* **2013**, *103*, 460–471. [CrossRef]

44. Hallen-Adams, H.E.; Wenner, N.; Kuldau, G.A.; Trail, F. Deoxynivalenol biosynthesis-related gene expression during wheat kernel colonization by *Fusarium graminearum*. *Phytopathology* **2011**, *101*, 1091–1096. [CrossRef]

45. Diamond, M.; Reape, T.J.; Rocha, O.; Doyle, S.M.; Kacprzyk, J.; Doohan, F.M.; McCabe, P.F. The *Fusarium* mycotoxin deoxynivalenol can inhibit plant apoptosis-like programmed cell death. *PloS ONE* **2013**, *8*, 1–8.

46. Walter, S.; Nicholson, P.; Doohan, F.M. Action and reaction of host and pathogen during Fusarium head blight disease. *New Phytol.* **2010**, *185*, 54–66. [CrossRef]

47. Langevin, F.; Eudes, F.; Comeau, A. Effect of trichothecenes produced by *Fusarium graminearum* during *Fusarium* head blight development in six cereal species. *Eur. J. Plant Pathol.* **2004**, *110*, 735–746. [CrossRef]

48. Jansen, C.; Von Wettstein, D.; Schafer, W.; Kogel, K.H.; Felk, A.; Maier, F.J. Infection patterns in barley and wheat spikes inoculated with wild-type and trichodiene synthase gene disrupted *Fusarium graminearum*. *Proc. Natl. Acad. Sci. USA* **2005**, *102*, 16892–16897. [CrossRef]

49. Desmond, O.J.; Manners, J.M.; Stephens, A.E.; MaClean, D.J.; Schenk, P.M.; Gardiner, D.M.; Munn, A.L.; Kazan, K. The *Fusarium* mycotoxin deoxynivalenol elicits hydrogen peroxide production, programmed cell death and defence responses in wheat. *Mol. Plant Pathol.* **2008**, *9*, 435–445. [CrossRef]

50. Audenaert, K.; Callewaert, E.; Hofte, M.; De Saeger, S.; Haesaert, G. Hydrogen peroxide induced by the fungicide prothioconazole triggers deoxynivalenol (DON) production by Fusarium graminearum. *BMC Microbiol.* **2010**, *10*, 1–14. [CrossRef]

51. Ponts, N.; Pinson-Gadais, L.; Barreau, C.; Richard-Forget, F.; Ouellet, T. Exogenous H_2O_2 and catalase treatments interfere with Tri genes expression in liquid cultures of *Fusarium graminearum*. *FEBS Lett.* **2007**, *581*, 443–447. [CrossRef]

52. Ponts, N.; Pinson-Gadais, L.; Verdal-Bonnin, M.N.; Barreau, C.; Richard-Forget, F. Accumulation of deoxynivalenol and its 15-acetylated form is significantly modulated by oxidative stress in liquid cultures of *Fusarium graminearum*. *FEMS Microbiol. Lett.* **2006**, *258*, 102–107. [CrossRef]

53. Boutigny, A.L.; Atanasova-Penichon, V.; Benet, M.; Barreau, C.; Richard-Forget, F. Natural phenolic acids from wheat bran inhibit *Fusarium culmorum* trichothecene biosynthesis *in vitro* by repressing Tri gene expression. *Eur. J. Plant Pathol.* **2010**, *127*, 275–286. [CrossRef]

54. Boutigny, A.L.; Barreau, C.; Atanasova-Penichon, V.; Verdal-Bonnin, M.N.; Pinson-Gadais, L.; Richard-Forget, F. Ferulic acid, an efficient inhibitor of type B trichothecene biosynthesis and Tri gene expression in *Fusarium* liquid cultures. *Mycol. Res.* **2009**, *113*, 746–753. [CrossRef]

55. Atanasova-Penichon, V.; Pons, S.; Pinson-Gadais, L.; Picot, A.; Marchegay, G.; Bonnin-Verdal, M.N.; Ducos, C.; Barreau, C.; Roucolle, J.; Sehabiague, P.; *et al*. Chlorogenic acid and maize ear rot resistance: A dynamic study investigating *Fusarium graminearum* development, deoxynivalenol production, and phenolic acid accumulation. *Mol. Plant Microbe Interact.* **2012**, *25*, 1605–1616. [CrossRef]

56. Engelhardt, G.; Koeniger, M.; Preiss, U. Influence of wheat phenolic acids on *Fusarium* head blight resistance and deoxynivalenol concentration. *Mycotoxin Res.* **2002**, *18*, 100–103. [CrossRef]

57. Goswami, R.S.; Kistler, H.C. Pathogenicity and *in planta* mycotoxin accumulation among members of the *Fusarium graminearum* species complex on wheat and rice. *Phytopathology* **2005**, *95*, 1397–1404. [CrossRef]

58. Boddu, J.; Cho, S.; Kruger, W.M.; Muehlbauer, G.J. Transcriptome analysis of the barley-*Fusarium graminearum* interaction. *Mol. Plant Microbe Interact.* **2006**, *19*, 407–417. [CrossRef]

59. Bowles, D.; Lim, E.K.; Poppenberger, B.; Vaistij, F.E. Glycosyltransferases of lipophilic small molecules. *Annu. Rev. Plant Biol.* **2006**, *57*, 567–597. [CrossRef]

60. Coleman, J.O.D.; BlakeKalff, M.M.A.; Davies, T.G.E. Detoxification of xenobiotics by plants: Chemical modification and vacuolar compartmentation. *Trends Plant Sci.* **1997**, *2*, 144–151. [CrossRef]

61. Bolton, M.D. Primary metabolism and plant defense: Fuel for the fire. *Mol. Plant Microbe Interact.* **2009**, *22*, 487–497. [CrossRef]

62. Seifi, H.S.; Van Bockhaven, J.; Angenon, G.; Hofte, M. Glutamate metabolism in plant Disease and defense: Friend or foe? *Mol. Plant Microbe Interact.* **2013**, *26*, 475–485. [CrossRef]

63. Jiao, F.; Kawakami, A.; Nakajima, T. Effects of different carbon sources on trichothecene production and Tri gene expression by *Fusarium graminearum* in liquid culture. *FEMS Microbiol. Lett.* **2008**, *285*, 212–219. [CrossRef]

64. Guenther, J.C.; Hallen-Adams, H.E.; Bucking, H.; Shachar-Hill, Y.; Trail, F. Triacylglyceride metabolism by *Fusarium graminearum* during colonization and sexual development on wheat. *Mol. Plant Microbe Interact.* **2009**, *22*, 1492–1503. [CrossRef]

65. Romero-Puertas, M.C.; Perazzolli, M.; Zago, E.D.; Delledonne, M. Nitric oxide signalling functions in plant-pathogen interactions. *Cell. Microbiol.* **2004**, *6*, 795–803. [CrossRef]

66. Elzahaby, H.M.; Gullner, G.; Kiraly, Z. Effects of powdery mildew infection of barley on the ascorbate-glutathione cycle and other antioxidants in different host-pathogen interactions. *Phytopathology* **1995**, *85*, 1225–1230. [CrossRef]

67. Gardiner, D.M.; Kazan, K.; Praud, S.; Torney, F.J.; Rusu, A.; Manners, J.M. Early activation of wheat polyamine biosynthesis during *Fusarium* head blight implicates putrescine as an inducer of trichothecene mycotoxin production. *BMC Plant Biol.* **2010**, *10*. [CrossRef]

68. Gunnaiah, R.; Kushalappa, A.C.; Duggavathi, R.; Fox, S.; Somers, D.J. Integrated metabolo-proteomic approach to decipher the mechanisms by which wheat QTL (Fhb1) contributes to resistance against *Fusarium graminearum*. *PloS ONE* **2012**, *7*, 1–15.

69. Lysoe, E.; Seong, K.Y.; Kistler, H.C. The transcriptome of *Fusarium graminearum* during the infection of wheat. *Mol. Plant Microbe Interact.* **2011**, *24*, 995–1000. [CrossRef]

70. Gardiner, D.M.; Kazan, K.; Manners, J.M. Nutrient profiling reveals potent inducers of trichothecene biosynthesis in *Fusarium graminearum*. *Fungal Genet. Biol.* **2009**, *46*, 604–613. [CrossRef]

71. Chen, F.F.; Zhang, J.T.; Song, X.S.; Yang, J.; Li, H.P.; Tang, H.R.; Liao, Y.C. Combined metabonomic and quantitative real-time PCR analyses reveal systems metabolic changes of *Fusarium graminearum* induced by Tri5 gene deletion. *J. Prot. Res.* **2011**, *10*, 2273–2285. [CrossRef]

72. Carapito, R.; Hatsch, D.; Vorwerk, S.; Petkovski, E.; Jeltsch, J.M.; Phalip, V. Gene expression in *Fusarium graminearum* grown on plant cell wall. *Fungal Genet. Biol.* **2008**, *45*, 738–748. [CrossRef]

73. Tsuge, T.; Harimoto, Y.; Akimitsu, K.; Ohtani, K.; Kodama, M.; Akagi, Y.; Egusa, M.; Yamamoto, M.; Otani, H. Host-selective toxins produced by the plant pathogenic fungus *Alternaria alternata*. *FEMS Microbiol. Rev.* **2013**, *37*, 44–66. [CrossRef]

74. Brauc, S.; De Vooght, E.; Claeys, M.; Geuns, J.M.C.; Hofte, M.; Angenon, G. Overexpression of arginase in *Arabidopsis thaliana* influences defence responses against *Botrytis cinerea*. *Plant Biol.* **2012**, *14*, 39–45. [CrossRef]

75. Zhang, Y.J.; Fan, P.S.; Zhang, X.; Chen, C.J.; Zhou, M.G. Quantification of *Fusarium graminearum* in harvested grain by real-time polymerase chain reaction to assess efficacies of fungicides on *Fusarium* head blight, deoxynivalenol contamination, and yield of winter wheat. *Phytopathology* **2009**, *99*, 95–100. [CrossRef]

76. Magan, N.; Hope, R.; Colleate, A.; Baxter, E.S. Relationship between growth and mycotoxin production by *Fusarium* species, biocides and environment. *Eur. J. Plant Pathol.* **2002**, *108*, 685–690. [CrossRef]

77. Simpson, D.R.; Weston, G.E.; Turner, J.A.; Jennings, P.; Nicholson, P. Differential control of head blight pathogens of wheat by fungicides and consequences for mycotoxin contamination of grain. *Eur. J. Plant Pathol.* **2001**, *107*, 421–431. [CrossRef]

78. Gaurilcikiene, I.; Mankeviciene, A.; Suproniene, S. The effect of fungicides on rye and triticale grain contamination with *Fusarium* fungi and mycotoxins. *Zemdirbyste* **2011**, *98*, 19–26.

79. Pirgozliev, S.R.; Edwards, S.G.; Hare, M.C.; Jenkinson, P. Effect of dose rate of azoxystrobin and metconazole on the development of *Fusarium* head blight and the accumulation of deoxynivalenol (DON) in wheat grain. *Eur. J. Plant Pathol.* **2002**, *108*, 469–478. [CrossRef]

80. Zhang, Y.J.; Yu, J.J.; Zhang, Y.N.; Zhang, X.; Cheng, C.J.; Wang, J.X.; Hollomon, D.W.; Fan, P.S.; Zhou, M.G. Effect of carbendazim resistance on trichothecene production and aggressiveness of *Fusarium graminearum*. *Mol. Plant Microbe Interact.* **2009**, *22*, 1143–1150. [CrossRef]

81. Zhang, L.; Jia, X.; Chen, C.; Zhou, M. Characterization of carbendazim sensitivity and trichothecene chemotypes of *Fusarium graminearum* in Jiangsu Province of China. *Physiol. Mol. Plant Pathol.* **2013**, *84*, 53–60. [CrossRef]

82. Edwards, S.G.; Pirgozliev, S.R.; Hare, M.C.; Jenkinson, P. Quantification of trichothecene-producing *Fusarium* species in harvested grain by competitive PCR to determine efficacies of fungicides against *Fusarium* head blight of winter wheat. *Appl. Environ. Microbiol.* **2001**, *67*, 1575–1580. [CrossRef]

83. Haidukowski, M.; Pascale, M.; Perrone, G.; Pancaldi, D.; Campagna, C.; Visconti, A. Effect of fungicides on the development of *Fusarium* head blight, yield and deoxynivalenol accumulation in wheat inoculated under field conditions with *Fusarium graminearum* and *Fusarium culmorum*. *J. Sci. Food Agric.* **2005**, *85*, 191–198. [CrossRef]

84. Ioos, R.; Belhadj, A.; Menez, M.; Faure, A. The effects of fungicides on *Fusarium* spp. and *Microdochium nivale* and their associated trichothecene mycotoxins in French naturally-infected cereal grains. *Crop Prot.* **2005**, *24*, 894–902. [CrossRef]

85. Paul, P.A.; Lipps, P.E.; Hershman, D.E.; McMullen, M.P.; Draper, M.A.; Madden, L.V. Efficacy of triazole-based fungicides for *Fusarium* head blight and deoxynivalenol control in wheat: A multivariate meta-analysis. *Phytopathology* **2008**, *98*, 999–1011. [CrossRef]

86. Reverberi, M.; Ricelli, A.; Zjalic, S.; Fabbri, A.A.; Fanelli, C. Natural functions of mycotoxins and control of their biosynthesis in fungi. *Appl. Microbiol. Biotechnol.* **2010**, *87*, 899–911. [CrossRef]

87. Kulik, T.; Lojko, M.; Jestoi, M.; Perkowski, J. Sublethal concentrations of azoles induce Tri transcript levels and trichothecene production in *Fusarium graminearum*. *FEMS Microbiol. Lett.* **2012**, *335*, 58–67. [CrossRef]

88. Wyand, R.A.; Brown, J.K.M. Sequence variation in the CYP51 gene of *Blumeria graminis* associated with resistance to sterol demethylase inhibiting fungicides. *Fungal Genet. Biol.* **2005**, *42*, 726–735. [CrossRef]

89. Leroux, P.; Walker, A.S. Multiple mechanisms account for resistance to sterol 14 alpha-demethylation inhibitors in field isolates of *Mycosphaerella graminicola*. *Pest Manag. Sci.* **2011**, *67*, 44–59. [CrossRef]

90. Hamamoto, H.; Hasegawa, K.; Nakaune, R.; Lee, Y.J.; Makizumi, Y.; Akutsu, K.; Hibi, T. Tandem repeat of a transcriptional enhancer upstream of the sterol 14 alpha-demethylase gene (CYP51) in *Penicillium digitatum*. *Appl. Environ. Microbiol.* **2000**, *66*, 3421–3426. [CrossRef]

91. Liu, X.; Yu, F.; Schnabel, G.; Wu, J.B.; Wang, Z.Y.; Ma, Z.H. Paralogous cyp51 genes in *Fusarium graminearum* mediate differential sensitivity to sterol demethylation inhibitors. *Fungal Genet. Biol.* **2011**, *48*, 113–123. [CrossRef]

92. De Waard, M.A.; Andrade, A.C.; Hayashi, K.; Schoonbeek, H.J.; Stergiopoulos, I.; Zwiers, L.H. Impact of fungal drug transporters on fungicide sensitivity, multidrug resistance and virulence. *Pest Manag. Sci.* **2006**, *62*, 195–207. [CrossRef]

93. Loffler, J.; Einsele, H.; Hebart, H.; Schumacher, U.; Hrastnik, C.; Daum, G. Phospholipid and sterol analysis of plasma membranes of azole-resistant *Candida albicans* strains. *FEMS Microbiol. Lett.* **2000**, *185*, 59–63.

94. Audenaert, K.; Monbaliu, S.; Deschuyffeleer, N.; Maene, P.; Vekeman, F.; Haesaert, G.; De Saeger, S.; Eeckhout, M. Neutralized electrolyzed water efficiently reduces *Fusarium* spp. *in vitro* and on wheat kernels but can trigger deoxynivalenol (DON) biosynthesis. *Food Control* **2012**, *23*, 515–521. [CrossRef]

95. Becher, R.; Weihmann, F.; Deising, H.B.; Wirsel, S.G.R. Development of a novel multiplex DNA microarray for *Fusarium graminearum* and analysis of azole fungicide responses. *BMC Genomics* **2011**, *12*. [CrossRef]

96. Poppenberger, B.; Berthiller, F.; Lucyshyn, D.; Sieberer, T.; Schuhmacher, R.; Krska, R.; Kuchler, K.; Glossl, J.; Luschnig, C.; Adam, G. Detoxification of the *Fusarium* mycotoxin deoxynivalenol by a UDP-glucosyltransferase from *Arabidopsis thaliana*. *J. Biol. Chem.* **2003**, *278*, 47905–47914. [CrossRef]

97. Zwiers, L.H.; Stergiopoulos, I.; Gielkens, M.M.C.; Goodall, S.D.; De Waard, M.A. ABC transporters of the wheat pathogen *Mycosphaerella graminicola* function as protectants against biotic and xenobiotic toxic compounds. *Mol. Genet. Genomics* **2003**, *269*, 499–507. [CrossRef]

98. Gardiner, D.M.; Stephens, A.E.; Munn, A.L.; Manners, J.M. An ABC pleiotropic drug resistance transporter of *Fusarium graminearum* with a role in crown and root diseases of wheat. *FEMS Microbiol. Lett.* **2013**, *348*, 36–45. [CrossRef]

99. Gardiner, D.M.; Osborne, S.; Kazan, K.; Manners, J.M. Low pH regulates the production of deoxynivalenol by *Fusarium graminearum*. *Microbiol. Sgm* **2009**, *155*, 3149–3156. [CrossRef]

100. Merhej, J.; Richard-Forget, F.; Barreau, C. The pH regulatory factor Pad1 regulates Tri gene expression and trichothecene production in *Fusarium graminearum*. *Fungal Genet. Biol.* **2011**, *48*, 275–284. [CrossRef]

101. Ramirez, M.L.; Chulze, S.; Magan, N. Temperature and water activity effects on growth and temporal deoxynivalenol production by two Argentinean strains of *Fusarium graminearum* on irradiated wheat grain. *Int. J. Food Microbiol.* **2006**, *106*, 291–296. [CrossRef]

102. Magan, N.; Aldred, D.; Mylona, K.; Lambert, R.J.W. Limiting mycotoxins in stored wheat. *Food Add. Contam. Part A Chem.* **2010**, *27*, 644–650. [CrossRef]

103. Kokkonen, M.; Ojala, L.; Parikka, P.; Jestoi, M. Mycotoxin production of selected *Fusarium* species at different culture conditions. *Int. J. Food Microbiol.* **2010**, *143*, 17–25. [CrossRef]

104. Jiang, J.H.; Liu, X.; Yin, Y.N.; Ma, Z.H. Involvement of a velvet protein FgVeA in the regulation of asexual development, lipid and secondary metabolisms and virulence in *Fusarium graminearum*. *PloS ONE* **2011**, *6*, e28291. [CrossRef]

105. Merhej, J.; Urban, M.; Dufresne, M.; Hammond-Kosack, K.E.; Richard-Forget, F.; Barreau, C. The velvet gene, FgVe1, affects fungal development and positively regulates trichothecene biosynthesis and pathogenicity in *Fusarium graminearum*. *Mol. Plant Pathol.* **2012**, *13*, 363–374. [CrossRef]

106. Jiang, J.H.; Yun, Y.Z.; Liu, Y.; Ma, Z.H. FgVELB is associated with vegetative differentiation, secondary metabolism and virulence in *Fusarium graminearum*. *Fungal Genet. Biol.* **2012**, *49*, 653–662. [CrossRef]

Section 3:
Impacts of DON on Animals and Humans

Article

Deoxynivalenol in the Gastrointestinal Tract of Immature Gilts under *per os* Toxin Application

Agnieszka Waśkiewicz [1], **Monika Beszterda** [1], **Marian Kostecki** [1], **Łukasz Zielonka** [2], **Piotr Goliński** [1,*] **and Maciej Gajęcki** [2]

[1] Department of Chemistry, Poznań University of Life Sciences, Wojska Polskiego 75, Poznań 60-625, Poland; agat@up.poznan.pl (A.W.); monika.beszterda@up.poznan.pl (M.B.); marian.kostecki@up.poznan.pl (M.K.)

[2] Department of Veterinary Prevention and Feed Hygiene, University of Warmia and Mazury in Olsztyn, Olsztyn 10-719, Poland; lukasz.zielonka@uwm.edu.pl (Ł.Z.); gajecki@uwm.edu.pl (M.G.)

* Author to whom correspondence should be addressed; piotrg@up.poznan.pl; Tel.: +48-61-848-78-37; Fax: +48-61-848-78-24

Received: 7 November 2013; in revised form: 13 February 2014; Accepted: 17 February 2014; Published: 5 March 2014

Abstract: Deoxynivalenol is also known as vomitoxin due to its impact on livestock through interference with animal growth and acceptance of feed. At the molecular level, deoxynivalenol disrupts normal cell function by inhibiting protein synthesis via binding to the ribosome and by activating critical cellular kinases involved in signal transduction related to proliferation, differentiation and apoptosis. Because of concerns related to deoxynivalenol, the United States FDA has instituted advisory levels of 5 µg/g for grain products for most animal feeds and 10 µg/g for grain products for cattle feed. The aim of the study was to determine the effect of low doses of deoxynivalenol applied *per os* on the presence of this mycotoxin in selected tissues of the alimentary canal of gilts. The study was performed on 39 animals divided into two groups (control, C; $n = 21$ and experimental, E; $n = 18$), of 20 kg body weight at the beginning of the experiment. Gilts received the toxin in doses of 12 µg/kg b.w./day (experimental group) or placebo (control group) over a period of 42 days. Three animals from two experimental groups were sacrificed on days 1, 7, 14, 21, 28, 35 and 42, excluding day 1 when only three control group animals were scarified. Tissues samples were prepared for high performance liquid chromatography (HPLC) analyses with the application of solid phase extraction (SPE). The results show that deoxynivalenol doses used in our study, even when applied for a short period, resulted in its presence in gastrointestinal tissues. The highest concentrations of deoxynivalenol reported in small intestine samples ranged from 7.2 (in the duodenum) to 18.6 ng/g (in the ileum) and in large intestine samples from 1.8 (in transverse the colon) to 23.0 ng/g (in the caecum). In liver tissues, the deoxynivalenol contents ranged from 6.7 to 8.8 ng/g.

Keywords: deoxynivalenol; gastrointestinal tract; immature gilts; low doses; mycotoxicosis

---

1. Introduction

Toxigenic *Fusarium* species are common pathogens of wheat, triticale and other cereals worldwide [1–3]. Many of them attack a range of plant parts and stages, such as seedlings, heads, roots, stems and ears, resulting in severe reductions of grain yield, often ranging from 10% to 40% [4,5]. Consequently, when cereal plants are infected with fungi, there is a risk of grain contamination with secondary metabolites, *i.e.*, mycotoxins, and their subsequent transfer to feed and food [6,7]. A major problem associated with animal feed contaminated with mycotoxins is not acute disease, but rather the ingestion of low levels of toxins, which may cause an array of metabolic, physiologic and immunologic disturbances [8–10]. Major *Fusarium* toxins found under the climatic conditions of Poland are deoxynivalenol, zearalenone, moniliformin and fumonisins detected in cereals during plant growth and vegetation [5,6,11,12]. The presence of deoxynivalenol (DON) in agricultural crops is an increasingly common problem associated with the occurrence of *Fusarium* head blight (FHB) infection not only under temperate weather conditions, possibly because of the extensive application of "no-tillage farming", changing climate patterns and simultaneously, enhanced cultivation of host crops such as maize and wheat [13]. Numerous studies indicated that choosing a less susceptible cultivar is a powerful tool to ensure a low DON concentration in cereal grain even under highly infectious conditions. This strategy enables farmers to make use of the benefits of conservation tillage and, simultaneously, produce high quality grain [14].

Structurally, DON is a polar organic compound with the IUPAC name: 12,13-epoxy-3α,7α,15-trihydroxytrichothec-9-en-8-on [15,16]. The ketone position in C_8 is a characteristic of the class B trichothecenes, also the number and position of the hydroxyl and acetyl-ester groups can influence the relative toxicity within cells. In addition, via the epoxy group, this toxin is able to bind to the large subunit of eukaryotic ribosomes and interfere with peptidyl transferase, thus impairing initiation or elongation of peptide chains [17,18]. DON relative capacity to interfere with protein synthesis has been attributed to a combination of different factors: the rate of transport into cells, metabolism by cytosol enzymes, changes in affinity to the active binding site or capacity to interfere with protein synthesis.

The intensity of DON toxic effects depends on the dose, species, duration of consumption, toxin purity and the route of administration [19,20]. Data on genotoxic effects of trichothecenes are scarce and these toxins are classified to group 3 (inadequate evidence) by the International Agency on Cancer Research. Among various animal species, swine are known to be especially susceptible to DON and could therefore serve as a model for human sensitivity to this mycotoxin. DON absorption in pigs is rapid, reaching peak plasma concentrations within 15 to 30 min of dosing [18,21]. Up to 0.82 systemic absorption was recorded in pigs administered DON orally. Transient tissue distribution of DON in pigs occurs with an elimination half-life of 3.9 h and very limited accumulation in tissues. From numerous studies on laboratory animals and cell lines it was concluded that low dose exposure upregulates expression of cytokines, chemokines and inflammatory genes with a concurrent immune stimulation. In addition, low to moderate dose acute oral exposure to trichothecenes causes vomiting, diarrhea, gastroenteritis, anorexia, reduced weight gain, malabsorption of glucose, glutamine and 5-methyltetrahydrofolic acid, oesophageal perforation, circulatory shock and can ultimately lead to death [22–24]. DON affects the integrity of intestinal epithelium through alterations in cell morphology and differentiation in the barrier function. In turn, high dose exposure promotes leukocyte apoptosis with a concomitant immune suppression and severe damage to the lymphoid and epithelial cells of the gastrointestinal mucosa resulting in hemorrhage, endotoxemia and shock [17,25–27]. Other targets include bone marrow and thymus, which can contribute to generalized immunosuppression. Interestingly, these gut effects can occur in animals exposed to trichothecenes via inhalation. The United States Food and Drug Administration (USFDA) has established Advisory Levels for DON in grain and grain by-products based on the intended use [18]. For beef, feedlot cattle older than 4 months and chickens it is 10 mg/kg while for pigs and all other animal species it is 5 mg of DON per kg of fodder.

The gastrointestinal tract is the first barrier against feed contaminants as well as the first target for mycotoxins. The intestine is a preferential immune site where immunoregulatory mechanisms simultaneously defend the body against pathogens, but also maintain tissue homeostasis to avoid immune-mediated pathology in response to environmental challenges [28]. However, the toxin has been demonstrated to readily cross the epithelial barrier by a paracellular pathway [29]. Hence, the purpose of the present study was to examine the effect of *per os* DON exposure at NOAEL (no observed adverse effect level) doses on the absorption, accumulation and final presence of this toxin in tissues of the gastrointestinal tract of gilts, *i.e.*, the liver, the small and large intestine, using high performance liquid chromatography tools.

2. Results and Discussion

In the last two decades, many studies have described DON toxicity using diverse species as models, where DON consumption was shown to affect numerous physiological functions such as food intake, reproduction or immunity [30–32]. The presented experiment involved two groups of immature gilts (with body weight of up to 20 kg), which were orally administered deoxynivalenol at 12 µg/kg b.w. (group E, $n = 18$) or placebo (group C, $n = 21$) over a period of 42 days. The concentration of the toxin in the small intestine, large intestine and liver tissues was chromatographically analyzed. Until now, most performance and toxicological data in pigs have been obtained with medium to high doses of the toxin, *i.e.*, 2 to 10 mg/kg of feed [33,34]. A survey of 11,022 cereal samples from 12 European countries showed that 57% were positive for DON, with 7% containing DON concentrations of 750 µg/kg or greater. One of the most important physicochemical properties of DON is its ability to withstand high temperatures, which poses a risk of its occurrence in feed and food [35]. *Ipso facto*, it is known that feed and food processing may have no effect on DON concentrations in the final product. Additionally, it seems possible that differences in the feed form could modulate the bioavailability of DON, as it could affect the liberation of the toxin from the matrix and thereby influence residue concentration in the animal tissue. In this study, the daily DON intake during the whole period of the experiment reflects slight differences between animals because of DON doses converted into body weight and precisely administered by a small portion of the feed matrix.

2.1. DON Residues in Small Intestine Tissues

The effect of feed contamination with low doses of DON on the presence of the toxin was investigated in tissue preparations from the duodenum, jejunum and ileum. As described in Table 1, in this experiment, diets significantly affect tissue concentrations of the toxin. Under individual intoxication of DON with 12 µg/kg b.w. per day doses, the toxin content in small intestine tissues ranged from 0.00 (terms I and II) to 18.60 ng/g (term VI) (Table 1). At terms I and II no mycotoxins were detected in any samples analyzed. The highest DON concentrations were detected during terms of slaughter IV–VI, at 7.20 ng/g for the duodenum, 9.20 ng/g for the jejunum and 18.60 ng/g for the ileum samples, respectively. Only in the ileum samples the content of DON in diets for pigs demonstrated linear dose relationships to tissue concentration. However, these relationships were characterized by high inter-individual variation.

Table 1. Concentration levels of deoxynivalenol (DON) [ng/g] with determined homogenous groups ($\alpha = 0.05$) and the derived carry over factor of DON in the first section of gastrointestinal tract-small intestine of gilts fed diets containing NOAEL (12 µg/kg b.w./day) concentrations of toxin.

Days of the experiment	Total doses [µg/kg b.w.]	DON amounts in small intestine [ng/g] ± standard deviation					
		Duodenum	Carry over factor	Jejunum	Carry over factor	Ileum	Carry over factor
7 (term I)	84	0 [a] ± 0	0	0 [a] ± 0	0	0 [a] ± 0	0
14 (term II)	168	0 [a] ± 0	0	0 [a] ± 0	0	0 [a] ± 0	0
21 (term III)	252	6.71 [b] ± 0.22	0.027	1.80 [a,b] ± 0.64	0.007	0 [a] ± 0	0
28 (term IV)	336	7.20 [b] ± 0.16	0.021	9.20 [c] ± 2.54	0.027	9.20 [b] ± 4.65	0.027
35 (term V)	420	4.24 [b] ± 1.57	0.010	4.54 [b] ± 0.95	0.011	11.19 [b,c] ± 2.40	0.027
42 (term VI)	504	6.13 [b] ± 2.13	0.012	8.06 [c] ± 1.08	0.016	18.60 [c] ± 4.26	0.037

Note: The same symbols for average values in a particular column indicate no statistically significant differences in the level of DON concentrations within the analyzed samples (Tukey HSD test).

Dänicke and co-workers [36] indicated that only about 1% of ingested DON reaches the proximal parts of the small intestine (duodenum and jejunum) in pigs with maximum DON contents of 1.3 mg/kg freeze-dried ingesta when swine weighing 88 kg were fed 1.1 kg feed with a DON concentration of 4.2 mg/kg. However, also lower and fluctuating DON accumulations can lead to morphological modifications of the intestinal epithelium after DON exposure *in vivo* [37]. In an experiment by Goyarts and co-workers [38], 15 min after *per os* exposure DON was found in all serum samples. This early detection of DON in serum [39] showed that the absorption of DON was rapid and indicated that it could start in the stomach or in the upper part of duodenum [40]. Presented arguments may explain the observed gradual saturation of tissues, which eventually led to an increased DON concentration in successive sections of the alimentary tract in proportion to the passing time. In turn, we would also have to remember that these slight DON levels, while gradually saturating tissues did not cause any increase in DON concentrations in earlier sections of the alimentary tract with the passage of time. Such a situation could be attributed to the acquisition of food tolerance, consisting in the inhibition of inflammation processes in the alimentary tract, particularly the ileum and the descending colon [41]. This might be a certain "escape" mechanism of mycotoxins before the induction of the local immune system, similarly as it is the case with Tregs induction by classical pathogens [42] during chronic infections.

Using the obtained results, the carry-over factor was calculated as the quotient of toxin concentration in the tissue (μg/kg) by toxin concentration in the diet (μg/kg), included as the total administered doses in the defined term experiment, according to previous researchers [38,43,44]

The mean carry-over factors in the first and second terms amounted to 0, while the maximum values in that period of time were observed in samples from terms III (for the duodenum–0.027), IV (for the jejunum–0.027) and VI (for the ileum–0.037) (Table 1). For the duodenum and jejunum specimens the median carry-over factors decrease with increasing proportions of exposure to DON starting from term IV for the duodenum and term V for the jejunum. The summary carry-over factor for the duodenum is higher than for the jejunum and ileum, respectively. In a study by Goyarts and co-workers [38] the mean carry-over factor of DON and de-epoxy-DON, presented as the content of both substances in the specimen divided by the toxin content in the diet, for all swine decreased in the descending order from the bile (0.1046), kidneys (0.0151), liver (0.0057), serum (0.0023), muscle (0.0016) to back fat (0.0002).

The percentage distribution of DON in analyzed small intestine tissues is shown in Figure 1. Starting from date 3 the quantitative participation of the mycotoxin in different tissues changed. At term III the highest DON concentration was detected in the duodenum, while at term IV the same percentage of DON was recorded in the jejunum and ileum. At terms V and VI the highest DON contents were detected in the ileum samples at a simultaneous similar participation of DON in the duodenum and jejunum specimens.

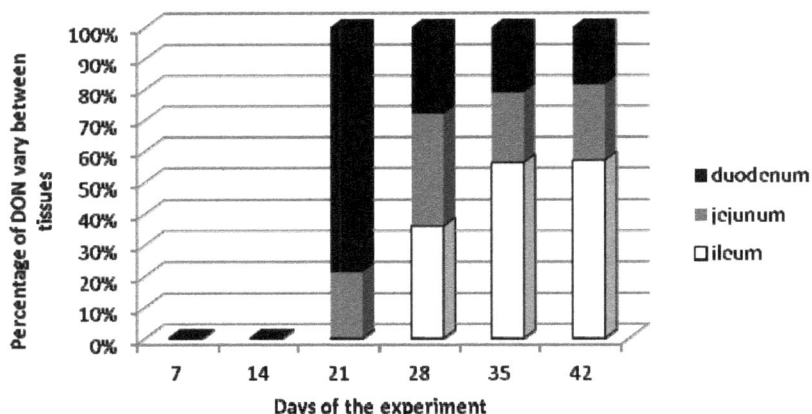

Figure 1. The percentage distribution of DON in tissues of small intestine (duodenum, jejunum, ileum).

2.2. DON Residue in Large Intestine and Liver Tissues

Using individual intoxication of DON the toxin was not detected in transverse and descending colon tissues collected in the first three dates of the experiment (Table 2). Up to term IV of sample collection the highest DON amounts were observed in the liver tissues (7.90 ng/g), and later in the ascending and descending colon sections (terms V and VI). The concentration of DON for different fragments of the gastrointestinal tract fell within the following ranges (in ng/g): for the liver from 6.70 (term VI) to 8.80 (term V), for the cecum from 0.0 (term I) to 20.5 (term V), for the transverse colon from 0.0 (terms I, II, III, V and VI) to 1.8 (term IV), and for the descending colon from 0.0 (terms I, II and III) to 20.00 (term VI). Averaging the recorded DON concentrations in tissues from all study terms the highest average level of the toxin was 8.70 ng/g in the cecum samples.

Table 2. DON concentration [ng/g] with determined homogenous groups ($\alpha = 0.05$) and the resulting carry-over factor of DON in the second section of the gastrointestinal tract, *i.e.*, the large intestine, of gilts fed diets containing NOAEL concentrations of DON.

Days of the experiment	Total doses [µg/kg b.w.]	DON amounts in large intestine [ng/g] ± standard deviation							
		Cecum	Carry over factor	Ascending colon	Carry over factor	Transverse colon	Carry over factor	Descending colon	Carry over factor
7 (term I)	84	0 [a] ± 0	0	0 [a] ± 0	0	0 [a] ± 0	0	0 [a] ± 0	0
14 (term II)	168	3.70 [a,b] ± 0.47	0.022	2.50 [a] ± 0.17	0.015	0 [a] ± 0	0	0 [a] ± 0	0
21 (term III)	252	6.42 [a,b] ± 0.22	0.025	3.10 [a] ± 1.56	0.012	0 [a] ± 0	0	0 [a] ± 0	0
28 (term IV)	336	9.81 [b] ± 4.08	0.029	7.44 [a] ± 0.92	0.022	1.80 [b] ± 0.18	0.005	6.75 [b] ± 2.53	0.020
35 (term V)	420	9.29 [b] ± 2.54	0.022	20.52 [b] ± 5.77	0.049	0 [a] ± 0	0	10.92 [b] ± 2.62	0.026
42 (term VI)	504	23.00 [c] ± 4.89	0.046	16.09 [b] ± 3.57	0.032	0 [a] ± 0	0	20.00 [c] ± 3.27	0.040

Note: The same symbols for average values in a particular column indicate no statistically significant differences in the level of DON concentrations within the analyzed samples (Tukey HSD test).

In a similar study the concentrations of DON decreased from bile > kidney > serum > liver = muscle in an experiment, where pigs received diets containing 0%, 25% and 50% of contaminated wheat containing 2.5 mg DON/kg [43].

The median carry-over factors for the large intestine samples showed significant differences. The maximum stability for the observed indicator was found for the cecum samples. The higher values of the carry-over factor were recorded in term VI for the cecum (0.046), term V for the ascending colon (0.049), term IV for the transverse colon and term VI for the descending colon (0.040) (Table 2).

The summary carry-over factor for the cecum was higher than for the ascending, descending and transverse colon, respectively. For the liver the median carry-over factors ranged from 0.013 to 0.081. The decreasing factor values were only observed in these tissue samples during the experimental periods (Table 3).

The percentage distribution of DON in the large intestine and liver tissue samples is shown in Figure 2. From term II to term III the highest percentages of the toxin with a downward trend were observed in the liver tissues, whereby starting from term IV the amounts of DON in the analyzed samples indicate similar levels in all tissues.

The observed situation is confirmed by the results of the authors' investigations [41], showing that the action of DON is multifaceted, e.g., it is an inhibitory factor for mRNA expression processes of the gene controlling the NOS-1 constitutive isomer and the NOS-2 inducible isomer. This causes specific changes in the functioning of the alimentary tract in gilts due to a reduced amount of NO, which is an inhibitor of neurotransmitters of the non-adrenergic and non-cholinergic systems [45,46]. As a consequence, low NO levels cause accelerated peristalsis of the oesophagus, stomach and intestines, inhibit stomach accommodation processes (relaxation, absorption, adaptation and successive contraction) and increased tension of intestinal sphincters, thus contributing to inhibition of gastric emptying and transfer of digesta in the intestines [47], *i.e.*, proliferation processes are activated [48]. All the presented situations are factors promoting enhanced biotransformation depending on energy resources of the organism [49], as well as better explain the cytotoxic properties of DON in the colon [50], which cytotoxicity in small doses is lower [51].

Table 3. Concentration levels of DON [ng/g] with determined homogenous groups ($\alpha = 0.05$) and the derived carry over factor of DON in liver of gilts fed diets containing NOAEL concentrations of DON.

Days of the experiment	Total doses [µg/kg b.w.]	DON amounts in liver [ng/g] ± standard deviation	
		Liver	Carry over factor
7 (term I)	84	6.79 [a] ± 1.28	0.081
14 (term II)	168	7.51 [a] ± 0.74	0.045
21 (term III)	252	7.90 [a] ± 2.87	0.031
28 (term IV)	336	7.78 [a] ± 3.35	0.023
35 (term V)	420	8.80 [a] ± 3.69	0.021
42 (term VI)	504	6.70 [a] ± 4.28	0.013

Note: The same symbols for average values in a particular column indicate no statistically significant differences in the level of DON concentrations within the analyzed samples (Tukey HSD test).

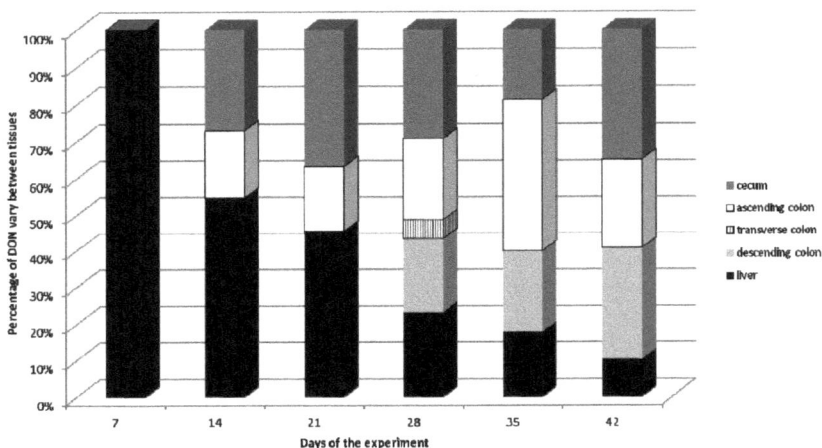

Figure 2. The percentage distribution of DON in large intestine (cecum, ascending colon, transverse colon and descending colon) and liver samples.

Up to now, exposure of swine to low doses of deoxynivalenol was investigated in the context of hematological, biochemical, immunological and histopathological parameters [9,52–54]. The toxicological principle of deoxynivalenol action may be similar in several organs, such as the lungs, liver or kidneys, and the toxin may cause damage and increase cell death in porcine intestinal cells, which is associated with metabolic stress [53]. Disturbed intestinal functions after DON ingestion seem to be connected with the fact that intestinal epithelial cells can be exposed to high luminal DON contents for a prolonged time following ingestion of contaminated feed [52,55]. An experiment on the cytotoxic effect of DON on a porcine intestinal cell line (IPEC-J2) demonstrated that the minimal effective dose to induce the cytotoxic effect was between 0.5 to 2.5 µM of DON solution in ethanol, whereas at a concentration of 10 µM this toxin caused cell damage, including rounding of cells, autolysis and cell loss from the monolayer. Such observations are confirmed by the currently unpublished results of a our study, from which it results that these low DON doses are only slightly cytotoxic to the local bacterial flora. It may be assumed that in the last segment of the large intestine (descending colon) the cytotoxic activity of DON is increased, but only at low doses [51] and the activity of fecal enzymes increases. In other words, a situation may occur promoting the occurrence of pathological changes in the final section of the porcine alimentary tract [56]. In addition, a significant decrease in ATP levels was observed at 48 h in a dose-dependent manner. Experimental acute DON toxicity in pigs includes digestive symptoms with emesis starting after 2 h following intravenous injection, together with increased defecation and diarrhea [57]. DON at high doses exceeding causes jejunal lesions, diagnosed after a 4-h exposure in an *ex-vivo* pig culture model including shortened and coalesced villi, lysis of enterocytes, and edema [58]. Similarly, at 4 mg/kg diet DON may cause corrugation in the fundic region of the stomach and after 5-week exposure to DON at 2.8 mg/kg diet, in the absence of body weight modulation in pigs, significant histopathological changes were observed [59]. Additionally, multifocal atrophy and villi fusion, apical necrosis of villi, cytoplasmatic vacuolation of enterocytes and edema of the lamina propria were observed in the jejunum and ileum of pigs administered DON [60]. In another study, Pinton and co-workers showed that DON causes a dose-dependent translocation of a pathogenic strain of *Escherichia coli* across the porcine IPEC-1 epithelial cell monolayers [61]. Similarly, an increased translocation of *Salmonella Thyphimurium* was observed in porcine IPEC-J2 exposed to low doses of DON [62].

Our experiments in combination with the results discussed above provide a significant contribution to the knowledge on DON toxicity and distribution in individual tissues of the porcine alimentary tract. This is the first report concerning quantitative analysis of deoxynivalenol in the gastrointestinal tract of gilts after *per os* toxin application and may be an important addition to knowledge concerning the effect of *per os* exposure to this toxin.

Summing up the presented results as well as studies conducted by other authors certain conclusions may be drawn. Accumulation of low DON exposure doses in the final section of the alimentary tract may contribute to specific situations, not necessarily pathological [48]. One of the may be connected with the cytotoxic property of DON. As a consequence, the count of microbial pathogens decreases, which constitute one of the proinflammatory factors in the large intestine [63]. On the other hand, exposure of animal organisms to low doses of the mycotoxin is a factor inhibiting mRNA expression of NOS controlling genes [41,46]. In this situation, we might propose a conclusion that low DON doses (below NOAEL) may have a therapeutic effect [51].

3. Experimental Section

All of the experimental procedures involving animals were carried out in compliance with the Polish legal regulations determining the terms and methods for performing experiments on animals (opinion of the local Ethics Committee for Animal Experimentation No. 88/2009).

3.1. Experimental Animals

This part of the study was conducted at the Department of Veterinary Prevention and Feed Hygiene, Faculty of Veterinary Medicine, the University of Warmia and Mazury in Olsztyn, Poland, using 39 clinically healthy female gilts. The mean initial body weight was ±20 kg. The gilts were housed in individual cages with an *ad libitum* access to water. The experimental diets were prepared locally and formulated according to the energy and amino acid requirements for piglets. The feed used in the present study was contaminated only with DON, which was confirmed by chromatography analysis of the feed matrix.

3.2. Experimental Design

A total of 39 animals were divided into two groups: control (n = 21) and experimental (n = 18). The animals from the experimental group were administered DON daily at a dose of 12 μg/kg body weight *per os* for 42 days. Analytical samples of the mycotoxin were applied in gelatin capsules before morning feeding. The animals from the control group were orally administered placebo (gelatin capsules without mycotoxin) for the same period of time as the experimental group. Amounts of toxin application were dependent on the body weight and updated weekly. Three animals from both groups (experimental and control) were sacrificed on days 1, 7, 14, 21, 28, 35 and 42 (a total of 6 gilts on each day), excluding day 1 when only three control group animals were sacrificed. Every week tissue samples were collected from the porcine gastrointestinal tract (the liver–left lobe, duodenum–first and middle sections, jejunum–middle section, ascending colon–middle section, descending colon–middle section).

3.3. Chemicals

The deoxynivalenol standard and organic solvents (HPLC grade) were purchased with a standard grade certificate from Sigma-Aldrich (Steinheim, Germany). All chemicals used for extraction and purification of DON were purchased from POCh (Gliwice, Poland). Water for the HPLC mobile phase was purified using a Milli-Q system (Millipore, Bedford, MA, USA).

3.4. Tissue Samples

Post-mortem samples of the following tissues: the duodenum, jejunum, ileum, cecum, ascending colon, transverse colon, descending colon and the liver were collected after rinsing with phosphate buffer. Pure tissues were stored at $-80\ ^{\circ}$C until the shredding and extraction procedure.

3.5. Extraction and Purification Procedure

Tissue fragments (2 g) were homogenized twice with methanol, then centrifuged (4000 rpm) and evaporated. The residue was diluted in 20 mL water and 20 mL 0.2 M CH_3COONa and applied on the top of immunoaffinity columns according to the manufacturer's recommendations (IAC, VICAM, Watertown, MA, USA). The toxin was eluted with a mixture of acetonitrile: water (82:12, v/v). The eluate was evaporated to dryness at 40 $^{\circ}$C under a stream of nitrogen. Dry residue was stored at $-20\ ^{\circ}$C until HPLC analyses.

3.6. HPLC Analysis of Deoxynivalenol

The chromatographic system consisted of a Waters 2695 high-performance liquid chromatograph (Waters, Milford, CT, USA) with a Waters 2996 Photodiode Array Detector with a Nova Pak C-18 column (300 mm $\times$ 3.9 mm) for DON (λ_{max} = 224 nm) analysis. Quantification of the toxin was performed by measuring the peak areas at the retention time according to the relevant calibration curve. The limit of detection was 0.001 µg/g for DON.

3.7. Statistical Analysis

Data on DON concentrations in various tissue samples were evaluated for statistically significant differences employing the Tukey-test. Values are presented as medians and ranges. All statistical analyses were conducted using the Statistica Package for Social Science program.

4. Conclusions

After 7-day exposure to DON (term I) among tissue samples the liver was the only organ where DON could be detected. Median DON concentrations decreased in the order from the ileum > duodenum > jejunum for the small intestine specimens and the cecum > ascending colon > liver > descending colon > transverse colon for the large intestine and liver samples. Maximum carry-over rates reached the following levels: 0.037 (term VI) for the ileum (the small intestine), 0.049 (term V) for the ascending colon and 0.046 (term VI) for the cecum (the large intestine) and 0.081 (term I) for the liver. In conclusion, the results show how dose and exposure time may affect the distribution of DON in the gastrointestinal tract of gilts after *per os* toxin application.

Acknowledgments: The study was supported by the Polish Ministry of Science and Higher Education (PMSHE) Project No. 12-0080-10/2010. The authors highly acknowledge the technical assistance by Ewa Rymaniak during the course of the experiments.

Conflicts of Interest: The authors declare no conflict of interest.

References

1. Gräfenhan, T.; Patrick, S.K.; Roscoe, M.; Trelka, R.; Gaba, D.; Chan, J.M.; McKendry, T.; Clear, R.M.; Tittlemier, S.A. *Fusarium* damage in cereal grains from Western Canada. 1. Phylogenetic analysis of moniliformin-producing *Fusarium* species and their natural occurrence in mycotoxin-contaminated wheat, oats, and rye. *J. Agric. Food Chem.* **2013**, *61*, 5425–5437. [CrossRef]

2. Juan, C.; Ritieni, A.; Mañes, J. Occurrence of *Fusarium* mycotoxins in Italian cereal and cereal products from organic farming. *Food Chem.* **2013**, *141*, 1747–1755. [CrossRef]

3. Slikova, S.; Gavurnikova, S.; Sudyova, V.; Gregova, E. Occurrence of deoxynivalenol in wheat in Slovakia during 2010 and 2011. *Toxins* **2013**, *5*, 1353–1361. [CrossRef]

4. Wiśniewska, H.; Stępień, Ł.; Waśkiewicz, A.; Beszterda, M.; Góral, T.; Belter, J. Toxigenic *Fusarium* species infecting wheat heads in Poland. *Cent. Eur. J. Biol.* **2013**, *9*, 163–172.

5. Goliński, P.; Waśkiewicz, A.; Wiśniewska, H.; Kiecana, I.; Mielniczuk, E.; Gromadzka, K.; Kostecki, M.; Bocianowski, J.; Rymaniak, E. Reaction of winter wheat (*Triticum aestivum* L.) cultivars to infection with *Fusarium* spp.: Mycotoxin contamination in grain and chaff. *Food Add. Contam.* **2010**, *27*, 1015–1024. [CrossRef]

6. Waśkiewicz, A.; Gromadzka, K.; Wiśniewska, H.; Goliński, P. Accumulation of zearalenone in genotypes of spring wheat after inoculation with *Fusarium culmorum*. *Cereal Res. Commun.* **2008**, *36*, 401–404.

7. Waśkiewicz, A.; Morkunas, I.; Bednarski, W.; Mai, V.-C.; Formela, M.; Beszterda, M.; Wiśniewska, H.; Goliński, P. Deoxynivalenol and oxidative stress indicators in winter wheat inoculated with *Fusarium graminearum*. *Toxins* **2014**, *6*, 575–591. [CrossRef]

8. Grenier, B.; Bracarense, A.-P.F.L.; Schwartz, H.E.; Lucioli, J.; Cossalter, A.-M.; Moll, W.-D.; Schatzmayr, G.; Oswald, I.P. Biotranformation approaches to alleviate the effects induced by *Fusarium* mycotoxins in swine. *J. Agric. Food Chem.* **2013**, *61*, 6711–6719. [CrossRef]

9. Gajęcka, M.; Rybarczyk, L.; Jakimiuk, E.; Zielonka, Ł.; Obremski, K.; Zwierzykowski, W.; Gajęcki, M. The effect of experimental long-term exposure to low-dose zearalenone on uterine histology in sexually immature gilts. *Exp. Toxicol. Pathol.* **2012**, *64*, 537–542. [CrossRef]

10. Girish, C.K.; MacDonald, E.J.; Scheinin, M.; Smith, T.K. Effects of feedborne *Fusarium* mycotoxins on brain regional neurochemistry of turkeys. *Poult. Sci.* **2008**, *87*, 1295–1302. [CrossRef]

11. Wegulo, S.N. Factors influencing deoxynivalenol accumulation in small grain cereals. *Toxins* **2012**, *4*, 1157–1180.

12. SCOOP. Collection of Occurrence Data of *Fusarium* Toxins in Food and Assessment of Dietary Intake by the Population of EU Member States. Reports on Tasks for Scientific Cooperation. 2003. Available online: http://ec.europa.eu/food/fs/scoop/task3210.pdf (accessed on 31 October 2013).

13. Waśkiewicz, A.; Irzykowska, L.; Bocianowski, J.; Koralewski, Z.; Kostecki, M.; Weber, Z.; Goliński, P. Occurrence of *Fusarium* fungi and mycotoxins in marketable asparagus spears. *Polish J. Environ. Stud.* **2010**, *19*, 219–225.

14. Koch, H.-J.; Pringas, C.; Maerlaender, B. Evaluation of environmental and management effects on *Fusarium* head blight infection and deoxynivalenol concentration in the grain of winter wheat. *Eur. J. Agron.* **2006**, *24*, 357–366. [CrossRef]

15. Sobrowa, P.; Adam, V.; Vasatkova, A.; Beklova, M.; Zeman, L.; Kizek, R. Deoxynivalenol and its toxicity. *Interdiscip. Toxicol.* **2010**, *3*, 94–99.

16. Nagy, C.M.; Fejer, S.N.; Berek, L.; Molnar, J.; Viskolcz, B. Hydrogen bondings in deoxynivalenol (DON) conformations—A density functional study. *J. Mol. Struct.* **2005**, *726*, 55–59. [CrossRef]

17. Döll, S.; Schrichx, J.A.; Dänicke, S.; Fink-Gremmels, J. Interactions of deoxynivalenol and lipopolysaccharides on cytokine excretion and mRNA expression in porcine hepatocytes and Kupffer cell enriched hepatocyte. *Toxicol. Lett.* **2009**, *190*, 96–105. [CrossRef]

18. Pestka, J.J. Deoxynivalenol: Toxicity, mechanisms and animal health risks. *Anim. Feed Sci. Technol.* **2007**, *137*, 283–298. [CrossRef]

19. Pestka, J.J. Deoxynivalenol: Mechanisms of action, human exposure, and toxicological relevance. *Arch. Toxicol.* **2010**, *84*, 663–679. [CrossRef]

20. Bonnet, M.S.; Roux, J.; Mounien, L.; Dallaporta, M.; Troadec, J.-D. Advances in deoxynivalenol toxicity mechanisms: The brain as a target. *Toxins* **2012**, *4*, 1120–1138. [CrossRef]

21. Pestka, J.J.; Smolinski, A.T. Deoxynivalenol: Toxicology and potential effects on humans. *J. Toxicol. Environ. Health B Crit. Rev.* **2005**, *8*, 39–69. [CrossRef]

22. Awad, W.A.; Aschenbach, J.R.; Setyabudi, F.M.C.S.; Razzazi-Fazeli, E.; Böhm, J.; Zentek, J. *In vitro* effects of deoxynivalenol on small intestinal D-glucose uptake and absorption of deoxynivalenol across the isolated jejunal epithelium of laying hens. *Poult. Sci.* **2007**, *86*, 15–20. [CrossRef]

23. Arnold, D.L.; McGuire, P.F.; Nera, E.A; Karpinski, K.F.; Bickis, M.G.; Zawidzka, Z.Z.; Fernie, S.; Vesonder, R.F. The toxicity of orally administered deoxynivalenol (vomitoxin) in rats and mice. *Food Chem. Toxicol.* **1986**, *24*, 935–941. [CrossRef]

24. Maresca, M.; Mahfoud, R.; Garmy, N.; Fantini, F. The mycotoxin deoxynivalenol affects nutrient absorption in human intestinal epithelial Cells. *J. Nutr.* **2002**, *132*, 2723–2731.

25. Pinton, P.; Tsybulskyy, D.; Lucioli, J.; Laffitte, J.; Callu, P.; Lyazhri, F.; Grosjean, F.; Bracarense, A.P.; Kolf-Clauw, M.; Oswald, I.P. Toxicity of deoxynivalenol and its acetylated derivatives on the intestine: Differential effects on morphology, barrier function, tight function proteins, and mitogen-activated protein kinases. *Toxicol. Sci.* **2012**, *130*, 180–190. [CrossRef]

26. Bensassi, F.; El Golli-Bennour, E.; Abid-Essefi, S.; Bouaziz, C.; Hajlaoui, M.R.; Bacha, H. Pathway of deoxynivalenol-induced apoptosis in human colon carcinoma cells. *Toxicology* **2009**, *264*, 104–109. [CrossRef]

27. Diesing, A.K.; Nossol, C.; Dänicke, S.; Walk, N.; Post, A.; Kahlert, S.; Rothkötter, H.J.; Kluess, J. Vulnerability of polarised intestinal porcine epithelial cells to mycotoxin deoxynivalenol depends on the route of application. *PLoS ONE* **2011**, *6*, e17472. [CrossRef]

28. Bouhet, S.; Oswald, I.P. The effects of mycotoxins, fungal food contaminants, on the intestinal epithelial cell-derived innate immune response. *Vet. Immunol. Immunopathol.* **2005**, *108*, 199–209. [CrossRef]

29. Sergent, T.; Parys, M.; Garsou, S.; Pussemier, L.; Schneider, Y.J.; Larondelle, Y. Deoxynivalenol transport across human intestinal Caco-2 cells and its effects on cellular metabolism at realistic intestinal concentrations. *Toxicol. Lett.* **2006**, *164*, 167–176. [CrossRef]

30. Seeling, K.; Dänicke, S.; Valenta, H.; van egmond, H.P.; Schothorst, R.C.; Jekel, A.A.; Lebzien, P.; Schollenberger, M.; Razzazi-Fazeli, E.; Flachowsky, G. Effects of *Fusarium* toxin-contaminated wheat and feed intake level on the biotransformation and carry-over of deoxynivalenol in dairy cows. *Food Addit. Contam.* **2006**, *23*, 1008–1020. [CrossRef]

31. Gutzwiller, A. Effects of deoxynivalenol (DON) in the lactation diet on the feed intake and fertility of sows. *Mycotoxin Res.* **2010**, *26*, 211–215. [CrossRef]

32. Collins, T.F.X.; Sprando, R.L.; Black, T.N.; Olejnik, N.; Eppley, R.M.; Hines, F.A.; Rorie, J.; Ruggles, D.I. Effects of deoxynivalenol (DON, vomitoxin) on in utero development in rats. *Food Chem. Toxicol.* **2006**, *44*, 747–757. [CrossRef]

33. Prelusky, D.B.; Gerdes, R.G.; Underhill, K.L.; Rotter, B.A.; Jui, P.Y.; Trenholm, H.L. Effects of low-level dietary deoxynivalenol on the hematological and clinical parameters of the pig. *Nat. Toxins* **1994**, *2*, 97–104. [CrossRef]

34. Swamy, H.V.L.N.; Smith, T.K.; MacDonald, E.J.; Boermans, H.J.; Squires, E.J. Effects of feeding a blend of grains naturally contaminated with *Fusarium* mycotoxins on growth and immunological measurements of starter pigs, and the efficacy of a polymeric glucomannan mycotoxin adsorbent. *J. Anim. Sci.* **2002**, *80*, 3257–3267.

35. Bretz, M.; Beyer, M.; Cramer, B.; Knecht, A.; Humpf, H.-U. Thermal degradation of the *Fusarium* mycotoxin deoxynivalenol. *J. Agric. Food Chem.* **2006**, *54*, 6445–6451. [CrossRef]

36. Dänicke, S.; Valenta, H.; Döll, S. On the toxicokinetics and the metabolism of deoxynivalenol (DON) in the pig. *Arch. Anim. Nutr.* **2004**, *58*, 169–180. [CrossRef]

37. Zielonka, Ł.; Wiśniewska, M.; Gajecka, M.; Obremski, K.; Gajecki, M. Influence of low doses of deoxynivalenol on histopathology of selected organs of pigs. *Pol. J. Vet. Sci.* **2009**, *12*, 89–95.

38. Goyarts, T.; Dänicke, S.; Valenta, H.; Ueberschär, K.H. Carry-over of *Fusarium* toxins (deoxynivalenol and zearalenone) from naturally contaminated wheat to pigs. *Food Addit. Contam.* **2007**, *24*, 369–380. [CrossRef]

39. Dänicke, S.; Brezina, U. Kinetics and metabolism of the Fusarium toxin deoxynivalenol in farm animals: Consequences for diagnosis of exposure and intoxication and carry over. *Food Chem. Toxicol.* **2013**, *60*, 58–75. [CrossRef]

40. Eriksen, G.S.; Pettersson, H.; Lindberg, J.E. Absorption, metabolism and excretion of 3-acetyl DON in pigs. *Arch. Tierernähr.* **2003**, *57*, 335–345.

41. Gajęcka, M.; Stopa, E.; Tarasiuk, M.; Zielonka, Ł.; Gajęcki, M. The expression of type-1 and type-2 nitric oxide synthase in selected tissues of the gastrointestinal tract during mixed mycotoxicosis. *Toxins* **2013**, *5*, 2281–2292. [CrossRef]

42. Silva-Campa, E.; Mata-Haro, V.; Mateu, E.; Hernández, J. Porcine reproductive and respiratory syndrome virus induces CD4+CD8+CD25+Foxp3+ regulatory T cells (Tregs). *Virology* **2012**, *430*, 73–80. [CrossRef]

43. Döll, S.; Dänicke, S.; Valenta, H. Residues of deoxynivalenol (DON) in pig tissue after feeding mash or pellet diets containing low concentrations. *Mol. Nutr. Food Res.* **2008**, *52*, 727–734. [CrossRef]

44. Dänicke, S.; Goyarts, T.; Döll, S.; Grove, N.; Spolders, M.; Flachowsky, G. Effects of the *Fusarium* toxin deoxynivalenol on tissue protein synthesis in pigs. *Toxicol. Lett.* **2006**, *165*, 297–311. [CrossRef]

45. Gupta, A.; Sharma, A.C. Despite minimal hemodynamic alterations endotoxemia modulates NOS and p38-MAPK phosphorylation via metalloendopeptidases. *Mol. Cell. Biochem.* **2004**, *265*, 4–56.

46. Grześk, E.; Grześk, G.; Koziński, M.; Stolarek, W.; Zieliński, M.; Kubica, J. Nitric oxide as a cause and a potential place therapeutic intervention in hypo responsiveness vascular in early sepsis. *Folia Cardiol.* **2011**, *6*, 36–43.

47. Castro, M.; Muñoz, J.M.; Arruebo, M.P.; Murillo, M.D.; Arnal, C.; Bonafonte, J.I.; Plaza, M.A. Involvement of neuronal nitric oxide synthase (nNOS) in the regulation of migrating motor complex (MMC) in sheep. *Vet. J.* **2012**, *192*, 352–358. [CrossRef]

48. Lucioli, J.; Pinton, P.; Callu, P.; Laffitte, J.; Grosjean, F.; Kolf-Clauw, M.; Oswald, I.P.; Bracarense, A.P.F.R.L. The food contaminant deoxynivalenol activates the mitogen activated protein kinases in the intestine: Interest of *ex vivo* models as an alternative to *in vivo* experiments. *Toxicon* **2013**, *66*, 31–36. [CrossRef]

49. Alonso-Pozos, I.; Rosales-Torres, A.M.; Avalos-Rodriguez, A.; Vergara-Onofre, M.; Rosado-Garcia, A. Mechanism of granulosa cell death during follicular atresia depends on follicular size. *Theriogenology* **2003**, *60*, 1071–1081. [CrossRef]

50. Waché, Y.J.; Valat, C.; Postollec, G.; Bougeard, S.; Burel, C.; Oswald, I.P.; Fravalo, P. Impact of deoxynivalenol on the intestinal microflora of pigs. *Int. J. Mol. Sci.* **2009**, *10*, 1–17.

51. Alassane-Kpembi, I.; Kolf-Clauw, M.; Gauthier, T.; Abrami, R.; Abiola, F.A.; Oswald, I.P.; Puel, O. New insights into mycotoxin mixtures: The toxicity of low doses of type B trichothecenes on intestinal epithelial cells is synergistic. *Toxicol. Appl. Pharmacol.* **2013**, *272*, 191–198. [CrossRef]

52. Awad, W.A.; Aschenbach, J.R.; Zentek, J. Cytotoxicity and metabolic stress induced by deoxynivalenol in the porcine intestinal IPEC-J2 cell line. *J. Anim. Physiol. Anim. Nutr.* **2012**, *96*, 709–716. [CrossRef]

53. Grenier, B.; Loureiro-Bracarense, A.-P.; Lucioli, J.; Pacheco, G.D.; Cossalter, A.-M.; Moll, W.-D.; Schatzmayr, G.; Oswald, I.P. Individual and combined effects of subclinical doses of deoxynivalenol and fumonisins in piglets. *Mol. Nutr. Food Res.* **2011**, *55*, 761–771. [CrossRef]

54. Weaver, A.C.; See, M.T.; Hansen, J.A.; Kim, Y.B.; de Souza, A.L.P.; Middleton, T.F.; Kim, S.W. The use of feed additives to reduce the effects of aflatoxin and deoxynivalenol on pig growth, organ health and immune status during chronic exposure. *Toxins* **2013**, *5*, 1261–1281. [CrossRef]

55. Awad, W.A.; Razzazi-Fazeli, E.; Böhm, J.; Ghareeb, K.; Zentek, J. Effect of addition of a probiotic microorganism to broiler diets contaminated with deoxynivalenol on performance and histological alterations of intestinal villi of broiler chickens. *Poult. Sci.* **2006**, *85*, 974–979. [CrossRef]

56. Arunachalam, C.; Doohan, F.M. Trichothecene toxicity in eukaryotes: Cellular and molecular mechanisms in plants and animals. *Toxicol. Lett.* **2013**, *217*, 149–158. [CrossRef]

57. Coppock, R.W.; Swanson, S.P.; Gelberg, H.B.; Koritz, G.D.; Hoffman, W.E.; Buck, W.B. Preliminary study of the pharmacokinetics and toxicopathy of deoxynivalenol (vomitoxin) in swine. *Am. J. Vet. Res.* **1985**, *46*, 169–174.

58. Kolf-Clauw, M.; Castellote, J.; Joly, B.; Bourges-Abella, N.; Raymond-Letron, I.; Pinton, P. Development of a pig jejunal explant culture for studying the gastrointestinal toxicity of the mycotoxin deoxynivalenol: Histopathological analysis. *Toxicol. in Vitro* **2009**, *23*, 1580–1584. [CrossRef]

59. D'Mello, J.P.F. Antinutritional factors and mycotoxins. In *Farm Animal Metabolism and Nutrition*; D'Mello, J.P.F., Ed.; CAB International: Wallingford, UK, 2000; pp. 383–403.

60. Bracarense, A.P.; Lucioli, J.; Grenier, B.; Drociunas Pacheco, G.; Moll, W.D.; Schatzmayr, G. Chronic ingestion of deoxynivalenol and fumonisin, alone or in interaction, induces morphological and immunological changes in the intestine of piglets. *Br. J. Nutr.* **2012**, *107*, 1776–1786. [CrossRef]

61. Pinton, P.; Nougayrede, J.P.; Del Rio, J.C.; Moreno, C.; Marin, D.E.; Ferrier, L. The food contaminant deoxynivalenol, decreases intestinal barrier permeability and reduces claudin expression. *Toxicol. Appl. Pharmacol.* **2009**, *237*, 41–48. [CrossRef]

62. Vandenbroucke, V.; Croubels, S.; Martel, A.; Verbrugghe, E.; Goossens, J.; van Deun, K. The mycotoxin deoxynivalenol potentiates intestinal inflammation by *Salmonella typhimurium* in porcine ileal loops. *PLoS ONE* **2011**, *6*, e23871.

63. Davila, A.-M.; Blachier, F.; Gotteland, M.; Andriamihaja, M.; Benetti, P.-H.; Sanz, Y.; Tomé, D. Re-print of "Intestinal luminal nitrogen metabolism: Role of the gut microbiota and consequences for the host". *Pharmacol. Res.* **2013**, *69*, 114–126. [CrossRef]

![toxins logo] *toxins*

MDPI

Article

Organ Damage and Hepatic Lipid Accumulation in Carp (*Cyprinus carpio* L.) after Feed-Borne Exposure to the Mycotoxin, Deoxynivalenol (DON)

Constanze Pietsch [1,*], **Carsten Schulz** [2,3], **Pere Rovira** [4], **Werner Kloas** [5] and **Patricia Burkhardt-Holm** [1,6]

[1] Man-Society-Environment, Department of Environmental Sciences, University of Basel, Vesalgasse 1, Basel CH-4051, Switzerland; patricia.holm@unibas.ch

[2] GMA Society/Association for Marine Aquaculture Ltd., Hafentörn 3, Büsum D-25761, Germany; cschulz@tierzucht.uni-kiel.de

[3] Christian Albrechts-University of Kiel, Institute for Animal Breeding and Husbandry, Olshausestr. 40, Kiel 24098, Germany

[4] Forest Sciences Centre of Catalonia (CTFC), Pujada del Seminari s/n, Solsona E-25280, Spain; pere.rovira@ctfc.es

[5] Department of Ecophysiology and Aquaculture, Leibniz-Institute of Freshwater Ecology and Inland Fisheries, Mueggelseedamm 310, Berlin D-12587, Germany; werner.kloas@igb-berlin.de

[6] Department of Biological Sciences, University of Alberta, CW 405 Biological Sciences Building, Edmonton, AB T6G 2E9, Canada

* Author to whom correspondence should be addressed; pietsch.constanze@gmail.com; Tel.: +41-61-256-0405; Fax: +41-61-256-0409.

Received: 3 January 2014; in revised form: 3 February 2014; Accepted: 6 February 2014; Published: 21 February 2014

Abstract: Deoxynivalenol (DON) frequently contaminates animal feed, including fish feed used in aquaculture. This study intends to further investigate the effects of DON on carp (*Cyprinus carpio* L.) at concentrations representative for commercial fish feeds. Experimental feeding with 352, 619 or 953 µg DON kg^{-1} feed resulted in unaltered growth performance of fish during six weeks of experimentation, but increased lipid peroxidation was observed in liver, head kidney and spleen after feeding of fish with the highest DON concentration. These effects of DON were mostly reversible by two weeks of feeding the uncontaminated control diet. Histopathological scoring revealed increased liver damage in DON-treated fish, which persisted even after the recovery phase. At the highest DON concentration, significantly more fat, and consequently, increased energy content, was found in whole fish body homogenates. This suggests that DON affects nutrient metabolism in carp. Changes of lactate dehydrogenase (LDH) activity in kidneys and muscle and high lactate levels in serum indicate an effect of DON on anaerobic metabolism. Serum albumin was reduced by feeding the medium and a high dosage of DON, probably due to the ribotoxic action of DON. Thus, the present study provides evidence of the effects of DON on liver function and metabolism.

Keywords: liver damage; oxidative stress; nutrient allocation; aquaculture

1. Introduction

Deoxynivalenol (DON) is a trichothecene mycotoxin that is commonly known to be produced by *Fusarium* fungi, but also fungal species, such as *Myrothecium*, *Cephalosporium*, *Verticimonosporium* and *Stachybotrys* [1]. The toxic effects of DON in mammals include diarrhoea, emesis and malabsorption of nutrients [2–4]. A recommended guidance value for DON in compound feed stuff of 5 mg kg^{-1} DON was established by the European Commission (2006/576/EC) [5]. As far as it is known, fish feeds do not exceed this level, although recent research has shown that DON frequently can be observed in fish feeds at concentrations of up to 825 µg kg^{-1} [6]. DON affects fish growth performance and health. For example, exposure to DON resulted in the reduction of growth performance in salmonids [7,8]. These investigations on salmonids also showed that histopathological changes and lesions in the liver of fish, including the altered appearance of hepatocytes, subcapsular edema and fat accumulation, occurred upon feeding with DON-contaminated diets at concentrations of 1.4 mg kg^{-1} and higher [8]. Lipid accumulation in liver tissue is a common problem in aquaculture and can be caused by parasites [9], inadequate nutrition [10–12], pesticides [13–15] and toxins [16]. Evidence for changes of nutritional status in fish has only been shown for rainbow trout treated with 2.6 mg kg^{-1} DON for 56 days [8], whereby crude protein values were reduced by DON compared to control fish. The effects of DON on the immune system of fish have only been shown for carp at even lower DON concentrations of 352 to 953 µg DON kg^{-1} feed [17].

In the last few years, many studies have been focused on fish health interactions with dietary management. In aquaculture, fish nutrition is critical, and the applied feed contributes up to more than 60% of the total production costs [18]. Thus, nutritional and economical optimization of dietary compositions for certain fish species supports the expanding aquaculture sector. The utilization of plant products in aquaculture feed has been increasing in the last few years tremendously, as the supply of conventional feed sources, e.g., fish meal, is limited, and prices are consequently rising. With increasing plant material utilization in aquaculture, the risk for feed borne exposure to DON increases. Therefore, further understanding of dietary DON on the health and nutritional value of carp (*Cyprinus carpio*), as an aquaculture species of high global relevance, is needed to guarantee fish health and a safe product for human nutrition.

The aim of the present study was therefore the evaluation of various DON feed dosages on the growth performance and health status of juvenile carp. The present study is a part of a feeding study in which stress and immune responses already have been evaluated and published elsewhere [17]. The analyses reported in the present study were conducted on preserved samples from this trial, to further elaborate the detrimental health effects of DON observed in the experiment. The main findings reported in the previous study were that fish fed the diet containing DON showed reduced immune parameters. The present study intended to investigate the effects of DON on liver condition and metabolism more closely.

2. Results and Discussion

2.1. Composition of the Diet and Growth Performance

The experimental diets were prepared as shown in Table 1. The inclusion of ingredients has been chosen to meet the nutritional requirements of carp [19]. The experimental diets differed only in their DON content.

The nutritional compositions of experimental diets are given in Table 2, showing that no nutritional differences between the diets occurred. The diets were formulated to be isonitrogenous (44.58%–45.91% crude protein) and isocaloric (22.26–22.51 MJ kg^{-1} dry matter).

After four weeks of feeding, two tanks containing six fish each were sampled, whereas two similar treated tanks containing 12 fish were fed the uncontaminated diet for a further two weeks and sampled thereafter. The growth performance, serum parameters and biochemical composition of all fish were analysed.

Table 1. The ingredients (percent of inclusion) used for preparations of experimental fish diets. DON, deoxynivalenol.

Ingredient	Basal Feed	Low DON	Medium DON	High DON
Fish meal	30.0	30.0	30.0	30.0
Blood meal	12.5	12.5	12.5	12.5
Casein	12.0	12.0	12.0	12.0
Dextrose	13.0	13.0	13.0	13.0
Potato starch	21.1	21.1	21.1	21.1
Fish oil	10.4	10.4	10.4	10.4
Vitamins[1]	0.5	0.5	0.5	0.5
Minerals[1]	0.5	0.5	0.5	0.5
DON (mg kg^{-1})	0	0.352	0.619	0.953

Notes: Vitamin and mineral mix (Spezialfutter Neuruppin-VM BM 55/13 no. 7318): vitamin A, 12,000 I.E.; vitamin D3, 1,600 I.E; vitamin E, 160 mg; vitamin K3, 6.4 mg; vitamin B1, 12 mg; vitamin B2, 16 mg; vitamin B6, 12 mg; vitamin B12, 26.4 µg; nicotinic acid, 120 mg; biotin, 800 µg; folic acid, 4.8 mg; pantothenic acid, 40 mg; inositol, 240 mg; vitamin C, 160 mg; antioxidants (BHT), 120 mg; iron, 100 mg; zinc, 24 mg; manganese, 16 mg; cobalt 0.8 mg; iodine, 1.6 mg; selenium, 0.08 mg.

Table 2. The composition of the experimental fish feeds. The values are given as the means ± SD of two independent determinations of the same feed batch. NFE, nitrogen-free extract.

Composition	Basal Feed	Low DON	Medium DON	High DON
Dry matter (% wet matter)	91.70 ± 0.01	91.58 ± 0.02	91.66 ± 0.05	91.68 ± 0.04
Crude protein (% dry matter)	44.58 ± 0.28	45.91 ± 0.17	45.14 ± 0.06	44.89 ± 0.48
Crude lipid (% dry matter)	14.61 ± 0.03	14.56 ± 0.07	14.63 ± 0.01	14.96 ± 0.20
NFE (% dry matter)	33.15 ± 0.36	32.13 ± 0.18	32.60 ± 0.01	32.47 ± 0.34
Crude ash (% dry matter)	7.66 ± 0.01	7.40 ± 0.03	7.64 ± 0.05	7.68 ± 0.05
Gross energy (MJ kg^{-1} dry matter)	22.51 ± 0.07	22.46 ± 0.01	22.26 ± 0.02	22.42 ± 0.01

Feeding carp the experimental diets at a restricted daily basis of 2% of body weight resulted in increased fish body weights after four weeks (Table 3). Compared to the initial weight of fish, significant weight gain was observed for the groups fed the DON-contaminated diets, but not for the control group. However, individual specific growth rates (SGR), calculated as shown in Equation (1) were not found to be different between treatment groups after four weeks of feeding (mean ± SEM: control fish: 1.31 ± 0.11; low dose: 1.13 ± 0.20; medium dose: 1.71 ± 0.31; high dose: 1.50 ± 0.06, respectively).

$$SGR = (\ln \text{ final weight} - \ln \text{ initial weight})/\text{days of experiments} \times 100 \qquad (1)$$

Table 3. The growth performance of experimental fish after four weeks of DON feeding; n = 12 each.

Growth Parameters	Basal Feed	Low DON	Medium DON	High DON
Initial Weight (g)	37.23 ± 6.31	36.36 ± 6.67	36.23 ± 8.00	34.46 ± 4.05
Final Weight (g)	48.38 ± 7.48	51.86 ± 5.50	52.71 ± 6.37	56.40 ± 10.63
Final Total Length (cm)	14.11 ± 0.61	14.64 ± 0.52	14.76 ± 0.66	14.62 ± 0.99
Final Condition Factor	0.016 ± 0.000	0.016 ± 0.000	0.016 ± 0.000	0.016 ± 0.000

Furthermore, condition factors were not different between treatment groups after four weeks of feeding and also after recovery for additional weeks (Tables 3 and 4). The final weight of fish after the recovery phase was also not significantly influenced by DON application (Table 4). Individual specific growth rates were also not different between treatment groups after the recovery phase (mean ± SEM: control fish: 1.26 ± 0.26; low dose: 1.62 ± 0.23; medium dose: 1.19 ± 0.30; high dose: 1.26 ± 0.25, respectively). However, the comparison of final weights to the initial weights at the start of the experiments showed a significant difference for the control group and the fish fed the low dose diet, but not for the fish fed the higher DON doses.

Table 4. The growth performance of experimental fish after four weeks of DON feeding with an additional two weeks of recovery; $n = 12$ each.

Growth Parameters	Basal Feed	Low DON	Medium DON	High DON
Initial Weight (g)	38.24 ± 5.51	33.46 ± 4.42	36.71 ± 5.01	38.18 ± 8.77
Final Weight (g)	65.58 ± 10.42	57.56 ± 8.36	50.99 ± 9.15	53.72 ± 9.83
Final Total Length (cm)	14.11 ± 0.61	14.64 ± 0.52	14.76 ± 0.66	14.62 ± 0.99
Final Condition Factor	0.015 ± 0.000	0.015 ± 0.000	0.015 ± 0.000	0.014 ± 0.000

Reduced intake of DON-contaminated feed leading to reduced weight gain was reported in mice [20], but not in the present study. Thus, it is unlikely that feed deprivation was a factor in the effects on carp metabolism, as fish were observed to ingest the entire feed ration. Nevertheless, Atlantic salmon showed reduced weight gain after 15 weeks of feeding of 3.7 mg DON per kilogram of feed, while rainbow trout revealed similar responses to 0.3 to 2.6 mg DON per kilogram of feed after 56 days of feeding [7,8]. In contrast, a previous study on zebrafish showed that treatment of fish with feed-borne DON concentrations of up to 3 mg per kilogram of feed did not result in effects on weight gain [21]. Thus, our results together with the study of Sanden *et al.* [21] suggest that weight gain is not a sensitive parameter when the effects of DON are investigated in cyprinids.

2.2. Histology

Prussian blue staining in liver tissue did not result in extensive staining of macrophages in liver tissues in the treated groups and in the control group. Therefore, results are not reported here. PAS reaction was positive in the liver of all fish. However, PAS staining did not indicate significant differences between control fish and DON-treated fish with respect to chrominance, RGB_{max} and luminosity values (Table 5).

Table 5. The histological estimation of glycogen in PAS-stained liver sections of experimental fish after four weeks of DON feeding; $n = 6$ each, calculated from five pictures from each slide.

Color Properties of Sections	Basal Feed	Low DON	Medium DON	High DON
Chrominance	50.1 ± 2.8	50.5 ± 1.2	46.7 ± 3.4	47.4 ± 3.0
RGB_{max}	135.1 ± 1.5	133.8 ± 2.4	129.9 ± 4.1	131.4 ± 2.4
Luminosity	23.7 ± 1.5	24.1 ± 0.9	23.1 ± 2.0	23.2 ± 1.7

However, the examination of haematoxylin and eosin (HE)-stained sections revealed significant differences between the condition of the liver tissue of control fish and DON-treated fish (Table 6). DON-treated fish showed significantly increased fat disposition (Figure 1) and severe hyperaemia, whereas no significant difference was found when tissue lesions and the degree of vacuolization were recorded. The observation of the dilatation of sinusoids revealed a significant difference between control fish and fish treated with the low and medium DON diet. In rainbow trout fed 1.4 mg kg^{-1} DON for 15 weeks, congestion in liver tissue was observed [8].

In addition, feeding 2.6 mg kg^{-1} DON in the same study led to fatty infiltration in liver. Histological changes, including fat deposition, have also been found in DON-treated carp in the present study. Liver damage was probably caused by the occurrence of oxidative stress (as indicated by the lipid peroxidation assay) together with apoptotic loss of cell integrity and disturbance of nutrient metabolism. Liver damage also often leads to the occurrence of increased liver fat content, as is reported for fish recovering from being fed the high-dose DON feed for four weeks (Table 6).

Table 6. The histological condition of haematoxylin and eosin (HE)-stained liver tissue after DON feeding and a two-week recovery phase (mean ± standard errors of six fish per treatment group; each fish was analysed by using 10 fields taken from two slides (0 = no alterations, 1 = mild alterations, 2 = moderate alterations, 3 = severe alterations); means with the same letter ([a] and/or [b]) are not significantly different from each other [significance tested with Mann-Whitney U-tests, $p < 0.05$)].

Histological Alteration	Basal Feed	Low DON	Medium DON	High DON
DON-treated:				
Lesions	0.03 ± 0.03	0.20 ± 0.13	0.14 ± 0.05	0.35 ± 0.15
Fat aggregation	0.92 ± 0.07 [a]	1.57 ± 0.28 [a,b]	1.94 ± 0.25 [b]	1.80 ± 0.19 [b]
Hyperaemia	1.00 ± 0.12 [a]	2.02 ± 0.26 [b]	1.90 ± 0.26 [b]	1.90 ± 0.23 [b]
Vacuolization	1.17 ± 0.06	1.77 ± 0.32	1.84 ± 0.32	1.85 ± 0.20
Dilation of sinusoids	1.20 ± 0.16 [a]	1.87 ± 0.07 [b]	1.82 ± 0.13 [b]	1.60 ± 0.16 [a,b]
Recovery:				
Lesions	0.02 ± 0.02	0.28 ± 0.15	0.12 ± 0.07	0.18 ± 0.09
Fat aggregation	0.50 ± 0.18 [a]	1.85 ± 0.26 [b]	1.47 ± 0.31 [a,b]	1.63 ± 0.29 [b]
Hyperaemia	1.26 ± 0.11 [a]	1.70 ± 0.17 [a,b]	1.37 ± 0.19 [a,b]	2.08 ± 0.23 [b]
Vacuolization	0.54 ± 0.22	1.45 ± 0.41	1.42 ± 0.38	1.48 ± 0.38
Dilation of sinusoids	1.44 ± 0.13 [a]	2.10 ± 0.12 [b]	1.78 ± 0.18 [a,b]	1.92 ± 0.11 [a,b]

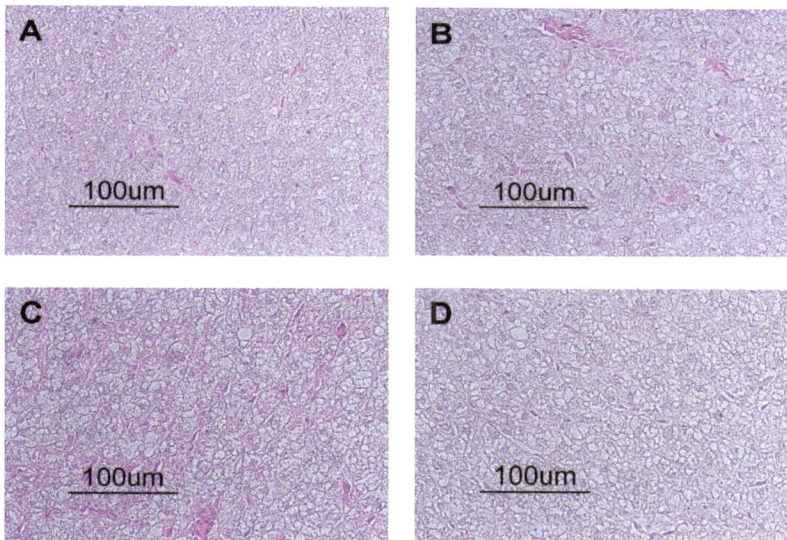

Figure 1. Histological sections of liver tissue from control fish (**A**) and DON-treated fish [low dose (**B**), medium dose (**C**) and high dose (**D**)].

2.3. Lipid Peroxidation

Measurement of lipid peroxidation indicated increased membrane damages in fish fed the diet with the high-dose DON compared to control fish (Figure 2A). This was no longer observed after the recovery phase of two weeks. A similar pattern was observed in spleen (Figure 2C). In contrast, lipid peroxidation in trunk kidney was reduced in the group treated with the high dose DON feed after four weeks of feeding and in all DON-treated fish after the recovery (Figure 2B).

Lipid peroxidation in liver was enhanced in fish treated with the high-dose feed after four experimental weeks and significantly reduced after the recovery phase compared to control fish (Figure 3A). Lipid peroxidation in muscle samples showed no differences, due to DON treatment (Figure 3B). Oxidative stress in these carp has already been indicated by the elevation of antioxidative enzymes in erythrocytes [17] and probably also contributed to the damage to several organs of DON-treated fish. Oxidative stress and lipid peroxidation due to the *Fusarium* toxins, deoxynivalenol and zearalenone, have frequently been shown in mammalian cell cultures and farm animals [22–24].

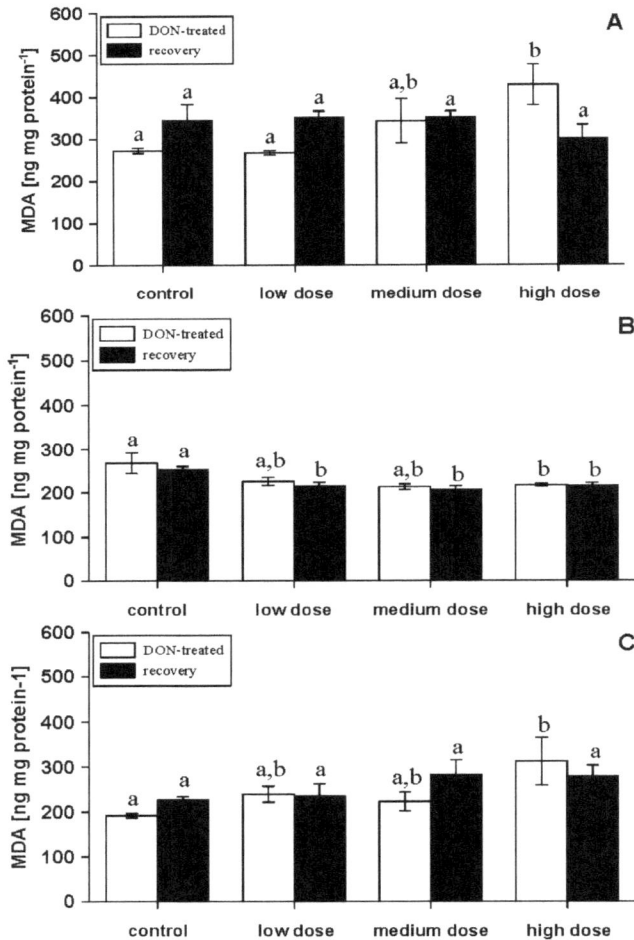

Figure 2. Lipid peroxidation measured as malondialdehyde (MDA; ng mg protein^{-1}) in different tissues [(**A**) head kidney, (**B**) trunk kidney, (**C**) spleen] in experimental fish with four weeks of DON feeding (DON-treated) and DON-fed fish with an additional two weeks of recovery (recovery); mean ± SEM; means with the same letter ([a] and/or [b]) are not significantly different from each other (significance tested with Mann-Whitney U-tests, $p < 0.05$).

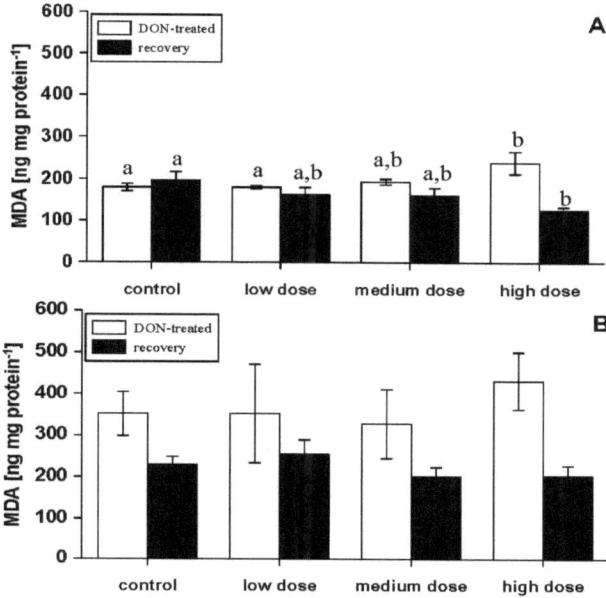

Figure 3. Lipid peroxidation measured as malondialdehyde (MDA; ng mg protein^{-1}) in different tissues [(**A**) liver, (**B**) white muscle] in experimental fish with four weeks of DON feeding (DON-treated) and DON-fed fish with an additional two weeks of recovery (recovery); mean ± SEM; means with the same letter ([a] and/or [b]) are not significantly different from each other (significance tested with Mann–Whitney U-tests, $p < 0.05$).

2.4. Measurement of LDH Activity in Different Tissues

The activity of the lactate dehydrogenase (LDH) showed differences in the kidneys after four weeks of DON feeding and also after additional two weeks of recovery but not in liver and spleen (Table 7). During DON feeding LDH activity was increased in the kidneys of all DON-fed fish compared to control fish which indicates increased anaerobic metabolism. However, a lower LDH activity in the recovery phase was observed between fish fed the highest DON diet and the control fish. The significant difference in LDH activity in muscle after four weeks of feeding the highest dose of DON was no longer observable after the recovery of two weeks. The observation that lactate levels are not increased in muscle samples although LDH activities are decreased in fish treated with the high dose DON diet, supports the hypothesis that lactate is transferred from white muscle via the blood stream before it accumulates.

2.5. Biochemical Body and Organ Composition

Investigation on whole body composition revealed that the fat content in whole body homogenates of fish treated with the medium- and high-dose DON diet was increased compared to control fish (Table 8), whereas in fish treated with the low dose DON diet, only a tendency for increased fat content was found ($p = 0.065$). Still, a significant relationship between the toxin concentrations in the experimental feeds and lipid content was found (Spearman correlation coefficient, 0.498, and significance, $p = 0.013$).

Table 7. Lactate dehydrogenase (LDH; mU mg protein^{-1}) activity in tissue homogenates after DON feeding and a two-week recovery phase; $n = 6$; mean ± SEM; means with the same letter ([a] and/or [b]) are not significantly different from each other (significance tested with Mann-Whitney U-tests, $p < 0.05$).

LDH Activity	Basal Feed	Low DON	Medium DON	High DON
DON-treated:				
head kidney	31.58 ± 4.77 [a]	71.55 ± 3.63 [b]	61.22 ± 5.11 [b]	51.89 ± 6.30 [b]
trunk kidney	34.34 ± 5.18 [a]	77.79 ± 3.95 [b]	66.23 ± 4.55 [b,c]	53.16 ± 6.84 [a,c]
spleen	2.64 ± 0.25	2.97 ± 0.47	2.90 ± 0.34	2.92 ± 0.43
liver	275.23 ± 61.86	370.96 ± 66.37	176.58 ± 40.47	264.67 ± 56.73
white muscle	38.28 ± 13.68 [a]	17.51 ± 5.77 [a,b]	20.86 ± 7.39 [a,b]	3.12 ± 1.81 [b]
Recovery:				
head kidney	51.21 ± 4.97 [a]	52.54 ± 6.65 [a]	57.12 ± 9.96 [a]	13.34 ± 2.77 [b]
trunk kidney	54.58 ± 6.08 [a]	65.33 ± 6.10 [a]	60.70 ± 13.15[a]	19.89 ± 4.50 [b]
spleen	10.88 ± 0.99	9.74 ± 0.90	7.92 ± 0.59	12.96 ± 2.86
liver	496.02 ± 118.64	669.79 ± 113.86	510.05 ± 106.11	344.31 ± 89.97
white muscle	17.73 ± 4.85	12.35 ± 4.57	22.29 ± 3.90	25.91 ± 3.92

After two additional weeks of recovery, all DON-treated fish showed lower lipid levels than the control fish (Table 9). These results were paralleled by significant differences in the energy content of the whole body homogenates. After recovery, the ash contents of the whole body homogenates were also found to be different in fish fed the high-dose DON diet compared to control fish (Table 9).

In the recovered fish, the ash content correlated with the lipid content negatively (Spearman correlation coefficient, −0.568; significance, $p = 0.002$). Moreover, a significant relationship between the toxin concentrations in the experimental feeds and the lipid and ash content was found (Spearman correlation coefficient, −0.792 and 0.568, and significance, $p < 0.000$ and $p = 0.007$, respectively) in fish after the recovery phase. This led to the investigation of similar parameters in samples of liver tissue and white dorsal musculature to allow a possible explanation for these observations.

Impairment of the intestinal nutrient uptake by DON was reported for mammalian systems [25–27]. Since the liver is an important metabolic organ that processes nutrients from feed, one would expect effects on the nutritional status of liver tissue if such impairment of intestinal nutrient uptake would also be present in the experimental carp.

Therefore, the biochemical composition of liver tissue of carp was analysed. Dry matter values of liver tissue were comparable to the values reported for other cyprinid species, such as freshwater major carp, *Catla catla* [28]. No significant differences in the dry matter of DON-treated fish compared to control fish were observed after four weeks of experiments, but an additional two weeks of recovery led to significantly different dry matter incorporations between fish fed the low-dose DON feed and the control fish (Table 10). Liver total lipid contents are higher than in the whole body homogenates. This seems feasible, since liver in carp is known to be important for the storage of lipids. Moreover, in fish treated with the highest concentration of DON, the lipid content in liver tissue was significantly increased after the recovery phase compared to control fish. This is in contrast to the whole body lipid levels of DON-treated fish after the recovery phase and indicates that liver damage was accompanied by hepatic lipid accumulation, which also influenced the lipid balance of the entire body.

Table 8. The whole body composition of experimental fish two weeks after DON feeding (recovery phase); $n = 6$; mean ± SEM; means with the same letter ([a] and/or [b]) are not significantly different from each other (significance tested with Mann-Whitney U-tests, $p < 0.05$).

Whole Body Composition	Basal Feed	Low DON	Medium DON	High DON
Crude lipid content (% dry matter)	24.83 ± 1.85 [a]	29.33 ± 1.36 [a,b]	32.17 ± 0.95 [b]	30.33 ± 0.95 [b]
Crude ash (% dry matter)	8.56 ± 1.22	8.72 ± 0.19	8.03 ± 1.01	9.37 ± 0.34
Energy content (MJ kg^{-1} dry matter)	24.32 ± 0.47 [a]	25.27 ± 0.34 [a,b]	25.78 ± 0.25 [b]	25.25 ± 0.31 [a,b]

Table 9. The whole body composition of experimental fish two weeks after DON feeding (recovery phase); $n = 6$; mean $\pm$ SEM; means with the same letter ([a] and/or [b]) are not significantly different from each other (significance tested with Mann-Whitney U-tests, $p < 0.05$).

Whole Body Composition	Basal Feed	Low DON	Medium DON	High DON
Crude lipid content (% dry matter)	33.17 $\pm$ 0.75 [a]	30.25 $\pm$ 0.48 [b]	29.60 $\pm$ 1.08 [b]	25.00 $\pm$ 2.71 [b]
Crude ash (% dry matter)	8.23 $\pm$ 0.35 [a]	8.46 $\pm$ 0.34 [a,b]	9.37 $\pm$ 0.34 [a,b]	9.98 $\pm$ 0.53 [b]
Energy content (MJ kg^{-1} dry matter)	25.63 $\pm$ 0.12 [a]	25.40 $\pm$ 0.19 [a,b]	25.02 $\pm$ 0.28 [a,b]	24.64 $\pm$ 0.23 [b]

Table 10. The composition of liver samples after four weeks of DON feeding and after the recovery phase of two weeks; $n = 6$; mean $\pm$ SEM; means with the same letter ([a] and/or [b]) are not significantly different from each other (significance tested with Mann-Whitney U-tests, $p < 0.05$).

Liver Composition	Basal Feed	Low DON	Medium DON	High DON
DON-treated:				
Dry matter (% wet matter)	25.9 $\pm$ 0.8	22.9 $\pm$ 2.2	25.6 $\pm$ 0.8	27.0 $\pm$ 1.4
Total lipids (% wet matter)	15.0 $\pm$ 1.3	14.5 $\pm$ 1.2	16.2 $\pm$ 1.7	17.3 $\pm$ 1.7
Free glucose (mM g^{-1} wet matter)	1.16 $\pm$ 0.27	0.87 $\pm$ 0.14	1.19 $\pm$ 0.08	0.91 $\pm$ 0.19
Lactate (mM g^{-1} wet matter)	1.75 $\pm$ 0.48	3.05 $\pm$ 1.29	3.77 $\pm$ 2.03	3.56 $\pm$ 1.87
AST (U mg protein^{-1})	90.3 $\pm$ 18.2	60.2 $\pm$ 20.5	81.7 $\pm$ 8.2	89.1 $\pm$ 16.6
ALT (U mg protein^{-1})	1.43 $\pm$ 0.14 [a]	2.61 $\pm$ 0.98 [a,b]	2.95 $\pm$ 0.51 [b]	2.73 $\pm$ 0.49 [a,b]
Ascorbate (μM g^{-1} wet matter)	81.7 $\pm$ 15.7	63.5 $\pm$ 8.3	71.5 $\pm$ 14.8	54.3 $\pm$ 8.8
Recovery:				
Dry matter (% wet matter)	25.1 $\pm$ 0.5 [a]	28.9 $\pm$ 1.4 [a]	27.5 $\pm$ 1.8 [a,b]	25.8 $\pm$ 1.7 [a,b]
Total lipids (% wet matter)	17.7 $\pm$ 1.2 [a]	19.2 $\pm$ 1.3 [a,b]	18.7 $\pm$ 0.8 [a,b]	23.6 $\pm$ 4.1 [b]
Free glucose (mM g^{-1} wet matter)	0.73 $\pm$ 0.10	0.96 $\pm$ 0.17	0.91 $\pm$ 0.12	0.99 $\pm$ 0.21
Lactate (mM g^{-1} wet matter)	1.36 $\pm$ 0.35 [a]	3.74 $\pm$ 1.13 [a,b]	6.48 $\pm$ 2.02 [b]	2.85 $\pm$ 0.80 [a,b]
AST (U mg protein^{-1})	98.5 $\pm$ 22.1	96.5$\pm$ 27.5	81.7 $\pm$ 8.2	99.7 $\pm$ 10.8
ALT (U mg protein^{-1})	2.47 $\pm$ 0.23	2.88 $\pm$ 0.22	2.79 $\pm$ 0.11	3.03 $\pm$ 0.26
Ascorbate (μM g^{-1} wet matter)	51.8 $\pm$ 6.5	58.8 $\pm$ 7.0	64.9 $\pm$ 7.5	62.2 $\pm$ 7.6

Liver glucose levels were not significantly different between treatment groups. It was suggested that amino acids are a superior energy sources to glucose for carp [29]. It has also been shown for carps under starvation that the conversion of lipid to glycogen in liver tissue was accompanied by an increase in blood glucose levels [30]. No changes of blood glucose (Table 11) and lipid (Table 10) in liver tissue were observed in DON-treated fish, which suggests that blood homeostasis was not subjected to a fasting-like status, due to an impairment of nutrient uptake in the intestine.

Liver glucose, lactate and ascorbate concentrations were not influenced by DON feeding and did not correlate with toxin concentrations in the experimental feeds. Although ascorbic acid has been shown to prevent the hemolytic action of DON on rat erythrocytes to some extent [22], ascorbate levels in carp remained unchanged, which indicated that ascorbate alone is not sufficient to prevent damage due to DON exposure. Liver alanine aminotransferase (ALT) activity was increased by treatment with the medium-dose DON diet for four weeks compared to control fish, while the other DON-treated groups showed no significant difference of this enzyme activity compared to the control group, probably due to higher individual variation (coefficient of variance (CV) for control fish of 24.6 *versus* CVs of 92.2, 42.3 and 51.0 for fish treated with the low-, medium- and high-dose, respectively, while the CVs for all fish in the recovery phase ranged from 9.5 to 23.2). ALT activity also correlated with the mycotoxin concentrations in feed (Spearman correlation coefficient, -0.409; significance, $p = 0.047$). This indicates damage to liver tissue, as has already been observed in carp under chemical stress [31]. In contrast, aspartate aminotransferase (AST) activity was not significantly influenced by DON feeding, which indicates that the rate of amino acid transformations via transamination is not influenced.

A considerable amount of the fish consists of white musculature, which shows low levels of myoglobin and is mostly used for burst swimming [32]. Swimming performance is known to lead to the utilization of nutrients from blood circulation and white and red musculature [33]. Several factors further influence the chemical body composition. For example, the genetic background of carp determined the dry matter content of the fillet, as has been shown for different crossbreds of common carp, accounting for 19% to 28% [34–36]. With respect to common mirror carps, a value for dry matter of 22.0% was observed [36]. The dry matter of the carp used in our study corresponds to this value and was only influenced by feeding the low-dose DON diet (Table 12).

As expected, the lipid content in carp muscle was found to be low in the present study, and it was not influenced by DON feeding. Different crossbreds of common carp showed lipid contents in the fillet of up to 9.9%, with mirror carps showing 2.4% lipid in fillet [36]. Low muscle lipid contents ranging from 0.5% to 2.6% have also been noted by another study [34]. However, slightly higher values, ranging from 2.9% to 5.2%, have been reported elsewhere [35,37]. Differences in the chemical composition of carp musculature can be due to the influence of rearing conditions [38,39], the influence of age [35,36,39] and differences in the composition of the diet [37,39–41], and these factors should be considered when the values for the present study are compared to other studies. For example, in the study of Steffens and Wirth [40], the addition of 10% different lipid sources in the diet led to 2.2 to 2.5% lipid in dorsal muscle of carp, which corresponds to the values in the present study.

The significant difference in the dry matter of white muscle of fish fed the low-dose diet compared to the control fish cannot be explained at the moment. Reductions of dry matter in the musculature can be caused by the depletion of tissue nutrient contents, which are compensated for by increasing the water content [29]. However, even after calculation of the dry weight lipid contents in white muscle, the samples do not show a significant influence of DON. Thus, the reason for this observation remains obscure.

Muscle glucose shows no differences of the DON treatment of fish. The glucose concentration in red musculature has been reported to be rather independent of blood glucose levels [42], but the aerobic glucose utilization was still assumed to be relying by approximately 30% on glucose in circulation. In contrast, white musculature in fish is known to largely depend on anaerobic glycogenolysis for energy liberation [43]. Lactate in carp white musculature ranged from four to 9 mM [44], even after exhaustive exercise. The lactate levels in muscle tissue in the present study were a bit lower and were not influenced by DON feeding. This might be due to the fact that even after severe hypoxic stress, lactate did not accumulate in white muscle, but was probably transferred out of the tissue [45].

Ascorbate levels in white muscle samples strongly depend on the supply via the diet and have been reported to be low in fish musculature [46]. This corresponds to the present study, although higher values have been reported for carp previously [47]. This study also reported an influence of the mycotoxin, sterigmatocystin, on ascorbate levels in the white musculature of carp, which was not observed after DON feeding of carp in the present study. Similar to the ascorbate levels in liver tissue, the lack of influence of DON on their concentration in muscle further indicates that ascorbate as an endogenous antioxidant does not prevent the detrimental effects of DON on carp.

2.6. Serum Parameter

Our study demonstrates a large variation of biochemical serum parameters, which are known to be affected by many endogenous and exogenous factors, such as age, health condition, nutrition or stress, including chemical stress [48,49].

Glucose levels in the experimental fish in the present study were comparable to values for unstressed carp reported previously [48,49]. Glucose concentrations in serum remained unchanged in fish treated with DON for four weeks (Table 11). A direct influence of DON on glucose metabolism has been shown in the human epithelial intestinal cell line, HT-29-D4, and in jejunum of chicken; however, cholesterol metabolism remained unchanged in this cell line [26,27]. No evidence for a similar impairment of glucose from the diet could be observed in the present study, although a possible decrease in blood glucose might also have been compensated for in DON-treated fish. Similar to other fish, carp are known to maintain blood glucose levels, even after prolonged starvation [29].

Serum lactate levels in control fish are comparable to values in unstressed fish in other studies [49,50]. However, lactate values in DON-treated carp were considerably higher. This was probably caused by an activation of gluconeogenesis to maintain levels of circulation glucose. Consequently, elevated serum glucose levels can be caused [49], which has also been observed in fish recovering for the treatment with the medium- and high-dose diet compared to the control fish (Table 11). Thus, it can be assumed that DON affects anaerobic metabolism in carp muscle.

Table 11. The serum parameter in DON-treated fish and in fish after a recovery of two weeks; $n = 6$ per group; mean $\pm$ SEM; means with the same letter ([a] and/or [b]) are not significantly different from each other (significance tested with Mann-Whitney U-tests, $p < 0.05$).

Serum Parameters	Basal Feed	Low DON	Medium DON	High DON
DON-treated:				
Free glucose (μM mL^{-1})	2.22 ± 0.22	2.37 ± 0.19	2.12 ± 0.09	2.53 ± 0.27
Lactate (μM mL^{-1})	8.04 ± 2.87 [a]	21.85 ± 5.78 [a,b]	20.50 ± 6.70 [a,b]	29.96 ± 8.44 [b]
LDH (mU mg protein^{-1})	15.7 ± 1.0 [a]	19.8 ± 1.8 [a,b]	18.6 ± 2.5 [a,b]	23.6 ± 2.1 [b]
SDH (mU mg protein^{-1})	11.2 ± 4.4 [a,b]	30.3 ± 4.8 [a]	23.1 ± 8.2 [a,b]	8.5 ± 2.1 [b]
AST (U mg protein^{-1})	7.1 ± 0.9 [a]	5.1 ± 0.2 [a,b]	4.8 ± 0.7 [b]	10.6 ± 3.8 [a,c]
ALT (U mg protein^{-1})	0.7 ± 0.2 [a]	0.7 ± 0.3 [a]	1.2 ± 0.6 [a]	0.2 ± 0.0 [b]
Total protein (mg mL^{-1})	23.05 ± 0.73	23.32 ± 0.64	26.22 ± 1.26	23.41 ± 0.78
Albumin (mg mL^{-1})	19.85 ± 1.49 [a]	15.14 ± 2.03 [a,b]	5.42 ± 2.40 [b]	6.92 ± 3.66 [b]
Recovery:				
Free glucose (μM mL^{-1})	1.66 ± 0.12 [a]	2.78 ± 0.28 [b]	2.84 ± 0.33 [b]	2.77 ± 0.25 [b]
Lactate (μM mL^{-1})	4.12 ± 1.08 [a]	34.87 ± 12.74 [b]	39.70 ± 8.35 [b]	19.83 ± 3.70 [b]
LDH (mU mg protein^{-1})	14.8 ± 1.4 [a]	18.8 ± 1.0 [b]	17.4 ± 0.9 [a,b]	21.9 ± 1.6 [b]
SDH (mU mg protein^{-1})	32.4 ± 8.8	16.1 ± 3.5	20.3 ± 5.4	34.5 ± 11.7
AST (U mg protein^{-1})	6.4 ± 0.8	5.1 ± 0.2	4.8 ± 0.7	7.9 ± 2.6
ALT (U mg protein^{-1})	0.4 ± 0.1 [a]	0.6 ± 0.1 [a]	0.7 ± 0.2 [a,b]	1.1 ± 0.2 [b]
Total protein (mg mL^{-1})	21.89 ± 0.73	22.04 ± 0.39	21.47 ± 1.86	29.86 ± 0.57
Albumin (mg mL^{-1})	15.32 ± 1.50	17.22 ± 1.74	16.94 ± 3.75	22.45 ± 2.40

Table 12. The composition of samples from dorsal white musculature after four weeks of DON feeding and after the recovery phase of two weeks; $n = 6$; mean $\pm$ SEM; means with the same letter ([a] and/or [b]) are not significantly different from each other (significance tested with Mann-Whitney U-tests, $p < 0.05$).

Composition of White Muscle	Basal Feed	Low DON	Medium DON	High DON
DON-treated:				
Dry matter (% wet matter)	22.4 ± 2.0 [a]	19.0 ± 0.7 [b]	20.9 ± 0.6 [a,b]	20.5 ± 0.4 [a,b]
Total lipids (% wet matter)	2.4 ± 0.4	2.2 ± 0.2	2.4 ± 0.1	2.1 ± 0.2
Free glucose (μM g^{-1} wet matter)	2.93 ± 0.39	3.30 ± 0.22	2.95 ± 0.29	2.76 ± 0.29
Lactate (mM g^{-1} wet matter)	3.54 ± 0.37	4.08 ± 0.29	4.56 ± 0.30	3.79 ± 0.21
Ascorbate (μM g^{-1} wet matter)	10.2 ± 0.5	10.0 ± 0.8	10.3 ± 0.5	9.5 ± 0.6
Recovery:				
Dry matter (% wet matter)	19.6 ± 0.9	19.7 ± 0.3	21.2 ± 0.6	19.7 ± 0.5
total lipids (% wet matter)	2.2 ± 0.3	2.5 ± 0.2	2.6 ± 0.2	2.9 ± 0.6
Free glucose (μM g^{-1} wet matter)	3.33 ± 0.34	3.12 ± 0.18	3.89 ± 0.40	3.32 ± 0.48
Lactate (mM g^{-1} wet matter)	3.03 ± 0.11	3.41 ± 0.48	2.95 ± 0.22	2.94 ± 0.31
Ascorbate (μM g^{-1} wet matter)	9.7 ± 0.3	9.3 ± 0.5	9.1 ± 0.7	10.2 ± 0.7

Total protein content in all carp used in the present study was slightly lower in most cases compared to previously reported values in other studies [33,49]. From this, it may be assumed that in general, the nutritional status of the fish was sufficient, since serum total protein levels are known to reflect the nutritional condition of carp [29]. The albumin concentration in the control group was comparable to previously reported values [33]. Although the total protein content of serum remained unchanged by DON feeding, albumin concentrations were significantly reduced in fish fed the medium dose and high dose diets for four weeks. This means that the ratio of albumin to total proteins in control fish of more than 80% is reduced in DON-treated fish to 66%, 21% and 31% in fish treated with the low-dose, medium-dose and high-dose diet, respectively. Which adaptations led to the maintenance of total protein levels, although albumin levels were reduced by DON treatment, remains unknown and should be investigated in future studies. Nevertheless, the effect on serum albumin levels was certainly caused by liver impairment that occurred upon DON treatment. That trichothecenes, including DON, are ribotoxic, targeting the 60S ribosomal subunit, and consequently, impairing protein synthesis and transcription, which leads to apoptosis, has already been reported for leukocytes and other actively proliferating eukaryotic cells of higher vertebrates [51].

In contrast to their activity in liver tissue (Table 10), the activities of AST and ALT in serum were significantly reduced by DON feeding in the medium-dose group and the high-dose group, respectively, compared to control fish (Table 11), and only ALT activity was increased in fish recovering from receiving the high-dose DON diet. This indicates that the rate of amino acid transformation via transamination is slowed down by DON. Lactate dehydrogenase (LDH) activity in serum was found to be increased in fish fed the high-dose feed for four weeks and after two weeks of recovery compared to control fish. Increased LDH in serum of carp, indicating membrane leakage in tissues, has also been shown after exposure to toxic concentrations of pesticides [52]. This parallels the increased lactate levels in DON-treated fish. SDH activity was increased in these fish compared to fish fed the low-dose DON diet. Significant increases of AST, ALT and LDH activities have been observed in carp that have been exposed to handling stress, regular exercise or toxic substances, such as microcystins and cyanide [33,53–56]. Although correlations between the toxin concentrations in feed and AST activities in serum were not significant in the present study during the feeding period, our results showed a significant correlation of ALT, LDH or SDH with the toxin concentrations in the experimental diets (Spearman correlation coefficient, −0.527, 0.466 and 0.475; significance, $p = 0.012$, 0.022 and 0.025, respectively). Moreover, after the same time period, albumin or lactate concentrations in serum correlated with the toxin concentration in the experimental feeds (Spearman correlation coefficient, −0.678 and 0.453; significance, $p = 0.000$ and 0.034, respectively). Similar results have also been obtained for the recovery phase for the correlation of ALT and LDH with the toxin concentrations previously applied (Spearman correlation coefficient, 0.433 and 0.555; significance, $p = 0.034$ and 0.005, respectively). Furthermore, glucose or lactate concentrations in serum correlated with the toxin concentration in the experimental feeds (Spearman correlation coefficient, 0.592 and 0.487; significance, $p = 0.002$ and 0.021, respectively). Thus, the activity of serum ALT and LDH together with lactate concentrations seems to be a sensitive indicator of the fish responses to DON. The symptoms of carp exposed to DON resemble the situation of freshwater snakehead fish, *Channa punctatus,* that were treated with sublethal concentrations of a carbamate pesticide, leading to increases of LDH and decreased SDH activity in several organs and hyperglycaemia and hyperlactaemia, which suggested that anaerobic metabolism was favored [57].

3. Experimental Section

3.1. Preparation of Feeds and Husbandry

Ingredients were chosen so that no cereals were included in the experimental diets to exclude cereal-based *Fusarium* toxin contamination of these diets. For the preparation of experimental feeds, all ingredients listed in Table 1 were mixed thoroughly [17]. The feeds were artificially contaminated by adding deoxynivalenol (DON, dissolved in ethanol; purity >98%, lot no. 011M4065V) to the fish oil during the feed preparation process, achieving different concentrations (low dose, 352 µg kg^{-1}, medium dose, 619 µg kg^{-1}, and high dose, 954 µg kg^{-1}, final feed, respectively) [17]. The preparation of the different diets was repeated three times in a pelletizer (L 14-175, Amandus Kahl, Reinbek, Germany) to allow the homogenous distribution of ingredients. The manufactured 4-mm pellets were allowed to cool down to room temperature for two hours before storage at 4 °C until use. The composition of the diets was analyzed by using standard methods. Experimental diets were analyzed for dry matter (DM) (105 °C, until constant weight), crude ash (550 °C, 2 h.), crude fat (Soxtec HT6, Tecator, Höganäs, Sweden) and crude protein content (N × 6.25; Kjeltec Auto System, Tecator, Höganäs, Sweden). Nitrogen-free extract and fibres (NFE) are summarized as shown in Equation (2).

$$\text{Nitrogen free extract + fibre, (NFE)} = 100 - (\% \text{ protein} + \% \text{ fat} + \% \text{ ash}) \qquad (2)$$

Carp were raised from eggs in our facilities and used for the experiments at 12–16 cm in total length. The fish were kept at a 16 h light/8 h dark photoperiod at 25 ± 0.2 °C (mean ± SD) and acclimatized to the tanks that were integrated into a flow-through system providing 6 L fresh and conditioned water per hour per tank prior to the experiments for three weeks.

3.2. Chemicals

All chemicals were obtained from Sigma-Aldrich (Buchs, Switzerland), unless indicated otherwise.

3.3. Experimental Feeding Design

The prepared pellets (4 mm in diameter) were given at 2% of body mass once every day to juvenile carp, which were separated into four different feeding groups (control, low dose, medium dose and high dose) with 6 fish each in quadruplicate 54-L tanks for each treatment. The maintenance of optimal rearing conditions (dissolved oxygen, water temperature, pH, conductivity) was controlled during the entire experiments [17]. All experimental procedures have been approved by the cantonal veterinarian authorities of Basel-Stadt (Basel, Switzerland) under permission number 2410. Feed amounts per tank were adjusted to the increased weight on a weekly basis. Uptake of feed was observed in all groups within less than 30 min after offering the experimental diets. Fish were fed the experimental diets for four weeks, and one half of the fish were sampled by using all fish from two tanks of each treatment group. All remaining fish were fed the uncontaminated control diet for a further two weeks before termination of the experiment, in order to investigate the possible reversal of the DON effects. Sampling of fish included blood sampling, recording of weight and length, as well as sampling of individual organs. The calculation of condition factors was achieved according to Equation (3).

$$\text{Condition factor} = \text{weight}/(\text{length})^3 \qquad (3)$$

3.4. Histological Determination of Glycogen and Histopathological Scoring

Histological assessments were conducted on liver ($n = 6$ fish per treatment level per sampling day). Tissues were automatically processed (TP1020 tissue processor, Leica Microsystems AG, Switzerland), and at least six sections per fish (3 µm-thick) were mounted on microscope slides and stained with haematoxylin and eosin (HE). Additionally, sections were stained with PAS or Prussian blue to analyze glycogen content and to detect changes in tissue iron content related to erythrocyte turnover. For histopathological examination (Nikon Eclipse 400 microscope), sections were examined in detail at 400× magnification. Quantitative analyses, as described below, were conducted on digital images, which were taken with a Nikon DXM 1200 F digital camera and Nikons ACT-1 software V2.63. Damage to liver tissue was estimated by histopathological scoring in 10 HE-stained sections per fish using semi-quantitative assessments of the severity (0 = no alterations, 1 = mild alterations, 2 = moderate alterations, 3 = severe alterations) according to the suggestions of Zodrow et al. [58]. Glycogen content of 5 PAS-stained sections per fish was analyzed by determining the luminosity and RGB$_{max}$. Luminosity can be deduced from the histogram settings in Adobe Photoshop. RGB$_{max}$ was achieved by lightening the green and blue channels in the RGB space. Chrominance was calculated by subtracting the minimal RGB value (which was obtained by darkening the minimal tones) from the RGB$_{max}$ (Adobe® Photoshop® CS3 Extended version 10.1).

3.5. Lipid Peroxidation Assay and LDH Activity Measurements

Tissue samples were homogenated in 19 volumes PBS containing 0.1% (w/v) butylated hydroxytoluene (BHT), homogenized using an UltraThurrax (IKA Werke, Staufen, Germany) for 10 s and centrifuged for 10 min at 10,000× g at 4 °C (Centrifuge 5415R, Eppendorf, Basel, Switzerland). The supernatant was used for the TBARs assay [59] with the following modifications. A volume of 40 µL of supernatant was mixed with 200 µL TBARs solution [60], containing 3.75 mg mL^{-1} thiobarbituric acid (TBA), 20% (w/v) trichloric acid (TCA), 9.1 µL mL^{-1} hydrochloric acid (37%),0.06% (w/v) BHT and 866.9 µL distilled water. Thereafter, samples were incubated at 70° C for 90 min, cooled to room temperature and centrifuged at 16,000× g for 15 min at room temperature. In parallel, standards containing 0 to 3200 nM malondialdehyde (MDA) were prepared. Optical densities of all samples were read at 532 nm (Infinite M200, Tecan Group Ltd., Männedorf, Switzerland). Aliquots of the tissue homogenate were also used for the lactate dehydrogenase (LDH) assay and protein determinations. The latter were conducted using the bicinchoninic acid (BCA) assay (Sigma), according to the manufacturer's protocol. The activity of LDH in tissue homogenates and serum samples was measured according to Bergmeyer [61]. In short, 164 µL NADH solution (0.244 mmol L^{-1} in Tris-NaCl solution (Tris, 81.3 mmol L^{-1}; NaCl, 203.2 mmol L^{-1} pH 7.2)) were mixed with 33 µL pyruvate solution (9.76 mmol L^{-1} in Tris-NaCl solution). The reaction was started by the addition of a 20 µL sample, and an absorption decrease at a wavelength of 339 nm was recorded for 10 min (Infinite M200, Tecan Group Ltd., Männedorf, Switzerland).

3.6. Nutrient Allocation in Fish

Fish from each group were killed by an overdose of anaesthetic, cut into small pieces with scissors and blended. Homogenates were then dried at 105 °C for 24 h and dry mass was noted. Fat content in dried homogenates was analysed in duplicate by petroleum ether extraction using a Soxhlet apparatus (Soxtec System HT, Tecator, Sweden).

The nutritional status of snap-frozen liver and muscle samples was analysed using different methods, as follows. To amounts of 100 to 250 mg, 1 mL distilled water was added. Liver samples were homogenated with an UltraThurrax (IKA Werke, Staufen, Germany) for 10 s, and the muscle samples were homogenized manually by using a glass potter (Wheaton™ Potter-Elvehjem Tissue Grinders, purchased from Fisher Scientific, Reinach, Switzerland). For the determination of total lipid content, samples were extracted with chloroform-methanol (2:1) containing 0.01% BHT as an antioxidant, according to the method of Bligh and Dyer [62], followed by analyses of total lipids with the sulfo-phospho-vanillin method [63]. A standard curve was prepared with olive oil dissolved in ethanol.

The energy content of dried homogenates was analysed using an IKA C 200 bomb calorimeter. By this method, the dried sample is wrapped in combustible paper and placed in a sealed iron bomb, where its explosive combustion is unleashed by an electric flash, under an O_2-saturated atmosphere. The bomb is placed in a water bath, which absorbs the heat generated during the combustion; the increase in the temperature of water relates to the heat generated during the combustion. The device is calibrated with a standard compound, benzoic acid (energy released upon combustion: 26,460 Joules g^{-1}). Obtained values (in Joules g^{-1} dry matter) were converted to values relative to the fresh weight of fish (% g fish^{-1}).

3.7. Measurement of Ascorbate

Ascorbate concentrations were analyzed in medium samples and cell extracts according to the method by Vislisel *et al.* [64]. Tissue samples were homogenized in 19 volumes of PBS, after which methanol and diethylenetriaminepentaacetic acid (DTPA) were added to achieve final concentrations of 60% (v:v) and 250 µM, respectively. Samples were centrifuged at 16,000 rpm for 2 min at 4 °C (Centrifuge %427 R, Eppendorf, Basel, Switzerland). Samples of 40 µL from the supernatant were used for the assay. Ascorbate in the samples was first oxidized to dehydroascorbate by the addition of 40 µL tempol (4-hydroxy-2,2,6,6-tetramethyl-piperidinyloxy, 2.3 mM in 2 M sodium acetate buffer) per well, followed by short shaking and the addition of 25 µL o-phenylenediamine (OPDA, 5.5 mM in 2 M sodium acetate buffer). Fluorescence emission values were recorded at 450 nm immediately using a plate reader (Infinite M200, Tecan Group Ltd., Männedorf, Switzerland) with excitation at 345 nm. Standard curves were prepared with ascorbate diluted to 14 different concentrations ranging from 0 to 150 µM using the methanol-water mixture containing DPTA.

3.8. Preparation of Serum Samples and Determination of Glucose and Lactate

Serum was immediately prepared from blood samples taken with heparinised syringes. Serum samples were stored at $-80\,^{\circ}$C until analyses. Tissue homogenates were centrifuged at 10,000× g for 10 min, and the supernatant was used for the analysis of glucose and lactate. Glucose was analysed according to the glucose oxidase method. Therefore, samples were mixed with sodium acetate buffer (2 M, pH 5.5) containing 0.1 mg mL^{-1} o-dianisidin, 4 U mL^{-1} glucose oxidase and 2.54 purpurgallin units of peroxidase (from horseradish) per mL. Plates were incubated at 37 $^{\circ}$C for 30 min, and the reaction was stopped by the addition of 100 µL 12 N sulphuric acid to each well. Optical densities were read at 540 nm (Infinite M200, Tecan Group Ltd., Männedorf, Switzerland).

Lactate in serum and tissue samples was determined according to Maughan [65] using hydrazine buffer (1.1 mM, pH = 9.0) and 227 µM NAD, and 125 U LDH from porcine heart per well. After incubation for 30 min at room temperature, optical densities were measured at 339 nm and lactate concentrations were calculated from a standard curve prepared with serial dilutions from a lactate standard (998 $\pm$ 6 mg L^{-1}).

3.9. Measurement of Total Protein and Albumin in Serum

Total protein contents were analysed from a diluted serum sample using the bicinchoninic acid (BCA) assay (Sigma), according to the manufacturer's protocol. Albumin was determined using the bromocresol green (BCG) method, as described by Doumas *et al.* [66], with the following modifications: BCG was solubilised in 0.1 N NaOH and diluted with distilled water and succinate buffer (0.1 M, pH 4.0) at a ratio of 1:3 (v/v). This BCG working solution was added to 25 mL of serum, followed by incubation for 10 min at room temperature and measurement of optical densities at 628 nm (Infinite M200, Tecan Group Ltd., Männedorf, Switzerland). A standard curve was prepared using essential globulin-free bovine serum albumin.

3.10. Measurement of AST, ALT and SDH Activity

Activities of aspartate aminotransferase (AST) and alanine aminotransferase (ALT) from samples of 20 µL of serum or tissue homogenate were determined according to the modified methods described by Casillas *et al.* [67] after incubation for 30 min at room temperature by monitoring NADH oxidation at 339 nm (Infinite M200, Tecan Group Ltd., Männedorf, Switzerland). Sorbitol dehydrogenase (SDH) activity in serum samples was measured as described by Bergmeyer [68] using fructose as the substrate. Absorption changes at a wavelength of 339 nm were recorded for 10 min (Infinite M200, Tecan Group Ltd., Männedorf, Switzerland).

3.11. Statistics

Data are presented as the mean $\pm$ standard error of the mean (SEM), unless indicated otherwise. Coefficients of variance were calculated as shown in Equation (4).

$$\text{Coefficient of variance } (CV) = \text{standard deviation/mean} \times 100 \qquad (4)$$

The effects of the treatments were determined by the comparison of treatment groups to controls using non-parametrical Mann-Whitney U-tests (SPSS 9.0 for Windows). Relationships between parameters were evaluated using Spearman correlation tests. A *p*-value of <0.05 was accepted as being statistically significant.

4. Conclusions

The histological alterations in carp livers of DON-treated fish suggest that the fish may face a metabolic crisis caused by tissue damage. The results indicate that oxidative stress leading to lipid peroxidation is involved in the detrimental effects of DON feeding. Taken together, the chemical body composition of the experimental fish in the present study was influenced by the abovementioned factors and gives strong evidence for an influence of DON on the nutritional status of carp. Similar metabolic disorders, including impairment of hepatic metabolism of fats due to oxidative stress, have also been observed in pesticide-treated carp [15]. Nevertheless, the specific biological action and molecular action of DON on liver function in fish is still unclear and needs to be elucidated in further studies.

Acknowledgments: The authors are grateful to Christian Michel, Simon Herzog, Heidi Schiffer, Nicole Seiler-Kurth Michael Schlachter and Florian Nagel for additional help in the laboratory work. We would furthermore like to thank Konstantin Bayer from the Anatomical Institute in Basel for providing the infrastructure for the preparation of histological samples and Sven Dänicke (Friedrich-Loeffler-Institute, Braunschweig, Germany) for the analysis of DON in the experimental feeds.

Conflicts of Interest: The authors declare no conflict of interest.

References

1. Grove, J.F. Non-macrocyclic trichothecenes, Part 2. *Prog. Chem. Org. Nat. Prod.* **2000**, *69*, 1–70.
2. Rotter, B.A. Invited review: Toxicology of deoxynivalenol (vomitoxin). *J. Toxicol. Environ. Health A* **1996**, *48*, 1–34. [CrossRef]
3. Arunachalam, C.; Doohan, F.M. Trichothecene toxicity in eukaryotes: Cellular and molecular mechanisms in plants and animals. *Toxicol. Lett.* **2013**, *217*, 149–158. [CrossRef]
4. Maresca, M. From the gut to the brain: Journey and pathophysiological effects of the food-associated trichothecene mycotoxin deoxynivalenol. *Toxins* **2013**, *5*, 784–820. [CrossRef]
5. European Commission. Commission Recommendation (EC) No 576/2006 of 17 August 2006 on the presence of deoxynivalenol, zearalenone, ochratoxin A, T-2 and HT-2 and fumonisins in products intended for animal feeding. *Off. J. Eur. Union* **2006**, *L229*, 7.
6. Pietsch, C.; Kersten, S.; Burkhardt-Holm, P.; Valenta, H.; Dänicke, S. Occurrence of deoxynivalenol and zearalenone in commercial fish feed—An initial study. *Toxins* **2013**, *5*, 184–192. [CrossRef]
7. Döll, S.; Valenta, H.; Baardsen, G.; Möller, P.; Koppe, W.; Stubhaug, I.; Dänicke, S. Effects of increasing concentrations of deoxynivalenol, zearalenone and ochratoxin A in diets for Atlantic salmon (*Salmo salar*) on performance, health and toxin residues. In Proceedings of the Abstracts of the 33rd Mycotoxin Workshop, Freising, Germany, 30 May–1 June 2011.
8. Hooft, J.M.; Elmor, A.E.H.I.; Encarnação, P.; Bureau, D.P. Rainbow trout (*Oncorhynchus mykiss*) is extremely sensitive to the feed-borne *Fusarium* mycotoxin deoxynivalenol (DON). *Aquaculture* **2011**, *311*, 224–232. [CrossRef]
9. Paperna, I. Diseases caused by parasites in the aquaculture of warm water fish. *Ann. Rev. Fish* **1991**, *1*, 155–194. [CrossRef]
10. Serrano, J.A.; Nematipour, G.R.; Gatlin, D.M., III. Dietary protein requirement of the red drum (*Sciaenops ocellatus*) and relative use of dietary carbohydrate and lipid. *Aquaculture* **1992**, *101*, 283–291. [CrossRef]
11. Shimeno, S.; Kheyyali, D.; Shikata, T. Metabolic response to dietary lipid to protein ratios in common carp. *Fish. Sci.* **1995**, *61*, 977–980.
12. Russell, P.M.; Davies, S.J.; Gouveia, A.; Tekinay, A.A. Influence of dietary starch source on liver morphology in juvenile cultured European sea bass (*Dicentrarchus labrax* L.). *Aquacult. Res.* **2001**, *32*, 306–314. [CrossRef]
13. Rojik, I.; Nemcsók, J.; Boross, L. Morphological and biochemical studies on liver, kidney and gill of fishes affected by pesticides. *Acta Biol. Hungarica* **1983**, *34*, 81–92.

14. Shakoori, A.R.; Mughal, A.L.; Iqbal, M.J. Effects of sublethal doses of fenvalerate (a synthetic pyrethroid) administered continuously for four weeks on the blood, liver, and muscles of a freshwater fish, *Ctenopharyngodon idella*. *Bull. Environ. Contam. Toxicol.* **1996**, *57*, 487–494. [CrossRef]

15. Xu, W.N.; Liu, W.B.; Shao, X.P.; Jiang, G.Z.; Li, X.F. Effect of trichlorfon on hepatic lipid accumulation in crucian carp (*Carassius auratus gibelio*). *J. Aquatic Anim. Health* **2012**, *24*, 185–194. [CrossRef]

16. Li, L.; Xie, P.; Chen, J. *In vivo* studies on toxin accumulation in liver and ultrastructural changes of hepatocytes of the phytoplanktivorous bighead carp i.p.-injected with extracted microcystins. *Toxicon* **2005**, *46*, 533–545. [CrossRef]

17. Pietsch, C.; Michel, C.; Kersten, S.; Valenta, H.; Dänicke, S.; Schulz, C.; Kloas, W.; Burkhardt-Holm, P. *In vivo* effects of deoxynivalenol (DON) on innate immune responses of carp (*Cyprinus carpio* L.). *Food Chem. Toxicol.* **2014**. under review.

18. Jamu, D.M.; Ayinla, O.A. Potential for the development of aquaculture in Africa. *NAGA World Fish Center Quarterly* **2003**, *26*, 9–13.

19. NRC (National Research Council). *Nutrient Requirements of Fish*; National Academy Press: Washington, DC, USA, 1993.

20. Gouze, M.E.; Laffitte, J.; Rouimi, P.; Loiseau, N.; Oswald, I.P.; Galtier, P. Effect of various doses of deoxynivalenol on liver xenobiotic metabolizing enzymes in mice. *Food Chem. Toxicol.* **2006**, *44*, 476–483. [CrossRef]

21. Sanden, M.; Jørgensen, S.; Hemre, G.-I.; Ørnsrud, R.; Sissener, N.H. Zebrafish (*Danio rerio*) as a model for investigating dietary toxic effects of deoxynivalenol contamination in aquaculture feeds. *Food Chem. Toxicol.* **2012**, *50*, 4441–4448. [CrossRef]

22. Rizzo, A.F.; Atroshi, F.; Hirvi, T.; Saloniemi, H. The hemolytic activity of deoxynivalenol and T-2 toxin. *Nat. Toxins* **1992**, *1*, 106–110. [CrossRef]

23. Kouadio, J.H.; Mobio, T.A.; Baudrimont, I.; Moukha, S.; Dano, S.D.; Creppy, E.E. Comparative study of cytotoxicity and oxidative stress induced by deoxynivalenol, zearalenone or fumonisin B1 in human intestinal cell line Caco-2. *Toxicology* **2005**, *213*, 56–65. [CrossRef]

24. Borutova, R.; Faix, S.; Placha, I.; Gresakova, L.; Cobanova, K.; Leng, L. Effects of deoxynivalenol and zearalenone on oxidative stress and blood phagocytic activity in broilers. *Arch. Anim. Nutr.* **2008**, *62*, 303–312. [CrossRef]

25. Hunder, G.; Schümann, K.; Strugala, G.; Gropp, J.; Fichtl, B.; Forth, W. Influence of subchronic exposure to low dietary deoxynivalenol, a trichothecene mycotoxin, on intestinal absorption of nutrients in mice. *Food Chem. Toxicol.* **1991**, *29*, 809–814. [CrossRef]

26. Maresca, M.; Mahfoud, R.; Garmy, N.; Fantini, J. The Mycotoxin deoxynivalenol affects nutrient absorption in human intestinal epithelial cells. *J. Nutr.* **2002**, *132*, 2723–2731.

27. Awad, W.A.; Aschenbach, J.R.; Setyabudi, F.M.C.S.; Razzazi-Fazeli, E.; Bohm, J.; Zentek, J. *In vitro* effects of deoxynivalenol on small intestinal D-glucose uptake and absorption of deoxynivalenol across the isolated jejunal epithelium of laying hens. *Poult. Sci.* **2007**, *86*, 15–20. [CrossRef]

28. Hassan, M.; Chatha, S.A.S.; Tahira, I.; Hussain, B. Total lipids and fatty acid profile in the liver of wild and farmed *Catla catla* fish. *Grasas Y Aceites* **2010**, *61*, 52–57. [CrossRef]

29. Navarro, I.; Gutiérrez, J. Fasting and Starvation. In *Biochemistry and Molecular Biology of Fishes*; Hochachka, P.W., Mommsen, T.P., Eds.; Elsevier: Amsterdam, The Netherlands, 1995; Chapter 17; Volume 4, pp. 393–434.

30. Nagai, M.; Ikeda, S. Carbohydrate metabolism in fish. I. Effect of starvation and dietary composition on the blood glucose level and the hepatopancreatic glycogen and lipid content in carp. *Nippon Suisan Gakkaishi* **1971**, *37*, 404–409. [CrossRef]

31. Han, J.; Cai, H.; Wang, J.; Liu, G. Detrimental effects of metronidazole on the liver of freshwater common carp (*Cyprinus carpio* L.). *Bull. Environ. Contam. Toxicol.* **2013**, *91*, 444–449. [CrossRef]

32. Johnston, I.A.; Davison, W.; Goldspink, G. Energy metabolism of carp swimming muscles. *J. Comp. Physiol.* **1977**, *114*, 203–216. [CrossRef]

33. Varga, D.; Molnár, T.; Balogh, K.; Mézes, M.; Hancz, C.; Szabó, A. Adaptation of common carp (*Cyprinus carpio* L.) to regular swimming exercise II. metabolism. muscle phospholipid fatty acid composition and lipid peroxide status. *Poult. Fish Wildl. Sci.* **2013**, *1*, 106–111.

34. Masurekar, V.B.; Pai, S.R. Observations on the fluctuations in protein, fat and water content in *Cyprinus carpio* (Linn.) in relation to the stages of maturity. *Indian J. Fish* **1979**, *26*, 217–224.

35. Mares, J.; Palikova, M.; Kopp, R.; Navratil, S.; Pikula, J. Changes in the nutritional parameters of muscles of the common carp (*Cyprinus carpio*) and the silver carp (*Hypopthalmichthys molitrix*) following environmental exposure to cyanobacterial water bloom. *Aquacult. Res.* **2009**, *40*, 148–156. [CrossRef]

36. Buchtova, H.; Svobodova, Z.; Kocour, M.; Velisek, J. Chemical composition of fillets of mirror crossbreds common carp (*Cyprinus carpio* L.). *Acta Vet. Brno* **2010**, *79*, 551–557. [CrossRef]

37. Hadjinikolova, L. The influence of nutritive lipid sources on the growth and chemical and fatty acid composition of carp (*Cyprinus carpio* L.). *Arch. Polish Fish* **2004**, *12*, 111–119.

38. Mráz, J.; Pickova, J. Differences between lipid content and composition of different parts of fillets from crossbred farmed carp (*Cyprinus carpio*). *Fish Physiol. Biochem.* **2009**, *35*, 615–623. [CrossRef]

39. irković, M.; Trbović, D.; Ljubojević, D.; Đorđević, V. Meat quality of fish farmed in polyculture in carp ponds in Republic od Serbia. *Tehnologija Mesa* **2011**, *1*, 106–121.

40. Steffens, W.; Wirth, M. Influence of nutrition on the lipid quality of pond fish: Common carp (*Cyprinus carpio*) and tench (*Tinca tinca*). *Aquaculture* **2007**, *15*, 313–319. [CrossRef]

41. Ljubojević, D.; Ćirković, M.; Đorđević, V.; Puvača, N.; Trbović, D.; Vukadinov, J.; Plavša, N. Fat quality of marketable fresh water fish species in the Republic of Serbia. *Czech. J. Food. Sci.* **2013**, *31*, 445–450.

42. West, T.G.; Brauner, C.J.; Hochachka, P.W. Muscle glucose utilization during sustained swimming in the carp (*Cyprinus carpio*). *Am. J. Physiol.* **1994**, *267*, R1226–R1234.

43. Crabtree, B.; Newsholme, E.A. The activities of phosphorylase, hexokinase, phosphofructokinase, lactate dehydrogenase and the glycerol 3-phosphate dehydrogenases in muscles from vertebrates and invertebrates. *Biochem. J.* **1972**, *126*, 49–58.

44. Van Ginneken, V.; Boot, R.; Murk, T.; van den Thillart, G.; Balm, P. Blood plasma substrates and muscle lactic-acid response after exhaustive exercise in common carp and trout: Indications for a limited lactate-shuttle. *Anim. Biol.* **2004**, *54*, 119–130. [CrossRef]

45. Driedzic, W.R.; Hochachka, P.W. The unanswered question of high anaerobic capabilities of carp white muscle. *Can. J. Zool.* **1975**, *53*, 706–712. [CrossRef]

46. Nettleton, J.A.; Exler, J. Nutrients in wild and farmed fish and shellfish. *J. Food Sci.* **1992**, *57*, 257–260. [CrossRef]

47. Abdelhamid, A.M. Effect of sterigmatocystin contaminated diets on fish performance. *Arch. Anim. Nutr. Berlin* **1988**, *38*, 833–846.

48. Svobodova, Z.; Vykusova, B.; Modra, H.; Jarkovsky, J.; Smutna, M. Haematological and biochemical profile of harvest- size carp during harvest and post-harvest storage. *Aquacult. Res.* **2006**, *37*, 959–965. [CrossRef]

49. Hoseini, S.M.; Ghelichpour, M. Effects of pre-sampling fasting on serum characteristics of common carp (*Cyprinus carpio* L.). *Int. J. Aquatic Biol.* **2013**, *1*, 6–13.

50. Pottinger, T.G. Changes in blood cortisol, glucose and lactate in carp retained in anglers' keepnets. *J. Fish Biol.* **1998**, *53*, 728–742.

51. Shifrin, V.I.; Anderson, P. Trichothecene mycotoxins trigger a ribotoxic stress response that activates c-Jun N-terminal kinase and p38 mitogen-activated protein kinase and induces apoptosis. *J. Biol. Chem.* **1999**, *274*, 13985–13992. [CrossRef]

52. Asztalos, B.; Nemcsok, J. Effect of pesticides on the LDH activity and isoenzyme pattern of carp (*Cyprinus carpio* L.) sera. *Comp. Biochem. Physiol.* **1985**, *82C*, 217–219.

53. Carbis, C.R.; Rawlin, G.T.; Grant, P.; Mitchell, G.F.; Anderson, J.W.; McCauley, I. A study of feral carp, *Cyprinus carpio* L., exposed to *Microcystis aeruginosa* at Lake Mokoan, Australia, and possible implications for fish health. *J. Fish Dis.* **1997**, *20*, 81–91.

54. Li, X.-Y.; Chung, I.-K.; Kim, J.-I.; Lee, J.-A. Subchronic oral toxicity of microcystin in common carp (*Cyprinus carpio* L.) exposed to *Microcystis* under laboratory conditions. *Toxicon* **2004**, *44*, 821–827. [CrossRef]

55. Dobšíková, R.; Svobodová, Z.; Bláhová, J.; Modrá, H.; Velíšek, J. The effect of transport on biochemical and haematological indices of common carp (*Cyprinus carpio* L.). *Czech J. Anim. Sci.* **2009**, *54*, 510–518.

56. Sadati, F.; Shahsavani, D.; Baghshani, H. Biochemical alterations induced by sublethal cyanide exposure in common carp (*Cyprinus carpio*). *J. Biol. Environ. Sci.* **2013**, *7*, 65–69.

57. Sastry, K.V.; Siddiqui, A.A. Chronic toxic effects of the carbamate pesticide sevin on carbohydrate metabolism in a freshwater snakehead fish, *Channa punctatus. Toxicol. Lett.* **1982**, *14*, 123–130. [CrossRef]

58. Zodrow, J.M.; Stegeman, J.J.; Tanguay, R.L. Histological analysis of acute toxicity of 2,3,7,8-tetrachlorodibenzo-p-dioxin (TCDD) in zebrafish. *Aquatic Toxicol.* **2004**, *66*, 25–38. [CrossRef]

59. Rau, M.A.; Whitaker, J.; Freedman, J.H.; Di Giulio, R.T. Differential susceptibility of fish and rat liver cells to oxidative stress and cytotoxicity upon exposure to prooxidants. *Comp. Biochem. Physiol. Part C* **2004**, *137*, 335–342.

60. Holt, S.; Gunderson, M.; Joyce, K.; Nayini, N.R.; Eyster, G.F.; Garitano l, A.M.; Zonia, C.; Krause, G.S.; Aust, S.D.; White, B.C. Myocardial tissue iron delocalization and evidence for lipid peroxidation after two hours of ischemia. *Ann. Emerg. Med.* **1986**, *15*, 1155–1159.

61. Bergmeyer, H.U. Lactate dehydrogenase. In *Methods of Enzymatic Analysis*, 2nd ed.; Academic Press: London, UK, 1974; pp. 574–579.

62. Bligh, E.G.; Dyer, W.J. A rapid method of total lipid extraction and purification. *Can. J. Biochem. Physiol.* **1959**, *37*, 911–917. [CrossRef]

63. Frings, C.S.; Fendley, T.W.; Dunn, R.T.; Queen, C.A. Improved determination of total serum lipids by the sulfo-phospho-vanillin reaction. *Clin. Chem.* **1972**, *18*, 673–674.

64. Vislisel, J.M.; Schafer, F.Q.; Buettner, G.R. A simple and sensitive assay for ascorbate using a plate reader. *Anal. Biochem.* **2007**, *365*, 31–39. [CrossRef]

65. Maughan, R.J. A simple, rapid method for the determination of glucose, lactate, pyruvate, alanine, 3-hydroxybutyrate and acetoacetate on a single 20-µL blood sample. *Clin. Chim. Acta* **1982**, *122*, 231–240. [CrossRef]

66. Doumas, B.T.; Watson, W.; Biggs, H.G. Albumin standards and the measurement of serum albumin with bromcreasol green. *Clin. Chim. Acta* **1971**, *31*, 87–96. [CrossRef]

67. Casillas, E.; Sundquist, J.; Ames, W.E. Optimization of assay conditions for, and the selected tissue distribution of, alanine aminotransferase and aspartate aminotransferase of English sole, Parophrys vetulus Girard. *J. Fish Biol.* **1982**, *21*, 197–204. [CrossRef]

68. Bergmeyer, H.U. Sorbitol dehydrogenase. In *Methods of Enzymatic Analysis*, 3rd ed.; Weinheim, V.C., Ed.; Academic Press: London, UK, 1974; pp. 569–573.

Article

Deoxynivanelol and Fumonisin, Alone or in Combination, Induce Changes on Intestinal Junction Complexes and in E-Cadherin Expression

Karina Basso, Fernando Gomes and Ana Paula Loureiro Bracarense *

Laboratory of Animal Pathology, Veterinary Medecine Department, Universidade Estadual de Londrina, Rodovia Celso Garcia Cid, Km 380, PO Box 10.011, Londrina, Paraná 86057-970, Brazil; karinavet.basso@gmail.com (K.B.); fercg01@hotmail.com (F.G.)

* Author to whom correspondence should be addressed; ana.bracarense@pq.cnpq.br; Tel.: +55-043-3371-4485; Fax: +55-043-3371-4714.

Received: 11 October 2013; in revised form: 10 November 2013; Accepted: 13 November 2013; Published: 28 November 2013

Abstract: Fusariotoxins such as fumonisin B1 (FB1) and deoxynivalenol (DON) cause deleterious effects on the intestine of pigs. The aim of this study was to evaluate the effect of these mycotoxins, alone and in combination, on jejunal explants from piglets, using histological, immunohistochemical and ultrastructural assays. Five 24-day old pigs were used for sampling the explants. Forty-eight explants were sampled from each animal. Explants were incubated for 4 hours in culture medium and medium containing FB1 (100 µM), DON (10 µM) and both mycotoxins (100 µM FB1 plus 10 µM DON). Exposure to all treatments induced a significant decrease in the normal intestinal morphology and in the number of goblet cells, which were more severe in explants exposed to DON and both mycotoxins. A significant reduction in villus height occurred in groups treated with DON and with co-contamination. Expression of E-cadherin was significantly reduced in explants exposed to FB1 (40%), DON (93%) and FB1 plus DON (100%). The ultrastructural assay showed increased intercellular spaces and no junction complexes on enterocytes exposed to mycotoxins. The present data indicate that FB1 and DON induce changes in cell junction complexes that could contribute to increase paracellular permeability. The *ex vivo* model was adequate for assessing intestinal toxicity induced by exposure of isolated or associated concentrations of 100 µM of FB1 and 10 µM of DON.

Keywords: *Fusarium* spp.; mycotoxins; gut; cell permeability; adherens junction

---

1. Introduction

Mycotoxins are secondary metabolites produced by several fungal genera. They act as natural contaminants and are commonly found in grains and fresh foods of vegetable origin. It is estimated that 25% of the grains worldwide are contaminated with these substances [1]. A survey including 7049 samples collected in Europe, Asia and Americas revealed that about 64% and 59% of raw materials and finished feed samples contained fumonisins (FB) and deoxynivalenol (DON), respectively. The mean levels were 1 mg/kg (maximum of 49 mg/kg) for DON and 2 mg/kg (maximum of 77 mg/kg) for FB [2].

The fumonisin B1 (FB1) corresponds to 70% of fumonisins and is produced by *Fusarium verticillioides* [3]. The toxicity of FB1 has been proved in several animal species, resulting in different effects such as acute pulmonary edema in pigs, leukoencephalomalacia in horses, liver cancer in rats and esophageal cancer in humans [4]. Data on the mechanisms of action of FB1 on the digestive system are scarce, and routes of action on the intestinal epithelium are poorly understood. Exposure to FB1 induces a reduction in cell number due to decrease in cell proliferation associated with increased apoptotic index, as well as a decrease in transepithelial electrical resistance, indicating changes in intestinal integrity [5]. *In vivo* studies in piglets demonstrated that acute and chronic ingestion of feed contaminated with fumonisin led to a significant increase in hepatic [6] and intestinal lesions such as atrophy and fusion of villi, and decreased E-cadherin expression [7]. Piglets exposed to this mycotoxin showed higher bacterial translocation to various organs [8], favoring the proliferation of opportunistic bacteria in the gut [9].

The fusariotoxin deoxynivalenol (DON) frequently contaminate corn and wheat, being a risk to human and animal health [10–13]. Exposure of intestinal explants to DON causes morphological changes in a dose-dependent manner, such as flattening of enterocytes; villi atrophy and increased apoptotic index [14]. One effect of this mycotoxin on the intestine is the reduction in the expression of proteins cell junctions as claudin-4 [15,16], E-cadherin and occludin [7], resulting in changes in paracellular and transcellular permeability, favoring penetration of pathogens [13,15–17].

The ultrastructural evaluation may help in understanding the pathophysiology of injury; however studies on the effects of mycotoxins on intestinal ultrastructure are scarce [18,19]. There is no data in the literature on ultrastructural changes induced by exposure of the bowel to fumonisins and deoxynivalenol. Health regulations only consider the effects of mono-contamination, but multi-contamination is a phenomenon often observed in natural contamination of feed [6]. The available data indicate that simultaneous intake of FB1 and DON induces an additive immunosuppressive effect as compared with exposure to a single toxin [6,7]. The need for more research into additive, synergistic or antagonistic effects in multi-contamination is necessary; however, *in vivo* studies are costly and involve bioethical issues. Thus, the use of alternative models which mimic the organic systems of interest is extremely interesting. The efficacy of *ex vivo* model for assessing the effects of exposure to DON on the intestine has been proven in previous studies [14,20]. The model is also appropriate to examine the expression of protein junctions of enterocytes [21]. Considering the need to broaden knowledge about the results of interactions between multiple mycotoxins and the limited available data, the aim of this study was to assess the effects of exposure to FB1 and DON, alone and in combination, with emphasis on E-cadherin expression and ultrastructural changes, using the *ex vivo* model of intestinal explant.

2. Results

2.1. Histological Analysis

The main histological changes observed in the control group were edema of lamina propria, mild cell degeneration and villi atrophy (Figure 1A). In explants exposed to FB1, flattening and focal loss of apical enterocytes, moderate fusion and villi atrophy were observed (Figure 1B). In the group treated with DON, the changes are similar to the FB1 group; however the intensity of the lesions was more severe and a reduction in villi number was also verified (Figure 1C). Explants exposed to both mycotoxins showed severe changes characterized by reduction in villi number, villi fusion and atrophy, besides lysis of enterocytes (Figure 1D). A significant decrease in histological score was observed in all explants treated with mycotoxins when compared to control group (Figure 1G). The reduction was 18.8%, 37% and 37.5% for exposure to FB1, DON and FB1 plus DON, respectively. The morphometric analysis showed a significant decrease in villi height in the explants treated with DON and FB1 plus DON ($p \leq 0.05$) compared to the control group (Figure 1H).

The explants exposed to all treatments showed a significant change in the number of goblet cells ($p \leq 0.05$) in comparison to control explants. The reduction was more pronounced in the groups treated with FB1 plus DON (98%) and DON (63.4%) (Figure 1I).

2.2. Expression of E-Cadherin

The immunohistochemical analysis was performed to evaluate the expression of the adherens junction protein, E-cadherin. We observed a significant decrease in expression of E-cadherin in the group treated with FB1 (40%) and DON (93%) compared to control (Figure 1E,F). The group treated with both mycotoxins showed no immunostaining for E-cadherin.

2.3. Ultrastructural Analysis

Cell viability and the integrity of the complex of enterocyte junctions in explants exposed to fusariotoxins were evaluated by ultrastructural analysis. Control explants showed a continuous monolayer of enterocytes lining the gut. The luminal membrane of enterocytes presented preserved microvilli, while the apical membrane between adjacent intestinal cells formed intercellular junctions and desmosomes (Figure 2A,B). The cytoplasm was homogeneous with no changes in organelles. The nuclei contained a low concentration of heterochromatin and large, distinct nucleoli. Goblet cells were observed between enterocytes, showing secretory granules throughout the length of the villi. On the other hand, explants exposed to mycotoxins presented increased intercellular spaces, decrease in the size and number of microvilli (Figure 2C,D). Junction complexes were not visualized between the enterocytes. The nucleus showed marginated heterochromatin and occasionally apoptotic corpuscles. The presence of cytoplasmic projections, vacuoles within the cytoplasm, cellular debris and desquamated cells in the lumen were observed mainly in the explants incubated with DON (Figure 2E). The group exposed to both mycotoxins showed so severe changes that impaired ultrastructural evaluation. In this group we observed only cellular debris with no preserved enterocytes.

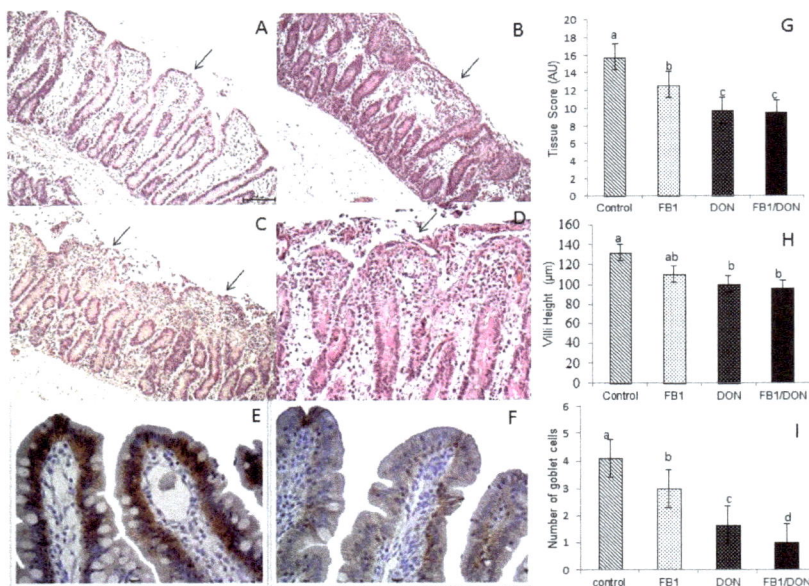

Figure 1. Effects of individual and combined exposition of jejunal explants to fumonisin and deoxynivalenol on histology. Explants were exposed to culture medium (▥) or culture medium with fumonisin B1 (FB1) (▧), deoxynivalenol (DON) (▤) or FB1 + DON (■). (**A**) Control explants. Edema of the lamina propria and mild villi atrophy (arrow); (**B**) FB1-exposed explant. Moderate fusion (arrow) and villous atrophy; (**C**) DON-exposed explant. Severe loss of apical enterocytes, fusion and atrophy (arrow); (**D**) FB1 + DON-exposed explant. Lysis of intestinal epithelium, villi atrophy, fusion (arrow) and cell debris. HE. Bar 100 μm; (**E**) Control explant showing a strong and homogeneous E-cadherin expression. Bar 20 μm; (**F**) DON-exposed explant showing reduced expression of E-cadherin. Bar 20 μm; (**G**) Tissue scores of pig intestinal explants exposed to FB1, DON and both mycotoxins; (**H**) Villi height in pig intestinal explants treated with FB1, DON and FB1 + DON; (**I**) Number of goblet cells per villus of pig intestinal explants treated with FB1, DON and FB1 + DON. Values are means with their standard deviation of the mean represented by vertical bars (n 5 animals). Mean values with unlike letters were significantly different ($p \leq 0.05$). AU = Arbitrary Units.

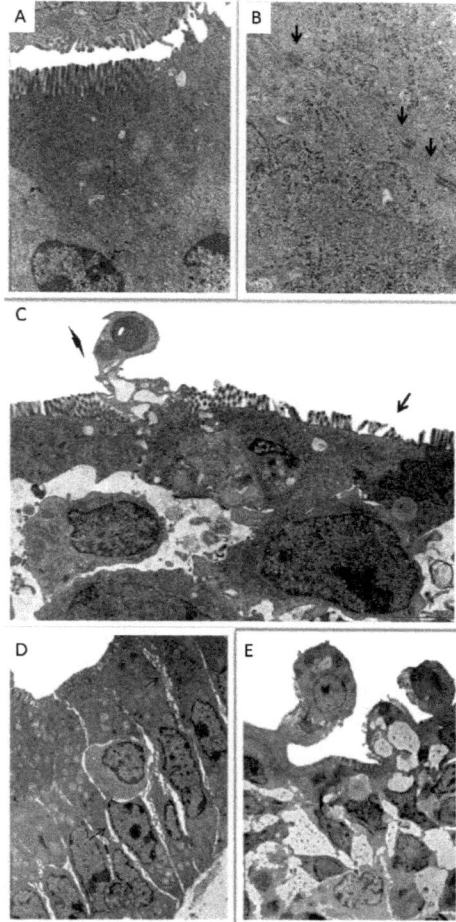

Figure 2. Effects of individual and combined exposition of jejunal explants to fumonisin and deoxynivalenol on ultrastructure. (**A**) Control explant. Enterocytes with normal morphology of microvilli and cytoplasm. Bar 500 nm; (**B**) Control explant. Enterocytes with junction complexes (arrow) and glycogen granules scattered in the cytoplasm. Bar 2 nm; (**C**) FB1-exposed explant. Focal loss of apical enterocytes (arrow head) and loss of microvilli (arrow). Bar 2 nm; (**D**) DON-exposed explant. Increased intercellular space (arrow) and loss of junction complexes. Bar 5 nm; (**E**) DON-exposed explant. Vacuoles within cytoplasm, membrane blebs and loss of apical enterocytes. Bar 10 nm.

3. Discussion

The integrity of the gut is dependent on the maintenance of various factors, including enterocyte and mucus layer integrity, as well as preservation and functionality of epithelial junctions cells [21]. There is increasing evidence that the intestinal epithelium is repeatedly exposed to mycotoxins, and at a higher concentration than other tissues [22]. The ingestion of fusariotoxins may induce changes on intestinal morphology and local immunity [7] affecting the barrier function of the gut. The effects of mycotoxins on gastrointestinal tract can diverge because the bioavailability of these fungal compounds is very diverse and differs between animal species [22]. In this study, we used an *ex vivo* model to demonstrate that fusariotoxins alone or in combination induced significant morphological and ultrastructural changes in jejunal explants from pigs.

The gut, as the first tissue to have contact with food contaminants, is considered a target organ for the action of mycotoxins [23]. In this experiment, we verified that exposition to DON, FB1 or both mycotoxins induced a significant decrease in intestinal score. Changes were more severe in explants incubated with DON and DON plus FB1. The main histological changes were lysis of enterocytes and villi atrophy and fusion. These histological findings are similar to those described in pigs chronically fed with FB1 and DON [7,24–26], demonstrating that the *ex vivo* model is suitable for assessing the effects of toxic substances on the intestine. It is interesting to observe that in both *ex vivo* and *in vivo* models a good correlation was reported for histological changes and phosphorylation of mitogen-activated kinases on intestine, even with differences in time of mycotoxin exposure (hours and days, respectively) [20,26]. In explants the morphological changes induced by fusariotoxins are more severe. Probably, the severity of the lesions on the explants is associated with a direct exposure to the toxin, whereas in the *in vivo* model the bioavailability of mycotoxins differs. In addition, the explants were submitted to a relative degree of hypoxia, a fact that can contribute to the intensity of histological changes.

It is worth noting that even with a short period of incubation with mycotoxins both toxins were able to cause significant injury on intestinal tissue. Deoxynivalenol induces a ribotoxic stress, compromising protein synthesis and triggering apoptosis [27], which may explain the microscopic changes observed in this study. Fumonisin leads to accumulation of sphinganine and sphingosine products that are cytotoxic, pro-apoptotic and cell growth inhibitors [28].

Changes in villi height imply in reduced absorption of nutrients [22]. In this study we verified a significant decrease in villi height in explants exposed to DON (24%) and DON plus FB1 (27%). Similar results were reported in pigs fed chronically diets contaminated with DON [7]. The reduction in villi height induced by DON is related to an increase in enterocyte apoptosis (data not shown). DON induces phosphorylation of mitogen-activated protein kinases (MAPKs) resulting in cell apoptosis [26]. In addition, DON is also able to reduce cell proliferation *in vitro* [27].

Regarding the effects of FB1 on the intestine, the results are conflicting. Doses of 30 mg/kg and 2.8 mol/Kg caused villus atrophy and fusion [29]; however, ingestion of a single dose of 83 mg/kg FB1 [3] or of 6 mg/kg chronically [7] induced no changes in villi height. In this study, explants exposed to FB1 showed no significant changes in villi height (reduction of 16.4%) compared to control explants.

Goblet cell density reflects mucus production, an important protective intestinal barrier that prevents adhesion of pathogens to enterocytes [30]. In this experiment, explants exposed to all treatments showed a significant change in the number of goblet cells. We observed a reduction that was more severe in DON and FB1 plus DON treatments. Chronic ingestion of FB1 induced no changes in globet cell density in piglets (6 mg/Kg FB) [7], whereas in poultry (300 mg/kg) a significant increase was observed [31]. These differences are probably related to the doses of FB1 used. The explants were exposed to 100 μM, which is equivalent to a dose of 144 mg/kg of FB1. Also, pigs are much more susceptible to fumonisins than poultry [22]. Reduction in the number of goblet cells after chronic exposure to DON intestinal has already been reported in *in vivo* studies [7]. In a previous study, jejunal explants exposed to DON and FB1 exhibited a significant increase in caspase-3 immunostaining, indicating cell apoptosis (unpublished data). We hypothesize that the changes in goblet cell staining are also related to increased apoptosis induced by mycotoxins exposition. On the other hand, changes in goblet cells staining could be associated with two other mechanisms, expulsion of mucin granules and necrosis. DON induces a ribotoxic stress [27] that results in decreased synthesis of proteins such as mucin, while FB1 induces apoptosis by activation of the caspase-3 pathway [32]. The explants exposed to FB1 plus DON showed a drastic reduction in the number of goblet cells suggesting an additive effect of these mycotoxins on globet cells.

A decrease in the expression of proteins cell junction and the induction of apoptosis are some of the factors associated with the deleterious effects of mycotoxins on the intestine [15]. Junction cell proteins as occludin, claudin and E-caderin have a fundamental role in maintaining epithelial architecture [33]. In this experiment we observed a decrease in E-cadherin expression, mainly in explants exposed to DON and to both mycotoxins. It is known that pigs chronically fed diets contaminated with DON, FB1 or FB1 plus DON present a significant decrease in E-cadherin and occludin expression [7]. In pigs' intestinal cells exposed to DON, the changes in claudin-4 expression, a junction protein, were related to phosphorylation of ERK 1/2, a mitogen activated protein kinase [16]. The molecular mechanisms underlying FB1 exposition and changes in junction proteins remain to be investigated. The reduction in the expression of E-cadherin reflects the ability of mycotoxins to alter epithelial integrity; however future studies evaluating other proteins that comprise the junctional complex can establish the types of changes on the epithelium and the courses of action of mycotoxins. A decreased expression of E-cadherin is a factor that contributes to increased paracellular permeability and facilitates the invasion of pathogens.

One of the aims of this study was to characterize the ultrastructural changes on the intestine induced by DON and FB1 with emphasis on cell junctions. To the best of our knowledge, this is the first study reporting the effects of DON and FB1 on intestinal ultrastructure. In this study we observed an increase in intercellular spaces and disappearance of junction structures induced by both mycotoxins. Increase in paracellular permeability has been reported in Caco-2 cells exposed to DON through inhibition of protein synthesis [34]. In addition, IPEC cells incubated with FB1 showed a decrease in transepithelial electrical resistance [5]. Our results indicate that DON and FB1 induce severe changes in junction complexes and cell structure increasing intestinal permeability and decreasing intestinal absorption. The ultrastructural findings observed on explants exposed to mycotoxins also strengthen those observed on immunohistochemical evaluation of E-cadherin expression reinforcing the changes on junction structures.

4. Material and Methods

4.1. Animals

Five piglets (Landrace × Large White × Duroc) of 24 days old (6.3 kg, ± 0.8) were used to collect the explants. The animals were weaned at 21 days of age, and then subjected to a standard diet after weaning in separate bays. At 24 days the piglets were euthanized with an intravenous injection of sodium pentobarbital (40 mg/kg of body weight). The institutional Ethics Committee for Animal Experimentation approved all animal procedures.

4.2. Culture of Explants and Exposure to FB1 and DON

The jejunum was chosen because in prior studies this region was shown to be more sensitive to the effects of mycotoxins [16]. Fragments of 5 cm of medial jejunum were sampled immediately after euthanasia, and washed with buffered saline (PBS) and opened longitudinally. The explants were collected with the aid of a biopsy punch (8 mm) and placed in six-well plates (EasyPath, São Paulo, Brazil), filled with 3 ml of agar and containing DMEM (GIBCO, NY, USA) plus fetal bovine serum (10%), glutamine (0.2 mL/L), gentamicin (0.5 mg/mL) and penicillin/streptomycin (10 mL/L).

The explants were deposited with the mucosa facing upwards (3 explants/well) and incubated at 37 °C in the presence of FB1 (F1147, Sigma-Aldrich, São Paulo Brazil) (100 μM) or DON (D0156, Sigma-Aldrich, São Paulo, Brazil) (10 μM) alone and FB1 (100 μM) plus DON (10 μM) for four hours. The doses used in this experiment are based in previous studies *in vitro* [35] and *ex vivo* [20] that have shown toxic effects with these concentrations. The dose of FB1 (100 μM) corresponds to an intake of 144 mg/kg of feed, while the dose of DON (10 μM) corresponds to an ingestion of 3 mg/kg of contaminated feed. From each animal 48 explants were sampled and submitted to the different treatments (24 explants/animal were used for immunohistological analysis and the other 24 for ultrastructural analysis). A total of 240 explants were evaluated by immunohistological and ultrastructural methods. For each treatment three explants immersed in culture medium without mycotoxins (control group) were incubated.

4.3. Histological and Morphometric Analysis

After the incubation period, explants were fixed in 10% buffered formalin solution, dehydrated in increasing alcohols and embedded in paraffin. Sections of 5μm were stained with hematoxylin and eosin (HE) for histological analysis. The histological changes were evaluated using an adapted tissue score [15] in which the intensity and severity of lesions were considered. The maximum score (22 points) indicates the overall integrity of the intestine. The criteria used to determine the score were morphology and number of villi, morphology of enterocytes and microvilli, presence of cellular debris, interstitial edema, lymphatic dilation and cellular necrosis. The lesion score was calculated by taking into account the degree of severity (severity factor) and the extent of each lesion (according to intensity or observed frequency, scored from 0 to 3). For each lesion, the score of the extent was multiplied by the severity factor.

To evaluate the density of goblet cells, sections were stained with alcian-blue. The number of goblet cells was counted in 10 randomly villi, using a 10 × magnification. Goblet cells present in the crypts were not counted. For morphometric evaluation the heights of 10 intestinal villi were measured randomly with the aid of the Motic Image Plus 2.0 software using a 10 × magnification.

4.4. Immunohistochemical Analysis

The expression of E-cadherin was evaluated using specific monoclonal antibody (clone 4A2C7, Zymed, Carlsbad, CA, USA) and the proportion of the intestinal section expressing E-cadherin was estimated. Each sample was assessed as showing either normal or reduced staining. Normal staining was considered when homogeneous and strong basolateral membrane staining of the enterocytes was detected. Heterogeneous and weak staining was considered to indicate reduced expression. The results are reported as the percentage of animals showing strong/homogeneous immunoreactivity to E-cadherin.

Tissue sections were deparaffinised with xylene and dehydrated through a graded ethanol series. Heat-mediated retrieval was done by heating the sections immersed in citrate buffer (pH 6.0) in a microwave oven (750 W) for 15 min. Endogenous peroxidase activity was blocked by incubation in methanol-hydrogen peroxide solution. The sections were incubated overnight at 4 °C with the primary antibody (diluted 1:50). The secondary antibody (Nichirei Biosciences, Tokyo, Japan) was applied followed by the addition of a chromogen (3,3′-diaminobenzidine). Finally, tissue sections were counterstained with hematoxylin and mounted on coverslips using a synthetic resin. All reactions were accompanied with negative and positive controls of the reaction according to the manufacturer.

4.5. Ultrastructural Analysis

The explants exposed to the different treatments were submitted to transmission electron microscopy. After the incubation period samples were fixed in Karnovsky modified solution and post-fixed in 1% osmium tetroxide. After complete sequential dehydration, the samples were embedded in epoxy resin and maintained for 3 days at 60 °C for polymerization. Ultrathin sections were stained with uranyl acetate and lead citrate and analyzed by TEM (model FEI Tecnai 12).

4.6. Statistical Analysis

The data used for statistical analysis were represented as means with their standard deviation. The experimental design used in the present study was entirely randomized with six replicates (each explant representing one replicate). Oneway analysis of variance (ANOVA) followed by a multiple comparison procedure (Tukey test) was used for statistical analysis. Data of the expression of E-cadherin were subjected to Fisher's exact test. The lack of normality or homogeneity of goblet cell number leads to the use of the non-parametric Kruskal-Wallis and Dunn tests. Differences were considered statistically significant at $p \leq 0.05$. Statistical analyses were performed with BioStat 5.0 (Belém, Pará, Brazil) software package. Ultrastructure parameters were subjected to a descriptive analysis.

5. Conclusions

In conclusion, results of the present study indicate that 10 µM of DON and 100 µM of FB1 alone or in combination induce significant damage on intestinal tissue. An association between exposition to these mycotoxins and reduced villi height, globet cell density, E-cadherin expression and junction complexes was also demonstrated. Taken together, these results support that DON and FB1 could increase enterocyte paracellular permeability and promote major changes on intestinal barrier function.

Acknowledgments: This study was supported by a grant from CNPq (474691/2012-8), Brazil. Karina Basso and Ana Paula Loureiro Bracarense were supported by a research fellowship financed by CNPq. We thank Celia Guadalupe Andrade for her help with ultrastructural analysis.

Conflicts of Interest: The authors declare no conflict of interest.

References

1. Cast, I. *Mycotoxins—Risks in Plant, Animal and Human Systems*; Task Force Report, No.139; Council for Agricultural Science and Technology: Ames, IA, USA, 2003; pp. 1–191.
2. Rodrigues, I.; Naehrer, K. A three-year survey on the worldwide occurrence of mycotoxins in feedstuffs and feed. *Toxins* **2012**, *4*, 663–675. [CrossRef]
3. Dilkin, P.; Direito, G.; Simas, M.M.S.; Mallmann, C.A.; Corrêa, B. Toxicokinetics and toxicological effects of single oral dose of fumonisin B1 containing *Fusarium verticillioides* culture material in weaned piglets. *Chemico-Biol. Interact.* **2010**, *185*, 157–160. [CrossRef]
4. Minami, L.; Meirelles, P.G.; Hirooka, E.Y.; Ono, E.Y.S. Fumonisinas: Efeitos toxicológicos, mecanismo de ação e biomarcadores para avaliação da exposição. *Semina* **2004**, *25*, 207–224.
5. Bouhet, S.; Hourcade, E.; Loiseau, N.; Fikry, A.; Martinez, S.; Roselli, M.; Galtier, P.; Mengheri, E.; Oswald, I.P. The mycotoxin fumonisin B1 alters the proliferation and the barrier function of porcine intestinal epithelial cells. *Toxicol. Sci.* **2004**, *77*, 165–171.
6. Grenier, B.; Bracarense, A.P.F.R.L.; Lucioli, J.; Pacheco, G.D.; Cossalter, A.M.; Moll, W.D.; Schatzmayr, G.; Oswald, I.P. Individual and combined effects of subclinical doses of deoxynivalenol and fumonisin in piglets. *Mol. Nutr. Food Res.* **2011**, *55*, 761–771. [CrossRef]
7. Bracarense, A.P.; Lucioli, J.; Grenier, B.; Pacheco, G.D.; Moll, W.D.; Schatzma, Y.R.G.; Oswald, I.P. Chronic ingestion of deoxynivalenol and fumonisin, alone or in interaction, induces morphological and immunological changes in the intestine of piglets. *Br. J. Nutr.* **2012**, *107*, 1776–1786. [CrossRef]
8. Oswald, I.P.; Desautels, C.; Lafftte, J.; Fournout, S.; Peress, Y.; Odin, M.; Bars, L.E.; Bars, J.; Fairbrother, J.M. Mycotoxin fumonisin B1 increases intestinal colonization by pathogenic Escherichia coli in pigs. *Appl. Environ. Microbiol.* **2003**, *69*, 5870–5874.
9. Maresca, M.; Yahi, N.; Younès-Sakr, L.; Boyron, M.; Caporiccio, B.; Fantini, J. Both direct and indirect effects account for the pro-inflammatory activity of enteropathogenic mycotoxins on the human intestinal epithelium: Stimulation of interleukin-8 secretion, potentiation of interleukin-1 β effect and increase in the transepithelial passage of commensal bacteria. *Toxicol. Appl. Pharmacol.* **2008**, *22*, 884–892.
10. Pestka, J.J. Deoxynivalenol: Mechanisms of action, human exposure, and toxicological relevance. *Arch. Toxicol.* **2010**, *84*, 663–679. [CrossRef]
11. Döll, S.; Dänicke, S. The fusarium toxins deoxynivalenol (DON) and zearalenone (ZON) in animal feeding. *Prev. Vet. Med.* **2011**, *102*, 132–145. [CrossRef]
12. Maresca, M. From the gut to the brain: Journey and pathophysiological effects of the food-associated trichothecene mycotoxin deoxynivalenol. *Toxins* **2013**, *5*, 784–820.
13. Maresca, M.; Fantini, J. Some food-associated mycotoxins as potential risk factors in humanspredisposed to chronic intestinal inflammatory diseases. *Toxicon* **2010**, *56*, 282–294. [CrossRef]
14. Kolf-clauw, M.; Castellote, J.; Joly, B.; Bourges-Abella, N.; Raymond-Letron, I.; Pinton, P.; Oswald, I.P. Development of a pig jejunal explant culture for studying the gastrointestinal toxicity of the mycotoxin deoxynivalenol: Histopathological analysis. *Toxicol. Vitro* **2009**, *23*, 1580–1584. [CrossRef]
15. Pinton, P.; Nougayrède, J.P.; Del Rio, J.C. The food contaminant deoxynivalenol, decreases intestinal barrier permeability and reduces claudin expression. *Toxicol. Appl. Pharmacol.* **2009**, *237*, 41–48.

16. Pinton, P.; Braicu, C.; Nougayrede, J.; Lafitte, J.; Taranu, I.; Oswald, I.P. Deoxynivalenol impairs porcine intestinal barrier function and decreases the protein expression of claudin-4 through a mitogen-activated protein kinase-dependent mechanism. *J. Nutr.* **2010**, *140*, 1956–1962.

17. Vandenbroucke, V.; Croubels, S.; Martel, A.; Verbrugghe, E.; Goossens, J.; Deun, K.V.; Boyen, F.; Thompson, A.; Shearer, N.; de Backer, P.; *et al.* The mycotoxin deoxynivalenol potentiates intestinal inflammation by *Salmonella* Typhimurium in porcine ileal loops. *PloS One* **2011**, *6*, e23871. [CrossRef]

18. Witloc, D.R.; Waytt, R.D.; Ruff, M.D. Morphological changes in the avian intestine induced by citrinin and lack of effect of aflatoxin and T-2 toxin as seen with scanning electron microscopy. *Toxicon* **1977**, *15*, 41–44. [CrossRef]

19. Obremski, K.; Gajecka, M.; Zielonka, L.; Jakimiuk, E.; Gajecki, M. Morphology and ultrastructure of small intestine mucosa in gilts with zearalenone mycotoxicosis. *Pol. J. Vet. Sci.* **2005**, *8*, 301–307.

20. Lucioli, J.; Pinton, P.; Callu, P.; Laffitte, J.; Grosjean, F.; Kolf-Clauw, M.; Oswald, I.P.; Bracarense, A.P.F.R.L. The food contaminant deoxynivalenol activates the mitogen activated protein kinases in the intestine: Interest of *ex vivo* models as an alternative to *in vivo* experiments. *Toxicon* **2013**, *66*, 31–36.

21. Randall, K.; Turton, J.; Foster, J.R. Explant culture of gastrointestinal tissue: A review of methods and applications. *Cell. Biol. Toxicol.* **2011**, *27*, 267–284. [CrossRef]

22. Grenier, B.; Applegate, T.J. Modulation of intestinal functions following mycotoxin ingestion: Meta-analysis of published experiments in animals. *Toxins* **2013**, *5*, 396–430. [CrossRef]

23. Bouhet, S.; Oswald, I.P. The intestine as a possible target for fumonisin toxicity. *Mol. Nutr. Food. Res.* **2007**, *51*, 925–931. [CrossRef]

24. Lessard, M.; Boudry, G.; Sève, B.; Oswald, I.P.; Lallès, J.P. Intestinal physiology and peptidase activity in male pigs are modulated by consumption of corn culture extracts containing fumonisins. *J. Nutr.* **2009**, *139*, 1303–1307. [CrossRef]

25. Zielonka, L.; Wiśniewska, M.; Obremski, K.; Gajecki, M. Influence of low doses of deoxynivalenol on histopathology of selected organs of pigs. *Pol. J. Vet. Sci.* **2009**, *12*, 89–95.

26. Pinton, P.; Tsybulskyy, D.; Lucioli, J.; Laffitte, J.; Callu, P.; Lyazhri, F.; Grosjean, F.; Bracarense, A.P.F.R.L.; Kolf-Clauw, M.; Oswald, I.P. Toxicity of deoxynivalenol and its acetylated derivatives on the intestine: Differential effects on morphology, barrier function, tight junctions proteins and mitogen-activated protein kinases. *Toxicol. Sci.* **2012**, *130*, 180–190. [CrossRef]

27. Bae, H.K.; Pestka, J.J. Deoxynivalenol induces p38 interaction with the ribosome in monocytes and macrophages. *Toxicol. Sci.* **2008**, *105*, 59–66. [CrossRef]

28. Voss, K.A.; Smith, G.M.; Hascheck, W.M. Fumonisins: Toxicokinetics, mechanism of action and toxicity. *Animal Feed Sci. Technol.* **2007**, *137*, 299–325. [CrossRef]

29. Dilkin, P.; Hassegawa, R.; Reis, T.A.; Mallmann, C.A.; Corrêa, B. Intoxicação experimental de suínos por fumonisinas. *Cienc. Rural.* **2004**, *34*, 175–191. [CrossRef]

30. McGuckin, M.A.; Lindén, S.K.; Sutton, P.; Florin, T.H. Mucin dynamics and enteric pathogens. *Nat. Rev. Microbiol.* **2011**, *9*, 265–278. [CrossRef]

31. Brown, T.; Rottinghaus, G.; Williams, M. Fumonisin mycotoxicosis in broilers: Performances and pathology. *Avian Diseases* **1992**, *36*, 450–454.

32. Goope, N.V.; Sharma, R.P.H. Fumonisin B1-induced apoptosis is associated with delayed inhibition of protein kinase C, nuclear factor-kappa B and tumor necrosis factor alpha in LLC-PK1 cells. *Chem-Biol. Interact.* **2003**, *146*, 131–145. [CrossRef]

33. Gartner, L.P.; Hiatt, J.L. Sistema Digestivo. In *Tratado de Histologia em cores*, 3rd ed.; Elsevier: Rio de Janeiro, Brasil, 2007; pp. 387–416.

34. De Walle, J.V.; Sergent, T.; Piront, N.; Toussaint, O.; Schneider, Y.J.; Larondelle, Y. Deoxynivalenol affects *in vitro* intestinal epithelial cell barrier integrity through inhibition of protein synthesis. *Toxicol. Appl. Pharmacol.* **2010**, *15*, 291–298.

35. Marin, D.E.; Gouze, M.E.; Taranu, I.; Oswald, I.P. Fumonisin B1 alters cell cycle progression and interleukin-2 synthesis in swine peripheral blood mononuclear cells. *Mol. Nutr. Food Res.* **2007**, *51*, 1406–1412. [CrossRef]

toxins

MDPI

Article

The Expression of Type-1 and Type-2 Nitric Oxide Synthase in Selected Tissues of the Gastrointestinal Tract during Mixed Mycotoxicosis

Magdalena Gajęcka [1],*, Ewa Stopa [2], Michał Tarasiuk [3], Łukasz Zielonka [1] and Maciej Gajęcki [1]

[1] Department of Veterinary Prevention and Feed Hygiene, Faculty of Veterinary Medicine, University of Warmia and Mazury in Olsztyn, ul. Oczapowskiego 13/29, Olsztyn 10-718, Poland; lukaszz@uwm.edu.pl (L.Z.); gajecki@uwm.edu.pl (M.G.)

[2] Veterinary Clinic, Ewa Stopa DVM, ul. Dąbrowskiego 15, Iława 14-200, Poland; ewastopa@autograf.pl

[3] BIOMIN Polska Sp. z o.o., ul. Grochowska 16, Warszawa 04-217, Poland; michal.tarasiuk@biomn.net

* Author to whom correspondence should be addressed; mgaja@uwm.edu.pl; Tel.: +48-89-523-32-37; Fax.: +48-89-523-36-18.

Received: 11 October 2013; in revised form: 12 November 2013; Accepted: 18 November 2013; Published: 22 November 2013

Abstract: The aim of the study was to verify the hypothesis that intoxication with low doses of mycotoxins leads to changes in the *m*RNA expression levels of nitric oxide synthase-1 and nitric oxide synthase-2 genes in tissues of the gastrointestinal tract and the liver. The experiment involved four groups of immature gilts (with body weight of up to 25 kg) which were orally administered zearalenone in a daily dose of 40 μg/kg BW (group Z, $n = 18$), deoxynivalenol at 12 μg/kg BW (group D, $n = 18$), zearalenone and deoxynivalenol (group M, $n = 18$) or placebo (group C, $n = 21$) over a period of 42 days. The lowest *m*RNA expression levels of nitric oxide synthase-1 and nitric oxide synthase-2 genes were noted in the sixth week of the study, in particular in group M. Our results suggest that the presence of low mycotoxin doses in feed slows down the *m*RNA expression of both nitric oxide synthase isomers, which probably lowers the concentrations of nitric oxide, a common precursor of inflammation.

Keywords: mycotoxins; low doses; zearalenone; deoxynivalenol; gene expression; nitric oxide synthases

1. Introduction

The liver and intestinal mucosa are subjected to high antigen loads and highly varied antigen molecules ingested with feed. Foodborne microorganisms colonize the digestive tract [1]. The intestinal barrier function involves a series of mechanisms which control absorption through the mucous membrane and prevent intestines against harmful substances. In a stable physiological state, small numbers of antigens are transported across the intestinal barrier which can quickly identify and eliminate foreign intruders that pose a health risk. Two types of immune responses are involved in this process: innate (nonspecific) immunity which is the body's first line of defense against infections and adaptive (specific) immunity which is further subdivided into cellular and humoral immunity. If the intestinal barrier function is disrupted by harmful substances ingested with food, such as mycotoxins [2,3], other antigen molecules or microorganisms, the structural continuity of the mucous membrane may be disrupted and the number of signaling molecules may be altered, leading to various pathological states. In a normal host, equilibrium is restored when the pathological factor is eliminated, but in susceptible hosts, intestinal permeability may be increased which, in extreme cases, can lead to chronic inflammations [4].

Nitric oxide (NO), first described as an endothelium-derived relaxing factor (EDRF), plays various roles in basic life functions of an organism. NO is one of the smallest known biologically active molecules. In stable physiological states, it acts as an intracellular signaling molecule, and at higher concentrations, it plays the role of an autocrine and paracrine molecule which regulates the immune response [5] or a factor which controls the peristalsis of the entire gastrointestinal tract [6]. NO levels are determined by the activity of various nitric oxide synthase (NOS) isomers [7,8]. Three basic forms of NOS have been identified to date: NOS-1, NOS-2 and NOS-3. They were previously classified as: NOS-1–nNOS (neuronal NOS) or NOS-I; NOS-2–iNOS (inducible NOS) or NOS-II; and NOS-3–eNOS (endothelial NOS) or NOS-III [6]. Mammalian cells contain two constitutive NOS enzymes—NOS-1 and NOS-3 which synthesize NO in response to an increase in calcium ion (Ca^{2+}) concentrations inside a cell [9]. In some cases, those enzymes can be activated independently in response to stress.

The activity of the inducible isoform of NOS-2 is not determined by intracellular concentrations of Ca^{2+}, but by calmodulin binding. Higher Ca^{2+} concentrations inside a cell increase calmodulin levels and supports calmodulin binding with NOS-1 and NOS-3, leading to momentary intensification of NO synthesis [10]. Unlike isoforms NOS-1 and NOS-3, NOS-2 can bind with calmodulin even at low intracellular concentrations of Ca^{2+}. NOS-2 activity is not affected by changes in Ca^{2+} levels, which prolongs synthesis [11] and increases local NO concentrations in the cell in comparison with the remaining NOS isoforms [12–14].

NOS-2 is found mainly in the central nervous system, peripheral nervous system, skeletal muscles, pancreatic islets, endometrium and the macula densa. NOS-2 modulates the transmission of neural signals, regulates nephron functions and controls gastrointestinal peristalsis [6]. NO produced by NOS-2 also acts as a neurotransmitter, in particular in the non-adrenergic non-cholinergic (NANC) nervous system. NOS-1 is activated in the first hours after cytokine release. Successive periods are characterized by a predominance of NOS-2 which synthesizes significant amounts of NO [15].

NOS-2 is also observed in macrophages, cardiac muscle, liver, smooth muscles and vascular endothelium where it is synthesized in response to endogenous and exogenous agents such as bacterial lipopolysaccharides, pro-inflammatory cytokines (IL-1β, IL-4, IL-6, IL-8 and IL-10 INF-γ, TNF-α) [15] and allergens [5,16,17], including zearalenone (ZEN), deoxynivalenol (DON) and mixtures thereof [18,19].

Mycotoxicosis caused by DON and/or ZEN reduced the number of mucus producing cells and decreased glycocalyx secretion in pigs exposed to low doses of those mycotoxins [20]. Intoxication with higher doses of ZEN alone exerted different effects by stimulating the activity of goblet cells and mucinogen granules [21]. Maresca and Fantini [2] demonstrated that intestinal mucosa is also affected by other mycotoxins which can contribute to inflammatory processes and uncontrolled proliferation of mucosal cells. The mechanisms responsible for mycotoxins' effects on mucus production have not yet been identified. Trichothecenes and estrogenic hormones were found to inhibit protein synthesis, and ZEN could be included in this group of mycotoxins [2].

The effects of ZEN and DON on local enteric immunity have not been investigated to date, but it can be assumed that ZEN is a potential and DON is a definite immunosuppressive agent [2]. Those mycotoxins participate in immunosuppressive processes that lead mainly to intestinal inflammations. Hyperadditive synergistic interactions contribute to the above [22]. Some mycotoxins have direct or indirect pro-inflammatory effects, and they exacerbate the existing inflammations [23]. They can provoke inflammations indirectly by modifying intestinal permeability and contributing to antigen transfer from the intestines, or directly by stimulating the release of pro-inflammatory cytokines from the intestinal epithelium [23]. Expression levels, in particular the expression of NOS-2, should be determined to evaluate the immunomodulative properties of the analyzed mycotoxins.

The objective of this study was to determine the effect of ZEN, DON and mixtures thereof, administered at low doses to gilts, on the expression levels of NOS-1 and NOS-2 proteins in different sections of the porcine gastrointestinal tract.

2. Materials and Methods

All of the experimental procedures involving animals were carried out in compliance with Polish legal regulations determining the terms and methods for performing experiments on animals (opinion of the Local Ethics Committee for Animal Experimentation No. 88/N of 16 December 2009).

The experiment was conducted at the Department of Veterinary Prevention and Feed Hygiene, Faculty of Veterinary Medicine, University of Warmia and Mazury in Olsztyn, Poland, on 75 clinically healthy gilts with initial body weight of 25 ± 2 kg. The gilts were penned in groups with ad libitum access to water. Administered feed was tested for the presence of mycotoxins: ZEA, α-ZEL and DON. Mycotoxin levels in the diets were estimated by common separation techniques with the use of immunoaffinity columns (Zearala-Test™ Zearalenone Testing System, G1012, VICAM, Watertown, USA and DON-Test™ DON Testing System, VICAM, Watertown, USA) and high performance liquid chromatography (HPLC) (Hewlett Packard, type 1050 and 1100) [24] with fluorescent and/or UV detection techniques.

The animals were divided into three experimental groups (Z, D and M; $n = 18$ in each group) and one control group (C; $n = 21$). Group Z animals were orally administered ZEN at 40 µg/kg BW, group D animals were orally administered DON at 12 µg/kg BW, and group M animals were orally administered a mixture of ZEN and DON (40 µg ZEN/kg BW + 12 µg DON/kg BW). Group C pigs were fed a placebo. In all experimental groups, mycotoxins were administered at doses below NOAEL values [22]. Both mycotoxins were synthesized and standardized by the Department of Chemistry of the Poznań University of Life Sciences under the supervision of Professor Piotr Goliński. The experiment covered a period of 42 days. Three animals from each experimental group were sacrificed on days 1, 7, 14, 21, 28, 35 and 42 (a total of 12 gilts on each day), excluding day 1 when only three control group animals were scarified.

2.1. Reagents

Analytical samples of the studied mycotoxins were administered *per os* daily in gelatin capsules before the morning feeding. Mycotoxin samples were diluted in 300 µL 96% ethyl alcohol (96% ethyl alcohol, SWW 2442-90, Polskie Odczynniki Chemiczne SA) to obtain the required doses (subject to body weight). The resulting solutions were stored at room temperature for 12 h to evaporate the solvent. The animals were weighed every seven days to update mycotoxin doses for each gilt.

2.2. mRNA Isolation and cDNA Synthesis

Tissue samples were collected from the porcine gastrointestinal tract (liver (L)—left lobe, duodenum (DU)—first and middle sections, jejunum (J)—middle section, ascending colon (AC)—middle section, descending colon (DC)—middle section) and rinsed in PBS. Tissue samples were placed in RNAlater®solution (Invitrogen, Carlsbad, California, USA) to stabilize RNA. They were incubated at 4 °C for 24 h and frozen at −80 °C. RNA was isolated from the porcine digestive tract with the use of the Total RNA Mini Plus kit (A&A Biotechnology, Gdynia, Polska). Tissue samples of 200 g were weighed and homogenized in TissueLyser II (Qiagen, USA). Total RNA was extracted in accordance with the manufacturer's protocol. RNA concentrations and purity were determined in the Nano Vue spectrophotometer (GE Health Care, Buckinghamshire UK). RNA quality was evaluated by electrophoresis in 2% agarose gel. The resulting total RNA was dissolved in an aqueous solution and stored at −80 °C until analysis by reverse transcription PCR (RT-PCR). RT-PCR was performed with the use of Fermentas reagents (Lithuania). The volume of the RNA solution containing 5 µg of total RNA was determined, and it was supplemented to 12.5 µg through the addition of RNase-free water and 1 µg of Oligo(dT)$_{18}$ (0.5 µg) primers (Fermentas, Lithuania). The resulting mixture was incubated at −65 °C for 5 minutes and cooled on ice. 4 µL of $5 \times$ RT buffer, 0.5 µL of 20U RNase inhibitor (RiboLock™ RNase Inhibitor) (Fermentas, Lithuania), 2 µL of dNTP mix, 10 mM each (Fermentas, Lithuania), and 1 µL of 200U reverse transcriptase (RevertAid™ Transcriptase) (Fermentas, Lithuania)

199

were added. The resulting mixture was incubated at 42 °C for 60 min, and the reaction was terminated at 70 °C for 10 min. The reverse transcription reaction was carried out in the Personal Mastercycler Eppendorf thermocycler (Hamburg, Germany). The cDNA synthesis reaction mixture was used in Real-Time PCR.

2.3. Real-Time PCR

Real-Time PCR was performed in the Genomics and Transcriptomics Laboratory of the Department of Animal Anatomy, Faculty of Veterinary Medicine of the University of Warmia and Mazury in Olsztyn. Real-Time PCR was carried out to determine the mRNA expression levels of NOS-2 and NOS-1 genes in reference to the GAPDH gene. Specific primers for the above genes were designed in the Pick Primer Blast application based on the following sequences: NOS-2–NM_001143690.1, NOS-1–XM_003132898.3 and GAPDH–NM_001206359.1 (Table 1). Reaction tubes were filled with 25 μL of the reaction mix, and the assay was performed in the 7500 Fast Real-Time PCR System thermocycler (Applied Biosystems, Carlsbad, California USA) under the following conditions: initial denaturation −10 min/95 °C, followed by 40 cycles: denaturation −15 s/95 °C, primer annealing −1 min/60 °C. The following reagents were used: 12.5 μL of FastStart Universal SYBR Green Master (Rox) (Roche, Vaud, Switzerland), 10.5 μL of nuclease-free water, 1 μL of cDNA, 1 μL of the primer mix, 5 μM each (forward + reverse). All samples were analyzed in duplicates.

Table 1. Real-time PCR primers.

Control gene	Primer sequence (sense and anti-sense)	Primer annealing temperature (°C)	Amplicon length (bp)	Gene bank No.
NOS-1	f: CCATGGCCGCCGATGTCCTC r: CGGTTGTCATCCCTCAGCCTGC	60 °C	109 bp	XM_003132898.3
NOS-2	f: CTCCAGGTGCCCACGGGAAA r: TGGGGATACACTCGCCCGCC	60 °C	117 bp	NM_001143690.1
GAPDH	f: TTCCACCCACGGCAAGTT r: GGCCTTTCCATTGATGACAAG	60 °C	69 bp	NM_001206359.1

2.4. Statistical Analysis

The data were grouped based on: (i) the administered mycotoxins relative to experimental dates and (ii) experimental dates relative to the analyzed mycotoxins. Differences between the administered mycotoxins relative to experimental dates and differences between experimental dates relative to the analyzed mycotoxins were analyzed. The results were processed statistically in the Statistica application. Differences between groups (mycotoxin or date) were determined by ANOVA. The equality of group variances was tested by the Brown-Forsythe test. When ANOVA revealed significant differences between groups ($P < 0.01$—highly significant differences, $0.01 < P < 0.05$—significant differences, $P > 0.05$—no differences), Tukey's HSD test was used to determine differences between specific groups.

3. Results

The results of statistical analysis of mRNA expression levels of NOS-1 genes in selected sections of the porcine gastrointestinal tract and the liver on different days of the experiment relative to the GAPDH gene are presented in Figure 1A–D. Differences at $P < 0.05$ were reported: on sampling date I—between group M and group C in tissue J (Figure 1A), between groups Z and D and group C in tissue DC (Figure 1B); on sampling date V—between group C and the remaining groups in tissue DC (Figure 1C); on sampling date VI—between group C and the remaining groups in tissue J (Figure 1D). Differences at $P < 0.01$ were observed only on date I between group D and group C in tissue J (Figure 1A).

The results of statistical analysis of mRNA expression levels of NOS-2 genes in selected sections of the gastrointestinal tract and the liver of gilts intoxicated with mycotoxins on different days of the experiment are presented in Figure 1E–F. Differences at $P < 0.05$ were reported between group Z and groups C and M on date VI in tissue DU (Figure 1E). Differences at $P < 0.01$ were observed on date I between group Z and groups C and D, and between group M and groups C and D in tissue L (Figure 1F).

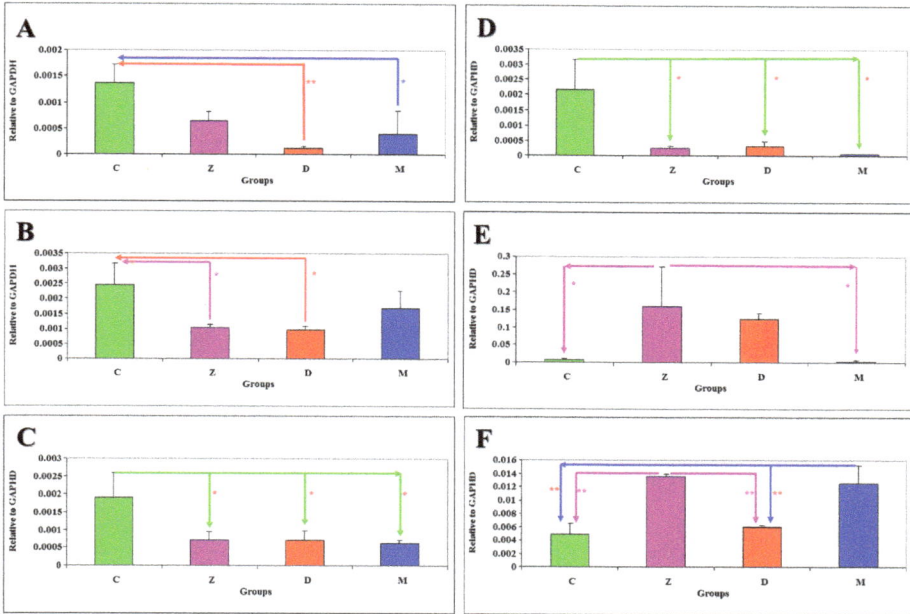

Figure 1. The results of statistical analysis of *m*RNA expression levels of nitric oxide synthase (NOS)-1 and NOS-2 genes in selected sections of the porcine gastrointestinal tract and the liver on different days of the experiment relative to the GAPDH gene are presented in: (**A**) Jejunum NOS-1 (sampling date I); (**B**) Descending colon NOS-1 (sampling date I); (**C**) Descending colon NOS-1 (sampling date V); (**D**) Jejunum NOS-1 (sampling date VI); (**E**) Duodenum NOS-2 (sampling date VI); (**F**) Liver NOS-2 (sampling date I).

The results of statistical analysis of *m*RNA expression levels of NOS-1 genes in selected sections of the gastrointestinal tract and the liver of gilts intoxicated with mycotoxins in different groups throughout the entire experiment are presented in Figure 2A–D. Differences at $P < 0.05$ were reported: in group C in tissue DC between date I and dates IV and VI (Figure 2A); in group Z—in tissue DU between dates I and V and between dates IV and VI (Figure 2B) and in tissue AC between date III and dates IV and V (Figure 2C), in tissue DC between date IV and dates I and II (Figure 2A); in group D—in tissue DC between date III and dates IV and V and between date II and dates IV and VI (Figure 2A), in tissue L between dates II and V and between dates IV and VI (Figure 2D); in group M—in tissue DC between date I and VI (Figure 2A). Differences at $P < 0.01$ were observed: in group Z—in tissue DU between dates IV and V (Figure 2B); in group D—in tissue DC between dates III and VI (Figure 1A) and in tissue L between dates V and VI (Figure 2D).

The results of statistical analysis of *m*RNA expression levels of NOS-2 genes in selected sections of the gastrointestinal tract and the liver of gilts intoxicated with myxotoxins in different groups throughout the entire experiment are presented in Figure 2E–H. Differences at $P < 0.05$ were reported: in group Z—in tissue DU between date VI and the remaining dates (Figure 2E), in tissue L between dates VI and V (Figure 2F); in group M—in tissue DC between date II and dates III, IV and VI (Figure 2G). Differences at $P < 0.01$ were observed: in all experimental groups in tissue J between date IV and the remaining dates (Figure 2H); in group M—in tissue L between date II and dates I and V (Figure 2F). A high number of differences were verging on the statistically significant level ($0.1 > P > 0.05$), representing statistical tendencies.

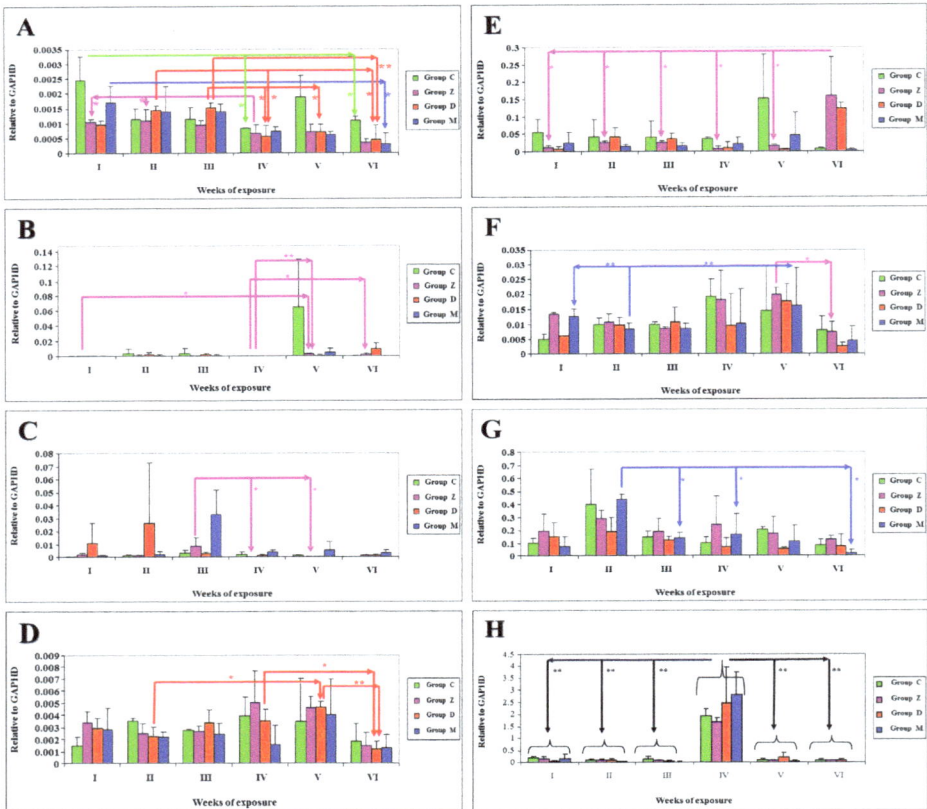

Figure 2. The results of statistical analysis of mRNA expression levels of NOS-1 genes in selected sections of the gastrointestinal tract and the liver of gilts intoxicated with mycotoxins in different groups throughout the entire experiment are presented in: (**A**) Descending colon (NOS-1); (**B**) Duodenum (NOS-1); (**C**) Ascending colon (NOS-1); (**D**) Liver (NOS-1); (**E**) Duodenum (NOS-2); (**F**) Liver (NOS-2); (**G**) Descending colon (NOS-2); (**H**) Jejunum (NOS-2).

The results of statistical analysis of *m*RNA expression levels of NOS-1 and NOS-2 genes in selected sections of the gastrointestinal tract and the liver of gilts intoxicated with mycotoxins on different experimental dates (Figure 1A–F) point to a decreasing trend in the *m*RNA expression levels of NOS-1 genes (Figure 1A–D) in all experimental groups in comparison with group C, between group M and groups Z and D, and between group Z and group D. A similar trend was noted in the *m*RNA expression levels of NOS-2 genes (Figure 1E–F), but an increase in *m*RNA expression was noted in group D in a comparison with group Z.

The values of NOS-1 gene expression in selected sections of the gastrointestinal tract and the liver of gilts intoxicated with mycotoxins in different groups throughout the experiment (Figure 2A–D) were not uniformly distributed, and the median for selected tissues in different groups was determined at: in group C—0.012336 for DU, 0.001267 for J, 0.001316 for AC, 0.001425 for DC and 0.002838 for

L; in group Z—0.001275 for DU, 0.000346 for J, 0.002249 for AC, 0.000795 for DC and 0.003281 for L; in group D—0.002531 for DU, 0.000262 for J, 0.007142 for AC, 0.000940 for DC and 0.002988 for L; in group M—0.001160 for DU, 0.000163 for J, 0.008002 for AC, 0.001026 for DC and 0.002392 for L. The lowest quartile of gene expression values was observed on date VI in 7 cases (3 × in DC and L in groups Z, D and M, and 1× in AC in group C) and on date IV in 6 cases (4 × in DU and 1 × in DC in group C and in AC in group Z). The highest quartile of gene expression values was noted on date V in 6 cases (3 × in DU in groups C, Z and M, 2 × in L in groups D and M, 1 × in J in group D), on date III in 5 cases (4 × in AC and 1 × in DC in group D) and on date I in 4 cases (2 × in J in groups Z and M and 2 × in DC in groups C and M).

The values of NOS-2 gene expression in selected sections of the porcine gastrointestinal tract and the liver across groups throughout the experiment (Figure 2E–H) were not uniformly distributed, and the median for selected tissues in different groups was determined at: in group C—0.055680 for DU, 0.416960 for J, 0.068257 for AC, 0.173322 for DC and 0.011045 for L; in group Z—0.041299 for DU, 0.349643 for J, 0.105470 for AC, 0.204706 for DC and 0.006676 for L; in group D—0.037921 for DU, 0.486018 for J, 0.062965 for AC, 0.109968 for DC and 0.009398 for L; in group M—0.021214 for DU, 0.500869 for J, 0.077001 for AC, 0.158163 for DC and 0.009989 for L. The highest quartile of gene expression values was noted: in DC on date II and in J on date IV in all groups, and in L in groups Z, D and M on date V. The lowest quartile of gene expression values was observed: on date VI in 12 of 20 identified cases (3 × in J and DC in groups C, Z and M; 3 × in L in groups Z, D and M; 2 × in DU in groups C and M, 1 × in AC in group Z); on date V in 4 cases (2 × in AC in groups Z and M, 1 × in DU and 1 × in DC in group D). All median values of *m*RNA expression levels of NOS-2 genes were higher than the median values of NOS-1 expression, excluding DU median values in group D.

4. Discussion

The presence of pathogens in digest can modulate the local immune response. The degree of immunomodulation is determined by the dose of the pathogenic substance, such as a mycotoxin. On the other hand, the clinical picture of intoxication does not reveal such a tendency, which is consistent with the low dose hypothesis proposed by Vandenberg *et al.*, [25]. The induction of regulatory T cells (Treg) is observed. This mechanism is deployed by pathogens, and not only, is to escape the immune response [26], and it observed during chronic and persistent infections. Very little is known about the possible effects of mycotoxins which can be ingested in small doses with feed over prolonged periods of time (monodiet) or on a regular basis.

Reactive oxygen species are activated before Treg induction [14]. They include NO, one of the main signaling molecules in the immune system. NO is active at the place of release or it may be transferred across mucous membranes into surrounding tissues. It is one of the most active smooth muscle relaxants. NO is catalyzed from arginine by NOS. Its activity is strictly determined by its concentrations at the site of the reaction which, in turn, are dependent on the expression levels of specific NOS isomers. At high concentrations, NO exerts proinflammatory and cytotoxic effects directly or indirectly via its active derivatives (nitrite, nitrogen trioxide, peroxynitrite, nitrosoperoxycarbonate). At low, physiological concentrations, NO regulates homeostasis in circulatory, respiratory [27] and immune systems and controls nerve conduction [28,29].

The results of our study investigating the *m*RNA expression levels of NOS-1 and NOS-2 genes in gilts administered very low doses of individual mycotoxins (ZEN and DON) or the mycotoxin mix for 42 days are partially consistent with published data. Significant differences (at $P < 0.05$ and $P < 0.01$) were observed in the *m*RNA expression levels of NOS-1 genes in selected sections of the porcine gastrointestinal tract and the liver on different days of the experiment (Figure 1A–F), in particular in tissues J and DC on dates I and VI. A decreasing trend in median values was reported in comparison with group C and between groups administered DON. A similar trend was observed in the *m*RNA expression levels of NOS-2 genes, but the noted values were significantly higher in comparison with NOS-1 expression. A drop in the *m*RNA expression levels of both genes was observed in groups Z,

D and M. This suggests that low doses of ZEN, DON or the mycotoxin mix in feed (below NOAEL values) inhibit both NOS isomers, which probably lowers NO synthesis. The results shown in Figure 1 could suggest that ZEN was the inhibitory factor at the initial stage of the experiment, but the longer the exposure to ZEN, the decrease in NOS-1 mRNA expression was leveled out, whereas the drop in NOS-2 mRNA expression was more pronounced in group Z.

Similar statistical tendencies were noted with respect to the *m*RNA expression levels of NOS-1 and NOS-2 genes in selected sections of the gastrointestinal tract and the liver of gilts intoxicated with mycotoxins in different groups throughout the entire experiment (Figure 2E–H). A decreasing trend was intensified over time and in groups administered the mycotoxin/mycotoxins. The only exceptions were tissue sections where significant differences were observed, mostly NOS-2 expression values in tissues DU, J, AC and DC.

Our findings indicate that both mycotoxins inhibit the *m*RNA expression of genes controlling constitutive isomer NOS-1 and inducible isomer NOS-2. It can be expected that low doses of ZEN or DON administered with feed (below NOAEL value) over prolonged periods of time can modify gastrointestinal functions due to a decrease in the concentrations of NO which inhibits NANC transmitters [30,31]. Low levels of NO can probably speed up peristalsis in the esophagus, stomach and intestines, inhibit gastric accommodation responses (receptive and adaptive relaxation, antral contraction) and increase anal sphincter pressure, thus slowing down gastric emptying and digesta passage through the intestines [6,32].

In consequence, decreased *m*RNA expression of the NOS-1 gene, released in the enteric nervous system, should slow down intestinal peristalsis and sphincter function. A drop in the *m*RNA expression of the NOS-2 gene should decrease intestinal permeability and inhibit bowel secretion [33]. The above conclusions were formulated by directly extrapolating from the symptoms caused by high mycotoxin doses to those caused by low doses. However, recent research findings provide evidence contrary to this assumption (low dose hypothesis). The influence of very low doses (below NOAEL values) of any hormonal agent which modulates tissue activity could be different. Undesirable hormonally active substances [25] such as ZEN, which can act as a signaling molecule, could serve as an example.

Interestingly, a decrease in the *m*RNA expression of genes controlling both NOS isomerases was particularly noted in distal sections of the digestive tract. The explanations can be provided for the above observation. Firstly, mycotoxins have antibacterial properties, and they decrease the populations of pathogenic microorganisms which are one of the key pro-inflammatory agents and which stimulate NO production [34]. Secondly, when administered at low doses, the analyzed mycotoxins inhibit the *m*RNA expression of both NOS genes, which could have therapeutic implications [31,35] at high levels but not at physiological levels of NO. The presence of low doses of ZEN and DON (below NOAEL value) in feed inhibits inflammatory processes in the digestive tract, in particular in tissues J, AC and DC. The above could be attributed to the "escape" of signaling molecules from local and systemic immune activation, similar to that observed when Tregs are induced by bacterial pathogens [26] in chronic infections.

Acknowledgments: The authors would like to thank Małgorzata Chmielewska, and Katarzyna Łosiewicz, for their help in laboratory analyses. The experiment was supported by research grant No. NR12-0080-10 of the Polish National Center for Research and Development.

Conflicts of Interest: The authors declare no conflict of interest.

References

1. Xu, X.; Xu, P.; Ma, C.; Tang, J.; Zhang, X. Gut microbiota, host health, and polysaccharides. *Biotechnol. Adv.* **2013**, *31*, 318–337.
2. Maresca, M.; Fantini, J. Some food-associated mycotoxins as potential risk factors in humans predisposed to chronic intestinal inflammatory diseases. *Toxicon* **2010**, *56*, 282–294. [CrossRef]
3. Pinton, P.; Guzylack, L.; Kolf-Clauw, M.; Oswald, I.P. Effects of some fungal toxins, the trichothecenes, on the epithelial intestinal barrier. *Curr. Immunol. Rev.* **2012**, *8*, 193–208. [CrossRef]

4. Groschwitz, K.R.; Hogan, S.P. Intestinal barrier function: Molecular regulation and disease pathogenesis. *J. Allergy Clin. Immunol.* **2009**, *124*, 3–20. [CrossRef]
5. Lechner, M.; Lirk, P.; Rieder, J. Inducible nitric oxide synthase (iNOS) in tumor biology: The two sides of the same coin. *Semin. Cancer Biol.* **2005**, *15*, 277–289. [CrossRef]
6. Castro, M.; Muñoz, J.M.; Arruebo, M.P.; Murillo, M.D.; Arnal, C.; Bonafonte, J.I.; Plaza, M.A. Involvement of neuronal nitric oxide synthase (nNOS) in the regulation of migrating motor complex (MMC) in sheep. *Vet. J.* **2012**, *192*, 352–358. [CrossRef]
7. Sessa, W.C. The nitric oxide synthase family of proteins. *J. Vasc. Res.* **1994**, *31*, 131–143. [CrossRef]
8. Dusting, G.J. Nitric oxide in cardiovascular disorders. *J. Vasc. Res.* **1995**, *32*, 143–161. [CrossRef]
9. Arnal, J.F.; Dinh-Xuanb, A.T.; Pueyoc, M.; Darbladea, B.; Ramia, J. Endothelium-derived nitric oxide and vascular physiology and pathology. *Cell Mol. Life Sci.* **1999**, *55*, 1078–1087. [CrossRef]
10. Bielewicz, J.; Kurzepa, J.; Łagowska-Lenard, M.; Bartosik-Psujek, H. The novel views on the patomechanism of ischemic stroke. *Wiad. Lek.* **2010**, *63*, 213–220.
11. Losada, A.P.; Bermúdez, R.; Faílde, L.D.; Quiroga, M.I. Quantitative and qualitative evaluation of iNOS expression in turbot (Psetta maxima) infected with Enteromyxum scophthalmi. *Fish Shellfish Immun.* **2012**, *32*, 243–248. [CrossRef]
12. Mehl, M.; Daiber, A.; Herold, S.; Shoun, H.; Ullrich, V. Peroxynitrite reaction with heme proteins. *Nitric Oxide* **1999**, *3*, 142–152. [CrossRef]
13. Stuehr, D.J.; Santolini, J.; Wang, Z.Q.; Wei, C.C.; Adak, S. Update on mechanism and catalytic regulation in the NO synthases. *J. Biol. Chem.* **2004**, *279*, 36167–36170.
14. Ługowski, M.; Saczko, J.; Kulbacka, J.; Banaś, T. Reactive oxygen and nitrogen species. *Pol. Merk. Lek.* **2011**, *31*, 313–317.
15. Alderton, W.K.; Cooper, C.E.; Knowles, R.G. Nitric oxide synthases: Structure, function and inhibiton. *Biochem. J.* **2001**, *357*, 593–615. [CrossRef]
16. Sokołowska, M.; Włodek, L. Good and bad sides of nitric oxide. *Folia Cardiol.* **2001**, *8*, 467–477.
17. Tokuhara, K.; Hamada, Y.; Tanaka, H.; Yamada, M.; Ozaki, T.; Matsui, K.; Kamiyama, Y.; Nishizawa, M.; Ito, S.; Okumura, T. Rebamipide, anti-gastric ulcer drug, up-regulates the induction of iNOS in proinflammatory cytokine-stimulated hepatocytes. *Nitric Oxide* **2008**, *18*, 28–36. [CrossRef]
18. Gajęcki, M.; Gajęcka, M.; Jakimiuk, E.; Zielonka, Ł. Feedingstuffs and human health. *Pol. J. Food. Nutr. Sci.* **2005**, *14*, 7–13.
19. Zielonka, Ł.; Obremski, K.; Gajęcka, M.; Rybarczyk, L.; Jakimiuk, E.; Gajęcki, M. An evaluation of selected indicators of immune response in pigs fed a diet containing deoxynivalenol, T-2 toxin and zearalenone. *Bull. Vet. Inst. Pulawy* **2010**, *54*, 631–635.
20. Obremski, K.; Zielonka, Ł.; Gajęcka, M.; Jakimiuk, E.; Bakuła, T.; Baranowski, M.; Gajęcki, M. Histological estimation of the small intestine wall after administration of feed containing deoxynivalenol, T-2 toxin and zearalenone in the pig. *Pol. J. Vet. Sci.* **2008**, *11*, 339–345.
21. Obremski, K.; Gajęcka, M.; Zielonka, Ł.; Jakimiuk, E.; Gajęcki, M. Morphology and ultrastructure of small intestine mucosa in gilts with zearalenone mycotoxicosis. *Pol. J. Vet. Sci.* **2005**, *8*, 301–307.
22. Boermans, H.J.; Leung, M.C.K. Mycotoxins and the pet food industry: Toxicological evidence and risk assessment. *Int. J. Food. Microbiol.* **2007**, *119*, 95–102. [CrossRef]
23. Maresca, M.; Yahi, N.; Younès-Sakr, L.; Boyron, M.; Caporiccio, B.; Fantini, J. Both direct and indirect effects account for the proinflammatory activity of enteropathogenic mycotoxins on the human intestinal epithelium: Stimulation of interleukin-8 secretion, potentiation of interleukin-1beta effect and increase in the transepithelial passage of commensal bacteria. *Toxicol. Appl. Pharmacol.* **2008**, *228*, 84–92. [CrossRef]
24. Obremski, K.; Gajęcki, M.; Zwierzchowski, W.; Bakuła, T.; Apoznański, J.; Wojciechowski, J. The level of zearalenone and α-zearalenol in the blood of gilts after feeding them of feed with a low content of zearalenone. *J. Ani. Feed Sci.* **2003**, *12*, 529–538.
25. Vandenberg, L.N.; Colborn, T.; Hayes, T.B.; Heindel, J.J.; Jacobs, D.R.; Lee, D.-H.; Shioda, T.; Soto, A.M.; vom Saal, F.S.; Welshons, W.V.; *et al.* Hormones and endocrine-disrupting chemicals: Low-dose effects and nonmonotonic dose responses. *Endocr. Rev.* **2012**, *33*, 378–455. [CrossRef]
26. Silva-Campa, E.; Mata-Haro, V.; Mateu, E.; Hernández, J. Porcine reproductive and respiratory syndrome virus induces CD4+CD8+CD25+Foxp3+ regulatory T cells (Tregs). *Virology* **2012**, *430*, 73–80. [CrossRef]

27. Ziętkowski, Z.; Ziętkowska, E.; Bodzenta-Łukaszyk, A. Exhaled nitric oxide measurements in the diagnosis of respiratory diseases. *Alerg. Astma Immun.* **2009**, *14*, 215–222.

28. Nussler, A.K.; Billiar, T.R. Inflammation, immunoregulation and inducible nitric oxide synthase. *J. Leukoc. Biol.* **1993**, *54*, 171–178.

29. Kroncke, K.D.; Fehsel, K.; Kolb-Bachofen, W. Inducible nitric oxide synthase and its product nitric oxide, a small molecule with complex biological activities. *Biol. Chem.* **1995**, *376*, 327–343.

30. Gupta, A.; Sharma, A.C. Despite minimal hemodynamic alterations endotoxemia modulates NOS and p38-MAPK phosphorylation via metalloendopeptidases. *Mol. Cell Biochem.* **2004**, *265*, 4–56.

31. Grześk, E.; Grześk, G.; Koziński, M.; Stolarek, W.; Zieliński, M.; Kubica, J. Nitric oxide as a cause and a potential place therapeutic intervention in hyporesponsiveness vascular in early sepsis. *Folia Cardiol.* **2011**, *6*, 36–43.

32. Bennett, M.R. Non-adrenergic non-cholinergic (NANC) transmission to smooth muscle: 35 years on. *Prog. Neurobiol.* **1997**, *52*, 159–195. [CrossRef]

33. Dijkstra, G.; van Goor, H.; Jansen, P.L.; Moshage, H. Targeting nitric oxide in the gastrointestinal tract. *Curr. Opin. Invest. Drugs* **2004**, *5*, 529–536.

34. Davila, A.-M.; Blachier, F.; Gotteland, M.; Andriamihaja, M.; Benetti, P.-H.; Sanz, Y.; Tomé, D. Re-print of "Intestinal luminal nitrogen metabolism: Role of the gut microbiota and consequences for the host". *Pharmacol. Res.* **2013**, *69*, 114–126. [CrossRef]

35. Yang, E.J.; Yim, E.Y.; Song, G.; Kim, G.O.; Hyun, C.G. Inhibition of nitric oxide production in lipopolysaccharide-activated RAW 264.7 macrophages by Jeju plant extracts. *Interdisc. Toxicol.* **2009**, *2*, 245–249.

toxins

Review

Effect of Deoxynivalenol and Other Type B Trichothecenes on the Intestine: A Review

Philippe Pinton [1,2] and Isabelle P. Oswald [1,2,*]

[1] INRA (Institut National de la Recherche Agronomique), UMR1331, Toxalim, Research Centre in Food Toxicology, Toulouse F-31027, France; Philippe.Pinton @toulouse.inra.fr

[2] Université de Toulouse, Institut National Polytechnique, UMR1331, Toxalim, Toulouse F-31000, France

* Author to whom correspondence should be addressed; Isabelle.Oswald@toulouse.inra.fr; Tel.: +33-561-285-480; Fax: +33-561-285-145.

Received: 20 December 2013; in revised form: 28 March 2014; Accepted: 9 May 2014; Published: 21 May 2014

Abstract: The natural food contaminants, mycotoxins, are regarded as an important risk factor for human and animal health, as up to 25% of the world's crop production may be contaminated. The _Fusarium_ genus produces large quantities of fusariotoxins, among which the trichothecenes are considered as a ubiquitous problem worldwide. The gastrointestinal tract is the first physiological barrier against food contaminants, as well as the first target for these toxicants. An increasing number of studies suggest that intestinal epithelial cells are targets for deoxynivalenol (DON) and other Type B trichothecenes (TCTB). In humans, various adverse digestive symptoms are observed on acute exposure, and in animals, these toxins induce pathological lesions, including necrosis of the intestinal epithelium. They affect the integrity of the intestinal epithelium through alterations in cell morphology and differentiation and in the barrier function. Moreover, DON and TCTB modulate the activity of intestinal epithelium in its role in immune responsiveness. TCTB affect cytokine production by intestinal or immune cells and are supposed to interfere with the cross-talk between epithelial cells and other intestinal immune cells. This review summarizes our current knowledge of the effects of DON and other TCTB on the intestine.

Keywords: barrier function; food-contaminant; immune response; intestinal lesions; mycotoxins

---

1. Introduction

Mycotoxins are structurally diverse fungal metabolites that can contaminate a variety of dietary components consumed by animals and humans. It is estimated that 25% of the world's crop production is contaminated by mycotoxins during the pre-harvest period, transport, processing or storage [1]. The major mycotoxin-producing fungal genera are _Aspergillus_, _Fusarium_ and _Penicillium_, mainly producing aflatoxins, zearalenone, trichothecenes, fumonisins, ochratoxins and ergot alkaloids.

Among the mycotoxins produced by the _Fusarium_ genus, the broad family of trichothecenes (TCT) is extremely prevalent. They represent the most diverse chemical group of all the mycotoxins, and their molecular weights range between 200 and 500 Da. All TCT possess a sesquiterpenoid structure with or without a macrocyclic ester or an ester-ether bridge between C-4 and C-15. They contain a common 12,13-epoxytrichothecene group responsible for their cytotoxicity and a 9,10-double bond with various side chain substitutions. The non-macrocyclic TCT constitute two groups: Type A, including T-2 toxin, HT-2 toxin, neosolaniol and diacetoxyscirpenol (DAS), while the Type B group contains a ketone and includes fusarenon-X (FUS-X), nivalenol (NIV) and deoxynivalenol (DON) and its 3-acetyl and 15-acetyl derivatives (3- and 15-ADON) (Figure 1). The number and position of the hydroxyl and acetyl-ester groups can influence the relative toxicity within eukaryotic cells. Their relative capacity to interfere with protein synthesis has been attributed to a combination of different factors: the rate of transport into cells, metabolism by cytosol enzymes, changes in affinity for the active binding site or the ability to interfere with protein synthesis [2].

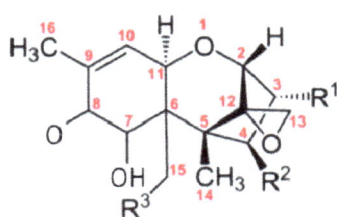

TCTB: Groups present in the R1, R2 and R3 position, respectively:

deoxynivalenol (OH, H, OH)

nivalenol (OH, OH, OH)

fusarenon-X (OH, OAc, OH)

3-acetyldeoxynivalenol (OAc, H, OH)

15-acetyldeoxynivalenol (OH, H, OAc)

Figure 1. Chemical structure of Type B trichothecenes. TCTB, Type B trichothecenes.

Deoxynivalenol and other Type B TCT (TCTB) are commonly found in cereals, such as wheat, rye, barley, oats and corn, all over the world [2–4]. These toxins are resistant to milling, processing and heating and, therefore, readily enter the food chain [5]. The total intake of DON in microgram per kilogram of body weight per day has been estimated to reach from 0.78 in an African diet to 2.4 in a Middle Eastern diet [6]. Intoxications following the consumption of foodstuffs contaminated with TCT have occurred in both humans and animals, with large numbers of people and livestock being affected [4]. Many outbreaks of acute human disease involving nausea, vomiting, gastro-intestinal upset, dizziness, diarrhea and headache have been reported in Asia [7,8]. These outbreaks have been attributed to the consumption of *Fusarium*-contaminated grains and, more recently, to the presence of DON at reported concentrations of 3–93 mg/kg in grain for human consumption [6].

More than 40 countries have introduced regulatory or guideline levels for DON in food and feed. In the USA, the Food and Drug Administration (FDA) has established an advisory level of 1 ppm of DON for bran, flour and germ targeted for human consumption [9]. The European Commission decided to limit the level of DON in food from 0.2 to 1.75 mg/kg for cereals and derived products, depending on the exposed population, and in feed from 0.9 to 5 mg/kg for complementary and complete feedstuffs, depending on the species [10,11].

Following the ingestion of contaminated food or feed, intestinal epithelial cells may be exposed to high concentrations of toxicants, potentially affecting intestinal functions [12]. The intestinal epithelium is a single layer of cells lining the gut lumen that acts as a selective filter, allowing the translocation of essential dietary nutrients, electrolytes and water from the intestinal lumen into the circulation. It also constitutes the largest and most important barrier to prevent the passage from the external environment into the organism of harmful intraluminal substances, including foreign antigens, microorganisms and their toxins [13]. The establishment of the epithelial monolayer by intestinal epithelial cells is dependent upon a considerably high degree of intracellular and intercellular organization. Within each epithelial cell, structural integrity is maintained by the presence of a complex cytoskeletal network of microfilaments playing a crucial role in maintaining cellular polarity and in supporting points of cell-cell contact. The interaction and contact between adjacent intestinal epithelial cells of the monolayer is mediated by distinct junctions, including tight junctions, desmosomes and adherens junctions [14]. The function as a selective permeable barrier places the mucosal epithelium at the center of interactions between the mucosal immune system and luminal contents, which includes dietary antigens and microbial products [15]. The intestine is a privileged immune site, where immunoregulatory mechanisms simultaneously defend the body against pathogens, but also preserve tissue homeostasis to avoid immune-mediated pathology in response to environmental challenges.

This review summarizes the consequences of exposure to DON and other TCTB on histopathological intestinal lesions, on the potential disruption of the intestinal barrier function and on the active role of the intestinal mucosa in immune responsiveness. We will focus on the data obtained in humans, laboratory animals, poultry and pigs, as this latter species can be regarded as a good model for man [16,17].

2. DON and Other TCTB Reduce Growth

The reduction in weight gain as a consequence of reduced feed consumption is strongly associated with the exposure of farm animals to DON, with pigs being one of the most sensitive species. Current regulatory standards for DON in foods are based on its ability to cause growth suppression [18]. While DON is considered one of the least lethal TCT, its anorexic and emetic potencies are equal to, or greater than, those reported from the more acutely toxic TCT, such as T2-toxin [19]. Recently, the anorectic potencies of TCTB were compared in mice, following intraperitoneal or oral exposure: NIV and FUS-X were shown to have greater effects than DON, which had similar effects as 3- and 15-ADON [20]. The ability of DON to induce anorexia may be the consequence of the dysregulation of various signaling pathways. DON may act at different levels to induce impaired growth and weight gain, including on neuroendocrine signaling, immune responses, growth hormone or a central neuronal network. The involvement of neuroendocrine factors, such as serotonin, has been proposed [21,22]. Serotonin is produced and released by the enterochromaffin cells in the gut, acts as a paracrine on the enteric nervous system [23] and can affect the secretion of both anorexigenic or orexigenic hormones [24]. The impact of DON on immune responses can also affect feed consumption, because the activation of proinflammatory cytokines is recognized as a cause of anorexia [24]. Indeed, these toxins induce several suppressors of cytokine signaling (SOCS) [25] that impair growth hormone signaling by suppressing two growth-related proteins, the hepatic insulin-like growth factor acid-labile subunit (IGFALS) and insulin-like growth factor 1 (IGF1), as demonstrated in mice [25]. In this species, DON also induces the release of the satiety hormones, peptide YY (PYY) and cholecystokinin (CCK), proposed as critical mediators of DON-induced anorexia [26]. The impact of DON exposure on the central regulation of energy balance has been studied in mice and pigs [27–29]. Recently, Girardet *et al.* [29] showed that in addition to its peripheral action, DON can reach the brain after *per os* administration and act centrally on the anorexigenic/orexigenic balance.

In conclusion, most data describing the effects of DON on food intake were obtained in mice or in pig, and they point out both central and peripheral neuroendocrine control mechanisms. Neuroendocrine factors and proinflammatory cytokines drive the anorexigenic effect of DON. Recent experiments obtained only in rodent demonstrated that anorexia is induced rapidly within a few minutes following DON ingestion. Complementary studies are needed to evaluate if the mechanisms involved in anorexia are similar in rodents and other species.

3. DON and Other TCTB Affect Nutrient Absorption

The intestinal epithelium mediates the selective permeability from the intestinal lumen into the circulation of essential dietary nutrients, electrolytes and water through two major routes: transcellular permeability, generally associated with solute transport through the epithelial cells and predominantly regulated by selective transporters for amino acids, electrolytes, short-chain fatty acids and sugars; paracellular permeability, associated with transport via the space between epithelial cells and regulated by intercellular membrane junctional complexes [13]. The intestinal epithelium is a recognized target for NIV and FUS-X with acute effects, such as impaired sugar and electrolyte absorption [30]. The impaired absorption of nutrients may participate in the effect of TCT on animal growth [31]. The impacts of DON and other TCTB on nutrient absorption or transport at the intestinal level are summarized in Table 1.

3.1. Humans

In the human intestinal epithelial cell line, HT-29-clone D4, exposure to 10 µM of DON affects the activities of intestinal transporters: inhibition of the D-glucose/D-galactose sodium-dependent transporter (SGLT1), of the D-fructose transporter, glucose transporter-5 (GLUT5), and of active and passive L-serine transporters has been observed. The transport of palmitate was increased, whereas the uptake of cholesterol was not affected by the mycotoxin. At high concentrations (100 µM), SGLT1 activity was inhibited, whereas the activities of all the other transporters were increased [32].

3.2. Rodents

Exposure of mice to DON at 10 mg/kg for six weeks did not modulate the absorption of water, L-leucine, L-tryptophan, iron or D-glucose. However, a slight, but significantly reduced, transfer of glucose was observed. Furthermore, a significant decrease (up to 50%) in the transfer, as well as tissue accumulation of 5-methyltetrahydrofolic acid in the jejunal segment was observed. These findings indicate that subchronic ingestion of DON can impair the intestinal transfer and uptake of nutrients [33].

Table 1. Effect of TCTB exposure on nutrient absorption. DON, deoxynivalenol; NIV, nivalenol; 15-ADON, 15-acetyl derivative of DON; FUS-X, fusarenon-X.

Toxin	Animal species	Concentration and duration of exposure	Effects on nutrients absorption	References
DON	Human HT-29 cell line (*in vitro*)	10 µM 48 h	Inhibition of D glucose/D galactose transporters	[32]
			Inhibition of D-fructose transporter Inhibition of the active L-serine transporter	
			Inhibition of active and passive L-serine transport	
			Increase in palmitate transport	
DON	Mouse (*in vivo*)	10 mg/kg feed 6- weeks	Reduced weight gain	[33]
			Decreased transfer of glucose	
			Decreased jejunal transfer and tissue accumulation of 5-methyltetrahydro folic acid	
	Poultry (*ex vivo*)	33 µM 30 min	Inhibition of jejunal Na$^+$-amino acid co-transport	[34]
	Poultry (*ex vivo*)	33 µM 30 and 45 min	Decrease in jejunal glucose uptake	[35]
NIV	Poultry (*ex vivo*)	33 µM 30 min	Decrease in jejunal glucose uptake	[31]
15-ADON	Poultry (*ex vivo*)	33 µM 30 min	Decrease in jejunal glucose uptake	[31]
FUS-X	Poultry (*ex vivo*)	33 µM 30 min	No obvious effect	[31]

3.3. Farm Animals

Several studies have investigated the effects of DON on farm animals, with most of the data obtained in chickens. The electrophysiological properties of chicken intestinal mucosa exposed to DON were evaluated using isolated jejunum fragments in Ussing chambers [34]. Intestinal transport was determined by changes in the short-circuit current (Isc), as a measure of ion transmembrane flux, in the middle segment of the jejunum of broilers. The addition of D-glucose produced an increase in the Isc, and this was reversed by different TCTB, including DON [31].

The Isc was decreased by the addition of L-proline on the luminal side of the isolated mucosa after DON treatment, an effect that could be attributed to a strong inhibition of the L-proline/sodium-dependent transporter by DON [34].

Cotransporters are specialized membrane proteins using electrochemical gradients across the membrane for transporting sugars, amino acids and ions. The inhibition of Na^+ transport and Na^+-D-glucose co-transport are important mechanisms of DON toxicity in the intestine of chickens [34,36,37]. Indeed, DON treatment (33.7 μM) decreased glucose uptake almost as efficiently as phlorizin, a specific inhibitor of the sodium-dependent glucose cotransporter, SGLT-1 [35].

When comparing the different TCTB, the activity of the glucose co-transporter appears to be more sensitive to DON, NIV and 15-ADON than to FUS-X in the jejunum of broilers [31].

In Ross broilers fed either a basal, low DON or high DON diet (0.26, 1.68 and 12.21 mg/kg of DON, respectively), a progressive decrease in the relative density (weight:length) of the small intestine with increasing time of exposure was observed, which could be correlated with a decrease in villus height in the small intestine. The Isc of the jejunal epithelium was reduced in birds fed the high DON diet [38]. Recently, morphometric analysis of duodenal sections of hybrid turkey poults, fed for three weeks with DON at 5.2 mg/kg demonstrated a significant reduction in villus height and apparent villus surface area [39]. Using a global transcriptomic approach, Dietrich *et al.* [40] identified, in the jejunum of broilers fed for 23 days with DON at 2.5 or 5 mg/kg, a downregulation of genes involved in the nutrient uptake into jejunal cells: *SLC2A5*, which facilitates glucose and fructose transport, *SLC27A4*, involved in the palmitate transport and *SLC16A1*, involved in monocarboxylate uptake.

In conclusion, the data obtained by *in vitro* and *ex vivo* experiments performed in different species (human, mouse and poultry) show that DON affects the absorption of amino acids and sugars in intestinal epithelial cells. These studies indicate that DON affects key nutrients transporters. The precise mechanisms of action of DON and other TCT is still unknown, but these toxicants could act on the transporter proteins themselves and also on other constituents of the "transportsome", such as regulatory molecules or scaffold proteins.

4. DON and Other TCTB Induce Intestinal Lesions

Chronic exposure to TCTB induces digestive problems, reduced food intake and food refusal. The most striking clinical sign is the alteration of growth performance, most often related to decreased feed intake and decreased weight gain (Table 2). The reporting of intestinal lesions has been inconsistent and not systematically correlated with the clinical signs. Animal species differ in their susceptibility to these toxins. For example, as far as DON is concerned, the animal species can be ranked in the following order: pigs > mice > rats > poultry ≈ ruminants [2].

Among different hypotheses, the interaction of mycotoxin-microbiota can be proposed to explain, at least in part, the differences between species. The microbiota is thought to play important roles in the maturation of the intestinal and immune systems, in the nutrition of the host and, finally, in its protection against pathogenic micro-organisms and hazardous chemicals/xenobiotics, including TCTB. Differences in the localization of the gut bacteria able to convert toxic DON into its non-toxic de-epoxide metabolite, DOM-1, prior to or after the small intestine can have a major effect on the bioavailability of ingested TCTB. On this basis, animals can be divided into two groups: polygastric animals and birds with a high bacterial content located both before and after the small intestine; and monogastric species (including humans, pigs and rodents) with a high bacterial content located only after the small intestine, *i.e.*, in their colon [41].

Table 2. Intestinal lesions reported after TCTB exposure.

Toxin	Animal species	Concentration and duration of exposure	Intestinal Lesions	References
		Repeated exposure to TCTB (dietary or gavage)		
DON	Pig	0.75–4.2 mg/kg 3–5 weeks	Edema and congestion	[42–44]
		0.7–5.8 mg/kg 4 weeks	Slight to moderate inflammation and congestion of intestinal mucosa. Slight to moderate degeneration of lymphoid cells in Peyer's patches and in lymph nodes	[45]
		4 mg/kg	Corrugations in the fundic region (stomach)	[46]
		2–3 mg/kg 4 weeks	Corrugations in jejunum	[47]
		2.8 mg/kg	Multifocal atrophy and villus fusion, Apical necrosis of villi,	
		5 weeks	Cytoplasmatic vacuolation of enterocytes, Edema of lamina propria Decrease in villus height Decrease in the number of goblet cells in the jejunum and the ileum	[48,49]
	Rat	10 mg/kg 4 weeks	Alteration in villus architecture of the jejunum (increased villus fusion and shorter villus length). Increased apoptosis score in jejunal epithelial cells in association with higher number of mitotic cells and crypt fission	[50]
DON 15-ADON	Pig	DON 2.3 mg/kg	Reduction in villus height greater in presence of DON + 15-ADON compared to DON	[51]
		DON 1.2 mg/kg+ 15-ADON 0.9 mg/kg 4 weeks	Histological scores of the jejunum lower in animals fed DON + 15-ADON compared to DON	
		Ex vivo short-term exposure to TCTB		
DON 3-ADON 15-ADON	Pig explants	10 µM 4 h	Flattened and coalescent villi Lyses of enterocytes Interstitial edema and apoptosis 15-ADON >> DON = 3-ADON	[51]

The gut is identified as one of the target organs for TCTB, but no particular intestinal segment appears more sensitive than others [52,53]. The difference in sensitivity between species may be explained by differences in absorption, distribution, metabolism and elimination of DON [2]. The exact mechanism of the cellular entry of TCTB is not well characterized, and one can speculate that differences exist between species, leading to a differential sensitivity to these compounds. Moreover, the biotransformation of TCT by the detoxifying enzymes in the liver is highly variable between species. In the rat or pig liver, there is no de-epoxidation of DON, while these species are able to carry out glucurono-conjugation.

4.1. Humans

Long-term exposure to *Fusarium* toxins has been associated with an increased incidence of esophageal cancer in China. DON and also Fumonisin B1 were suspected to be a risk factor [54,55], though the role of these toxins needs to be clarified [2].

4.2. Rodents

In mice, after two years of exposure to NIV at a 30 mg/kg dietary concentration, the survival rate was generally higher in the NIV- treated animals than in the controls, where naturally occurring tumors, mostly lymphomas, were observed. No intestinal lesions were observed after two years of NIV exposure [56].

212

In weaning rats, after 28 days of exposure to DON-contaminated feed at 10 mg/kg, a decreased feed intake was observed, associated with a reduced weight gain. Lesions were observed in the jejunum: the villus architecture was altered, with an increase in villus fusion and a shorter villus length. The apoptotic score was increased in jejunal epithelial cells and associated with a greater number of mitotic cells and crypt fission [50].

4.3. Pigs

Congestion and erosions of the gastric and intestinal mucosae have been described following chronic DON exposure in pigs [42–45]. At 4 mg/kg of diet, DON may cause corrugations in the fundic region of the stomach [46], which were also observed at the jejunal level with lower doses of toxin (2–3 mg/kg diet) [47]. After five weeks of exposure to DON at 2.8 mg/kg of the diet, in the absence of changes in the pigs' body weight, significant histological changes were observed. Indeed, multifocal atrophy and villus fusion, apical necrosis of villi, cytoplasmatic vacuolation of enterocytes and edema of lamina propria were detected in the jejunum and ileum of DON-treated pigs [48]. A significant decrease in villus height was observed in the jejunum, probably reflecting a change in the balance between epithelial cell proliferation and apoptosis [49], while no difference in crypt depth was observed in any intestinal region. The number of goblet cells that synthesize and secrete mucin, involved in gut barrier function, decreased significantly in the jejunum and the ileum of piglets fed DON [49]. Similarly, a significant reduction in villus height and in the number of goblet cells was observed in pig jejunal explants exposed to 10 µM DON for four hours [57]. When comparing the effects of DON and its acetylated derivatives, Pinton *et al.* [51] observed that the reduction in villus height was greater in animals receiving DON + 15-ADON than in the animals receiving feed contaminated only with DON. The histological scores of the jejunum reflecting the main histological changes were lower in animals fed with DON + 15-ADON compared with animals fed DON.

To conclude, comparable macroscopic lesions of the intestinal epithelium are observed in rodents and pig exposed to TCTB. The exact mechanisms leading to these lesions are not characterized; it is especially important to delineate the effect of DON on: (i) the alteration of intestinal epithelial cells; (ii) the repair of these cells; and (iii) the coalescence and shortening of villi. Future research should focus on the differential effects of TCT on the potential target cells in the intestinal epithelium (crypts and stem cells *vs.* villus and differentiated cells). The effect of TCT on the factors that control the balance between cell proliferation, differentiation and cell death should also be analyzed.

5. DON and Other TCTB Alter Intestinal Barrier Function

The toxicity of DON and other TCTB is partially explained by the ability of these compounds to bind to eukaryotic ribosomes [58] and to rapidly activate the mitogen-activated protein kinases (MAPKs) via a process termed the "ribotoxic stress response". The MAPK cascades are central signaling pathways that regulate a wide variety of stimulated cellular processes, including proliferation, differentiation, apoptosis and stress response [59].

At present, the four different MAPK cascades identified are named according to their MAPK components: extracellular signal-regulated kinase 1 and 2 (ERK1/2), c-Jun N-terminal kinase (JNK), p38 and ERK5. Two possible upstream signal transducers for the DON-induced MAPK activation are the double-stranded RNA-activated protein kinase (PKR) and the hematopoietic cell kinase (Hck), a Src-family tyrosine kinase [24]. The consequence of this activation is an increase in proinflammatory gene expression, and the downstream effects include anorexia, reduced weight gain, immune stimulation, tissue injury and apoptosis. DON modulates cytokine and chemokine gene expression [60]. Highly dividing cells, such as intestinal epithelial cells or immune cells, are especially sensitive to TCTB, and the exposure of intestinal epithelial cells to these toxins may alter their ability to proliferate and to ensure a proper barrier function.

5.1. Effects on Cell Proliferation and Differentiation

In order to maintain an effective barrier function, the intestinal epithelium rapidly regenerates entirely in approximately one week, throughout life. Mature cells derived from intestinal stem cells migrate upwards along the crypt-villus axis towards the tip of the villus, gradually differentiating as they come closer to the tip [61]. Several studies have investigated the effects of mycotoxins on intestinal epithelial cell proliferation and on intestinal morphology (Table 3).

5.1.1. Effects on Cell Growth

The effect of TCT on intestinal epithelial cell growth has mainly been studied in the two human cell lines, Caco-2 and HT-29 [32,62,63]. When treated with a range of concentrations of DON (84 nM to 84 μM), Caco-2 cells showed a reduction in protein synthesis, proliferation and survival [64]. Dividing Caco-2 cells were found to be more sensitive compared to differentiated cells [65]. The greater sensitivity of proliferating cells is probably due to the capacity of the toxin to inhibit protein synthesis and, subsequently, nucleic acid synthesis [66]. DON was also demonstrated to decrease the cell proliferation in the porcine intestinal epithelial cell line, IPEC-1, whereas the acetylated derivatives exhibited differential effects. Indeed, 3-ADON was less toxic and 15-ADON was equally toxic as DON [51]. In the IPEC-J2 porcine intestinal epithelial cell line, the cytotoxicity of DON was correlated with an increase in lactate dehydrogenase release and decrease in ATP content [67]. In the human intestinal cell line HT-29, the cytotoxic effect of DON was not correlated with the induction of the heat shock protein, Hsp 70, or with the generation of reactive oxygen species, but was associated with a fragmentation of DNA and the activation of the apoptotic molecules, p53 and caspase-3 [68]. In intestinal cells from rat species (IEC-6 cell line), DON at 10 μM reduced the viability and induced apoptosis, independently of any cell cycle arrest, but involving caspase-3 activation [69].

Table 3. Effect of TCTB exposure on intestinal barrier function.

Toxin	Animal species	Concentration and duration of exposure	Effects on barrier function	References
		In vivo approach		
DON	Mouse	acute exposure 25 mg/kg bw (one gavage)	Increase in 4 kDa dextran permeability Effect on the distribution pattern of claudin 1, 3 and 3 tight junction proteins in small intestine	[70]
	Rat	chronic exposure 2 mg/kg feed 28 days	Decrease in transepithelial electrical resistance (TEER) Increase in 4 kDa dextran permeability	[50]
		In vitro approach: intestinal epithelial cell lines		
DON	Human HT-29 cell line	2 to 50 μM 24 h 10 μM 0–24 h	Dose dependent inhibition of cell viability (IC$_{50}$ = 10 μmol/L) Time dependent: Increase in total DNA damage Increase in p53 protein level Increase in caspase-3 activity	[68]
	Human Caco-2 cell line	84 μM 24 h	Decreased survival rate of 40%	[64]
	Human HT-29 cell line	0.13 to 0.7 μM 6 to 15 d	Decrease in brush border enzyme activity Decrease in protein content Decrease in transepithelial electrical resistance (TEER) Increase in lucifer yellow permeability	[62]
	Human Caco-2 cell line	30 μM 48 h	Decrease in TEER Decrease in claudin-4 tight junction proteins Increase in 4 kDa dextran permeability	[71]
	Human Caco-2 cell line	10 μM 12 h	Increase in *E coli* K12 translocation	[72]
	Human Caco-2 cell line	1.7 to 17 μM 24 h	Decrease in claudin-4 tight junction proteins	[73]

Table 3. *Cont.*

Toxin	Animal species	Concentration and duration of exposure	Effects on barrier function	References
DON	Porcine IPEC-1 cell line	10 to 50 µM 48 h	Decrease in TEER Decrease in claudin-3 and 4 tight junction proteins Increase in 4 kDa dextran permeability Increase in *E coli* 28C translocation	[71]
DON 3-ADON 15-ADON	Porcine IPEC-1 cell line	10 to 30 µM	Decrease in TEER and increase in 4 kDa dextran permeability 15-ADON >> DON > 3-ADON	[51]
		24 to 48 h	Decrease in claudin-3 and -4 tight junction proteins expression 15-ADON >> DON = 3-ADON	
DON	Porcine IPEC-J2 cell line	2.5 to 10 µM 24 h	Decrease in cell viability Increase of lactate dehydrogenase release Decrease in ATP content	[67]
Ex vivo approaches				
DON	Porcine tissue (Ussing chamber)	20 to 50 µM 2 h	Increase in 4 kDa dextran permeability	[71]
	Porcine tissue (jejunal explants)	1 to 10 µM 4 h	Shortened and coalescent villi, lysis of enterocytes, edema	[48]

As the intestine is potentially exposed to mixtures of mycotoxins [74,75], Alassane-Kpembi *et al.* evaluated interactions caused by co-exposure to TCTB on proliferating Caco-2 cells [76]. Using the MTT test and neutral red uptake, the authors observed that binary or ternary mixtures show synergistic effects when toxins were at low concentrations (cytotoxic effect between 10% and 40%) and additive or nearly additive effects at higher concentrations (cytotoxic effect around 50%).

5.1.2. Effects on Cell Differentiation

Using scanning electron microscopy, Kasuga *et al.* [62] demonstrated on differentiated Caco-2 cells that the formation of the brush border and the expression of two membrane-associated hydrolases related to enterocyte differentiation were affected by DON in a dose-dependent manner [62].

In broiler chicks, the ingestion of DON-contaminated feed produced an alteration in the small intestinal morphology, especially in the duodenum and jejunum, where the villi were shorter and thinner [77].

5.2. Effects on Barrier Functions

Polarized cells form strong barriers through the development of tight junctions between them. The intercellular tight junction is the rate-limiting barrier in the paracellular pathway for permeation by ions and larger solutes [78]. The investigations concerning the effects of TCTB on the intestinal barrier functions are only just beginning.

5.2.1. Effects on TEER

The transepithelial electrical resistance (TEER) of cell monolayers can be considered a good indicator of the degree of organization of the tight junctions within the cell monolayer and epithelial integrity [79]. Several studies have investigated the effect of TCTB on the TEER of intestinal epithelial cell lines (Table 3). In three different human intestinal epithelial cell lines, HT-29, Caco-2 and T84, DON was found to induce a dose-dependent decrease in the TEER [32,62,63,71,72,80]. The same effect was observed in the porcine intestinal epithelial cell lines IPEC-1 and IPEC-J2 [71,81]. Interestingly, IPEC-1 cells showed greater sensitivity to DON compared with Caco-2 [71]. Several hypotheses can explain this higher sensitivity of IPEC-1 cells [71]. Firstly, Caco-2 cells were obtained from an adenocarcinoma, whereas IPEC-1 cells were derived from normal newborn piglets [82]. Secondly, even if Caco-2 cells express many morphological and biochemical characteristics of small intestine [83], they are derived from the colon. By contrast, IPEC-1 cells were obtained from jejunum and ileum. Thirdly, these two cell lines are from different species (pig *versus* human), and among animal species, pig is the species most sensitive to DON. However, it is difficult to assess the susceptibility of humans [2]. Interestingly, a differential sensitivity to DON of the TEER of intestinal cells from the IPEC-J2 cell line has been observed depending on the route of application. Indeed, following basolateral exposure, the TEER was significantly decreased compared to apical exposure [81].

Differential effects were observed in the decrease of the TEER of IPEC-1 cells after a 24-h exposure to 10 µM of DON or acetylated derivatives, and the toxins were ranked in the following order of toxicity: 15-ADON >> DON > 3-ADON [51]. Similar data were obtained using human Caco-2 cells, confirming the differential effects of toxins of a close chemical structure [84].

5.2.2. Effects on Intestinal Permeability

The observed reduction in the TEER induced by trichothecenes can be due to an alteration of the tight junction barrier properties, but also to an effect on the plasma membrane, such as alterations in transcellular ion transport [85]. It is thus of interest to determine the effect of these toxins on the paracellular permeability of a tracer, such as lucifer yellow or dextran (Table 3).

Kasuga *et al.* [62] demonstrated a significant increase in the permeability of lucifer yellow in human Caco-2 and human T84 cells treated with DON. Similarly, we observed that DON increased the paracellular permeability of human Caco-2 cells and porcine IPEC-1 cells to 4 kDa dextran in a time and dose-dependent manner [71]. Akbari *et al.* [70] observed during 24 h that DON induced the dysfunction of the epithelial barrier of a Caco-2 cells by measuring the decline in impedance values. DON and its acetylated derivatives exhibit differential effects on 4 kDa dextran permeability of IPEC-1 cells, and after a 24 h exposure to 10 µM of toxin, they were ranked in the following order of toxicity: 15-ADON >> DON > 3-ADON [51].

This effect of DON on paracellular permeability was confirmed in rats chronically exposed to DON at 2 mg/kg of feed during 28 days. Pieces of jejunum mounted in an Ussing chamber showed a decrease of the TEER associated with an increase of permeability to 4 kDa dextran [50]. In pig explants mounted in Ussing chambers and exposed to DON *ex vivo*, we observed a two-fold increase in the paracellular passage of FITC-dextran across intestinal tissue treated with 20 µM and 50 µM of DON, when compared to untreated ones [71].

The numerous pores present in the basement membrane of the intestinal villi are essential for the communication of enterocytes with cells in the lamina propria. An 11-week exposure of pigs to DON at 2.2 to 2.9 mg/kg of feed led to an increase in the pore number in jejunum and potentially improved the antigen sampling in the intestinal epithelium [86].

5.2.3. Effects on Bacterial Translocation

The impaired intestinal integrity could lead to the entry of luminal antigens and bacteria that are normally restricted to the gut lumen by the intestinal barrier function (Table 3). We observed that DON induces a dose-dependent translocation of a pathogenic strain of *Escherichia coli* across the porcine IPEC-1 epithelial cell monolayers [71]. An increased translocation of *Salmonella typhimurium* was observed in porcine IPEC-J2 exposed to low doses of DON, with undifferentiated cells being more sensitive than the differentiated ones [87]. Maresca *et al.* [72] demonstrated that among other mycotoxins, DON allowed the transepithelial passage of apically added non-invasive commensal bacteria across human Caco-2 cell monolayers. However, in this case, no modification of paracellular permeability evaluated by the TEER measurement or tracer flux was observed. Such an increase in the bacterial passage through intestinal epithelial cells after DON treatment could have major implications for human health in terms of sepsis and inflammation. In mice, DON-contaminated diet accelerates *S. enteritidis* infection [88] and transiently increases the severity of reovirus infection [89].

5.2.4. Mode of Action

The mechanism underlying the trichothecene-induced impairment of the intestinal barrier function has been poorly investigated. The effect of DON on bacterial translocation could be related to the ability of this toxin to specifically decrease the expression of claudin proteins. Indeed, we have observed that, in porcine intestinal epithelial cell monolayers, the increased permeability was accompanied by a specific reduction in the expression of claudins. This increased permeability was also noted in pig explants treated with DON, and a reduction of claudin expression was described in the jejunum of piglets exposed to DON-contaminated feed [71]. This reduction of epithelial integrity through inhibition of the claudin-4 protein synthesis was observed in DON-exposed Caco-2 cells [73]. This decrease was not due to diminished transcription or increased degradation and was also observed for a tight junction-independent protein, *i.e.*, intestinal alkaline phosphatase [73]. As well as claudins, E-cadherin also plays a fundamental role in maintaining the epithelial architecture, and its expression is decreased in pig jejunal explants exposed to 10 µM of DON for 4 h [57]. The MAPK signaling pathway could be involved in the regulation of tight junction protein expression. Indeed, the ribotoxic stress induced by trichothecenes leads to the activation of members of the family of Src tyrosine kinases implicated as upstream regulators of a large number of intracellular signaling pathways [90]. They most likely represent critical signals that precede MAPK activation and the induction of resultant downstream responses [91]. In our study, we observed that the MAPK p44/42 ERK activation, induced by DON treatment, decreased the expression of claudin in correlation with a reduction in the barrier function of the intestine evaluated by TEER and paracellular permeability [92]. A recent study indicates that 6 h following the exposure of mice to 25 mg DON/kg bw, the distribution pattern of claudins 1, 2 and 3 was affected. In addition, the increase in the paracellular permeability, evaluated by the measure of FITC-dextran in the serum of mice, strengthened the hypothesis that the tight junction protein network is a target of DON [70]. We proposed a potential mechanism to explain the loss of the barrier function of intestinal epithelial cells following DON exposure, mediated by MAPK and claudin. The correlation between claudin 4 decreased expression and the MAPK activation was not observed in Caco-2 cells [73], suggesting a differential mechanism of action between porcine untransformed and human transformed intestinal epithelial cell lines.

Recently, we showed that the differential effects of DON and its acetylated derivatives on the intestinal barrier function were correlated, at the molecular level, with the exacerbated capacity of 15-ADON to activate MAPK ERK1/2, p38 and JNK, both in the intestinal cell line, explants and the jejunum from exposed animals, at a lower dose than DON and 3-ADON, and to decrease the expression of the tight junction proteins, claudin 3 and 4 [51].

DON cannot only interact with epithelial cells on the apical side during intestinal passage and absorption, but following absorption in the stomach and upper small intestine, detectable concentrations of DON can be found in blood serum, potentially exposing epithelial cells from their basolateral side. Diesing *et al.* [81] demonstrated a differential decrease in the tight junction protein, claudin 3, in the IPEC-J2 cell line, depending on the route of application, whereas the protein, ZO-1, was unaffected by the treatment. In addition, using a comparative global genomic approach, they showed that the apical and basolateral challenges to epithelial cell layers trigger different gene response profiles paralleled with a higher susceptibility towards basolateral challenge. The genes regulated were involved in metabolism, genetic or environment information processing and cellular processes [93].

The data summarized in this paragraph, obtained in different species (mouse, pig, human) and using different models (cell cultures, explants, *in vivo* experiments) confirm that one of the main target of DON in the intestine is the tight junction protein network. The modulation of claudin proteins by DON correlated with the increased intestinal permeability observed *in vitro*, *ex vivo* and *in vivo*. In terms of human or animal health, the consequences of the DON-induced increase in intestinal permeability still have to be determined.

6. Genotoxic Effects of DON and Other TCTB

The data on the genotoxic effects of trichothecenes are scarce, and these toxins are classified in Group 3 (inadequate evidence) by the International Agency on Cancer Research [94]. As far as the intestine is concerned, the genotoxic potential of NIV and FUS-X were evaluated *in vitro* on the human intestinal epithelial cell line, Caco-2. In differentiated post-confluent cells, a short exposure (3 h) to NIV or FUS-X did not cause any DNA damage, whereas DNA damage was observed after 24 h or 72 h [66]. In HT-29 cells exposed to DON, Bensassi *et al.* [68] demonstrated that the increase in DNA damage was induced in a time-dependent manner. Interestingly, in mice exposed orally and intraperitoneally to NIV (50% of the LD_{50}), Tsuda *et al.* [95] observed that NIV was genotoxic in the gastrointestinal tract, with the colon mucosa being preferentially damaged.

7. DON and Other TCTB Modulate the Intestinal Immune Response

The intestinal immune response involves the coordinated action of both immune (dendritic cells, macrophages, lymphocytes) and non-immune cells, including epithelial cells. Monocytes, macrophages, dendritic cells, as well as T- and B-lymphocytes can be cellular targets of DON and other TCT. Low to moderate toxin concentrations upregulate the expression of cytokines, chemokines and genes, inducing an inflammatory response, both transcriptionally and post-transcriptionally [60]. Not only immune cells, but also intestinal epithelial cells produce cytokines, crucial for the recruitment and activation of the immune system, including TGF-α, IL-1, IL-10, IL-15 and IL-18 [96]. Other cytokines, such as IL-1-α or β, IL-6, IL-8, TNF-α, MCP-1, CCL20 and GM-CSF, are also expressed by normal epithelial cells and are markedly upregulated in response to microbial infections. Intestinal epithelial cells also drive the development of dendritic cells: they release thymic stromal lymphopoietin that inhibits IL-12 production by dendritic cells and TGF-β and retinoic acid involved in the development of tolerogenic dendritic cells. In addition, thymic stromal lymphopoietin favors the release of a proliferation-inducing ligand (APRIL) and the B-cell activation factor of the TNF family (BAFF) by intestinal epithelial cell-conditioned dendritic cells and supports IgA class switching directly in the lamina propria [97,98].

7.1. DON and Other TCTB Induce Intestinal Inflammation

7.1.1. Modulation of the Cytokine Production in Intestinal Tissue by TCTB

The effect of DON or TCTB on cytokine production in intestinal tissue has been evaluated in different studies (Table 4). The increase in the expression of IL-1β, IL-8, MCP1 and IL-6 in pig intestinal loops exposed to *Salmonella typhimurium* was potentiated when DON was co-exposed with the bacteria [87]. The consequence of the DON intake could be an increase in the susceptibility to *Salmonella typhimurium*, with a subsequent potentiation of the inflammatory response in the gut.

Table 4. DON modulates cytokine production by intestinal epithelial cells.

Toxin	Species/model	Concentration and duration exposure	Cytokine modulation	References
	In vitro approach: intestinal epithelial cell lines			
DON	Human intestine 407 and Caco-2 cell lines	0–3.3 µM 12 h [94]	↗ IL-8	[72,80,99]
		0–10 µM, 12 h [64]		
		0–16.9 µM, 48 h [66]		
	Porcine IPEC-J2 cell line	0.5 µM, 48 h	↗ IL-1b, IL-6, IL-8,	[100]
		2 µM, 48 h	↘ IL-1a, MCP1	
			↗ IL-1a, IL-1b, IL-6, IL-8, TNFa, MCP1	
	Human intestine 407 cell line	24 h pre-exposure to LPS endotoxin	↘ IL-8	[101]
		1.7 µM, 12 h		
	Ex vivo and *in vivo* approaches			
DON	Porcine jejunal explants (*ex vivo*)	10 µM, 24 h	↗ IL-21, IL-22, IL-23	[102]
			↘ FoxP3, RALDH1	
	Porcine intestinal loops (*in vivo*)	0–3.3 µM, 6 h	↗ IL-1b, IL-8, MCP1, IL-6	[87]
	Broiler chickens (*in vivo*)	10 mg DON/kg, 35 d	↘ IL-1β, IFN-g, TGFBR1	[103]
			→ TNF-α, IL-8, NF-$\kappa\beta$,	

In porcine jejunal explants, 10 μM of DON led to an increase in the expression of IL-6, IL-23 and IL-1β, but did not affect the expression of TGF-β and strongly repressed FoxP3 and RALDH1. These data suggest that in this model, DON mainly drives the intestinal immune system towards a Th17 response elicited by the Th17 helper lymphocytes, recently described as important mediators of the mucosal immunity, the defense against extracellular pathogens and autoimmunity [102]. Besides these direct effects, DON also potentiates the effects of pro-inflammatory stimuli, such as TLR-4 ligands, lipopolysaccharide (LPS) and bacteria on immune cells [104–106]. Exposure to LPS is common and can occur through infections, via gastrointestinal translocation of gut microflora, due to inflammatory bowel diseases, or gut injury [107]. In mice, simultaneous exposure of subtoxic intravenous doses of LPS and dietary DON caused a sequential elevation of IL1-β overexpression and severe apoptotic depletion of lymphoid tissue [108,109]. However, such a mechanism needs to be demonstrated in intestinal lymphocytes during dietary exposure to LPS and TCT.

A recent study demonstrated a downregulation of the expression of IL-1β, IFN-g and transforming growth factor beta receptor I (TGFBR1) in broiler chickens fed for 35 days with DON at 10 mg/kg feed, but no changes in TNF-α, IL-8 and NF-κB in the jejunum of the animals [103].

7.1.2. Modulation of the Cytokine Production in Intestinal Epithelial Cells by TCTB

DON provokes intestinal inflammation *in vivo* [110], which results from a direct effect on the production of pro-inflammatory cytokines, especially IL-8, by intestinal epithelial cells. IL-8 is an early marker of the inflammatory process and is a potent chemo-attractant for leukocytes and T-lymphocytes underlying gut epithelial cells. IL-8 also enhances cell proliferation and controls the repair processes during injury of the intestinal mucosa or cytotoxic stress [111]. Several studies have shown that DON stimulates the secretion of IL-8 in various human intestinal epithelial cell lines [72,80,99]. Indeed, after exposure of Caco-2 cells to DON, a dose-dependent increase in IL-8 secretion through an NF-kB activity mechanism was observed. This effect was amplified upon pro-inflammatory stimulation, showing that DON exposure could cause or exacerbate intestinal inflammation. Moreover, Maresca *et al.* [72] have shown that direct IL-8 secretion from differentiated Caco-2 cells in response to DON is dependent on the ribotoxic-associated activation of PKR, NF-kB and p38. By contrast, DON-induced IL-8 secretion in human embryonic epithelial intestine 407 cells (Int407) was dependent on the activation of MAPK ERK1/2, but not on the activation of p38. This difference is probably due to the maturation status of the cells: differentiated mature Caco-2 cells *vs.* undifferentiated Int407 cells [99]. Moreover, DON modulates the production of several pro-inflammatory cytokines following a 48-h treatment of IPEC-J2 cells. A 2-μM exposure upregulates IL-1α, IL-1β, IL-6, IL-8, TNFα and MCP1, whereas a 0.5-μM exposure upregulates IL-1β, IL-6, IL-8 and downregulates IL-1 and MCP1 [100]. The consequence of the production of pro-inflammatory cytokines is the modulation of the intestinal tight junction barrier, potentially favoring an increased translocation of luminal antigens [112].

The detection of bacteria by intestinal epithelial cells, which induces IL-8 secretion, is known to be mediated through the interaction of bacteria flagella with the cellular Toll-like receptor 5 [113]. An indirect pro-inflammatory effect of mycotoxins could result from an alteration of the intestinal barrier function, allowing the transepithelial passage of non-invasive commensal bacteria. Indeed, high doses of DON (around 100 μM) compromise tight junctions and allow the transepithelial passage of apically added non-invasive commensal bacteria [72]. In the same study, the authors showed that DON at 1 and 10 μM potentiates the effects of basolaterally added bacteria on the secretion of IL-8 by human intestinal epithelial cells.

The inflammation processes act to maintain tissue homeostasis, but as some cytokines are potent mediators of potentially damaging tissue responses, several mechanisms exist to ensure that the effects of these cytokines are restricted [114]. Recently, DON exposure was demonstrated to suppress BAFF gene expression via the induction of SOCS3 in human enterocytes. As TCT are ribosomal stress agents, their ingestion could exert adverse effects on the regulation of BAFF, a vital cytokine for B-cell development [115].

Moon *et al.* [101] showed that human epithelial cells are less responsive to DON-induced IL-8 production after pre-exposure to the endotoxin LPS. As the intestine of newborn infants becomes established with the normal microflora, their epithelium can recognize the external bacteria components. After further constitutive experience of the commensals and their endotoxins, the epithelium becomes hypo-responsive to the normal microflora and controls its associated physiological inflammation [116]. The mechanism of the hypo-production of IL-8 in intestinal epithelial cells is the extended production of the DON-induced proliferator-activated receptor γ (PPAR-γ) after pre-exposure to endotoxin. DON increased PPAR-γ gene expression, which was transiently maintained, but endotoxin pre-exposure extends the duration of DON-induced PPAR-γ expression, thus sensitizing cells to induce an extended PPAR-γ in response to DON treatment. Constitutively-expressed PPAR-γ on the intestinal epithelial surface may trigger the tolerance to the normal microflora and its associated inflammation [117]. For example, the impaired expression of PPAR-γ has been observed in inflammatory bowel diseases. The results demonstrated by Moon *et al.* [101] suggest that there is a potential risk of mucosal inflammation after DON exposure in young infants compared with endotoxin-tolerant adults.

7.2. DON and TCTB May Interfere with the Intestinal Homeostasis

As described previously, epithelial cells act as initiators, mediators and regulators in innate and adaptive immune responses, as well as in the transition from innate immunity to adaptive immunity. Dendritic cells collaborate as sentinels against foreign particulate antigens by building a transepithelial interacting cellular network. During inflammatory and immune responses, epithelial cells express pattern-recognition receptors to trigger a host defense response and interact with dendritic cells to regulate antigen sensitization and release cytokines to recruit effector cells [118]. Moreover, macrophages and T-cells are also able to modulate dendritic cell functions [119]. The effects of TCTB on the different cell types involved in gut homeostasis can explain various pathologies associated with the ingestion of mycotoxin contaminated food.

Firstly, DON can potentiate the effect of IL1-β on IL-8 secretion and increase the transepithelial passage of commensal bacteria [71,72]. IL-8 has been implicated in many chronic diseases, ranging from inflammatory bowel disease [120,121] to rheumatoid arthritis [122]. Then, in addition to potentially exacerbating established intestinal inflammation, this mycotoxin may thus participate in the induction of sepsis and intestinal inflammation *in vivo* [72]. Indeed, inflammatory bowel diseases, such as Crohn's disease, are generally associated with the presence of adherent-invasive bacteria [123]. A hypothesis would be that at least in some cases, the ingestion of food contaminated with mycotoxins could be involved in inducing inflammatory bowel diseases [63,71,72].

The induction of proinflammatory cytokines, such as IL-6 by macrophages, plays a pivotal role, as they are directly linked to the differentiation of B-cells and to the stimulation of IgA secretion [124]. Prolonged feeding of DON causes a dramatic elevation in total serum IgA in mice. Moreover, dietary exposure to DON or NIV selectively upregulates membrane IgA-bearing cells in mouse Peyer's patches [124]. IL-6 is critical to mucosal IgA immunity based both on its differentiative effects on IgA-committed B-cells and its production in the gut by macrophages and T-cells [125]. *In vivo* and *in vitro*, DON upregulates the IL-6 expression that drives the differentiation of IgA-committed B-cells to IgA secretion [126–128], mimicking the early stage of human IgA nephropathy. It will be interesting to study the implication of TCT on the release of soluble molecules, such as thymic stromal lymphopoietin (TSLP), retinoic acid and TNF-β by intestinal epithelial cells. Indeed, TSLP is shown to favor the release of BAFF and APRIL by conditioned dendritic cells and induces IgA switching directly in the lamina propria. The disruption of gut homeostasis induced by TCT could explain the modification of mucosal IgA responses, as well as the diverse immune-mediated pathology observed in response to these fungal toxins.

8. Conclusions

The intestinal mucosa is the first biological barrier encountered by natural toxins, and consequently, it could be exposed to high amounts of dietary toxins. An increasing number of studies demonstrate that intestinal epithelial cells are targets for food contaminants, including mycotoxins [129–131].

In this review, we summarize the data concerning the ingestion of DON and other TCTB. These toxins induce intestinal pathologies in humans and animals, including necrosis of the intestinal epithelium. They also disturb the barrier function, potentially leading to the increased translocation of pathogens and an increased susceptibility to enteric infectious diseases. DON modulates the immune responsiveness of the intestinal mucosa, may interact in the cross-talk between epithelial cells and intestinal immune cells and could represent a predisposing factor to inflammatory diseases [102,132]. In farm and laboratory animals, dietary exposure to DON decreases growth performances. This is due to a local effect of the toxin altering the structure of the epithelium, reducing nutrient absorption by the enterocytes and modulating hormone production by enterochromaffin cells. A central effect of DON has also been described, involving the regulation of growth hormone production by inhibitors of cytokine signaling and the direct action of the toxin on the central neuronal network.

One important research field for the future will concern the impact of DON and other TCTB on the intestinal microbiota. Indeed, these toxins may directly target the microbiota [133]. In addition, the mucosal exposure to ribotoxic stress and the subsequent inflammatory responses may alter bacterial composition and, thus, reduce the microbial diversity.

As mentioned in this review, the effects of DON and other TCTB have been investigated in different species, including man, laboratory animals, poultry and pigs. Among these species, pigs are of particular concern for at least two reasons: (i) due to the cereal-rich diet, pigs can be exposed to a high level of toxins; and (ii) the pig is one of the most sensitive species. In addition, because of the similarities in the intestinal tract, pigs can be considered as a good model for humans. In this species, several complementary approaches have been developed to investigate the effects of DON and other TCTB on the intestine. *In vivo* trials and cell culture models were used to study the long-term exposure to mycotoxins. Intestinal loops and explants enable multiple exposure conditions of the entire intestinal tissue to be investigated, but these models are limited to short-term exposure. The data obtained using these complementary approaches all show an impact of DON and other TCTB on the intestine, confirming the validity of the pig to investigate the effect of these toxins. Surprisingly, despite the available tool of this animal species, very few data have been obtained on the effect of DON on the intestine of rodents. This might be due to the fact that they are not very sensitive to this toxin.

The concentrations of toxins used in the *in vivo* trials presented in this review are, most of the time, in accordance with plausible levels of contamination. Similarly, in the different models of study of the intestine, cell culture or explants, the range of concentration used (generally 5–30 µM) is in accordance with these plausible levels [63]. Under these conditions of realistic exposure, TCTB produce deleterious effects on intestinal morphology and/or function. As highlighted recently by Maresca [41], the differences between the doses of DON affecting cell functions and the doses of DON susceptible to being present [134] in relation to the actual provisional maximum tolerable daily intake represent a low safety factor. Thus, DON can represent a risk to human health, mainly because of its effect on the intestinal and immune systems.

The data presented in this review focused on the effects of DON on the intestine. *Fusarium* species are able to synthesize, in addition to DON, other related toxins, among them the acetylated derivatives, the toxicity of which has been mainly investigated through *in vitro* experiments. Masked forms of mycotoxins, such as deoxynivalenol-3-β-D-glucoside (DON-3G), that result from the detoxification metabolism in plants, are an emerging problem [135]. The occurrence and toxicity of this new DON-metabolite are poorly documented, as well as its possible hydrolysis to the parent mycotoxin in the intestine.

The toxicity of DON also needs to be addressed in the context of mycotoxin mixtures. Indeed, the co-occurrence of mycotoxins is likely to arise due to at least three different reasons: (i) most fungi are able to produce a number of mycotoxins simultaneously; (ii) food commodities can be contaminated by several fungi; and (iii) a complete diet is made up of various different food commodities [75]. Unfortunately, the toxicity of combinations of mycotoxins cannot always be predicted based upon their individual toxicities. Recent data suggest that the type of interaction depends not only on the type of toxin and their ratio, but also on the concentration of the toxin-mixture at a constant ratio [76]. More research is needed to understand the impact of mycotoxin combinations and to determine when synergistic interactions occur. These data are needed to assess the health risk due to the exposure of multi-mycotoxin contaminated food and feed [74].

Acknowledgments: This work was supported by European Commission projects-Knowledge Based Bio-Economy (2007-222690-2, "MYCORED" and 2008-227549, "INTERPLAY") and by the French Agence Nationale de la Recherche project-"DON and Co". The authors are solely responsible for the work described in this article, and their opinions are not necessarily those of the European Union. We thank John Woodley for language editing.

Author Contributions: Philippe Pinton selected the publications for illustrating the impact of trichothecenes on the different functions of the intestine and prepared the majority of the manuscript drafts. Isabelle Oswald has a research background in mycotoxins. She provided the main inputs in the structure of the manuscript and provided critical feedback.

Conflicts of Interest: The authors declare no conflict of interest.

References

1. Oswald, I.P.; Marin, D.E.; Bouhet, S.; Pinton, P.; Taranu, I.; Accensi, F. Immunotoxicological risk of mycotoxins for domestic animals. *Food Addit. Contam.* **2005**, *22*, 354–360. [CrossRef]

2. Pestka, J.J.; Smolinski, A.T. Deoxynivalenol: Toxicology and potential effects on humans. *J. Toxicol. Environ. Health B Crit. Rev.* **2005**, *8*, 39–69. [CrossRef]

3. Chu, F.S. *Mycotoxins-Occurrence and Toxic Effect. Encyclopedia of Human Nutrition*; Sadler, M., Strain, J.J., Caballero, B., Eds.; Academic Press: New York, NY, USA, 1998; pp. 858–869.

4. Larsen, J.C.; Hunt, J.; Perrin, I.; Ruckenbauer, P. Workshop on trichothecenes with a focus on DON: Summary report. *Toxicol. Lett.* **2004**, *153*, 1–22. [CrossRef]

5. Sugita-Konishi, Y.; Park, B.J.; Kobayashi-Hattori, K.; Tanaka, T.; Chonan, T.; Yoshikawa, K.; Kumagai, S. Effect of cooking process on the deoxynivalenol content and its subsequent cytotoxicity in wheat products. *Biosci. Biotechnol. Biochem.* **2006**, *70*, 1764–1768. [CrossRef]

6. Canady, R.; Coker, R.; Rgan, S.; Krska, R.; Kuiper-Goodman, T.; Olsen, M.; Pestka, J.; Resnik, S.; Schlatter, J. Deoxynivalenol. Safety evaluation of certain mycotoxins in food. *WHO Food Addit. Ser.* **2001**, *47*, 420–555.

7. Bhat, R.V.; Beedu, S.R.; Ramakrishna, Y.; Munshi, K.L. Outbreak of trichothecene mycotoxicosis associated with consumption of mould-damaged wheat production in Kashmir Valley, India. *Lancet* **1989**, *1*, 35–37.

8. Luo, Y. Fusarium toxins contamination of cereals in China. In Proceedings of the 7th International IUPAC Symposium on Mycotoxins and Phycotoxins, Tokyo, Japan, 16–19 August 1988; Aibara, K., Kumagai, S., Ohtsubo, K., Yoshizawa, T., Eds.; Japanese Association of Mycotoxicology: Tokyo, Japan, 1988; pp. 97–98.

9. Food and Drugs Administration. Guidance for Industry and FDA: Advisory Levels for Deoxynivalenol (DON) in Finished Wheat Products for Human Consumption and Grains and Grain By-Products used for Animal Feed. FDA: Rockville, MD, USA. Available online: http://www.fda.gov/downloads/Food/GuidanceRegulation/UCM217558.pdf (accessed on 15 May 2014).

10. European-Union. Commission Recommendation of 17 August 2006 on the Presence of Deoxynivalenol, Zearalenone, Ochratoxin A, T-2 and HT-2 and Fumonisins in Products Intended for Animal Feeding. Available online: http://eur-lex.europa.eu/LexUriServ/LexUriServ.do?uri=OJ:L:2006:229:0007:0009:EN: PDF (accessed on 15 May 2014).

11. European-Union. COUNCIL REGULATION (EC) No 463/2005 of 16 March 2005 terminating the partial interim review of the anti-dumping measures applicable to imports of certain tube or pipe fittings, of iron or steel, originating, inter alia, in Thailand. Available online: http://eur-lex.europa.eu/legal-content/EN/TXT/PDF/?uri=CELEX:32005R0463&from=EN (accessed on 15 May 2014).

12. Bouhet, S.; Oswald, I.P. The effects of mycotoxins, fungal food contaminants, on the intestinal epithelial cell-derived innate immune response. *Vet. Immunol. Immunopathol.* **2005**, *108*, 199–209. [CrossRef]

13. Groschwitz, K.R.; Hogan, S.P. Intestinal barrier function: Molecular regulation and disease pathogenesis. *J. Allergy Clin. Immunol.* **2009**, *124*, 3–20. [CrossRef]

14. Pitman, R.S.; Blumberg, R.S. First line of defense: The role of the intestinal epithelium as an active component of the mucosal immune system. *J. Gastroenterol.* **2000**, *35*, 805–814. [CrossRef]

15. Turner, J.R. Intestinal mucosal barrier function in health and disease. *Nat. Rev. Immunol.* **2009**, *9*, 799–809. [CrossRef]

16. Nejdfors, P.; Ekelund, M.; Jeppsson, B.; Westrom, B.R. Mucosal *in vitro* permeability in the intestinal tract of the pig, the rat, and man: Species- and region-related differences. *Scand. J. Gastroenterol.* **2000**, *35*, 501–507.

17. Rothkotter, H.J.; Sowa, E.; Pabst, R. The pig as a model of developmental immunology. *Hum. Exp. Toxicol.* **2002**, *21*, 533–536. [CrossRef]

18. Flannery, B.M.; Wu, W.; Pestka, J.J. Characterization of deoxynivalenol-induced anorexia using mouse bioassay. *Food Chem. Toxicol.* **2011**, *49*, 1863–1869. [CrossRef]

19. Rotter, B.A.; Prelusky, D.B.; Pestka, J.J. Toxicology of deoxynivalenol (vomitoxin). *J. Toxicol. Environ. Health* **1996**, *48*, 1–34.

20. Wu, W.; Flannery, B.M.; Sugita-Konishi, Y.; Watanabe, M.; Zhang, H.; Pestka, J.J. Comparison of murine anorectic responses to the 8-ketotrichothecenes 3-acetyldeoxynivalenol, 15-acetyldeoxynivalenol, fusarenon X and nivalenol. *Food Chem. Toxicol.* **2012**, *50*, 2056–2061. [CrossRef]

21. Fioramonti, J.; Dupuy, C.; Dupuy, J.; Bueno, L. The mycotoxin, deoxynivalenol, delays gastric emptying through serotonin-3 receptors in rodents. *J. Pharmacol. Exp. Ther.* **1993**, *266*, 1255–1260.

22. Girish, C.K.; MacDonald, E.J.; Scheinin, M.; Smith, T.K. Effects of feedborne fusarium mycotoxins on brain regional neurochemistry of turkeys. *Poult. Sci.* **2008**, *87*, 1295–1302.

23. Li, Y. Sensory signal transduction in the vagal primary afferent neurons. *Curr. Med. Chem.* **2007**, *14*, 2554–2563. [CrossRef]

24. Pestka, J.J. Deoxynivalenol: Mechanisms of action, human exposure, and toxicological relevance. *Arch. Toxicol.* **2010**, *84*, 663–679. [CrossRef]

25. Amuzie, C.J.; Shinozuka, J.; Pestka, J.J. Induction of suppressors of cytokine signaling by the trichothecene deoxynivalenol in the mouse. *Toxicol. Sci.* **2009**, *111*, 277–287. [CrossRef]

26. Flannery, B.M.; Clark, E.S.; Pestka, J.J. Anorexia induction by the trichothecene deoxynivalenol (vomitoxin) is mediated by the release of the gut satiety hormone peptide YY. *Toxicol. Sci.* **2012**, *130*, 289–297. [CrossRef]

27. Gaige, S.; Bonnet, M.; Tardivel, C.; Pinton, P.; Trouslard, J.; Jean, A.; Guzylack, L.; Troadec, J.; Dallaporta, M. c-Fos immunoreactivity in the pig brain following deoxynivalenol intoxication: Focus on NUCB2/nesfatin-1 expressing neurons. *NeuroToxicology* **2013**, *34*, 135–149. [CrossRef]

28. Girardet, C.; Bonnet, M.S.; Jdir, R.; Sadoud, M.; Thirion, S.; Tardivel, C.; Roux, J.; Lebrun, B.; Mounien, L.; Trouslard, J.; *et al.* Central inflammation and sickness-like behavior induced by the food contaminant deoxynivalenol: A PGE2-independent mechanism. *Toxicol. Sci.* **2011**, *124*, 179–191. [CrossRef]

29. Girardet, C.; Bonnet, M.S.; Jdir, R.; Sadoud, M.; Thirion, S.; Tardivel, C.; Roux, J.; Lebrun, B.; Wanaverbecq, N.; Mounien, L.; *et al.* The food-contaminant deoxynivalenol modifies eating by targeting anorexigenic neurocircuitry. *PLoS One* **2011**, *6*, e26134. [CrossRef]

30. Kumagai, S.; Shimizu, T. Effects of fusarenon-X and T-2 toxin on intestinal absorption of monosaccharide in rats. *Arch. Toxicol.* **1988**, *61*, 489–495. [CrossRef]

31. Awad, W.A.; Razzazi-Fazeli, E.; Bohm, J.; Zentek, J. Effects of B-trichothecenes on luminal glucose transport across the isolated jejunal epithelium of broiler chickens. *J. Anim. Physiol. Anim. Nutr.* **2008**, *92*, 225–230. [CrossRef]

32. Maresca, M.; Mahfoud, R.; Garmy, N.; Fantini, J. The mycotoxin deoxynivalenol affects nutrient absorption in human intestinal epithelial cells. *J. Nutr.* **2002**, *132*, 2723–2731.

33. Hunder, G.; Schumann, K.; Strugala, G.; Gropp, J.; Fichtl, B.; Forth, W. Influence of subchronic exposure to low dietary deoxynivalenol, a trichothecene mycotoxin, on intestinal absorption of nutrients in mice. *Food Chem. Toxicol.* **1991**, *29*, 809–814. [CrossRef]

34. Awad, W.A.; Rehman, H.; Bohm, J.; Razzazi-Fazeli, E.; Zentek, J. Effects of luminal deoxynivalenol and L-proline on electrophysiological parameters in the jejunums of laying hens. *Poult. Sci.* **2005**, *84*, 928–932. [CrossRef]

35. Awad, W.A.; Aschenbach, J.R.; Setyabudi, F.M.; Razzazi-Fazeli, E.; Bohm, J.; Zentek, J. *In vitro* effects of deoxynivalenol on small intestinal D-glucose uptake and absorption of deoxynivalenol across the isolated jejunal epithelium of laying hens. *Poult. Sci.* **2007**, *86*, 15–20. [CrossRef]

36. Awad, W.A.; Bohm, J.; Razzazi-Fazeli, E.; Hulan, H.W.; Zentek, J. Effects of deoxynivalenol on general performance and electrophysiological properties of intestinal mucosa of broiler chickens. *Poult. Sci.* **2004**, *83*, 1964–1972. [CrossRef]

37. Awad, W.A.; Bohm, J.; Razzazi-Fazeli, E.; Zentek, J. *In vitro* effects of deoxynivalenol on electrical properties of intestinal mucosa of laying hens. *Poult. Sci.* **2005**, *84*, 921–927. [CrossRef]

38. Yunus, A.W.; Blajet-Kosicka, A.; Kosicki, R.; Khan, M.Z.; Rehman, H.; Bohm, J. Deoxynivalenol as a contaminant of broiler feed: Intestinal development, absorptive functionality, and metabolism of the mycotoxin. *Poult. Sci.* **2012**, *91*, 852–861. [CrossRef]

39. Devreese, M.; Girgis, G.N.; Tran, S.T.; de Baere, S.; de Backer, P.; Croubels, S.; Smith, T.K. The effects of feed-borne Fusarium mycotoxins and glucomannan in turkey poults based on specific and non-specific parameters. *Food Chem. Toxicol.* **2014**, *63*, 69–75. [CrossRef]

40. Dietrich, B.; Neuenschwander, S.; Bucher, B.; Wenk, C. Fusarium mycotoxin-contaminated wheat containing deoxynivalenol alters the gene expression in the liver and the jejunum of broilers. *Animal* **2013**, *6*, 278–291.

41. Maresca, M. From the gut to the brain: Journey and pathophysiological effects of the food-associated mycotoxin Deoxynivalenol. *Toxins* **2013**, *5*, 784–820. [CrossRef]

42. Pollmann, D.S.; Koch, B.A.; Seitz, L.M.; Mohr, H.E.; Kennedy, G.A. Deoxynivalenol-contaminated wheat in swine diets. *J. Anim. Sci.* **1985**, *60*, 239–247.

43. Trenholm, H.L.; Hamilton, R.M.; Friend, D.W.; Thompson, B.K.; Hartin, K.E. Feeding trials with vomitoxin (deoxynivalenol)-contaminated wheat: Effects on swine, poultry, and dairy cattle. *J. Am. Vet. Med. Assoc.* **1984**, *185*, 527–531.

44. Harvey, R.B.; Kubena, L.F.; Huff, W.E.; Corrier, D.E.; Clark, D.E.; Phillips, T.D. Effects of aflatoxin, deoxynivalenol, and their combinations in the diets of growing pigs. *Am. J. Vet. Res.* **1989**, *50*, 602–607.

45. Cote, L.M.; Beasley, V.R.; Bratich, P.M.; Swanson, S.P.; Shivaprasad, H.L.; Buck, W.B. Sex-related reduced weight gains in growing swine fed diets containing deoxynivalenol. *J. Anim. Sci.* **1985**, *61*, 942–950.

46. D'Mello, J.P.F. Antinutritional factors and mycotoxins. In *Farm Animal Metabolism and Nutrition*; D'Mello, J.P.F., Ed.; CAB International: Wallingford, UK, 2000; pp. 383–403.

47. Kolf-Clauw, M; Oswald, I.P. Toxalim, Research Centre in Food Toxicology, Toulouse, France. Effect of DON on the intestine in pig. 2013.

48. Kolf-Clauw, M.; Castellote, J.; Joly, B.; Bourges-Abella, N.; Raymond-Letron, I.; Pinton, P.; Oswald, I.P. Development of a pig jejunal explant culture for studying the gastrointestinal toxicity of the mycotoxin deoxynivalenol: Histopathological analysis. *Toxicol. Vitro* **2009**, *23*, 1580–1584. [CrossRef]

49. Bracarense, A.P.; Lucioli, J.; Grenier, B.; Drociunas Pacheco, G.; Moll, W.D.; Schatzmayr, G.; Oswald, I.P. Chronic ingestion of deoxynivalenol and fumonisin, alone or in interaction, induces morphological and immunological changes in the intestine of piglets. *Br. J. Nutr.* **2012**, *107*, 1776–1786. [CrossRef]

50. Payros, D; Oswald, I.P. Toxalim, Research Centre in Food Toxicology, Toulouse, France. Effect of DON on the intestine in rat. 2014.

51. Pinton, P.; Tsybulskyy, D.; Lucioli, J.; Laffitte, J.; Callu, P.; Lyazhri, F.; Grosjean, F.; Bracarense, A.P.; Kolf-Clauw, M.; Oswald, I.P. Toxicity of deoxynivalenol and its acetylated derivatives on the intestine: Differential effects on morphology, barrier function, tight junctions proteins and MAPKinases. *Toxicol. Sci.* **2012**, *130*, 180–190. [CrossRef]

52. Awad, W.A.; Razzazi-Fazeli, E.; Bohm, J.; Zentek, J. Influence of deoxynivalenol on the D-glucose transport across the isolated epithelium of different intestinal segments of laying hens. *J. Anim. Physiol. Anim. Nutr.* **2007**, *91*, 175–180. [CrossRef]

53. Zielonka, L.; Wisniewska, M.; Gajecka, M.; Obremski, K.; Gajecki, M. Influence of low doses of deoxynivalenol on histopathology of selected organs of pigs. *Pol. J. Vet. Sci.* **2009**, *12*, 89–95.

54. Yoshizawa, T.; Yamashita, A.; Luo, Y. Fumonisin occurrence in corn from high- and low-risk areas for human esophageal cancer in China. *Appl. Environ. Microbiol.* **1994**, *60*, 1626–1629.

55. Luo, Y.; Yoshizawa, T.; Katayama, T. Comparative study on the natural occurrence of Fusarium mycotoxins (trichothecenes and zearalenone) in corn and wheat from high- and low-risk areas for human esophageal cancer in China. *Appl. Environ. Microbiol.* **1990**, *56*, 3723–3726.

56. Ohtsubo, K.; Ryu, J.C.; Nakamura, K.; Izumiyama, N.; Tanaka, T.; Yamamura, H.; Kobayashi, T.; Ueno, Y. Chronic toxicity of nivalenol in female mice: A 2-year feeding study with Fusarium nivale Fn 2B-moulded rice. *Food Chem. Toxicol.* **1989**, *27*, 591–598. [CrossRef]

57. Basso, K.; Gomes, F.; Bracarense, A.P. Deoxynivanelol and fumonisin, alone or in combination, induce changes on intestinal junction complexes and in e-cadherin expression. *Toxins* **2013**, *5*, 2341–2352. [CrossRef]

58. Ueno, Y. *Trichothecenes: Chemical, Biological, and Toxicological Aspects. Trichothecenes*; Ueno, Y., Ed.; Elsevier Press: Amsterdam, The Netherlands, 1983; pp. 135–146.

59. Plotnikov, A.; Zehorai, E.; Procaccia, S.; Seger, R. The MAPK cascades: Signaling components, nuclear roles and mechanisms of nuclear translocation. *Biochim. Biophys. Acta* **2011**, *1813*, 1619–1633. [CrossRef]

60. Pestka, J.J. Mechanisms of deoxynivalenol-induced gene expression and apoptosis. *Food Addit. Contam.* **2008**, *25*, 1128–1140. [CrossRef]

61. Booth, C.; Potten, C.S. Gut instincts: Thoughts on intestinal epithelial stem cells. *J. Clin. Invest.* **2000**, *105*, 1493–1499. [CrossRef]

62. Kasuga, F.; Hara-Kudo, Y.; Saito, N.; Kumagai, S.; Sugita-Konishi, Y. In vitro effect of deoxynivalenol on the differentiation of human colonic cell lines Caco-2 and T84. *Mycopathologia* **1998**, *142*, 161–167. [CrossRef]

63. Sergent, T.; Parys, M.; Garsou, S.; Pussemier, L.; Schneider, Y.J.; Larondelle, Y. Deoxynivalenol transport across human intestinal Caco-2 cells and its effects on cellular metabolism at realistic intestinal concentrations. *Toxicol. Lett.* **2006**, *164*, 167–176. [CrossRef]

64. Instanes, C.; Hetland, G. Deoxynivalenol (DON) is toxic to human colonic, lung and monocytic cell lines, but does not increase the IgE response in a mouse model for allergy. *Toxicology* **2004**, *204*, 13–21. [CrossRef]

65. Bony, S.; Carcelen, M.; Olivier, L.; Devaux, A. Genotoxicity assessment of deoxynivalenol in the Caco-2 cell line model using the Comet assay. *Toxicol. Lett.* **2006**, *166*, 67–76. [CrossRef]

66. Bony, S.; Olivier-Loiseau, L.; Carcelen, M.; Devaux, A. Genotoxic potential associated with low levels of the Fusarium mycotoxins nivalenol and fusarenon X in a human intestinal cell line. *Toxicol. Vitro* **2007**, *21*, 457–465. [CrossRef]

67. Awad, W.A.; Aschenbach, J.R.; Zentek, J. Cytotoxicity and metabolic stress induced by deoxynivalenol in the porcine intestinal IPEC-J2 cell line. *J. Anim. Physiol. Anim. Nutr.* **2012**, *96*, 709–716. [CrossRef]

68. Bensassi, F.; El Golli-Bennour, E.; Abid-Essefi, S.; Bouaziz, C.; Hajlaoui, M.R.; Bacha, H. Pathway of deoxynivalenol-induced apoptosis in human colon carcinoma cells. *Toxicology* **2009**, *264*, 104–109. [CrossRef]

69. Bianco, G.; Fontanella, B.; Severino, L.; Quaroni, A.; Autore, G.; Marzocco, S. Nivalenol and deoxynivalenol affect rat intestinal epithelial cells: A concentration related study. *PLoS One* **2012**, *7*, e52051.

70. Akbari, P.; Braber, S.; Gremmels, H.; Koelink, P.J.; Verheijden, K.A.; Garssen, J.; Fink-Gremmels, J. Deoxynivalenol: A trigger for intestinal integrity breakdown. *FASEB J.* **2014**, in press.

71. Pinton, P.; Nougayrede, J.P.; del Rio, J.C.; Moreno, C.; Marin, D.E.; Ferrier, L.; Bracarense, A.P.; Kolf-Clauw, M.; Oswald, I.P. The food contaminant deoxynivalenol, decreases intestinal barrier permeability and reduces claudin expression. *Toxicol. Appl. Pharmacol.* **2009**, *237*, 41–48. [CrossRef]

72. Maresca, M.; Yahi, N.; Younes-Sakr, L.; Boyron, M.; Caporiccio, B.; Fantini, J. Both direct and indirect effects account for the pro-inflammatory activity of enteropathogenic mycotoxins on the human intestinal epithelium: Stimulation of interleukin-8 secretion, potentiation of interleukin-1beta effect and increase in the transepithelial passage of commensal bacteria. *Toxicol. Appl. Pharmacol.* **2008**, *228*, 84–92. [CrossRef]

73. Van De Walle, J.; Sergent, T.; Piront, N.; Toussaint, O.; Schneider, Y.J.; Larondelle, Y. Deoxynivalenol affects *in vitro* intestinal epithelial cell barrier integrity through inhibition of protein synthesis. *Toxicol. Appl. Pharmacol.* **2010**, *245*, 291–298. [CrossRef]

74. Grenier, B.; Oswald, I.P. Mycotoxin co-contamination of food and feed: Meta-analysis of publications describing toxicological interactions. *World Mycotoxin J.* **2011**, *4*, 285–313. [CrossRef]

75. Streit, E.; Schatzmayr, G.; Tassis, P.; Tzika, E.; Marin, D.; Taranu, I.; Tabuc, C.; Nicolau, A.; Aprodu, I.; Puel, O.; et al. Current situation of mycotoxin contamination and co-occurrence in animal feed–focus on Europe. *Toxins* **2012**, *4*, 788–809. [CrossRef]

76. Alassane-Kpembi, I.; Kolf-Clauw, M.; Gauthier, T.; Abrami, R.; Abiola, F.A.; Oswald, I.P.; Puel, O. New insights into mycotoxin mixtures: The toxicity of low doses of Type B trichothecenes on intestinal epithelial cells is synergistic. *Toxicol. Appl. Pharmacol.* **2013**, *272*, 191–198. [CrossRef]

77. Awad, W.A.; Bohm, J.; Razzazi-Fazeli, E.; Ghareeb, K.; Zentek, J. Effect of addition of a probiotic microorganism to broiler diets contaminated with deoxynivalenol on performance and histological alterations of intestinal villi of broiler chickens. *Poult. Sci.* **2006**, *85*, 974–979. [CrossRef]

78. Madara, J.L. Regulation of the movement of solutes across tight junctions. *Annu. Rev. Physiol.* **1998**, *60*, 143–159. [CrossRef]

79. Harhaj, N.S.; Antonetti, D.A. Regulation of tight junctions and loss of barrier function in pathophysiology. *Int. J. Biochem. Cell Biol.* **2004**, *36*, 1206–1237. [CrossRef]

80. Van De Walle, J.; Romier, B.; Larondelle, Y.; Schneider, Y.J. Influence of deoxynivalenol on NF-kappaB activation and IL-8 secretion in human intestinal Caco-2 cells. *Toxicol. Lett.* **2008**, *177*, 205–214.

81. Diesing, A.K.; Nossol, C.; Danicke, S.; Walk, N.; Post, A.; Kahlert, S.; Rothkotter, H.J.; Kluess, J. Vulnerability of polarised intestinal porcine epithelial cells to mycotoxin deoxynivalenol depends on the route of application. *PLoS One* **2011**, *6*, e17472. [CrossRef]

82. Gonzalez-Vallina, R.; Wang, H.; Zhan, R.; Berschneider, H.M.; Lee, R.M.; Davidson, N.O.; Black, D.D. Lipoprotein and apolipoprotein secretion by a newborn piglet intestinal cell line (IPEC-1). *Am. J. Physiol.* **1996**, *271*, G249–G259.

83. Pinto, M.; Robine-Leon, S.; Appay, M. Enterocyte-like differentiation and polarization of the human colon carcinoma cell line Caco-2 in culture. *Biol. Cell* **1983**, *47*, 323–330.

84. Kadota, T.; Furusawa, H.; Hirano, S.; Tajima, O.; Kamata, Y.; Sugita-Konishi, Y. Comparative study of deoxynivalenol, 3-acetyldeoxynivalenol, and 15-acetyldeoxynivalenol on intestinal transport and IL-8 secretion in the human cell line Caco-2. *Toxicol. Vitro* **2013**, *27*, 1888–1895. [CrossRef]

85. Barrett, K.E. Positive and negative regulation of chloride secretion in T84 cells. *Am. J. Physiol.* **1993**, *265*, C859–C868.

86. Nossol, C.; Diesing, A.K.; Kahlert, S.; Kersten, S.; Kluess, J.; Ponsuksili, S.; Hartig, R.; Wimmers, K.; Danicke, S.; Rothkotter, H.J. Deoxynivalenol affects the composition of the basement membrane proteins and influences en route the migration of CD16 cells into the intestinal epithelium. *Mycotoxin Res.* **2013**, *29*, 245–254. [CrossRef]

87. Vandenbroucke, V.; Croubels, S.; Martel, A.; Verbrugghe, E.; Goossens, J.; van Deun, K.; Boyen, F.; Thompson, A.; Shearer, N.; de Backer, P.; *et al.* The mycotoxin deoxynivalenol potentiates intestinal inflammation by salmonella typhimurium in porcine ileal loops. *PLoS One* **2011**, *6*, e23871. [CrossRef]

88. Hara-Kudo, Y.; Sugita-Konoshi, Y.; Kasuga, F.; Kumagai, S. Effects of deoxynivalenol on *Salmonella enteritidis* infection. *Mycotoxins* **1996**, *42*, 51–55.

89. Li, M.; Cuff, C.F.; Pestka, J.J. Modulation of murine host response to enteric reovirus infection by the trichothecene deoxynivalenol. *Toxicol. Sci.* **2005**, *87*, 134–145.

90. Lowell, C.A. Src-family kinases: Rheostats of immune cell signaling. *Mol. Immunol.* **2004**, *41*, 631–643. [CrossRef]

91. Zhou, H.R.; Jia, Q.; Pestka, J.J. Ribotoxic stress response to the trichothecene deoxynivalenol in the macrophage involves the SRC family kinase Hck. *Toxicol. Sci.* **2005**, *85*, 916–926. [CrossRef]

92. Pinton, P.; Braicu, C.; Nougayrede, J.P.; Laffitte, J.; Taranu, I.; Oswald, I.P. Deoxynivalenol impairs porcine intestinal barrier function and decreases the protein expression of claudin-4 through a mitogen-activated protein kinase-dependent mechanism. *J. Nutr.* **2010**, *140*, 1956–1962. [CrossRef]

93. Diesing, A.K.; Nossol, C.; Ponsuksili, S.; Wimmers, K.; Kluess, J.; Walk, N.; Post, A.; Rothkotter, H.J.; Kahlert, S. Gene regulation of intestinal porcine epithelial cells ipec-j2 is dependent on the site of deoxynivalenol toxicological action. *PLoS One* **2012**, *7*, e34136. [CrossRef]

94. International Agency for Research on Cancer (IARC). Some Naturally Occurring Substances: Food Items and Constituents, Heterocyclic Aromatic Amines and Mycotoxins. Monographs on the evaluation of carcinogenic risks to humans. Available online: http://monographs.iarc.fr/ENG/Monographs/vol56/volume56.pdf (accessed on 15 May 2014).

95. Tsuda, S.; Kosaka, Y.; Murakami, M.; Matsuo, H.; Matsusaka, N.; Taniguchi, K.; Sasaki, Y.F. Detection of nivalenol genotoxicity in cultured cells and multiple mouse organs by the alkaline single-cell gel electrophoresis assay. *Mutat. Res.* **1998**, *415*, 191–200. [CrossRef]

96. Stadnyk, A.W. Intestinal epithelial cells as a source of inflammatory cytokines and chemokines. *Can. J. Gastroenterol.* **2002**, *16*, 241–246.

97. Rescigno, M.; di Sabatino, A. Dendritic cells in intestinal homeostasis and disease. *J. Clin. Invest.* **2009**, *119*, 2441–2450. [CrossRef]

98. Chen, F.; Ma, Y.; Xue, C.; Ma, J.; Xie, Q.; Wang, G.; Bi, Y.; Cao, Y. The combination of deoxynivalenol and zearalenone at permitted feed concentrations causes serious physiological effects in young pigs. *J. Vet. Sci.* **2008**, *9*, 39–44. [CrossRef]

99. Moon, Y.; Yang, H.; Lee, S.H. Modulation of early growth response gene 1 and interleukin-8 expression by ribotoxin deoxynivalenol (vomitoxin) via ERK1/2 in human epithelial intestine 407 cells. *Biochem. Biophys. Res. Commun.* **2007**, *362*, 256–262. [CrossRef]

100. Wan, L.Y.; Turner, P.C.; El-Nezami, H. Individual and combined cytotoxic effects of Fusarium toxins (deoxynivalenol, nivalenol, zearalenone and fumonisins B1) on swine jejunal epithelial cells. *Food Chem. Toxicol.* **2013**, *57*, 276–283. [CrossRef]

101. Moon, Y.; Yang, H.; Park, S.H. Hypo-responsiveness of interleukin-8 production in human embryonic epithelial intestine 407 cells independent of NF-kappaB pathway: New lessons from endotoxin and ribotoxic deoxynivalenol. *Toxicol. Appl. Pharmacol.* **2008**, *231*, 94–102.

102. Cano, P.M.; Seeboth, J.; Meurens, F.; Cognie, J.; Abrami, R.; Oswald, I.P.; Guzylack-Piriou, L. Deoxynivalenol as a new factor in the persistence of intestinal inflammatory diseases: An emerging hypothesis through possible modulation of Th17-mediated response. *PLoS One* **2013**, *8*, e53647.

103. Ghareeb, K.; Awad, W.A.; Soodoi, C.; Sasgary, S.; Strasser, A.; Bohm, J. Effects of feed contaminant deoxynivalenol on plasma cytokines and mRNA expression of immune genes in the intestine of broiler chickens. *PLoS One* **2013**, *8*, e71492.

104. Islam, Z.; Gray, J.S.; Pestka, J.J. p38 Mitogen-activated protein kinase mediates IL-8 induction by the ribotoxin deoxynivalenol in human monocytes. *Toxicol. Appl. Pharmacol.* **2006**, *213*, 235–244. [CrossRef]

105. Mbandi, E.; Pestka, J.J. Deoxynivalenol and satratoxin G potentiate proinflammatory cytokine and macrophage inhibitory protein 2 induction by Listeria and Salmonella in the macrophage. *J. Food Prot.* **2006**, *69*, 1334–1339.

106. Zhou, H.R.; Harkema, J.R.; Yan, D.; Pestka, J.J. Amplified proinflammatory cytokine expression and toxicity in mice coexposed to lipopolysaccharide and the trichothecene vomitoxin (deoxynivalenol). *J. Toxicol. Environ. Health A* **1999**, *57*, 115–136. [CrossRef]

107. Van Leeuwen, P.A.; Boermeester, M.A.; Houdijk, A.P.; Ferwerda, C.C.; Cuesta, M.A.; Meyer, S.; Wesdorp, R.I. Clinical significance of translocation. *Gut* **1994**, *35*, S28–S34.

108. Islam, L.N.; Nabi, A.H. Endotoxins of enteric pathogens modulate the functions of human neutrophils and lymphocytes. *J. Biochem. Mol. Biol.* **2003**, *36*, 565–571. [CrossRef]

109. Islam, Z.; Pestka, J.J. Role of IL-1(beta) in endotoxin potentiation of deoxynivalenol-induced corticosterone response and leukocyte apoptosis in mice. *Toxicol. Sci.* **2003**, *74*, 93–102. [CrossRef]

110. Azcona-Olivera, J.I.; Ouyang, Y.; Murtha, J.; Chu, F.S.; Pestka, J.J. Induction of cytokine mRNAs in mice after oral exposure to the trichothecene vomitoxin (deoxynivalenol): Relationship to toxin distribution and protein synthesis inhibition. *Toxicol. Appl. Pharmacol.* **1995**, *133*, 109–120. [CrossRef]

111. Maheshwari, A.; Lacson, A.; Lu, W.; Fox, S.E.; Barleycorn, A.A.; Christensen, R.D.; Calhoun, D.A. Interleukin-8/CXCL8 forms an autocrine loop in fetal intestinal mucosa. *Pediatr. Res.* **2004**, *56*, 240–249. [CrossRef]

112. Al-Sadi, R.; Boivin, M.; Ma, T. Mechanism of cytokine modulation of epithelial tight junction barrier. *Front. Biosci.* **2009**, *14*, 2765–2778. [CrossRef]

113. Bambou, J.C.; Giraud, A.; Menard, S.; Begue, B.; Rakotobe, S.; Heyman, M.; Taddei, F.; Cerf-Bensussan, N.; Gaboriau-Routhiau, V. In vitro and ex vivo activation of the TLR5 signaling pathway in intestinal epithelial cells by a commensal Escherichia coli strain. *J. Biol. Chem.* **2004**, *279*, 42984–42992. [CrossRef]

114. Hopkins, S.J. The pathophysiological role of cytokines. *Leg Med. Tokyo* **2003**, *5*, S45–S57. [CrossRef]

115. Do, K.H.; Choi, H.J.; Kim, J.; Park, S.H.; Kim, K.H.; Moon, Y. SOCS3 regulates BAFF in human enterocytes under ribosomal stress. *J. Immunol.* **2013**, *190*, 6501–6510. [CrossRef]

116. Borka, K.; Kaliszky, P.; Szabo, E.; Lotz, G.; Kupcsulik, P.; Schaff, Z.; Kiss, A. Claudin expression in pancreatic endocrine tumors as compared with ductal adenocarcinomas. *Virchows Arch.* **2007**, *450*, 549–557. [CrossRef]

117. Eun, C.S.; Han, D.S.; Lee, S.H.; Paik, C.H.; Chung, Y.W.; Lee, J.; Hahm, J.S. Attenuation of colonic inflammation by PPARgamma in intestinal epithelial cells: Effect on Toll-like receptor pathway. *Dig. Dis. Sci.* **2006**, *51*, 693–697. [CrossRef]

118. Oswald, I.P. Role of intestinal epithelial cells in the innate immune defence of the pig intestine. *Vet. Res.* **2006**, *37*, 359–368. [CrossRef]

119. Denning, T.L.; Wang, Y.C.; Patel, S.R.; Williams, I.R.; Pulendran, B. Lamina propria macrophages and dendritic cells differentially induce regulatory and interleukin 17-producing T cell responses. *Nat. Immunol.* **2007**, *8*, 1086–1094. [CrossRef]

120. Alzoghaibi, M.A.; Walsh, S.W.; Willey, A.; Yager, D.R.; Fowler, A.A., 3rd; Graham, M.F. Linoleic acid induces interleukin-8 production by Crohn's human intestinal smooth muscle cells via arachidonic acid metabolites. *Am. J. Physiol. Gastrointest. Liver Physiol.* **2004**, *286*, G528–G537. [CrossRef]

121. Banks, C.; Bateman, A.; Payne, R.; Johnson, P.; Sheron, N. Chemokine expression in IBD. Mucosal chemokine expression is unselectively increased in both ulcerative colitis and Crohn's disease. *J. Pathol.* **2003**, *199*, 28–35. [CrossRef]

122. Georganas, C.; Liu, H.; Perlman, H.; Hoffmann, A.; Thimmapaya, B.; Pope, R.M. Regulation of IL-6 and IL-8 expression in rheumatoid arthritis synovial fibroblasts: The dominant role for NF-kappa B but not C/EBP beta or c-Jun. *J. Immunol.* **2000**, *165*, 7199–7206.

123. Martin, H.M.; Campbell, B.J.; Hart, C.A.; Mpofu, C.; Nayar, M.; Singh, R.; Englyst, H.; Williams, H.F.; Rhodes, J.M. Enhanced Escherichia coli adherence and invasion in Crohn's disease and colon cancer. *Gastroenterology* **2004**, *127*, 80–93. [CrossRef]

124. Pestka, J.J. Deoxynivalenol-induced IgA production and IgA nephropathy-aberrant mucosal immune response with systemic repercussions. *Toxicol. Lett.* **2003**, *140–141*, 287–295. [CrossRef]

125. Beagley, K.W.; Eldridge, J.H.; Lee, F.; Kiyono, H.; Everson, M.P.; Koopman, W.J.; Hirano, T.; Kishimoto, T.; McGhee, J.R. Interleukins and IgA synthesis. Human and murine interleukin 6 induce high rate IgA secretion in IgA-committed B cells. *J. Exp. Med.* **1989**, *169*, 2133–2148. [CrossRef]

126. Shi, Y.; Pestka, J.J. Attenuation of mycotoxin-induced IgA nephropathy by eicosapentaenoic acid in the mouse: Dose response and relation to IL-6 expression. *J. Nutr. Biochem.* **2006**, *17*, 697–706. [CrossRef]

127. Yan, D.; Zhou, H.R.; Brooks, K.H.; Pestka, J.J. Potential role for IL-5 and IL-6 in enhanced IgA secretion by Peyer's patch cells isolated from mice acutely exposed to vomitoxin. *Toxicology* **1997**, *122*, 145–158. [CrossRef]

128. Pinton, P.; Accensi, F.; Beauchamp, E.; Cossalter, A.M.; Callu, P.; Grosjean, F.; Oswald, I.P. Ingestion of deoxynivalenol (DON) contaminated feed alters the pig vaccinal immune responses. *Toxicol. Lett.* **2008**, *177*, 215–222. [CrossRef]

129. Bouhet, S.; le Dorze, E.; Peres, S.; Fairbrother, J.M.; Oswald, I.P. Mycotoxin fumonisin B1 selectively down-regulates the basal IL-8 expression in pig intestine: *in vivo* and *in vitro* studies. *Food Chem. Toxicol.* **2006**, *44*, 1768–1773. [CrossRef]

130. McLaughlin, J.; Padfield, P.J.; Burt, J.P.; O'Neill, C.A. Ochratoxin A increases permeability through tight junctions by removal of specific claudin isoforms. *Am. J. Physiol. Cell Physiol.* **2004**, *287*, C1412–C1417. [CrossRef]

131. Sergent, T.; Dupont, I.; Jassogne, C.; Ribonnet, L.; van der Heiden, E.; Scippo, M.L.; Muller, M.; McAlister, D.; Pussemier, L.; Larondelle, Y.; *et al.* CYP1A1 induction and CYP3A4 inhibition by the fungicide imazalil in the human intestinal Caco-2 cells-comparison with other conazole pesticides. *Toxicol. Lett.* **2009**, *184*, 159–168. [CrossRef]

132. Maresca, M.; Fantini, J. Some food-associated mycotoxins as potential risk factors in humans predisposed to chronic intestinal inflammatory diseases. *Toxicon* **2010**, *56*, 282–294. [CrossRef]

133. Wache, Y.J.; Valat, C.; Postollec, G.; Bougeard, S.; Burel, C.; Oswald, I.P.; Fravalo, P. Impact of deoxynivalenol on the intestinal microflora of pigs. *Int. J. Mol. Sci.* **2009**, *10*, 1–17.

134. Sirot, V.; Fremy, J.M.; Leblanc, J.C. Dietary exposure to mycotoxins and health risk assessment in the second French total diet study. *Food Chem. Toxicol.* **2013**, *52*, 1–11. [CrossRef]

135. Berthiller, F.; Crews, C.; Dall'asta, C.; Saeger, S.D.; Haesaert, G.; Karlovsky, P.; Oswald, I.P.; Seefelder, W.; Speijers, G.; Stroka, J. Masked mycotoxins: A review. *Mol. Nutr. Food Res.* **2013**, *57*, 165–186. [CrossRef]

Review

The Impact of *Fusarium* Mycotoxins on Human and Animal Host Susceptibility to Infectious Diseases

Gunther Antonissen [1,2,*], **An Martel** [2], **Frank Pasmans** [2], **Richard Ducatelle** [2], **Elin Verbrugghe** [2], **Virginie Vandenbroucke** [3], **Shaoji Li** [2], **Freddy Haesebrouck** [2], **Filip Van Immerseel** [2] and **Siska Croubels** [1,*]

[1] Department of Pharmacology, Toxicology and Biochemistry, Faculty of Veterinary Medicine, Ghent University, Salisburylaan 133, 9820 Merelbeke, Belgium

[2] Department of Pathology, Bacteriology and Avian Diseases, Faculty of Veterinary Medicine, Ghent University, Salisburylaan 133, 9820 Merelbeke, Belgium; An.Martel@UGent.be (A.M.); Frank.Pasmans@UGent.be (F.P.); Richard.Ducatelle@UGent.be (R.D.); Elin.Verbrugghe@UGent.be (E.V.); Shaoji.Li@UGent.be (S.L.); Freddy.Haesebrouck@UGent.be (F.H.); Filip.VanImmerseel@UGent.be (F.V.I.)

[3] Animal Health Care Flanders, Industrielaan 29, 8820 Torhout, Belgium; Virginie.Vandenbroucke@dgz.be

[*] Correspondence: Gunther.Antonissen@UGent.be (G.A.); Siska.Croubels@UGent.be (S.C.); Tel.: +32-9-264-74-34 (G.A.); +32-9-264-73-47 (S.C.); Fax: +32-9-264-74-97 (S.C.)

Received: 21 December 2013; in revised form: 16 January 2014; Accepted: 16 January 2014; Published: 28 January 2014

Abstract: Contamination of food and feed with mycotoxins is a worldwide problem. At present, acute mycotoxicosis caused by high doses is rare in humans and animals. Ingestion of low to moderate amounts of *Fusarium* mycotoxins is common and generally does not result in obvious intoxication. However, these low amounts may impair intestinal health, immune function and/or pathogen fitness, resulting in altered host pathogen interactions and thus a different outcome of infection. This review summarizes the current state of knowledge about the impact of *Fusarium* mycotoxin exposure on human and animal host susceptibility to infectious diseases. On the one hand, exposure to deoxynivalenol and other *Fusarium* mycotoxins generally exacerbates infections with parasites, bacteria and viruses across a wide range of animal host species. Well-known examples include coccidiosis in poultry, salmonellosis in pigs and mice, colibacillosis in pigs, necrotic enteritis in poultry, enteric septicemia of catfish, swine respiratory disease, aspergillosis in poultry and rabbits, reovirus infection in mice and Porcine Reproductive and Respiratory Syndrome Virus infection in pigs. However, on the other hand, T-2 toxin has been shown to markedly decrease the colonization capacity of *Salmonella* in the pig intestine. Although the impact of the exposure of humans to *Fusarium* toxins on infectious diseases is less well known, extrapolation from animal models suggests possible exacerbation of, for instance, colibacillosis and salmonellosis in humans, as well.

Keywords: deoxynivalenol; fumonisin; *Fusarium* mycotoxins; human; infectious diseases; mouse; pig; poultry; T-2 toxin; zearalenone

1. Introduction

Mycotoxins are toxic fungal metabolites that can contaminate a wide array of food and feed [1]. Mycotoxin-producing fungi can be classified into either field or storage fungi. Field fungi, such as the *Fusarium* species, produce mycotoxins on the crops in the field, whereas storage fungi, such as the *Aspergillus* and *Penicillium* species, produce mycotoxins on the crops after harvesting [2]. *Fusarium* fungi have traditionally been associated with temperate climatic conditions, since they require somewhat lower temperature for growth and mycotoxin production than, for example, the *Aspergillus* species [3]. The most toxicologically important *Fusarium* mycotoxins are trichothecenes (including deoxynivalenol (DON) and T-2 toxin (T-2)), zearalenone (ZEN) and fumonisin B1 (FB1).

Fusarium mycotoxins are capable of inducing both acute and chronic toxic effects. These effects are dependent on the mycotoxin type, the level and duration of exposure, the animal species that is exposed and the age of the animal [4]. Intake of high doses of mycotoxins may lead to acute mycotoxicoses, which are characterized by well-described clinical signs [5,6]. Exposure of pigs to high concentrations of DON causes abdominal distress, malaise, diarrhea, emesis and even shock or death. Exposure of pigs to fumonisins can lead to pulmonary edema due to cardiac insufficiency. In horses fumonisins can cause equine leukoencephalomalacia (ELEM) and target the brain [7]. Since these high contamination levels are rare in modern agricultural practice [8], this review will not discuss extensively their effect on animal or human health. Indeed, although the results of a global survey indicate that the *Fusarium* mycotoxins DON, fumonisins, and ZEN respectively contaminated 55%, 54% and 36% of feed and feed ingredients in the period 2004–2011, the majority of samples was found to comply with even the most stringent European Union regulations or recommendations on the maximal tolerable concentration (Table A3) [8]. Therefore, this review will focus on the effect of low to moderate doses of the major *Fusarium* mycotoxins.

Following oral intake of low to moderate amounts of these mycotoxins, the gastro-intestinal epithelial cell layer will be exposed first [9]. The intestinal mucosa acts as a barrier, preventing the entry of foreign antigens including food proteins, xenobiotics (such as drugs and toxins), commensal microbiota and pathogens into the underlying tissues [9,10]. The mucosal immunity, which consists of an innate and adaptive immune system, can be affected by *Fusarium* mycotoxins (Figure 1) [9,10]. An important component of the innate immune system are the intestinal epithelial cells, which are interconnected by tight junctions, and covered with mucus, produced by goblet cells [11]. By measuring the transepithelial electrical resistance (TEER), several *in vitro* and *ex vivo* studies indicate that DON and FB1 are able to increase the permeability of the intestinal epithelial layer of human, porcine and avian origin [12–14]. Also the viability and proliferation of animal and human intestinal epithelial cells can be negatively affected by *Fusarium* mycotoxins [9,15–20]. Their effect on mucus production is variable: co-exposure of low doses of DON, T-2 and ZEN reduces the number of goblet cells in pigs [21], but ZEN given alone at higher doses increases the activity of goblet cells [22]. Several mycotoxins are also able to modulate the production of cytokines *in vitro* and *in vivo* [9,23]. For example, DON increases the expression of TGF-β and IFN-γ in mice and fumonisins decrease the expression of IL-8 in an intestinal porcine epithelial cell line (IPEC-1) [9].

Fusarium mycotoxins can cross the intestinal epithelium and reach the systemic compartment [20,24], affecting the immune system. Exposure to these toxins can either result in immunostimulatory or immunosuppressive effects depending on the age of the host and exposure dose and duration [20,25]. Mycotoxin-induced immunomodulation may affect innate and adaptive immunity by an impaired function of macrophages and neutrophils, a decreased T- and B-lymphocyte activity and antibody production [23,25,26]. In addition to the effect of *Fusarium* mycotoxins on the animal or human host, these mycotoxins may alter the metabolism of the pathogen, which may alter the outcome of the infectious disease [27,28].

A wealth of research papers clearly indicate a negative influence of *Fusarium* mycotoxins on the intestinal function and immune system. Since the intestinal tract is also a major portal of entry to many enteric pathogens and their toxins, mycotoxin exposure could increase the animal susceptibility to these pathogens. Furthermore, mycotoxin-induced immunosuppression may also result in decreased animal or human host resistance to infectious diseases.

This review attempts to summarize the impact of *Fusarium* mycotoxin exposure on the animal and human host susceptibility to infectious diseases. More specifically, the effect of *Fusarium* mycotoxins on enteric, systemic and respiratory infectious diseases in livestock animals and animal models for human diseases are highlighted.

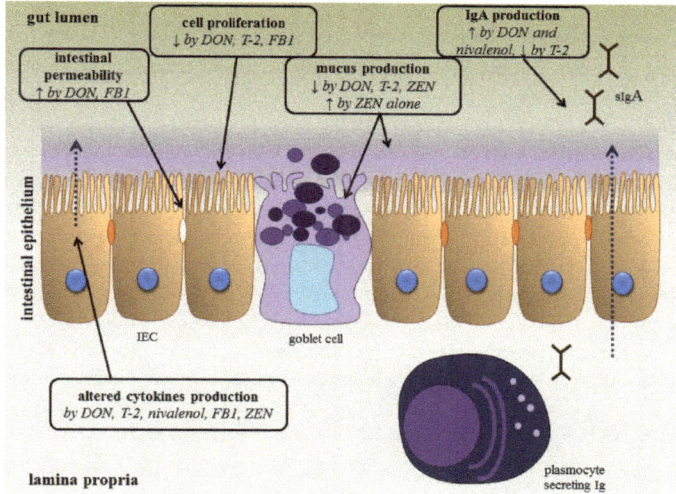

Figure 1. The effect of *Fusarium* mycotoxins on the intestinal epithelium. A variety of *Fusarium* mycotoxins alter the different intestinal defense mechanisms including epithelial integrity, cell proliferation, mucus layer, immunoglobulins (Ig) and cytokine production. (IEC: intestinal epithelial cell) (based on [9]).

2. Effect of *Fusarium* Toxins on Parasitic Diseases

Coccidiosis

Intestinal protozoa, including the coccidia (*Eimeria*, *Isospora*, *Cryptosporidium* and *Sarcosporidia*) and flagellates, are important infectious agents. Coccidiosis in poultry generally refers to the disease caused by the *Eimeria* species, and is still considered one of the most important enteric diseases affecting performance. These obligate intracellular parasites have an oral-fecal life cycle with developmental stages alternating between the external environment and the host [29].

Seven species of *Eimeria* (*E. acervulina*, *E. brunetti*. *E. maxima*, *E. mitis*, *E. necatrix*, *E. praecox* and *E. tenella*) are found in chickens [29]. The physical and biological characteristics, pathogenicity and immunogenicity depend on the species. Immunity to *Eimeria* is complex, multifactorial and influenced by both host and parasite [30].

Cell-mediated immunity, mainly evoked by the intraepithelial lymphocytes (IEL) and lymphocytes of the lamina propria, is the major protective immune component against avian coccidiosis [31,32]. The CD4$^+$ T-lymphocytes, IEL and macrophages are involved in the response against primary exposure to *Eimeria* [31], while CD8$^+$ T-lymphocytes and IFN-γ are important in the protective immune response against *Eimeria* infection [33]. Girgis *et al.* [34,35] showed a negative impact of diets naturally contaminated with *Fusarium* mycotoxins on the cell-mediated immune response against coccidiosis in broilers (Table A2). Following primary infection of broilers with *Eimeria*, *Fusarium* mycotoxins decreased the percentage of CD4$^+$ and CD8$^+$ T-cells in the jejunal mucosa [35]. In addition, feeding on a mycotoxin-contaminated diet lowered the blood levels of CD8$^+$ T-cells and monocytes, which could suggest an increased recruitment at the intestinal site of coccidial infection or a delayed replication necessary to replenish these subsets in the circulation [34,35]. Additionally, feeding on a *Fusarium* mycotoxin-contaminated diet increased IFN-γ gene expression in the cecal tonsils of *Eimeria*-challenged birds, however, without being linked to the apparent resistance to coccidial infection in terms of changes in oocyst yield [34]. The cecal tonsils constitute a lymphoid tissue in the cecum belonging to the gut-associated lymphoid tissue (GALT). Resistance to *Eimeria* infection is related to the expression of a set of interleukins rather than only IFN-γ and the up-regulation of the gene may not necessarily be associated with functional secretion [34]. Furthermore, it was shown that moderate levels of *Fusarium* mycotoxins negatively affect intestinal morphology and interfere with intestinal recovery from an enteric coccidial infection, indicated by a lower villus height and apparent villus area (Table A2) [36]. Although Girgis *et al.* [34,35] demonstrated that *Fusarium* mycotoxins impair the *Eimeria*-induced immune response, no effect was seen on fecal oocyst counts. Similarly, Békési *et al.* [37] showed no impact of a T-2 and ZEN-contaminated diet on *Cryptosporidium baileyi* oocyst excretion in broilers.

Research investigating the influence of mycotoxins on the animal susceptibility to infectious diseases focuses mainly on exposure to single major mycotoxins. Limited information about the impact of mycotoxin co-occurrence and plant metabolites of mycotoxins on this interaction is available. Nevertheless, Girgis *et al.* [34,35] showed that the combination of DON, 15-acetylDON (15-AcDON), ZEN and fumonisins alters the *Eimeria*-induced immune response. Interestingly, mycotoxin contamination of broiler feed may reduce the efficacy of the anti-coccidial treatment with lasalocid [38].

To conclude, *Fusarium* mycotoxins negatively affect the innate and adaptive cellular immune response against *Eimeria*, though without changing the oocyst yield. Further data of clinical coccidiosis lesion scoring is still needed in order to evaluate the effect of *Fusarium* mycotoxins on the severity of the disease.

3. Effect of *Fusarium* Toxins on Bacterial Diseases

3.1. Salmonellosis

Salmonellosis is an infection with the Gram-negative *Salmonella* bacterium, a facultative anaerobic, facultative intracellular microorganism of the *Enterobacteriaceae* family. The host—*Salmonella* interaction is complex, with a broad array of mechanisms used by the bacteria to overcome host defenses. Two important disease manifestations are differentiated, *i.e.*, gastroenteritis and enteric fever, caused by nontyphoidal and typhoidal *Salmonella* serovars, respectively [39].

Nontyphoidal *Salmonella* strains, such as *Salmonella* serovar Typhimurium and *Salmonella* serovar Enteritidis strains, infect a wide range of animal hosts, including pigs and poultry, without causing clinical symptoms in these animals. Infection in slaughter pigs and poultry can cause meat and egg contamination [39,40].

An infection with *Salmonella* generally occurs in three stages: the adhesion to the intestinal wall, the invasion of the gut wall and the dissemination to mesenteric lymph nodes and other organs. Via bacterial-mediated endocytosis, *Salmonella* invades the intestinal epithelial cells, after which the bacterium becomes enclosed within an intracellular phagosomal compartment (the *Salmonella*-containing vacuole (SCV)). After crossing the epithelial barrier, the bacterium is located predominantly in macrophages in the underlying tissue [39].

Feeding pigs a *Fusarium* mycotoxin-contaminated diet influences the intestinal phase of the pathogenesis of *Salmonella* Typhimurium infections as illustrated in Figure 2. Non-cytotoxic concentrations of DON and T-2 enhance intestinal *Salmonella* invasion and increase the passage of *Salmonella* Typhimurium across the epithelium (Table A1) [28,41]. Chronic exposure of specific pathogen-free pigs to naturally fumonisin-contaminated feed had no impact on *Salmonella* Typhimurium translocation [42]. Once *Salmonella* has invaded the intestinal epithelium, the innate immune system is triggered and the porcine gut will start to produce several cytokines [28,43]. Both *Fusarium* mycotoxins and *Salmonella* affect the innate immune system. Vandenbroucke *et al.* [27] showed that low concentrations of DON could potentiate the early intestinal immune response induced by *Salmonella* Typhimurium infection. Co-exposure of the intestine to DON and *Salmonella* Typhimurium resulted in increased expression of several cytokines, for instance, those responsible for the stimulation of the inflammatory response (TNF-α) and T-lymphocyte stimulation (IL-12) (Table A2). The authors suggested that the enhanced intestinal inflammation could be due to a DON-induced stimulation of *Salmonella* Typhimurium invasion in and translocation across the intestinal epithelium [27].

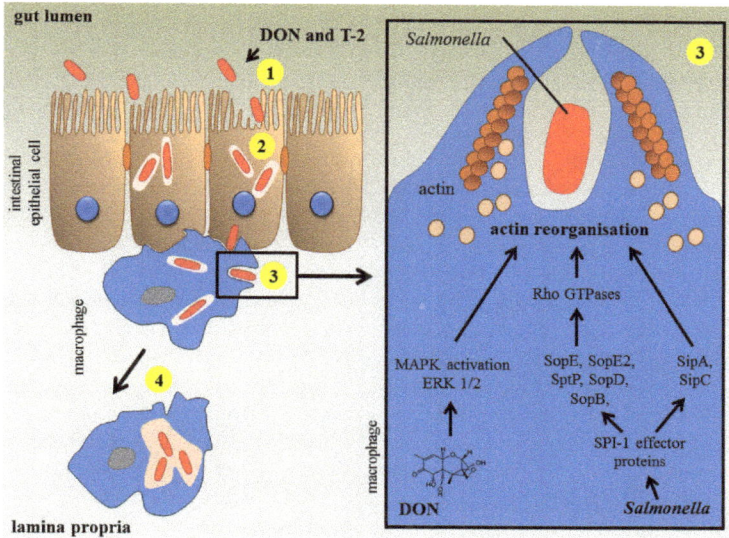

Figure 2. The impact of deoxynivalenol and T-2 toxin on a *Salmonella* Typhimurium infection in pigs. *In vitro*, deoxynivalenol (DON) and T-2 toxin (T-2) promote *Salmonella* invasion (1) and transepithelial passage (2) of IPEC-J2 cell layer. Subsequently, the bacterium can spread to the bloodstream using the host macrophage to establish the systemic infection. *In vitro*, DON and T-2 enhance *Salmonella* uptake (3) in porcine alveolar macrophages. The *Salmonella* invasion of macrophages coincides with membrane ruffling, caused by actin cytoskeletal changes. Activation of host Rho GTPases by the *Salmonella* pathogenicity island (SPI)-1 type 3 secretion system (T3SS) effector proteins SopB, SopE, SopE2 and SopD leads to actin cytoskeleton reorganization. After *Salmonella* internalization has occurred, the bacterium injects the effector protein SptP which promotes the inactivation of Rho GTPases. The bacterium can also modulate the actin dynamics of the host cell in a direct manner through the bacterial effector proteins SipA and SipC. The mycotoxin DON enhances the uptake of *Salmonella* in macrophages through activation of the mitogen-activated protein kinases (MAPK) extracellular signal-regulated kinases (ERK1/2) pathway, which induces actin reorganizations and membrane ruffles. DON and T-2 do not affect intracellular bacterial proliferation (4) (based on [41,44]).

Fusarium mycotoxins also affect the systemic part of the *Salmonella* Typhimurium infection in pigs. After the intestinal phase of the pathogenesis, *Salmonella* can spread to the bloodstream using the host macrophage to establish the systemic infection. However, in pigs the systemic part of *Salmonella* Typhimurium is poorly documented and colonization is mostly limited to the gastrointestinal tract [44]. After bacterial uptake by the macrophage, *Salmonella* can survive and even proliferate in this cell. Exposure of macrophages to non-cytotoxic concentrations of DON and T-2 promotes the uptake of *Salmonella* Typhimurium (Figure 2, Table A1). *Salmonella* entry in host cells involves a complex series of actin cytoskeletal changes. Macrophage invasion coincides with membrane ruffling, followed by bacterium uptake and formation of *Salmonella*-containing vacuole [41]. Vandenbroucke *et al.* [41] showed *in vitro* that DON enhances *Salmonella* Typhimurium engulfment, since low concentrations of DON modulate the cytoskeleton of macrophages through ERK1/2 F-actin reorganization resulting in an enhanced uptake of *Salmonella* Typhimurium in porcine alveolar macrophages (PAM) (Figure 2,

Table A1). Non-cytotoxic concentrations of the *Fusarium* mycotoxins DON and T-2 did not affect the intracellular proliferation of *Salmonella* Typhimurium in porcine macrophages (Figure 2) [28,41].

In addition to the effects of *Fusarium* mycotoxins on the host susceptibility to a *Salmonella* Typhimurium infection, these mycotoxins also modulate the bacterial metabolism. Although no effect of DON or T-2 on the growth of *Salmonella* Typhimurium is detected, DON and T-2 modulate the *Salmonella* gene expression [28,41]. The enhanced inflammatory effect following exposure to DON is more likely a result of the toxic effect of the mycotoxin on the intestine than on the bacterium [27]. Only high concentrations of DON increase the bacterial expression of regulators of *Salmonella* pathogenicity island (SPI)-1 and SPI-2, respectively *hilA* and *ssrA*. SPI-1 consists of genes coding for bacterial secretion systems necessary for invasion, while SPI-2 genes encode essential intracellular replication mechanisms [41]. For T-2 the toxic effects on the bacterium itself are probably more pronounced than the host cell-mediated effects resulting in a reduced *in vivo* colonization in pigs. Low concentrations of T-2 cause a reduced motility of *Salmonella* and a general down regulation of genes involved in *Salmonella* metabolism, genes encoding ribosomal proteins and SPI-1 genes [28].

Only limited information is available concerning the interaction between *Fusarium* mycotoxins and *Salmonella* Typhimurium infection in other animals. The currently available publications mainly focus on the interaction of T-2 and the systemic phase of a *Salmonella* Typhimurium infection. In T-2-challenged broiler chickens and mice an increased level of *Salmonella* Typhimurium-related organ lesions or mortality was seen (Table A2) [45–48]. Infection of mice with *Salmonella* Typhimurium results in systemic infection and a disease similar to that seen in humans after infection with *Salmonella* Typhi [49]. Increased mortality might be explained partly by the synergistic effects of bacterial lipopolysaccharide (LPS) and T-2 during the late phase of murine salmonellosis [50]. In addition to *Salmonella* Typhimurium, DON reduces the resistance to oral infection with *Salmonella* Enteritidis in mice by promoting translocation of *Salmonella* to mesenteric lymph node (MLN), liver and spleen (Table A2) [51].

Mouse and pig models are important animal models to investigate the impact of mycotoxins, infectious diseases and their combination on animal health [52,53]. Infection of mice with *Salmonella* Typhimurium is an important host–pathogen interaction model to investigate typhoid fever in humans. Moderate to high concentrations of T-2 have shown to increase *Salmonella*-induced mortality [46,47,50]. The pig is very similar to humans in terms of anatomic and physiologic characteristics such as size, digestive physiology, kidney structure and function, pulmonary vascular bed structure, coronary artery distribution, respiratory rates, cardiovascular anatomy and physiology, and immune response, and has been used to study various intestinal pathogens, including *Salmonella* and *Escherichia coli* [53]. The interaction between mycotoxins and *Salmonella* Typhimurium studied in a porcine model of infection, gives us relevant information concerning the impact of this interaction on human intestinal inflammation and immune response [27].

In conclusion, the exact outcome of co-exposure to *Fusarium* mycotoxins and *Salmonella* Typhimurium is difficult to predict. Published data show an influence of mycotoxin exposure on the bacterium, the host cells and the host–pathogen interaction. Depending on the characteristics of the mycotoxin exposure, one of these effects will determine the outcome of the interaction between *Fusarium* mycotoxins and *Salmonella* Typhimurium.

3.2. Colibacillosis

Escherichia coli is a Gram-negative, non-sporulating rod-shaped bacterium of the family *Enterobacteriaceae*. Although this bacterium is considered to be a normal component of the intestinal microbiota, it is frequently associated with both intestinal and extra-intestinal infections in humans and animals. A certain number of these strains possess particular combinations of virulence factors which enables them to cause disease. Clinical syndromes resulting from infection with these pathotypes include enteric/diarrheal disease, urinary tract infections and sepsis/meningitis.

The pathogenesis of *E. coli* infections depends on the pathotype involved and may include colonizing the intestinal mucosa, evasion of host defenses, multiplication, and induction of host damage [54,55].

Fusarium mycotoxins may influence the pathogenesis of *E. coli* infections in different animal species by stimulating intestinal colonization and translocation and negatively affecting the immune response. Feeding a diet contaminated with a moderate level of FB1 to pigs enhanced intestinal colonization and translocation of a septicemic *E. coli* (SEPEC) strain from the intestine to the systemic compartment. FB1-treatment resulted in a higher bacterial translocation to the mesenteric lymph nodes and lungs, and to a lesser extent to liver and spleen (Table A2) [56]. It was shown *in vitro* that DON increased the translocation of SEPEC over the intestinal epithelial cell monolayer (IPEC-1) (Table A1) [14].

Mycotoxins increase the calf susceptibility to shiga toxin or verotoxin-producing *E. coli* (STEC)-associated hemorrhagic enteritis. Recently, Baines *et al.* [57] showed that exposing calves of less than one month old to the combination of aflatoxin and fumonisins promoted STEC-associated hemorrhagic enteritis (Table A2) [57].

Feeding a FB1-contaminated diet to pigs negatively affects the mucosal immune response against an infection with enterotoxigenic *E. coli* (ETEC). Devriendt *et al.* [58] showed a prolonged intestinal infection of *E. coli* in pigs administered fumonisins for 10 consecutive days and subsequently challenged with *E. coli* (F4+ ETEC) (Table A2). Antigen-presenting cells (APCs) have an important role in the mucosal immune system by connecting the innate and adaptive immune response, through uptake of antigen in lamina propria, maturation and migration to GALT, and interaction with T cells. FB1 negatively affected the function of intestinal APCs by a reduced up-regulation of the major histocompatibility complex class II (MHC-II), cluster of differentiation (CD) 80/6 and IL-12p40 cytokine gene expression [58]. This altered function of APCs could therefore influence the *E. coli*-induced adaptive immune response [58,59]. Additionally, moniliformin and FB1 delayed systemic *E. coli* (avian pathogenic *E. coli*, APEC) clearance in broilers and turkeys after intravenous administration (Table A2) [60,61].

The results of these studies may also be valid for human infections since the gastro-intestinal tract of pigs and humans are very similar [58]. Infant diarrhea caused by enteropathogenic *E. coli* (EPEC) is known to be of major concern in developing countries and, for instance, enterohemorrhagic *E. coli* (EHEC) infections are a major worldwide public health hazard.

3.3. Necrotic Enteritis in Broilers

Necrotic enteritis (NE) is a disease in broilers caused by *Clostridium perfringens*. This Gram-positive spore-forming bacterium occurs naturally in the environment, feed and gastrointestinal tract of chickens and other animals [62,63]. NE is a complex, multifactorial enteric disease with many known and unknown factors influencing its occurrence and the severity of the outbreaks. The best-known predisposing factor is mucosal damage caused by coccidial pathogens [64]. Only *C. perfringens* strains expressing the NetB toxin are capable of inducing NE in broilers [65]. *C. perfringens* is auxotrophic for several amino acids, thus availability of these amino acids would allow extensive bacterial proliferation [63].

The intake of DON-contaminated feed is a predisposing factor for the development of necrotic enteritis in broiler chickens due to the negative influence on the epithelial barrier, and to an increased intestinal nutrient availability for clostridial proliferation. Recently, we [66] showed in an experimental subclinical NE infection model that chickens fed a diet contaminated with DON for three weeks were more prone to develop NE lesions compared to chickens on a control diet (Table A2). The negative effects of DON on the small intestinal barrier can lead to an impaired nutrient digestion and leakage of plasma amino acids into the intestinal lumen, providing the necessary growth substrate for extensive proliferation of *C. perfringens* [66].

3.4. Edwardsiella ictaluri Infection in Catfish

Edwardsiella ictaluri is a Gram-negative bacterium of the *Enterobacteriaceae* family. Bacillary Necrosis of *Pangasianodon* (BNP) caused by *E. ictaluri* is the most frequently occurring infectious disease in catfish [67]. Besides the Vietnamese freshwater production, also the American channel catfish (*Ictalurus punctatus*) industry suffers massively from *E. icatluri* infections which have been termed Enteric Septicemia of Catfish (ESC). BNP is characterized by multifocal irregular white spots of varying sizes on several organs including liver, spleen and kidney [68]. ESC in channel catfish may occur in an acute form characterized by enteritis and septicemia with rapid mortality, or in a chronic form, which is characterized by meningoencephalitis, open lesions on the cranial region and exophthalmia [69]. Mortality associated with the co-occurrence of *Fusarium* mycotoxins and *E. ictaluri* is difficult to predict in juvenile channel catfish. T-2 increased *E. ictaluri*-associated mortality [70], while moderate contamination of DON improved the survival of the channel catfish [71] (Table A2). Mycotoxin sensitivity differs between fish species. Rainbow trout, for example, are extremely sensitive to DON, while channel catfish are rather resistant [71,72]. Important data concerning the toxicity of the mycotoxin on the bacterium are lacking. Further investigation of the interaction between *Fusarium* mycotoxins and *E. ictaluri* will be necessary to evaluate the outcome.

3.5. Swine Respiratory Disease

Respiratory disease in pigs is often caused by the combined effects of multiple pathogens and predisposing factors [73]. Primary infections with bacteria such as *Actinobacillus pleuropneumoniae*, *Mycoplasma hyopneumoniae*, *Bordetella bronchiseptica* or viruses such as influenza virus and Porcine Reproductive and Respiratory Syndrome Virus (PRRSV), can predispose pigs to secondary pathogens such as *Pasteurella multocida* and *Trueperella pyogenes* [74]. Respiratory symptoms can vary depending on the pathogens involved. *M. hyopneumoniae* is the principal etiological agent responsible for enzootic pneumonia in pigs [75]. *M. hyopneumoniae* is an obligate symbiotic and host-specific bacterium, which is lacking a cell wall. This pathogen affects the respiratory mucosal clearance system by disrupting the celia on the epithelial surface and modulates the immune system of the respiratory tract. Consequently, *M. hyopneumoniae* predisposes animals to concurrent infections with other respiratory pathogens [75]. Dietary exposure to fumonisins induces pulmonary edema and may facilitate *M. hyopneumoniae* infection (Table A2) [76].

The progressive form of porcine atrophic rhinitis is often due to a combined infection with *B. bronchiseptica* and toxigenic *P. multocida* [73,77]. Dietary exposure to FB1 of piglets infected with both bacteria increases the risk of pneumonia and the severity of the pathological changes [73]. *P. multocida* type A is the most frequently occurring secondary pathogen that can cause pneumonic pasteurellosis [78]. Halloy *et al.* [74] showed that inoculation of piglets with *P. multocida* combined with an oral bolus of FB1 induced a cough and a lung inflammatory process characterized by an increased number of total cells, macrophages and lymphocytes in broncheo-alveolar lavage fluid (BALF). Lung lesions were more severe in these animals and consisted of subacute interstitial pneumonia [74].

4. Effect of *Fusarium* Toxins on Fungal Diseases

Aspergillosis

Aspergillus fumigatus is an ubiquitous saprophytic fungus found in soil, plant debris, and the indoor environment, including hospitals. This fungus is also an opportunistic pathogen. Inhalation of its conidia can cause life-threatening infections in the respiratory system of immunocompromised animals and humans. Respiratory macrophages are the first line of defense against inhaled *Aspergillus* conidia. T-2 impaired the phagocytotic activities of macrophages against *A. fumigatus* conidia in chickens and rabbits (Tables A1 and A2) [79,80]. However, the pro-inflammatory response of *A. fumigatus* infected chicken macrophages was increased by T-2 (Table A1) [80]. The effect of T-2 on the innate immune response against *Aspergillus* conidia is dual, which suggests that depending on the characteristics of the mycotoxin exposure and the animal, one of these effects will determine the outcome of this interaction.

5. Effect of Fusarium Toxins on Viral Diseases

5.1. Reovirus

Reovirus is a non-enveloped double-stranded RNA virus that has been isolated from the gastro-intestinal tract and respiratory tract of both humans and animals [81,82]. Enteric reoviruses cause mostly a mild and self-limiting infection [82]. Nevertheless, reovirus infections can be more severe, affecting, for example, the central nervous system in mice and rats [81]. Viral arthritis is the most frequent reovirus-associated disease in poultry, which is characterized by lameness and swellings affecting primarily tarsometatarsal joints and the feet [83–85].

Fusarium mycotoxins negatively affect the intestinal virus clearance in mice. Li *et al.* [82,86] showed that high concentrations of DON and T-2 suppress the host immune response to reovirus as evidenced by the inability to clear the virus from the intestine as well as by increased fecal shedding of the virus (Table A2). Trichothecene exposure increased the intestinal viral load, which could increase inflammation and discomfort to the host during the infection process. The increased fecal shedding could enhance virus dissemination among individuals [86]. Both mycotoxins decreased the cell-mediated viral clearance by suppressing the gene expression of IFN-γ in Peyer's Patches (PP) [82,86]. DON enhanced Th2 cytokine expression prior to and after reovirus infection, which potentiates the IgA and IgG responses to reovirus [82]. In contrast, T-2 suppressed reovirus-induced immunoglobulin responses [86]. The lack of a similar effect of Th2 cytokines by T-2 suggests inherent differences between both mycotoxins in their capacity to modulate cytokines during viral infection, although both mycotoxins belong to the class of trichothecenes [86]. Nevertheless, the intestinal clearance of reovirus was less efficient after T-2 exposure compared to DON [86]. Since reovirus infection in mice is used as a model for several enteric and respiratory viral infections in humans and other animals [81], these results could assume an impact of mycotoxins on host susceptibility to more virulent viruses.

5.2. Porcine Reproductive and Respiratory Syndrome Virus (PRRSV)

Porcine Reproductive and Respiratory Syndrome Virus (PRRSV) is an enveloped single-stranded RNA virus belonging to the family *Arteriviridae*, within the order *Nidovirales* [87]. Currently, PRRS is one of the most economic significant diseases in swine production [87,88]. The clinical symptoms, respiratory or reproductive, vary with the viral strain, the immune status of the herd, and management factors [88,89]. PRRSV is a highly infectious virus that replicates within the monocytes or macrophages with the lung being a predominant site of viral multiplication [89]. Exposure of piglets to FB1 increased the risk for PRRSV disease [90]. More severe histopathological lesions were observed when pigs were exposed to FB1 and subsequently inoculated with PRRSV. The authors suggest that FB1 causes immunosuppression, facilitating PRRSV to induce more severe lesions [89]. Given the importance of PRRSV in worldwide swine production and the frequent occurrence of fumonisins, research should be performed investigating this interaction also at lower doses of FB1.

6. Discussion

In recent years, research investigating the effects of *Fusarium* mycotoxins on the intestinal and immune functions has made substantial progress. However, only limited information is available on the interaction between mycotoxins and infectious diseases. The aforementioned literature data indicates that *Fusarium* mycotoxins may influence the animal and human host susceptibility to enteric, systemic and respiratory infectious diseases. Depending on host, pathogen and mycotoxin characteristics, exposure to *Fusarium* mycotoxins can generally exacerbate infectious diseases. On the other hand, T-2 has been shown to decrease the colonization capacity of *Salmonella* in the pig intestine. *Fusarium* mycotoxins may influence the host–pathogen interaction by negatively affecting the intestinal barrier function and the innate and adaptive immune response [9,23,26]. *Fusarium* mycotoxins affect the morphology and the barrier function of the intestinal layer [9], leading to increased translocation of different bacterial species including *Salmonella* enterica and *E. coli*, to the systemic compartment. The negative influence of these mycotoxins on the function of macrophages results in impaired phagocytosis of bacterial and fungal pathogens. However, also the adaptive immune response is targeted, demonstrated by the effect on gene expression of several cytokines, leading to an altered Th1 and Th2 response.

The economic impact of mycotoxins on animal production is generally considered to be mainly due to losses related to direct effects on animal health and trade losses related to grain rejection [91]. It is clear, however, that the indirect influence of myocotoxins on animal health, by enhancing infectious diseases, should also be taken into account. These effects, as reviewed here, occur even at low to moderate mycotoxin contamination levels of feed [8]. Some publications showed that these effects can even occur at contamination levels below the European guidance levels, suggesting that the legislation may not cover all deleterious health effects of mycotoxins.

Fusarium mycotoxins have various acute and chronic effects on humans [92]. DON could play a role in diseases such as inflammatory bowel disease (IBD) [20,93]. Taken into account conditions such as environmental, socio-economic and food production, it seems plausible that the risk for food-associated mycotoxin exposure is even higher in developing countries [94]. Besides the risk for acute mycotoxicosis in developing countries [95], results obtained in animals suggest that low to moderate concentrations of these mycotoxins could also influence human susceptibility to infectious diseases.

The effect of multi-mycotoxin contamination and of less well-known or emerging mycotoxins on the human or animal susceptibility to infectious diseases is rather unknown. Multi-mycotoxin contamination of feed is frequently occurring, raising the question on the impact on animal toxicity of this phenomenon [3]. Several in vitro and in vivo studies demonstrated an enhanced toxicity and more severe immune suppression compared to single mycotoxin contamination [96–98]. In addition, plant metabolites of mycotoxins may also be present in feed and are known as masked mycotoxins [99]. *Fusarium* fungi and infected plants may produce conjugated forms of, for instance, DON, such as 3-AcDON (3-acetylDON), 15-AcDON and DON-3G (DON-3-glucoside). Furthermore, mycotoxins can also be conjugated by certain food-processing techniques. These conjugated forms could have a direct toxic effect, or may be hydrolyzed to their precursor mycotoxin in the digestive tract of animals, resulting in higher exposure levels [100–102]. The influence of mycotoxin co-occurrence and masked mycotoxins on human and animal susceptibility to infectious diseases will be an important research question in the future.

Global warming and increasing world population of humans are further important issues. Climate changes may affect the global distribution of mycotoxigenic fungi and their mycotoxins [103,104], but also the distribution of infectious diseases [105]. Livestock farming will remain an important component of the global food supply in the future. Animal health, including the impact of mycotoxins and susceptibility to infectious diseases, will be important future topics to produce enough safe food for the entire human population.

In conclusion, *Fusarium* mycotoxins may alter the human and animal susceptibility to infectious diseases by affecting the intestinal health and the innate and adaptive immune system. Further research will be necessary to investigate the impact of mycotoxins on infectious diseases and to develop practical, economically justified, solutions to counteract mycotoxin contamination of feed and food, and its effects on human and animal health. Acknowledgments G. Antonissen was supported by a PhD fellowship from Biomin GmbH, Herzogenburg, Austria.

Acknowledgments: G. Antonissen was supported by a PhD fellowship from Biomin GmbH, Herzogenburg, Austria.

Conflicts of Interest: The authors declare no conflict of interest.

Appendix A

Table A1. Interaction between *Fusarium* mycotoxins and infectious diseases: *in vitro* approach.

Mycotoxin	Exposure dose	Exposure period	Cell line (host species)	Pathogen	Effect	Reference(s)
DON or T-2	>25 ng DON/mL or 5 ng T-2/mL; ≥0.75 µg DON/mL or ≥2.5 ng T-2/mL	24 h	undifferentiated IPEC[1]-J2; differentiated IPEC[1]-J2; (pig)	*Salmonella* Typhimurium	↑ invasion	[27,28]
DON or T-2	0.5 µg DON/mL or ≥1.0 ng T-2/mL	24 h	differentiated IPEC[1]-J2 (pig)	*Salmonella* Typhimurium	↑ translocation	[27,28]
DON or T-2	0.025 µg DON/mL or 1 ng T-2/mL	24 h	PAM[2] (pig)	*Salmonella* Typhimurium	↑ invasion	[28,41]
DON	5–50 µM (1.5–15 µg/mL)	48 h	IPEC[1]-J1 (pig)	*E. coli* (SEPEC)[3]	↑ translocation	[14]
T-2	0.001 µM	6 h	peritoneal macrophages (mouse)	*P. aeruginosa*[4]	↓ phagocytosis	[48]
T-2	0.01—0.05 µM	20 h	alveolar macrophages (rat)	*S. cerevisiae*[5]	↓ phagocytosis	[106]
T-2	0.1 µM	6 h	alveolar macrophages (rat)	*S. aureus*[6]	↓ phagocytosis	[106]
T-2	1–5 ng/mL; 2–5 ng/mL	24 h	HD-11 cell line[8] (chicken)	*A. fumigatus*[7]	↓ phagocytosis; ↑ immune response(A); ↑ germination	[80]

DON = deoxynivalenol; T-2 = T-2 toxin; [1] IPEC = Intestinal Porcine Epithelial Cell; [2] PAM = porcine alveolar macrophage; [3] septicemic *Escherichia coli*; [4] *Pseudomonas aeruginosa*; [5] *Saccharomyces cerevisiae*; [6] *Staphylococcus aureus*; [7] *Aspergillus fumigatus*; [8] chicken macrophages; (A) = increased gene expression of IL-1β, IL-6, CCLi1, CXCLi1, CXCLi2, IL-18 and IL-12β.

Table A2. The influence of *Fusarium* mycotoxins on infectious diseases in animals: *in vivo* approach.

Mycotoxin	Exposure dose	Exposure period	Animal species	Age	Pathogen	Effect: compared to negative control	Reference(s)
DON, 15-acetylDON, ZEN and fumonisins	6.5 mg DON, 0.44 mg 15-acetylDON, 0.59 mg ZEN and 0.37 mg fumonisins/kg feed	6 weeks	chicken (broiler)	1 day	E. maxima[1]	↓ percentage of CD4+ and CD8+ T cells in jejunal mucosa	[35]
DON, 15-acetylDON and ZEN	3.8 mg DON and 0.3 mg 15-acetylDON and 0.2 mg ZEN/kg feed	10 weeks	chicken (broiler)	1 day	E. acervulina[1], E. maxima[1], E. tenella[1]	↓ level of blood monocytes at end of challenge period; percentage of CD8+ T-cells not. Restored at end of recovery period; ↑ IFN-γ gene expression	[34]
DON, 15-acetylDON and ZEN	3.8 mg DON, 0.3 mg 15-acetylDON and 0.2 mg ZEN/kg feed	10 weeks	chicken (broiler)	1 day	E. acervulina[1], E. maxima[1], E. tenella[1]	↓ intestinal recovery: duodenal villus height and apparent villus surface area	[36]
DON	1 µg/mL	6 h	pig	5 weeks	Salmonella Typhimurium	synergistic ↑ gene expression IL-12, TNF-α, IL-1β, IL-8, MCP-1 and IL-6	[27]
T-2	15 and 83 µg/kg feed	23 days	pig	3 weeks	Salmonella Typhimurium	↓ colonization of the cecum	[28]
FB1 and FB2	8.6 mg FB1 and 3.2 mg FB2/kg feed	9 weeks	pig	4 weeks	Salmonella Typhimurium	synergistic transient effect digestive microbiota balance	[42]
T-2	2 mg/kg BW	2 days	chicken (broiler)	1 day	Salmonella Typhimurium	↑ mortality	[45]
T-2	1 mg/kg BW	3 weeks	mouse	5-6 weeks	Salmonella Typhimurium	↑ mortality	[46]
T-2	1 mg/kg BW	10 days	mouse	5-6 weeks	Salmonella Typhimurium	↑ bacteria-related organ lesions	[47]
T-2	2 mg/kg BW	s.a.	mouse	-	Salmonella Typhimurium	↑ mortality	[48]
DON	1 mg/L drinking water	3 weeks	mouse	7 weeks	Salmonella Enteritidis	↑ translocation to mesenteric lymph node, liver and spleen	[51]
FB1	150 mg/kg feed	6 weeks	Japanese quail	1 day	Salmonella Gallinarum	↓ clinical signs and mortality; ↓ blood lymphocyte number	[107]
FB1	0.5 mg/kg BW	6 days	pig	3 weeks	E. coli (SEPEC)[2]	↑ intestinal colonization; ↑ translocation to the mesenteric lymph node, lung, liver and spleen	[56]
FB1	1 mg/kg BW	10 days	pig	3-4 weeks	E. coli (ETEC)[3]	intestinal infection prolonged; impaired function of intestinal antigen presenting cells	[58]
fumonisins and aflatoxin	[a] 50–350 ng fumonisins/mL and 1–3 ng aflatoxin/mL	-	calf	<1 month	E. coli (STEC)[4]	↑ susceptibility to hemorrhagic enteritis	[57]
moniliformin	75–100 mg/kg feed	3 weeks	chicken (broiler)	0 day	E. coli (APEC)[5]	↓ bacterial clearance	[60]
moniliformin and FB1	100 mg moniliformin and 200 mg FB1/kg feed	3 weeks	turkey	0 day	E. coli[3] (APEC)[5]	↓ bacterial clearance	[61]
DON	4-5 mg/kg feed	3 weeks	chicken (broiler)	1 day	C. perfringens[6]	↑ number of chickens with necrotic enteritis	[66]
DON	5-10 mg/kg feed	10 weeks	channel catfish	juvenile	E. ictaluri[7]	↓ mortality	[71]
T-2	1-2 mg/kg	6 weeks	channel catfish	juvenile	E. ictaluri[7]	↑ mortality	[70]
FB1, FB2 and FB3	20 mg FB1, 3.5 mg FB2 and 1.9 mg FB3/kg feed	42 days	pig	3 days	M. hyopneumoniae[8]	↑ severity of the pathological changes	[76]
FB1	10 mg/kg feed	24 days	pig	3 days	B. bronchiseptica[9] and P. multocida[10] (type D)	↑ extent and severity of the pathological changes	[73]
FB1	0.5 mg/kg BW	7 days	pig	piglets	P. multocida[10] (type A)	↓ growth rate and ↑ coughing; ↑ total number of cells, number of macrophages and lymphocytes in BALF; ↑ gross pathological lesions and histopathological lesion of lungs	[74]
T-2	mg/mouse ≈ 3.3 mg/kg BW	20 days	mouse	adult	M. tuberculosis[11] (H37Rv-R-KM)	↑ bacterial count in spleen	[108]

Table A2. *Cont.*

Mycotoxin	Exposure dose	Exposure period	Animal species	Age	Pathogen	Effect: compared to negative control	Reference(s)
T-2	0.1 mg/mouse ≈ 3.3 mg/kg BW	20 days	mouse	adult	M. bovis 12	↓ mouse survival time	[108]
T-2	0.5 mg/kg BW	21 days	rabbit	–	A. fumigatus 13	↓ phagocytosis by alveolar macrophages	[79]
T-2	2 mg/kg BW	s.a.	mouse	–	P. aeruginosa 14	↓ phagocytosis by peritoneal macrophages ↓ viral clearance and ↑ fecal shedding↓	[48]
DON	25 mg/kg BW	s.a.	mouse	7–10 weeks	reovirus (serotype 1)	Th1 response by ↓ IFN-γ gene expression↑ intestinal IgA and ↑ Th 2 response: by ↑ IL-4, IL-6 and IL-10 gene expression	[82]
T-2	1.75 mg/kg BW	s.a.	mouse	7–10 weeks	reovirus (serotype 1)	↓ viral clearance and ↑ fecal shedding; ↓ Th1 response by ↓ IFN-γ gene expression	[86]
FB1	12 mg/kg BW	18 days	pig	1 month	PRRSV15	↑ histopathological lesions of lungs	[89]

DON = deoxynivalenol; T-2 = T-2 toxin; ZEN = zearalenone; FB1 = fumonisin B1; FB2 = fumonisin B2; FB3 = fumonisin B3; BW = bodyweight; a mycotoxin level detected in the hemorrhaged mucosa; s.a. = single administration; 1 Eimeria; 2 septicemic Escherichia coli; 3 enterotoxigenic Escherichia coli; 4 shiga toxin producing Escherichia coli; 5 avian pathogenic Escherichia coli; 6 Clostridium perfringens; 7 Edwardsiella ictaluri; 8 Mycoplasma hyopneumoniae; 9 Bordetella bronchiseptica; 10 Pasteurella multocida; 11 Mycobacterium tuberculosis; 12 Mycobacterium bovis; 13 Aspergillus fumigatus; 14 Pseudomonas aeroginosa; 15 PRRSV = Porcine Reproductive and Respiratory Syndrome Virus.

Table A3. European Union limits for foodstuffs for human consumption, feed material and finished feed for animals adapted from the European Commission Regulation No 1881/2006 [109] and the European Commission Recommendations 2006/576/EC [110] and 2013/165/EU [111].

Mycotoxin	Foodstuffs for human consumption/finished animal feed	Maximum levels (µg/kg)
	unprocessed cereals other than durum wheat, oats and maize	1250
	unprocessed durum wheat and oats	1750
	unprocessed maize, with the exception of unprocessed maize intended to be processed by wet milling	1750
	cereals intended for direct human consumption, cereal flour, bran and germ as end product marketed for direct human consumption, with the exception of foodstuffs listed in (1)	750
	pasta (dry)	750
	bread (including small bakery wares), pastries, biscuits, cereal snacks and breakfast cereals	500
	(1) processed cereal-based foods and baby foods for infants and young children	200
DON	*feed materials:* cereals and cereal products with the exception of maize by-products	8000
	maize by-products	12,000
	complementary and complete feedingstuffs: all animal species with the exception of (2)	5000
	(2) complementary and complete feedingstuffs for pigs	900
	(2) complementary and complete feedingstuffs for calves (<4 months), lambs and kids	2000

Table A3. *Cont.*

Mycotoxin	Foodstuffs for human consumption/finished animal feed	Maximum levels (µg/kg)
	unprocessed cereals other than maize	100
	unprocessed maize with the exception of unprocessed maize intended to be processed by wet milling	350
	cereals intended for direct human consumption, cereal flour, bran and germ as end product marketed for direct human consumption, with the exception of foodstuffs listed in [2]	75
	refined maize oil	400
	bread (including small bakery wares), pastries, biscuits, cereal snacks and breakfast cereals, excluding maize snacks and maize-based breakfast cereals	50
ZEN	[2] maize intended for direct human consumption, maize-based snacks and maize-bases breakfast cereals	100
	[2] processed cereal-based foods (excluding processed maize-based foods) and baby foods for infants and young children	20
	[2] processed maize-based foods for infants and young children	20
	feed materials:	
	cereals and cereal products with the exception of maize by-products	2000
	maize by-products	3000
	complementary and complete feedingstuffs:	
	complementary and complete feedingstuffs for piglets and gilts (young sows)	100
	complementary and complete feedingstuffs for sows and fattening pigs complementary and complete feedingstuffs for calves, dairy cattle, sheep (including lamb) and goats (including kids)	250
	complementary and complete feedingstuffs for calves, dairy cattle, sheep (including lamb) and goats (including kids)	500
	unprocessed maize with the exception of unprocessed maize intended to be processed by wet milling	4000
	maize intended for direct human consumption, maize-based foods for direct human consumption, with the exception of foodstuffs listed in [3]	1000
	[3] maize-based breakfast cereals and maize-based snacks	800
	[3] processed maize-based foods and baby foods for infants and young children	200
Fumonisins (sum FB1 + FB2)	*feed materials:*	
	maize and maize products	60,000
	complementary and complete feedingstuffs:	
	complementary and complete feedingstuffs for pigs, horses (*Equidae*), rabbits and pet animals	5000
	complementary and complete feedingstuffs for fish	10,000
	complementary and complete feedingstuffs for poultry, calves (<4 months), lambs and kids	20,000
	complementary and complete feedingstuffs for adult ruminants (>4 months) and mink	50,000

Table A3. *Cont.*

Mycotoxin	Foodstuffs for human consumption/finished animal feed	Maximum levels (µg/kg)
	unprocessed cereals:	
	barley (including malting barley) and maize	200
	oats (with husk)	1000
	wheat, rye and other cereals	100
	cereal grains for direct human consumption:	
	oats	200
	maize	100
	other cereals	50
Sum T-2 and HT-2	*cereal products for human consumption:*	
	oat bran and flaked oats	200
	cereal bran except oat bran, oat milling products other than oat bran and flaked oats, and maize milling products	100
	other cereal milling products	50
	breakfast cereals including formed cereal flakes	75
	bread (including small bakery wares), pastries, biscuits, cereal snacks, pasta	25
	cereal-based foods for infants and young children	15
	cereal products for feed:	
	oat milling products (husks)	2000
	other cereal products	500
	compound feed:	
	compound feed, with the exception of feed for cats	250

(DON = deoxynivalenol, ZEN= zearalenone, T-2= T-2 toxin, HT-2= HT-2 toxin, FB1 = fumonisin B1, FB2 = fumonisin B2)

References

1. Binder, E.M. Managing the risk of mycotoxins in modern feed production. *Anim. Feed Sci. Tech.* **2007**, *133*, 149–166. [CrossRef]
2. Filtenborg, O.; Frisvad, J.C.; Thrane, U. Moulds in food spoilage. *Int. J. Food Microbiol.* **1996**, *33*, 85–102. [CrossRef]
3. Placinta, C.; D'mello, J.; Macdonald, A. A review of worldwide contamination of cereal grains and animal feed with *Fusarium* mycotoxins. *Anim. Feed Sci. Tech.* **1999**, *78*, 21–37. [CrossRef]
4. D'mello, J.; Placinta, C.; Macdonald, A. *Fusarium* mycotoxins: A review of global implications for animal health, welfare and productivity. *Anim. Feed Sci. Tech.* **1999**, *80*, 183–205. [CrossRef]
5. Smith, T.K.; Diaz, G.; Swamy, H. Current Concepts in Mycotoxicoses in Swine. In *The Mycotoxin Blue Book*; Diaz, D.E., Ed.; Nottingham University Press: Nottingham, UK, 2005; pp. 235–248.
6. Devegowda, G.; Murthy, T. Mycotoxins: Their Effects in Poultry and Some Practical Solutions. In *The Mycotoxin Blue Book*; Diaz, D.E., Ed.; Nottingham University Press: Nottingham, UK, 2005; pp. 25–56.
7. Devreese, M.; de Backer, P.; Croubels, S. Overview of the most important mycotoxins for the pig and poultry husbandry. *Vlaams Diergeneeskundig Tijdschrift* **2013**, *82*, 171–180.
8. Streit, E.; Naehrer, K.; Rodrigues, I.; Schatzmayr, G. Mycotoxin occurrence in feed and feed raw materials worldwide-long term analysis with special focus on Europe and Asia. *J. Sci. Food Agric.* **2013**, *93*, 2892–2899. [CrossRef]
9. Bouhet, S.; Oswald, I.P. The effects of mycotoxins, fungal food contaminants, on the intestinal epithelial cell-derived innate immune response. *Vet. Immunol. Immun.* **2005**, *108*, 199–209. [CrossRef]
10. Oswald, I.P. Role of intestinal epithelial cells in the innate immune defence of the pig intestine. *Vet. Res.* **2006**, *37*, 359–368. [CrossRef]
11. Schenk, M.; Mueller, C. The mucosal immune system at the gastrointestinal barrier. *Best Pract. Res. CL GA* **2008**, *22*, 391–409. [CrossRef]
12. Maresca, M.; Mahfoud, R.; Garmy, N.; Fantini, J. The mycotoxin deoxynivalenol affects nutrient absorption in human intestinal epithelial cells. *J. Nutr.* **2002**, *132*, 2723–2731.
13. Sergent, T.; Parys, M.; Garsou, S.; Pussemier, L.; Schneider, Y.-J.; Larondelle, Y. Deoxynivalenol transport across human intestinal Caco-2 cells and its effects on cellular metabolism at realistic intestinal concentrations. *Toxicol. Lett.* **2006**, *164*, 167–176. [CrossRef]
14. Pinton, P.; Nougayrède, J.-P.; Del Rio, J.-C.; Moreno, C.; Marin, D.E.; Ferrier, L.; Bracarense, A.-P.; Kolf-Clauw, M.; Oswald, I.P. The food contaminant deoxynivalenol, decreases intestinal barrier permeability and reduces claudin expression. *Toxicol. Appl. Pharm.* **2009**, *237*, 41–48. [CrossRef]
15. Bouhet, S.; Hourcade, E.; Loiseau, N.; Fikry, A.; Martinez, S.; Roselli, M.; Galtier, P.; Mengheri, E.; Oswald, I.P. The mycotoxin fumonisin B1 alters the proliferation and the barrier function of porcine intestinal epithelial cells. *Toxicol. Sci.* **2004**, *77*, 165–171.
16. Awad, W.A.; Bohm, J.; Razzazi-Fazeli, E.; Zentek, J. Effects of feeding deoxynivalenol contaminated wheat on growth performance, organ weights and histological parameters of the intestine of broiler chickens. *J. Anim. Physiol. Anim. Nutr.* **2006**, *90*, 32–37. [CrossRef]
17. Yunus, A.W.; Blajet-Kosicka, A.; Kosicki, R.; Khan, M.Z.; Rehman, H.; Bohm, J. Deoxynivalenol as a contaminant of broiler feed: Intestinal development, absorptive functionality and metabolism of the mycotoxin. *Poult. Sci.* **2012**, *91*, 852–861. [CrossRef]
18. Hoerr, F.; Carlton, W.; Yagen, B. Mycotoxicosis caused by a single dose of T-2 toxin or diacetoxyscirpenol in broiler chickens. *Vet. Pathol.* **1981**, *18*, 652–664.
19. Awad, W.A.; Hess, M.; Twaruzek, M.; Grajewski, J.; Kosicki, R.; Böhm, J.; Zentek, J. The impact of the *fusarium* mycotoxin deoxynivalenol on the health and performance of broiler chickens. *Int. J. Mol. Sci.* **2011**, *12*, 7996–8012. [CrossRef]
20. Maresca, M. From the gut to the brain: Journey and pathophysiological effects of the food-associated trichothecene mycotoxin deoxynivalenol. *Toxins* **2013**, *5*, 784–820. [CrossRef]
21. Obremski, K.; Zielonka, Ł.; Gajecka, M.; Jakimiuk, E.; Bakuła, T.; Baranowski, M.; Gajecki, M. Histological estimation of the small intestine wall after administration of feed containing deoxynivalenol, T-2 toxin and zearalenone in the pig. *Pol. J. Vet. Sci.* **2008**, *11*, 339–345.

22. Obremski, K.; Gajecka, M.; Zielonka, L.; Jakimiuk, E.; Gajecki, M. Morphology and ultrastructure of small intestine mucosa in gilts with zearalenone mycotoxicosis. *Pol. J. Vet. Sci.* **2004**, *8*, 301–307.
23. Bondy, G.S.; Pestka, J.J. Immunomodulation by fungal toxins. *J. Toxicol. Env. Heal. B* **2000**, *3*, 109–143. [CrossRef]
24. Osselaere, A.; Devreese, M.; Goossens, J.; Vandenbroucke, V.; de Baere, S.; de Backer, P.; Croubels, S. Toxicokinetic study and absolute oral bioavailability of deoxynivalenol, T-2 toxin and zearalenone in broiler chickens. *Food Chem. Toxicol.* **2012**, *51*, 350–355.
25. Corrier, D. Mycotoxicosis: Mechanisms of immunosuppression. *Vet. Immunol. Immun.* **1991**, *30*, 73–87. [CrossRef]
26. Oswald, I.; Marin, D.; Bouhet, S.; Pinton, P.; Taranu, I.; Accensi, F. Immunotoxicological risk of mycotoxins for domestic animals. *Food Addit. Contam.* **2005**, *22*, 354–360. [CrossRef]
27. Vandenbroucke, V.; Croubels, S.; Martel, A.; Verbrugghe, E.; Goossens, J.; van Deun, K.; Boyen, F.; Thompson, A.; Shearer, N.; de Backer, P. The mycotoxin deoxynivalenol potentiates intestinal inflammation by *Salmonella* Typhimurium in porcine ileal loops. *PLoS One* **2011**, *6*. [CrossRef]
28. Verbrugghe, E.; Vandenbroucke, V.; Dhaenens, M.; Shearer, N.; Goossens, J.; de Saeger, S.; Eeckhout, M.; D'herde, K.; Thompson, A.; Deforce, D. T-2 toxin induced *Salmonella* Typhimurium intoxication results in decreased *Salmonella* numbers in the cecum contents of pigs, despite marked effects on *Salmonella*-host cell interactions. *Vet. Res.* **2012**, *43*, 1–18. [CrossRef]
29. Lillehoj, H.S.; Lillehoj, E.P. Avian coccidiosis. A review of acquired intestinal immunity and vaccination strategies. *Avian Dis.* **2000**, *44*, 408–425. [CrossRef]
30. Chapman, H.D.; Barta, J.R.; Blake, D.; Gruber, A.; Jenkins, M.; Smith, N.C.; Suo, X.; Tomley, F.M. A selective review of advances in coccidiosis research. *Adv. Parasitol.* **2013**, *83*, 93–171.
31. Lillehoj, H. Role of T lymphocytes and cytokines in coccidiosis. *Int. J. Parasitol.* **1998**, *28*, 1071–1081. [CrossRef]
32. Lillehoj, H.; Min, W.; Dalloul, R. Recent progress on the cytokine regulation of intestinal immune responses to Eimeria. *Poult. Sci.* **2004**, *83*, 611–623.
33. Lillehoj, H.; Kim, C.; Keeler, C.; Zhang, S. Immunogenomic approaches to study host immunity to enteric pathogens. *Poult. Sci.* **2007**, *86*, 1491–1500.
34. Girgis, G.N.; Sharif, S.; Barta, J.R.; Boermans, H.J.; Smith, T.K. Immunomodulatory effects of feed-borne *fusarium* mycotoxins in chickens infected with coccidia. *Exp. Biol. Med.* **2008**, *233*, 1411–1420. [CrossRef]
35. Girgis, G.N.; Barta, J.R.; Girish, C.K.; Karrow, N.A.; Boermans, H.J.; Smith, T.K. Effects of feed-borne *fusarium* mycotoxins and an organic mycotoxin adsorbent on immune cell dynamics in the jejunum of chickens infected with *Eimeria maxima*. *Vet. Immunol. Immun.* **2010**, *138*, 218–223. [CrossRef]
36. Girgis, G.; Barta, J.; Brash, M.; Smith, T. Morphologic changes in the intestine of broiler breeder pullets fed diets naturally contaminated with *fusarium* mycotoxins with or without coccidial challenge. *Avian Dis.* **2010**, *54*, 67–73. [CrossRef]
37. Békési, L.; Hornok, S.; Szigeti, G.; Dobos-Kovács, M.; Széll, Z.; Varga, I. Effect of F-2 and T-2 fusariotoxins on experimental *Cryptosporidium baileyi* infection in chickens. *Int. J. Parasitol.* **1997**, *27*, 1531–1536. [CrossRef]
38. Varga, I.; Ványi, A. Interaction of T-2 fusariotoxin with anticoccidial efficacy of lasalocid in chickens. *Int. J. Parasitol.* **1992**, *22*, 523–525. [CrossRef]
39. Andrews-Polymenis, H.L.; Bäumler, A.J.; McCormick, B.A.; Fang, F.C. Taming the elephant: *Salmonella* biology, pathogenesis, and prevention. *Infect. Immun.* **2010**, *78*, 2356–2369. [CrossRef]
40. Ohl, M.E.; Miller, S.I. *Salmonella*: A model for bacterial pathogenesis. *Annu. Rev. Med.* **2001**, *52*, 259–274. [CrossRef]
41. Vandenbroucke, V.; Croubels, S.; Verbrugghe, E.; Boyen, F.; De Backer, P.; Ducatelle, R.; Rychlik, I.; Haesebrouck, F.; Pasmans, F. The mycotoxin deoxynivalenol promotes uptake of *Salmonella* Typhimurium in porcine macrophages, associated with ERK1/2 induced cytoskeleton reorganization. *Vet. Res.* **2009**, *40*. [CrossRef]
42. Burel, C.; Tanguy, M.; Guerre, P.; Boilletot, E.; Cariolet, R.; Queguiner, M.; Postollec, G.; Pinton, P.; Salvat, G.; Oswald, I.P. Effect of low dose of fumonisins on pig health: Immune status, intestinal microbiota and sensitivity to *Salmonella*. *Toxins* **2013**, *5*, 841–864.

43. Skjolaas, K.; Burkey, T.; Dritz, S.; Minton, J. Effects of *Salmonella enterica* serovars Typhimurium (st) and Choleraesuis (sc) on chemokine and cytokine expression in swine ileum and jejunal epithelial cells. *Vet. Immunol. Immun.* **2006**, *111*, 199–209. [CrossRef]

44. Boyen, F.; Haesebrouck, F.; Maes, D.; van Immerseel, F.; Ducatelle, R.; Pasmans, F. Non-typhoidal *Salmonella* infections in pigs: A closer look at epidemiology, pathogenesis and control. *Vet. Microbiol.* **2008**, *130*, 1–19. [CrossRef]

45. Ziprin, R.; Elissalde, M. Effect of T-2 toxin on resistance to systemic *Salmonella* Typhimurium infection of newly hatched chickens. *Am. J. Vet. Res.* **1990**, *51*, 1869–1872.

46. Tai, J.; Pestka, J. Impaired murine resistance to *Salmonella* Typhimurium following oral exposure to the trichothecene t-2 toxin. *Food Cem. Toxicol.* **1988**, *26*, 691–698. [CrossRef]

47. Tai, J.-H.; Pestka, J. T-2 toxin impairment of murine response to *Salmonella* Typhimurium: A histopathologic assessment. *Mycopathologia* **1990**, *109*, 149–155. [CrossRef]

48. Vidal, D.; Mavet, S. *In vitro* and *in vivo* toxicity of t-2 toxin, a fusarium mycotoxin, to mouse peritoneal macrophages. *Infect. Immun.* **1989**, *57*, 2260–2264.

49. Mittrücker, H.-W.; Kaufmann, S. Immune response to infection with *Salmonella* Typhimurium in mice. *J. Leukocyte Biol.* **2000**, *67*, 457–463.

50. Tai, J.; Pestka, J. Synergistic interaction between the trichothecene T-2 toxin and *Salmonella* Typhimurium lipopolysaccharide in C3H/HeN and C3H/HeJ mice. *Toxicol. Lett.* **1988**, *44*, 191–200. [CrossRef]

51. Hara-Kudo, Y. Effects of deoxynivalenol on *Salmonella* Enteritidis infection. *Mycotoxins* **1996**, *1996*, 42, 51–56.

52. Rothkötter, H.; Sowa, E.; Pabst, R. The pig as a model of developmental immunology. *Hum. Exp. Toxicol.* **2002**, *21*, 533–536. [CrossRef]

53. Meurens, F.; Summerfield, A.; Nauwynck, H.; Saif, L.; Gerdts, V. The pig: A model for human infectious diseases. *Trends Microbiol.* **2012**, *20*, 50–57. [CrossRef]

54. Kaper, J.B.; Nataro, J.P.; Mobley, H.L. Pathogenic *Escherichia coli*. *Nat. Rev. Microbiol.* **2004**, *2*, 123–140. [CrossRef]

55. Nataro, J.P.; Kaper, J.B. Diarrheagenic *Escherichia coli*. *Clin. Microbiol. Rev.* **1998**, *11*, 142–201.

56. Oswald, I.P.; Desautels, C.; Laffitte, J.; Fournout, S.; Peres, S.Y.; Odin, M.; le Bars, P.; le Bars, J.; Fairbrother, J.M. Mycotoxin fumonisin B1 increases intestinal colonization by pathogenic escherichia coli in pigs. *Appl. Environ. Microb.* **2003**, *69*, 5870–5874. [CrossRef]

57. Baines, D.; Sumarah, M.; Kuldau, G.; Juba, J.; Mazza, A.; Masson, L. Aflatoxin, fumonisin and shiga toxin-producing *Escherichia coli* infections in calves and the effectiveness of celmanax/dairyman's choice applications to eliminate morbidity and mortality losses. *Toxins* **2013**, *5*, 1872–1895. [CrossRef]

58. Devriendt, B.; Verdonck, F.; Wache, Y.; Bimczok, D.; Oswald, I.P.; Goddeeris, B.M.; Cox, E. The food contaminant fumonisin B1 reduces the maturation of porcine CD11r1+ intestinal antigen presenting cells and antigen-specific immune responses, leading to a prolonged intestinal etec infection. *Vet. Res.* **2009**, *40*, 1–14.

59. Grenier, B.; Applegate, T.J. Modulation of intestinal functions following mycotoxin ingestion: Meta-analysis of published experiments in animals. *Toxins* **2013**, *5*, 396–430.

60. Li, Y.; Ledoux, D.; Bermudez, A.; Fritsche, K.; Rottinghaust, G. Effects of moniliformin on performance and immune function of broiler chicks. *Poult. Sci.* **2000**, *79*, 26–32.

61. Li, Y.; Ledoux, D.; Bermudez, A.; Fritsche, K.; Rottinghaus, G. The individual and combined effects of fumonisin B1 and moniliformin on performance and selected immune parameters in turkey poults. *Poult. Sci.* **2000**, *79*, 871–878.

62. Barbara, A.J.; Trinh, H.T.; Glock, R.D.; Glenn Songer, J. Necrotic enteritis-producing strains of *Clostridium perfringens* displace non-necrotic enteritis strains from the gut of chicks. *Vet. Microbiol.* **2008**, *126*, 377–382.

63. Timbermont, L.; Haesebrouck, F.; Ducatelle, R.; van Immerseel, F. Necrotic enteritis in broilers: An updated review on the pathogenesis. *Avian Pathol.* **2011**, *40*, 341–347. [CrossRef]

64. Williams, R. Intercurrent coccidiosis and necrotic enteritis of chickens: Rational, integrated disease management by maintenance of gut integrity. *Avian Pathol.* **2005**, *34*, 159–180. [CrossRef]

65. Keyburn, A.L.; Boyce, J.D.; Vaz, P.; Bannam, T.L.; Ford, M.E.; Parker, D.; di Rubbo, A.; Rood, J.I.; Moore, R.J. NetB, a new toxin that is associated with avian necrotic enteritis caused by *Clostridium perfringens*. *PLoS Pathog.* **2008**, *4*. [CrossRef]

66. Antonissen, G.; Van Immerseel, F.; Pasmans, F.; Ducatelle, R.; Haesebrouck, F.; Timbermont, L.; Verlinden, M.; Janssens, G.P.J.; Eeckhout, M.; de Saeger, S.; *et al.* Deoxynivalenol predisposes for necrotic enteritis by affecting the intestinal barrier in broilers. In Proceedings of the International Poultry Scientific Forum, Atlanta, Georgia, USA, 28–29 January 2013; pp. 9–10.

67. Crumlish, M.; Dung, T.; Turnbull, J.; Ngoc, N.; Ferguson, H. Identification of *Edwardsiella ictaluri* from diseased freshwater catfish, *Pangasius hypophthalmus* (sauvage), cultured in the mekong delta, Vietnam. *J. Fish Dis.* **2002**, *25*, 733–736. [CrossRef]

68. Ferguson, H.; Turnbull, J.; Shinn, A.; Thompson, K.; Dung, T.T.; Crumlish, M. Bacillary necrosis in farmed *Pangasius hypophthalmus* (sauvage) from the Mekong delta, Vietnam. *J. Fish Dis.* **2001**, *24*, 509–513. [CrossRef]

69. Newton, J.; Wolfe, L.; Grizzle, J.; Plumb, J. Pathology of experimental enteric septicaemia in channel catfish, *Ictalurus punctatus* (rafinesque), following immersion-exposure to *Edwardsiella ictaluri*. *J. Fish Dis.* **1989**, *12*, 335–347. [CrossRef]

70. Manning, B.B.; Terhune, J.S.; Li, M.H.; Robinson, E.H.; Wise, D.J.; Rottinghaus, G.E. Exposure to feedborne mycotoxins T-2 toxin or ochratoxin a causes increased mortality of channel catfish challenged with *Edwardsiella ictaluri*. *J. Aquat. Anim. Health* **2005**, *17*, 147–152. [CrossRef]

71. Manning, B.B.; Abbas, H.K.; Wise, D.J.; Greenway, T. The effect of feeding diets containing deoxynivalenol contaminated corn on channel catfish (*Ictalurus punctatus*) challenged with edwardsiella ictaluri. *Aquac. Res.* **2013**. [CrossRef]

72. Hooft, J.M.; Elmor, A.E.H.I.; Encarnação, P.; Bureau, D.P. Rainbow trout (*Oncorhynchus mykiss*) is extremely sensitive to the feed-borne *Fusarium* mycotoxin deoxynivalenol. *Aquaculture* **2011**, *311*, 224–232. [CrossRef]

73. Pósa, R.; Donkó, T.; Bogner, P.; Kovács, M.; Repa, I.; Magyar, T. Interaction of *Bordetella bronchiseptica*, *Pasteurella multocida*, and fumonisin B1 in the porcine respiratory tract as studied by computed tomography. *Can. J. Vet. Res.* **2011**, *75*, 176–182.

74. Halloy, D.J.; Gustin, P.G.; Bouhet, S.; Oswald, I.P. Oral exposure to culture material extract containing fumonisins predisposes swine to the development of pneumonitis caused by *Pasteurella multocida*. *Toxicology* **2005**, *213*, 34–44. [CrossRef]

75. Maes, D.; Segales, J.; Meyns, T.; Sibila, M.; Pieters, M.; Haesebrouck, F. Control of *Mycoplasma hyopneumoniae* infections in pigs. *Vet. Microbiol.* **2008**, *126*, 297–309. [CrossRef]

76. Pósa, R.; Magyar, T.; Stoev, S.; Glávits, R.; Donkó, T.; Repa, I.; Kovács, M. Use of computed tomography and histopathologic review for lung lesions produced by the interaction between Mycoplasma hyopneumoniae and fumonisin mycotoxins in pigs. *Vet. Pathol.* **2013**, *50*. [CrossRef]

77. Chanter, N.; Magyar, T.; Rutter, J.M. Interactions between *Bordetella bronchiseptica* and toxigenic *Pasteurella multocida* in atrophic rhinitis of pigs. *Res. Vet. Sci.* **1989**, *47*, 48–53.

78. Davies, R.L.; MacCorquodale, R.; Baillie, S.; Caffrey, B. Characterization and comparison of *Pasteurella multocida* strains associated with porcine pneumonia and atrophic rhinitis. *J. Med. Microbiol.* **2003**, *52*, 59–67. [CrossRef]

79. Niyo, K.; Richard, J.; Niyo, Y.; Tiffany, L. Effects of T-2 mycotoxin ingestion on phagocytosis of *Aspergillus fumigatus* conidia by rabbit alveolar macrophages and on hematologic, serum biochemical, and pathologic changes in rabbits. *Am. J. Vet. Res.* **1988**, *49*, 1766–1773.

80. Li, S.-J.; Pasmans, F.; Croubels, S.; Verbrugghe, E.; Van Waeyenberghe, L.; Yang, Z.; Haesebrouck, F.; Martel, A. T-2 toxin impairs antifungal activities of chicken macrophages against *Aspergillus fumigatus* conidia but promotes the pro-inflammatory responses. *Avian Pathol.* **2013**, *42*, 457–463. [CrossRef]

81. Nibert, M.; Furlong, D.; Fields, B. Mechanisms of viral pathogenesis. Distinct forms of reoviruses and their roles during replication in cells and host. *J. Clin. Invest.* **1991**, *88*, 727–934. [CrossRef]

82. Li, M.; Cuff, C.F.; Pestka, J. Modulation of murine host. Response to enteric reovirus infection by the trichothecene deoxynivalenol. *Toxicol. Sci.* **2005**, *87*, 134–145. [CrossRef]

83. Jones, R. Avian reovirus infections. *Rev. Sci. Tech. OIE* **2000**, *19*, 614–625.

84. Jones, R.; Kibenge, F. Reovirus-induced tenosynovitis in chickens: The effect of breed. *Avian Pathol.* **1984**, *13*, 511–528. [CrossRef]

85. Benavente, J.; Martínez-Costas, J. Avian reovirus: Structure and biology. *Virus Res.* **2007**, *123*, 105–119. [CrossRef]

86. Li, M.; Cuff, C.F.; Pestka, J.J. T-2 toxin impairment of enteric reovirus clearance in the mouse associated with suppressed immunoglobulin and IFN-γ responses. *Toxicol. Appl. Pharm.* **2006**, *214*, 318–325. [CrossRef]

87. Chand, R.J.; Trible, B.R.; Rowland, R.R. Pathogenesis of porcine reproductive and respiratory syndrome virus. *Curr. Opin. Virol.* **2012**, *2*, 256–263. [CrossRef]
88. Rowland, R.; Morrison, R. Challenges and opportunities for the control and elimination of porcine reproductive and respiratory syndrome virus. *Transbound. Emerg. Dis.* **2012**, *59*, 55–59. [CrossRef]
89. Ramos, C.M.; Martinez, E.M.; Carrasco, A.C.; Puente, J.H.L.; Quezada, F.; Perez, J.T.; Oswald, I.P.; Elvira, S.M. Experimental trial of the effect of fumonisin B and the PRRS virus in swine. *J. Anim. Vet. Adv.* **2010**, *9*, 1301–1310. [CrossRef]
90. Bane, D.P.; Neumann, E.J.; Hall, W.F.; Harlin, K.S.; Slife, R.L.N. Relationship between fumonisin contamination of feed and mystery swine disease-a case-control study. *Mycopathologia* **1992**, *117*, 121–124.
91. Wu, F. Measuring the economic impacts of *Fusarium* toxins in animal feeds. *Anim. Feed Sci. Tech.* **2007**, *137*, 363–374. [CrossRef]
92. Zain, M.E. Impact of mycotoxins on humans and animals. *J. Saudi Chem. Soc.* **2011**, *15*, 129–144. [CrossRef]
93. Maresca, M.; Fantini, J. Some food-associated mycotoxins as potential risk factors in humans predisposed to chronic intestinal inflammatory diseases. *Toxicon* **2010**, *56*, 282–294. [CrossRef]
94. Wagacha, J.; Muthomi, J. Mycotoxin problem in africa: Current status, implications to food safety and health and possible management strategies. *Int. J. Food Microbiol.* **2008**, *124*, 1–12. [CrossRef]
95. Wild, C.P.; Gong, Y.Y. Mycotoxins and human disease: A largely ignored global health issue. *Carcinogenesis* **2010**, *31*, 71–82. [CrossRef]
96. Wan, L.Y.M.; Turner, P.C.; El-Nezami, H. Individual and combined cytotoxic effects of *Fusarium* toxins (deoxynivalenol, nivalenol, zearalenone and fumonisins B1) on swine jejunal epithelial cells. *Food Chem. Toxicol.* **2013**, *57*, 276–283. [CrossRef]
97. Grenier, B.; Oswald, I. Mycotoxin co-contamination of food and feed: Meta-analysis of publications describing toxicological interactions. *World Mycotoxin J.* **2011**, *4*, 285–313. [CrossRef]
98. Grenier, B.; Loureiro-Bracarense, A.P.; Lucioli, J.; Pacheco, G.D.; Cossalter, A.M.; Moll, W.D.; Schatzmayr, G.; Oswald, I.P. Individual and combined effects of subclinical doses of deoxynivalenol and fumonisins in piglets. *Mol. Nutr. Food Res.* **2011**, *55*, 761–771. [CrossRef]
99. De Boevre, M.; di Mavungu, J.D.; Landschoot, S.; Audenaert, K.; Eeckhout, M.; Maene, P.; Haesaert, G.; De Saeger, S. Natural occurrence of mycotoxins and their masked forms in food and feed products. *World Mycotoxin J.* **2012**, *5*, 207–219. [CrossRef]
100. Nagl, V.; Schwartz, H.; Krska, R.; Moll, W.-D.; Knasmüller, S.; Ritzmann, M.; Adam, G.; Berthiller, F. Metabolism of the masked mycotoxin deoxynivalenol-3-glucoside in rats. *Toxicol. Lett.* **2012**, *213*, 367–373. [CrossRef]
101. Dall'Erta, A.; Cirlini, M.; Dall'Asta, M.; del Rio, D.; Galaverna, G.; Dall'Asta, C. Masked mycotoxins are efficiently hydrolysed by the human colonic microbiota, releasing their toxic aglycones. *Chem. Res. Toxicol.* **2013**, *26*, 305–312. [CrossRef]
102. Broekaert, N.; Devreese, M.; de Mil, T.; Fraeyman, S.; de Baere, S.; de Saeger, S.; de Backer, P.; Croubels, S. Development and validation of an LC-MS/MS method for the toxicokinetic study of deoxynivalenol and its acetylated derivatives in animal plasma. *Anal. Bioanal. Chem..* Submitted.
103. Paterson, R.R.M.; Lima, N. How will climate change affect mycotoxins in food? *Food Res. Int.* **2010**, *43*, 1902–1914. [CrossRef]
104. Magan, N.; Medina, A.; Aldred, D. Possible climate change effects on mycotoxin contamination of food crops pre-and postharvest. *Plant Pathol.* **2011**, *60*, 150–163. [CrossRef]
105. Shuman, E.K. Global climate change and infectious diseases. *N. Engl. J. Med.* **2010**, *362*, 1061–1063. [CrossRef]
106. Gerberick, G.F.; Sorenson, W.; Lewis, D. The effects of T-2 toxin on alveolar macrophage function *in vitro*. *Environ. Res.* **1984**, *33*, 246–260. [CrossRef]
107. Deshmukh, S.; Asrani, R.; Jindal, N.; Ledoux, D.; Rottinghaus, G.; Sharma, M.; Singh, S. Effects of fusarium moniliforme culture material containing known levels of fumonisin B1 on progress of *Salmonella* Gallinarum infection in Japanese quail: Clinical signs and hematologic studies. *Avian Dis.* **2005**, *49*, 274–280. [CrossRef]
108. Kanai, K.; Kondo, E. Decreased resistance to mycobacterial infection in mice fed a trichothecene compound (T-2 toxin). *Jpn. J. Med. Sci. Biol.* **1984**, *37*, 97.
109. European Commission. Commission Regulation (EC) of 19 December 2006 setting maximum levels for certain contaminants in foodstuffs no. 1881/2006. *Off. J. Eur. Union* **2006**, *L364*, 5–24.

110. European Commission. Commission Recommendation of 17 August 2006 on the presence of deoxynivalenol, zearalenone, ochratoxin A, T-2 and HT-2 and fumonisins in products intended for animal feeding (2006/576/EC). *Off. J. Eur. Union* **2006**, *L229*, 7–9.

111. European Commission. Commission Recommendation of 27 March 2013 on the presence of T-2 and HT-2 toxin in cereals and cereal products (2013/165/EU). *Off. J. Eur. Union* **2013**, *L91*, 12–15.

Section 4:
Remediation Strategies to Limit the Presence of DON

toxins

Article

Fusarium Head Blight Control and Prevention of Mycotoxin Contamination in Wheat with Botanicals and Tannic Acid

Hans-Rudolf Forrer *, Tomke Musa, Fabienne Schwab [†], Eveline Jenny, Thomas D. Bucheli, Felix E. Wettstein and Susanne Vogelgsang *

Agroscope, Institute for Sustainability Sciences, Reckenholzstrasse 191, 8046 Zurich, Switzerland; tomke.musa@agroscope.admin.ch (T.M.); eveline.jenny@agroscope.admin.ch (E.J.); thomas.bucheli@agroscope.admin.ch (T.D.B.); felix.wettstein@agroscope.admin.ch (F.E.W.)

* Correspondence: hans-rudolf.forrer@agroscope.admin.ch (H.-R.F.); susanne.vogelgsang@agroscope.admin.ch (S.V.); Tel.: +41-58-468-72-30 (H.-R.F.); +41-58-468-72-29 (S.V.)

† Present Address: Center for the Environmental Implications of Nanotechnology, Duke University, 121 Hudson Hall, Durham, NC 27708, USA; fabienne.schwab@duke.edu.

Received: 24 December 2013; in revised form: 5 February 2014; Accepted: 13 February 2014; Published: 26 February 2014

Abstract: Suspensions or solutions with 1% of Chinese galls (*Galla chinensis*, GC) or 1% of tannic acid (TA), inhibited germination of conidia or mycelium growth of *Fusarium graminearum* (FG) by 98%–100% or by 75%–80%, respectively, whereas dried bark from buckthorn (*Frangula alnus*, FA) showed no effect at this concentration. In climate chamber experiments where the wheat variety "Apogee" was artificially inoculated with FG and F. crookwellense (FCr) and treated with 5% suspensions of TA, GC and FA, the deoxynivalenol (DON) content in grains was reduced by 81%, 67% and 33%, respectively. In field experiments with two commercial wheat varieties and artificial or semi-natural inoculations, mean DON reductions of 66% (TA) and 58% (FA), respectively, were obtained. Antifungal toxicity can explain the high efficacies of TA and GC but not those of FA. The Fusarium head blight (FHB) and mycotoxin reducing effect of FA is probably due to elicitation of resistance in wheat plants. With semi-natural inoculation, a single FA application in the first half of the flowering period performed best. However, we assume that applications of FA at the end of ear emergence and a treatment, triggered by an infection period, with TA or GC during flowering, might perform better than synthetic fungicides.

Keywords: *Fusarium graminearum* (FG); antifungal; natural compound; phenolic; phytoalexin; elicitor; deoxynivalenol (DON); forecasting; organic

1. Introduction

Contamination of food and feed with mycotoxins is a major concern for growers and industry in small grain cereals and especially in maize production. Globally, *Fusarium graminearum* (FG) Schwabe (teleomorph *Gibberella zeae*) is the most prevalent Fusarium head blight (FHB) causing fungus and the main source of deoxynivalenol (DON) and zearalenone (ZEA) contamination of wheat [1], most probably because of its high genetic diversity and an increasing surface of maize cropping. The key factors of *F. graminearum* (FG) infections in wheat are maize or wheat as a previous crop, reduced or zero tillage and susceptible wheat varieties [2,3]. The reasons for increasing cases of such high risk situations are mostly of economical nature. In Switzerland, even by using the lowest susceptible wheat variety and intensive mechanical maize residue mulching treatments, the DON contamination has not often been reduced below the maximum limit of 1.25 mg kg^{-1} in unprocessed cereals [4] when wheat following grain maize was sown with reduced or zero tillage [5]. For control of FHB, the application of a fungicide during wheat anthesis (growth stage [GS] 61–69; Zadoks [6]) is often recommended. In field trials in the UK by Edwards and Godley [7], applications of the prothioconazole product Proline® at GS 65 resulted in an FHB and DON reduction in wheat of nearly 60%. In own field trials with artificial infections with *F. culmorum* and the application of triazole fungicides at GS 57 and 67, we observed DON reductions of 29% and 71%, respectively [8]. Apart from the choice of product to optimise the efficacy of a treatment, the timing of the application based on a forecasting system such as FusaProg [9] is helpful for sufficient control of FHB.

In European countries, except UK, FHB and DON contamination in organic wheat production is generally considered as less important as in conventional production [10–12]. With the trend to reduced soil conserving tillage and expanding maize cropping in organic wheat production [13,14], increasing problems must be expected and natural fungicides could help to reduce the risk of mycotoxin contaminations [15]. Botanicals and other natural antifungal agents have been shown to inhibit *Phytophthora infestans*, the cause of potato late blight [16–19], *Microdochium majus*, the cause of snow mold in wheat [20] or to be effective against food contaminating fungi such as *F. oxysporum*, *Alternaria alternata* und *Aspergillus* species [21]. With respect to choice of natural antifungal compounds, botanicals used in Chinese medicine and known for antioxidant and antimicrobial activity could be promising. For example, *Rheum palmatum* (RP) L. (Chinese rhubarb), *Frangula alnus* (FA) Mill. (buckthorn bark) and *Galla chinensis* (GC; Chinese gallnuts) are all rich in tannins and other phenolic compounds [19,20,22,23]. Tannins and tannic compounds are used in dietary and medicinal herbs with antioxidant and antimicrobial activity and have also been considered to prevent cancer [24–26]. As early as in 1913, Knudson [27] reported that tannic acid (TA; $C_{76}H_{52}O_{46}$) is even at low concentrations toxic to a large number of fungi. Furthermore, strong inhibition was also observed towards bacteria [28,29]. Antibiotic phenolic compounds were found in many plants and play constitutively or induced by elicitors a crucial role in the defense of plant diseases [30]. Examples of such induction of antibiotic phenolics in wheat are seed treatments with Chitosan®, a product based on chitin, and silicon [31,32]. Seed treatments with Chitosan® reduced seed borne FG incidence by more than 50% whereas silicon sprayed on wheat plants induced the formation of phenolics which in turn reduced powdery mildew (*Erysiphe graminis*) incidence on wheat leaves.

The main objective of this study was to evaluate the potential of TA and the botanicals GC, RP and FA to reduce head blight caused by FG and the DON concentration in integrated and organic wheat production. The particular aims were to investigate the effect of the antifungal botanicals (ABs): (1) on *in vitro* conidia germination and mycelial growth; (2) on FHB in wheat through FG and *F. crookwellense* (FCr) as well as DON and nivalenol (NIV) contamination in climate chamber; and (3) in field experiments from 2006 to 2010.

2. Results and Discussion

2.1. Isolate Specific Inhibition of Conidia Germination with TA

Based on re-isolations of *Fusarium* species from wheat grains of our first field experiment, we realised that we did not investigate four isolates of FG as originally planned, but with three FG and one FCr isolate. Since the FCr isolate is of Swiss origin and a potent NIV producer, we decided to continue with this combination of strains and both *Fusarium* species. The mean rates of conidia germination of the three FG isolates FG0407, FG0410, FG9915 and the FCr isolate FCr9703 (all single conidia) in the control treatment with water were 86%, 79%, 82% and 97%, respectively. With 0.19% Pronto® Plus (PrP), zero germination was observed from the isolates FG0407 and FG0410 and low germination rates of 7% and 1% for FG9915 and FCr9703, respectively. Corrected for a germination rate of 100% of the isolates in the control treatments, the effective concentration for 50% inhibition (EC_{50}) through TA at 0.125%, 0.25%, 0.5% and 1% was calculated. The EC_{50} with TA and the three FG isolates varied little between 0.43% and 0.46%. The EC_{50} for the FCr isolate was 0.55%, making it slightly, but nonetheless, significantly ($p < 0.05$) higher (Figure 1).

The EC_{50} with TA for the FG and FCr isolates was about 10 times higher than those reported for *Colletotrichum lindemuthianum* with extracts of cascalote (*Ceasalpinia cacalaco*) [33]. Cascalote is also a source for phenolics such as gallic and tannic acid. However, for *Microdochium majus* and suspensions with 0.1%–1.0% GC, similar efficacies as with TA and FG were observed [20].

2.2. Inhibition of Conidia Germination of FG0407 with TA and Botanicals

Based on the results and the narrow EC_{50} band for all isolates, we subsequently restricted the comparison of the conidia inhibiting effect of the three ABs and TA to one isolate, FG0407. Although it is not known whether all four isolates would react similar to FG0407, this isolate proved to be pathogenic and toxigenic in preliminary climate chamber trials. An application of PrP at 0.19% completely inhibited the germination of FG0407 (data not shown). Complete inhibition was also observed with TA and GC at 1%. Almost no effect was observed with the RP and the FA extracts at 1%. Elevated concentrations with 10% RP or FA reduced the germination rate down to about 20% or 80%, respectively (Figure 2).

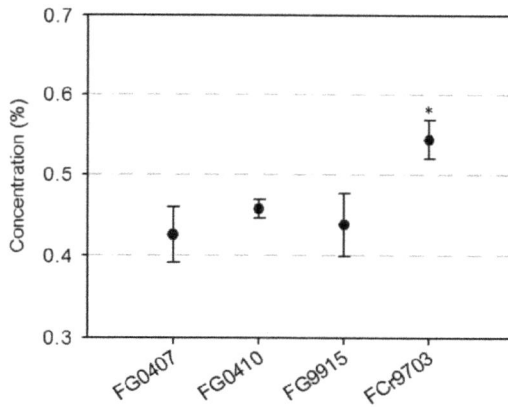

Figure 1. *In vitro* experiment: isolate specific inhibition of conidial germination with tannic acid (TA): concentration of TA needed for 50% inhibition (EC_{50}) of conidia germination from three *Fusarium graminearum* (FG) and one *F. crookwellense* (FCr) isolate/s. Mean values of the germination of each 60 conidia and the standard error of means. A "*" indicates a significant difference from the other isolates, according to a one way analysis of variance (ANOVA) and a Holm-Sidak post hoc test at $p < 0.05$.

2.3. Inhibition of Mycelial Growth of FG0407 with TA and Botanicals

Application of 2 mL of 1% TA and 1% GC suspensions in agar in Petri dishes significantly ($p < 0.001$) inhibited the mycelium growth of FG0407 by 80% or by 73%, respectively. A slight but not significant inhibition (13%) resulted with the RP treatment and almost none was observed after a treatment with FA (Figure 3).

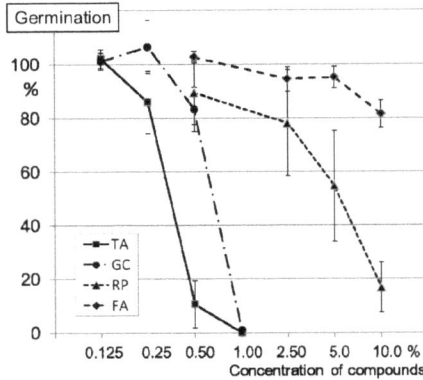

Figure 2. *In vitro* experiment: conidia germination of FG0407 with various concentrations of TA and botanicals. Solutions or aqueous extracts of TA and the botanicals *Galla chinensis* (GC), *Rheum palmatum* (RP) and *Frangula alnus* (FA) were applied. The control treatment with water was set to 100% germination. Each data point represents the relative mean of the germination of 90 conidia. Bars indicate the standard error of mean.

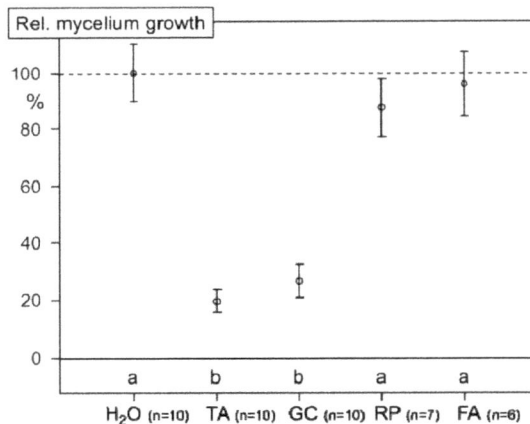

Figure 3. *In vitro* experiment: mycelial growth of FG0407 with TA and botanicals. Solutions of TA and suspensions of 1% of GC, RP and FA were applied on agar (2 mL suspension per Petri dish). The control treatment with water was set to 100% radial growth. For treatments labeled with the same letter, mean values are not statistically different ($p < 0.05$) according to a Tukey test.

In preliminary trials, we observed that agar solidification was not successful when concentrations of TA greater than 0.5% were incorporated. Therefore, all antifungal agents, including TA solutions, botanicals and PrP were applied onto the surface of the already solidified agar. The aqueous solutions were absorbed by the agar or were forced to evaporate in an air stream at room temperature. Certainly, with this procedure adapted for TA and our ABs, a direct comparison with other investigations employing agar incorporated agents is not possible. With our approach we can only estimate the EC of the agents needed to sufficiently reduce mycelial growth. Still, the primary aim of this experiment was to compare the effect of the selected agents and not to determine absolute values for toxic concentrations. If we assume that the 2 mL 1% GC poured on 20 g agar in our experiment (~45%

inhibition of FG) is distributed in the agar, the concentration corresponds approximately to those of 0.1% GC incorporated in agar for an assay with *M. majus* (~70%) [20]. Hence, we assume that the experimental set-ups are comparable. Nevertheless, a direct comparison is doubtful since different fungal species were investigated. This might explain the contrasting results with RP in the current study and those from an investigation with *P. infestans*: in our experiment with 1% RP, hardly no effect was observed, whereas in the study by Hu *et al.* [17], concentrations of 0.4% RP incorporated into the agar completely inhibited mycelial growth of *P. infestans* [17].

2.4. Climate Chamber Experiment—Reduction of Disease Severity and Mycotoxins in Artificially Inoculated Wheat

In these experiments, the effect of antifungal agents was examined on the artificially inoculated wheat cultivar "Apogee". A mixture of the same isolates as *in vitro* was used for the inoculation. Since the preparations with TA and the botanicals were acid, a tap water and acidified tap water control were used (Table 1).

Table 1. Description of treatments in climate chamber experiments with the spring wheat cultivar "Apogee" and in field experiments with the winter wheat cultivars "Runal" and "Levis". Type of experiments: A: climate chamber experiments; and B: field experiments with artificial inoculations (2006 and 2008–2010).

Treatment Nb.	Ingredient	Identifier	Application *vs.* inoculation		Concentration	pH	Type of experiment	
			before	after	%	±0.2	A	B
1	Tap water (control 1)	Water	×	-	-	7.8	×	×
2	Acidified water (c. 2)	ac-Water	×	-	-	4.0	×	-
3	Tannic acid	TA b i.	×	-	5	3.8	×	×
4	Tannic acid	TA a i.	-	×	5	3.8	×	-
5	Tannic acid	TA b+a i.	×	×	5	3.8	×	×
6	*Galla chinensis*	GC b+a i.	×	×	5	3.9	×	×
7	*Rheum palmatum*	RP b+a i.	×	×	5	5.0	×	×
8	*Frangula alnus*	FA b+a i.	×	×	5	5.2	×	×
9	Pronto® Plus	PrP b i.	×	-	0.375	8.2	×	×

b i./a i.: before/after inoculation; b+a i.: application one day before and after inoculation; pH: pH of water, suspensions with botanicals and PrP.

The artificial inoculation had a strong effect on all evaluated parameters. The best effect on disease inhibition and DON reduction was achieved with PrP resulting in a complete elimination of symptoms and 98% less DON compared with the control treatment (Figure 4). A significant effect ($p < 0.05$) was also observed with TA applications before and after inoculation (TA b+a i.) with a reduction of the disease severity from 59% down to 12% (Figure 4A). All botanicals, with the exception of FA and the treatment with TA before infection, reduced the DON contamination in the grains significantly by 67% to 80% (Figure 4B). An even stronger effect was observed for the yield and the thousand kernel weight (TKW). The "TA b+a i." treatment performed as excellent as PrP and gave a 77%–80% higher yield compared with the control treatments (Figure 4C,D). There was a strong correlation between disease, yield, TKW and DON (Table 2). In fact, the Spearman correlation coefficients between disease and DON were as high as 0.948 ($r^2 = 0.90$) (Table 2). Such strong relationships were also reported for the incidence of *F. poae* in grains and the NIV content in field experiments with artificial inoculations of different cultivars [34]. In the current study, the correlation between disease severity and NIV was substantially lower but still very high ($r^2 = 0.77$). The NIV and AcDON contents were analysed only in one of the two climate chamber experiments. In this experiment, the r^2 for DON and AcDON was 0.89, a strong indication that AcDON was produced by the FG isolates. Between NIV and DON, the r^2 was still remarkable with 0.57. Therefore, it can be assumed that NIV was produced by an important fraction of FG isolates and not only by the FCr isolate as originally expected. In these climate chamber experiments, the amount of harvested seed was only sufficient to analyse the mycotoxins but not to perform a seed health test.

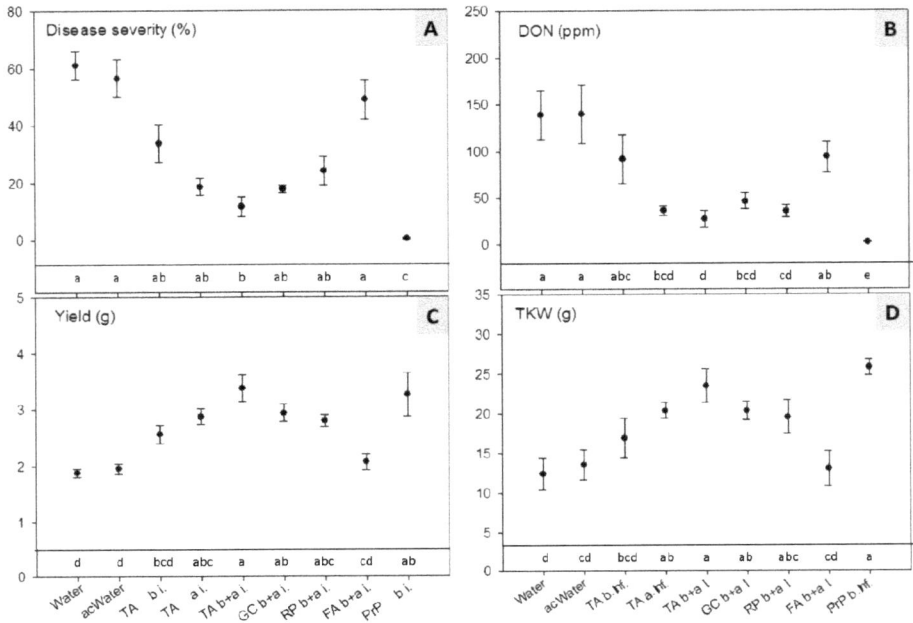

Figure 4. Climate chamber experiment: Effect of treatments with TA, GC, RP, FA and Pronto Plus® (PrP) on: (**A**) disease severity (% area with FHB symptoms); (**B**) deoxynivalenol (DON) content; (**C**) yield of three heads; and (**D**) thousand kernel weight (TKW) of the spring wheat cultivar "Apogee" after artificial inoculation with a mixture of three *F. graminearum* (FG) and one FCr isolate/s. Data are pooled results from two experiments with four replicates of each treatment. For treatments labeled with the same letter, mean values are statistically not different according to a Tukey test ($p < 0.05$). Error bars indicate the standard error of mean. Treatment abbreviations are as in Table 1.

In contrast to other studies showing good efficacies against potato late blight under field conditions [19], no significant effect was observed with buckthorn bark (FA) in the "Apogee" experiments. However, this result was not surprising since there was no effect of FA on *in vitro* FG conidia germination or on mycelial growth (Figures 2 and 3). Nevertheless, the good efficacy obtained with Chinese rhubarb (RP b+a i.), which showed little effects *in vitro*, but an *in vivo* performance that was comparable with the gallnut suspension (GC b+a i.) with strong effects *in vitro*, was not expected. One possible reason for this phenomenon could be an elicitation of defense mechanisms. For example, in experiments with RP and FA against downy mildew (*Plasmopara viticola*) on grapes (*Vitis vinifera*), a strong induction of resveratrol and other stilbenoids (polyphenols) was found [35]. In these experiments, TA and GC completely inhibited infections by *P. viticola*, but they did not induce the production of stilbenoids.

Table 2. Climate chamber experiment: Correlation between disease severities (% area with FHB symptoms), yield and mycotoxin contamination with data from two experiments with the spring wheat cultivar "Apogee". Analysis for nivalenol (NIV) and acetylated deoxynivalenol (Ac-DON) were conducted only in the second experiment. The Spearman correlation coefficients were all with $p < 0.001$; coefficients with $r > 0.8$ are indicated in bold.

Parameter	TKW ($n = 72$)	Disease severity ($n = 72$)	DON ($n = 72$)	NIV ($n = 36$)	AcDON ($n = 36$)
Yield	0.660	−0.737	−0.637	−0.560	−0.675
TKW	-	**−0.919**	**−0.932**	−0.664	**−0.814**
Disease severity	-	-	**0.948**	**0.877**	**0.907**
DON	-	-	-	0.752	**0.941**
NIV	-	-	-	-	0.786

2.5. Field Experiments with Artificial Inoculation—Reduction of FHB and Mycotoxins in Wheat

In 2006 and from 2008 to 2010, field experiments with artificial inoculations were performed with the two winter bread wheat cultivars "Runal" and "Levis". Artificial inoculations were in general conducted near mid anthesis (GS 63–65) together with applications for FHB control one day before and/or after the fungal inoculation. For logistic reasons, however, both cultivars were inoculated at the same day.

All treatments with antifungal compounds caused significant effects on FHB disease symptoms on wheat heads, yield and mycotoxin accumulation. The fungicide PrP showed the best performance and resulted in the highest yield ($p < 0.05$) with a mean increase of 37% for both cultivars compared with the water control throughout the four experimental years (Figure 5C). With the botanicals and TA, yield increases of 13%–23% (TA a+b i.) were achieved. With respect to FHB on heads (disease severity; Figure 5A), DON and NIV (Figure 5B,D), the TA treatment "TA b+a i." was as effective as PrP.

The higher yield caused by the synthetic fungicide could be explained by its broad spectrum for disease control. Apart from the excellent performance of TA and GC under field conditions, the efficacy of buckthorn bark (FA) was also highly remarkable, since preparations of FA were not only ineffective in the *in vitro* experiments but also in the climate chamber experiments with the cultivar "Apogee" (Figure 4). This finding may be explained by induction of self defense mechanisms with production of phenolic compounds as observed in the mentioned experiment on downy mildew of grapes [35] or with Chitosan® which reduced seed-borne FG by inducing the formation of phenolic acids and lignin [29]. The function of induced or constitutive phenolic compounds in disease resistance has been described for diverse pathogen-host interactions [30,36] and phenolic compounds may also play a role as resistance factors in the FHB wheat interaction [37]. The assumed lack in induction of FHB resistance in the cultivar "Apogee" through FA could be based on its overall high susceptibility to various diseases and a deleted QTL region of the chromosome 3BS FHB [38].

High correlations were observed for FHB severity on heads *versus* yield, FHB severity *versus* DON and *versus* NIV content as well as for yield *versus* DON and *versus* NIV content (Table 3). The high correlation between DON and NIV ($r = 0.893$) together with that between the FG incidence in grains and NIV indicates that NIV was also produced by the poly conidia FG isolates and not only by FCr.

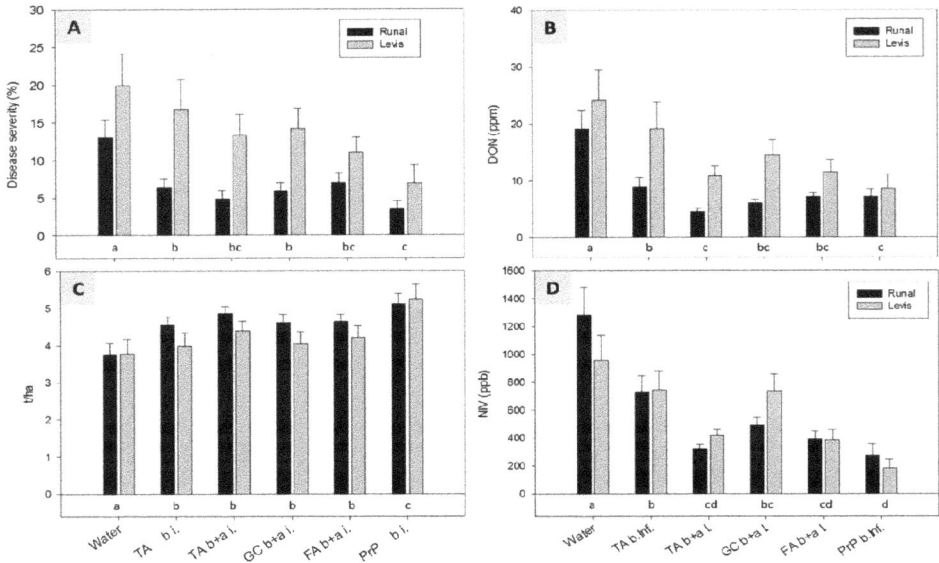

Figure 5. Field experiments with artificial inoculation: Effect of treatments with TA, GC, FA and PrP: (**A**) on the area of heads with FHB symptoms (disease severity); (**B**) on DON content; (**C**) on yield; and (**D**) on NIV content of the winter wheat cultivars "Runal" and "Levis" after artificial inoculation with a mixture of three FG and one FCr isolate/s. Data are pooled results from field experiments located at Zurich-Reckenholz in 2006 and from 2008 to 2010 (for choice of years, see Experimental Section). Bars with means of 16 values (four years and four replicates) and standard error of means. For treatments labeled by the same letter, mean values are statistically not different according to Tukey test ($p < 0.05$).

Table 3. Field experiments with artificial inoculations in 2006, 2007, 2009 and 2010 (for choice of years, see Experimental Section 3.6): Correlation between yield, FHB (% area with symptoms on heads) and mycotoxin concentrations. Spearman correlation coefficients for 178 observations and symbols for significance, NS: not significant, * = $p < 0.05$, ** = $p < 0.01$, and *** = $p < 0.001$. Coefficients with $r > 0.8$ are indicated in bold. ZEA: zearalenone.

Parameter	Disease severity	FG incidence	FCr incidence	DON	ZEA	NIV
Yield	**−0.868** ***	−0.680 ***	−0.168 *	**−0.814** ***	−0.112 NS	**−0.836** ***
Disease severity (FHB)	- -	0.703 ***	0.117 NS	**0.879** ***	0.333 ***	**0.849** ***
FG incidence	- -	- -	0.228 **	0.775 ***	0.348 ***	0.637 ***
FCr incidence	- -	- -	- -	0.254 **	0.133 NS	0.370 ***
DON	- -	- -	- -	- -	0.475 ***	**0.893** ***
ZEA	- -	- -	- -	- -	- -	0.316 ***

2.6. Field Experiments with Semi-Natural Inoculation—Reduction of FHB and Mycotoxins in Wheat

In commercial wheat fields with maize as the pre crop and zero or reduced tillage situations, DON values exceeded often the EU limit of 1.25 ppm in unprocessed cereals, even when growing varieties with low FHB susceptibility [5]. Field experiments with semi-natural inoculation represent an approach to mimic conditions as in commercial wheat fields. Such a field experiment was conducted in 2010. Our DON forecasting system FusaProg [9] was employed in order to optimise the timing of the applications. In the first half of wheat flowering, the system forecasted FG infection periods for 6 and 7 June 2010. Therefore, the treatments were applied as soon as possible on 7 and 8 June 2010. In the second half of the flowering period, no additional infection periods were registered. The fungicide PrP significantly increased the yield of "Levis" but not that of "Runal" (Figure 6A,B). For "Runal", all treatments except those with application of GC (1) and FA (2) significantly reduced the DON content (Figure 6C). For Levis, all treatments significantly reduced the DON content (Figure 6D). The good efficacy of a single application FA (FA (1); Figure 6C,D) on "Levis" and on "Runal" can again be best explained with the induction of plant defense mechanisms [35]. Differences in the efficacy between the two cultivars might be a result of different defense induction abilities of cultivars and/or effects of the growth stage as it has been reported in other investigations with the crops wheat and grapevine [29,37].

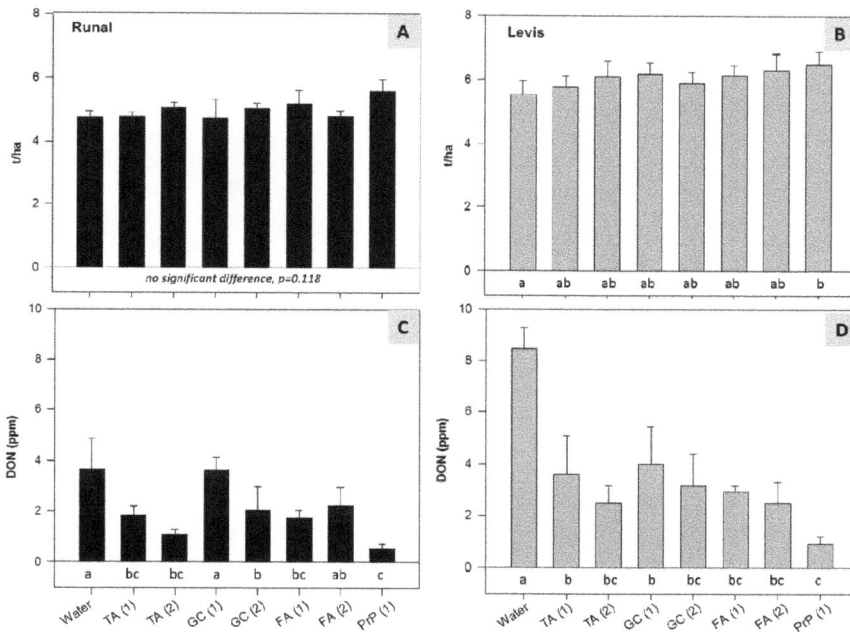

Figure 6. Field experiment with semi-natural inoculation (2010): Effect of treatments with TA, GC, FA and PrP on (**A,B**) yield and (**C,D**) DON content, in the winter wheat cultivars "Runal" and "Levis". (1): application on 7 June 2010; (2): applications on 7 and 8 June 2010. Data from field experiment at Zurich Reckenholz with infections originating from FG/FCr infected maize stubbles. Bars with mean values and standard error of means. For treatments labeled by the same letter, mean values are statistically not different according to Tukey test ($p < 0.05$).

3. Experimental Section

3.1. Fungal Isolates, Growth Conditions and Antifungal Agents

For *in vitro*, *in vivo* and field experiments, three isolates of FG, FG9915 (CBS 121291; CBS: Centraalbureau voor Schimmelcultures), FG0407 (CBS 121296), FG0410 (CBS 121292) and one isolate of FCr, FCr9703 (CBS 121293), were used. All isolates originate from wheat grains from eastern and midland regions of Switzerland. Starter cultures and fungal inoculum was produced in 9 cm diameter

Petri plates containing autoclaved (20 min, 121 °C) potato dextrose agar (PDA; 39 g L^{-1}, CM0139; Oxoid Ltd., Hampshire, UK). After inoculation, plates were incubated for 6–7 days at 19 ± 1 °C with a photoperiod of 12 h dark/12 h near-ultraviolet light. Conidia suspensions for *in vitro* tests, artificial infections in climate chambers and in the field were obtained by washing off the conidia from the cultures with deionised water with 0.125‰ Tween®20 (Sigma Aldrich, Buchs, Switzerland). Conidia concentrations were measured and adjusted to the desired concentration.

The following ABs were used: dried bark of FA (Frangulae corticis sicc norm), dried root of RP (Rhei radix pulv) both from Hänseler AG (Herisau, Switzerland), powder of GC galls (origin: Sichuan, China; purchased from Berg-Apotheke, Zurich, Switzerland) and the polyphenol TA (tannic acid powder, puriss; Sigma Aldrich). The plant material was finely ground with a centrifugal mill (mesh size 0.08 mm; Retsch ZM 200, Schieritz & Hauenstein AG, Arlesheim, Switzerland). The selection of the ABs was based on promising botanicals from earlier experiments, including investigations with *Phytophthora infestans* and *Microdochium majus* [17–20]. According to the results, GC was one of the best performing botanicals. In China, GC galls from *Rhus chinensis* Mill., induced by larvae of the aphid *Melaphis chinensis*, serve as raw material for industrial production of TA, hence, this compound was also integrated in the experiments of this study [17]. This high molecular weight polyphenolic has a broad field of applications and, for example, is used in human medicine for inhibition of melanogenesis in melanoma cells [39] and also for antidiarrheal effects [40]. The fungicide Pronto Plus® (PrP; active ingredients 25.5% spiroxamine, 13.6% tebuconazole) was integrated to compare the efficacy of the antifungal agents with good performing synthetic fungicides to control FG in wheat [7].

3.2. Isolate Specific Inhibition of Conidia Germination with TA

Microscope slides (76 mm × 26 mm) were placed in Petri plates onto moistened (2 mL sterile deionised water) filter papers (diameter 8.5 cm, Nr. 591, Schleicher & Schuell, München, Germany) and three water agar plugs (1 cm diameter) were placed on each slide. Each treatment consisted of two Petri plates, resulting in a total of six agar plugs. One droplet of 15 µL TA with concentrations of 0.125%, 0.25%, 0.5% and 1% were pipetted onto each agar plug and were allowed to evaporate for 20 min in a sterile bench. Sterile, deionised water served as the control treatment. Subsequently, for each isolate, one droplet consisting of 15 µL of the conidial suspensions with 3.3 × 10^4 conidia mL^{-1} was applied to the plugs. Petri plate lids were closed and plugs were incubated for 24 h at 10 °C and 70% relative humidity (RH) in the dark. Conidia were killed and stained with one drop of a solution with 0.19% PrP and 0.5% cotton blue. The germination rate was assessed on each plug with the aid of a light microscope (400 magnification) by determining the ratio of germinated conidia from a total of 30 conidia within three different visual fields. A conidium was assigned as germinated when the germination tube was longer than the width of the conidium. Based on the results from the five concentrations with six agar plugs, the concentration for a 50% inhibition of the germination of the conidia (EC$_{50}$) was calculated with the software ED50plus v1.0 [41]. The experiment was conducted twice and the results from the two experimental runs were pooled.

3.3. Inhibition of Conidia Germination of FG0407 with TA and Botanicals

Using a single conidium isolate of FG0407, we examined the efficacy of the ABs and the fungicide PrP. Preliminary trials showed that germinating conidia are hardly visible on agar containing powder particles. Hence, for this experiment, aqueous extracts of the botanicals were used as opposed to suspensions from botanicals in the *in vivo* and the field experiments. For each of the three botanicals, 10 g powder were suspended in 100 mL autoclaved deionised water and stirred for 3 h at ambient temperature. The aqueous extracts were subsequently filtered using fluted filters (diameter 15 cm, 520 A 1/2, Schleicher & Schuell).

TA and GC were tested at 0.125%, 0.25%, 0.5% and 1% (w/v). RP and FA were tested with aqueous extracts of 0.5%, 2.5%, 5% and 10% (w/v). The elevated concentrations for RP and FA were used, since in preliminary trials almost no effect was observed with RP and FA at 1%. Sterile, deionised water served as the control treatment. For each treatment, the conidia germination rate was determined as described above. The germination rate of the control treatment was set to 100% and the results from the other treatments were adjusted correspondingly. The experiment was repeated three times and the results from the three experimental runs were pooled.

3.4. Inhibition of Mycelial Growth of FG0407 with TA and Botanicals

Schott flasks with autoclaved PDA medium were placed in a water bath (60 °C) and while stirring amended with streptomycin sulphate (0.1 g L^{-1}). For each Petri plate, 20 g of agar was poured into Petri plates. After one to two days, 2 mL water (control), water with 1% powder of ABs or with TA were evenly spread over the agar surface with a sterilized spreader rod. Subsequently, the Petri dishes were opened and placed into a sterile bench for 2–3 h until the weight of the agar was reduced back to 20 g. For each treatment, five Petri plates were used. Using a cork borer, mycelial plugs (diameter 0.5 cm) were cut from the margin of seven day-old colonies of a single conidium isolate of FG407. For each PDA plate, one plug was placed in the center with the mycelial side facing the agar. Plates were incubated in the dark at 24 ± 1 °C and 70% RH for six days. Subsequently, radial growth was determined by measuring the diameter of the fungal colony at two positions (smallest and largest diameter) and calculating the average of both values. Data are presented as percentage growth of the aqueous controls. The experiment was conducted two times and the results were pooled.

3.5. Climate Chamber Experiment—Reduction of Disease Severity and Mycotoxins in Artificially Inoculated Wheat

For these experiments, poly conidia isolates of FG0407, FG0410, FG9915 and FCr9703 were used. Suspensions with powder of the ABs GC, RP and FA were applied to wheat plants without any filtration. The methods for cultivation of the spring wheat cultivar "Apogee" (*Triticum aestivum* L.), inoculation of fungi, disease assessment and harvest were conducted according to Vogelgsang *et al.* [42]. One modification to this protocol was that for inoculation, the pots were transferred for 24 h in a walk-in climate chamber at 19–20 °C with 90% RH, followed by 48 h with 85% RH and a dark and light period of 9 h and 15 h, respectively. In addition, the TKW of harvested grains was assessed using a "Contador" seed counter (Baumann Saatzuchtbedarf, Waldenburg, Germany). The experimental set-up was a randomised complete block design. Each treatment consisted of eight pots (handled as four replicates with each two pots) with three wheat plants in each pot. "Apogee" is a full dwarf hard red spring wheat developed for life support system in space and is highly susceptible towards FHB [38,43].

The botanical powders were suspended in water with 0.1‰ Greemax® (surfactant and adhesive emulsifier; Madora GmbH, Lörrach, Germany) [44] and stirred at room temperature for 2–3 h. All preparations and the water controls were applied with 37.5 mL per pot. TA and botanicals were applied with 5% suspensions, PrP with 0.375%. Two water controls with Greemax® (0.1‰) amended tap water, whereof one was acidulated with acetic acid to pH 4.0, were used (see Table 1). For application of the botanicals, the pots were placed on a turntable and the heads of the wheat plants were sprayed from all sides until run-off. After application, the wheat heads were allowed to dry during 2 h. Visual disease assessments were conducted three times within 7–14 days post-inoculation by counting spikelets from all heads in each pot with symptoms and estimating the percentage of the diseased area. The experiment was conducted twice and the results were pooled.

For determination and quantification of mycotoxins, liquid chromatography tandem mass spectrometry (LC-MS/MS) analysis was used. 10 g wheat flour was placed in a 100-mL flask and 40 mL of an acetonitrile/acetone/water mixture 50:25:25 (v/v) (all organic solvents from Scharlau Multisolvent, Sentmenat, Spain; water from Gradient A10, Millipore, Bedford, MA, USA) were added. Closed flasks were manually agitated until no larger wheat flour aggregates were visible. The extraction was conducted on a rotary shaker (Bühler SM-30, Hechingen, Germany) for 2 h (180 rev min^{-1}). Flour and solvent mixture were separated over a folded filter (Whatman 595½, Dassel, Germany), and the extract was collected in a 22-mL vial with a solid screw cap (Supelco, Bellefonte, PA, USA). Matrix components including lipids or fat were removed by cleaning 1 mL of extract over a 3 mL cartridge (Isolute, Uppsala, Sweden) filled with 0.15 g of celite (Fluka, 545 coarse, Buchs, Switzerland)/alox (Fluka, for chromatography, Buchs, Switzerland) 1:1 (w/w), wetted and precleaned with 2 mL of the same solvent mixture used for extraction. The resulting extract was collected in a 5-mL Reacti-vial (Supelco). After percolation of 1-mL extract, the cartridge was rinsed with 2 mL of solvent mixture and emptied by use of vacuum. The final volume of the cleaned extract (3 mL) was reduced at 40 °C to 0.4 mL with compressed air and transferred into a 2 mL high-performance liquid chromatography (HPLC)-vial. The Reacti-vial was rinsed with 0.4 mL water/methanol 90:10 during 10 s by the aid of a vortex (Scientific Industries, Bohemia, NY, USA) and transferred to the HPLC-vial as well. The final volume of the extract was adjusted with water/methanol 90:10 to 1 mL. The samples were stored in the dark at room temperature and were processed within 48 h. The LC-MS/MS analysis was performed on a Varian 1200-L system (Varian Inc., Walnut Creek, CA, USA). The analytes DON, NIV, acetylated-deoxynivalenol (Ac-DON: sum of 3-Ac- DON and 15-Ac-DON (all from R-Biopharm, Darmstadt, Germany) and ZEA (Sigma-Aldrich, St. Louis, MO, USA) were separated on a Polaris C18-A column (50 mm × 2.0 mm, 3 μm; Varian Inc., Walnut Creek, CA, USA). Operating the mass spectrometer in negative atmospheric pressure chemical ionization (APCI) mode, the analytes NIV, DON, Ac-DON and ZEA were detected with the next elution gradient: 0 min: 5% B (95% A); 1 min: 5% B; 4 min: 30% B; 5 min: 100% B; 12.5 min: 100% B; 13 min: 5% B; 20 min: 5% B. Eluent A consisted of water/methanol 95:5 (v/v) and eluent B of water/methanol 5:95 (v/v); both were buffered with 5 mL 1 M ammonium acetate per Liter (Fluka, Puriss P.A., Buchs, Switzerland). Each analyte was detected with two transitions (qualifier and quantifier) in multiple reactions monitoring (MRM). Analyte identification was confirmed using chromatographic retention time, correct mass of the mother ion, correct mass of the two daughter ions and agreement of the ratio of qualifier to quantifier with the calibration ($\pm$10%). For quantification, the method of matrix matching calibration was implemented to correct for eventual ion suppression. Recoveries for low (0.5 mg kg^{-1}) and high (2 mg kg^{-1}) spiked blank samples (n = 4) were between 86%–126% and 78%–107%, respectively. Method precision was in the range of 2%–12%, whereas instrument precision was between 2% and 10%. The limit of quantification for DON ranged between 0.09 mg kg^{-1} and 0.126 mg kg^{-1}, for ZEA between 0.008 mg kg^{-1} and 0.021 mg kg^{-1} and for NIV between 0.058 mg kg^{-1} and 0.185 mg kg^{-1}, depending on the sample series analyzed.

3.6. Field Experiments with Artificial Inoculation—Reduction of FHB and Mycotoxins in Wheat

Field experiments were conducted from the harvest years 2006–2010 with the two Swiss winter wheat cultivars "Runal" and "Levis". According to the Swiss national catalogue, the resistance for FHB of these bread wheat cultivars is considered to be "medium" and "medium to poor", respectively. The field experiments were carried out on the experimental farm of the Research Station Agroscope in Zurich-Reckenholz, Switzerland. The experiments comprised seven treatments (Table 1) with four replications each, using a Latin square design. The size of individual plots was 6.5 m × 2.6 m. Each plot was divided in two subplots for the two varieties sown in bands. Husbandry management was standard for the farm, except that no fungicides were applied.

For inoculation of the wheat plants, a suspension with a mixture of the four poly conidia isolates, FG0407, FG0410, FG9915 and FCr9703, was used. The suspension contained 2×10^5 conidia mL^{-1} with equal amounts of each isolate and 0.125‰ Tween 20. The conidia mixture and a water control with 0.125‰ Tween 20 were applied from both sides along the field with a volume of 750 L ha^{-1} using a knapsack sprayer (width 1.5 m, 3 bar, Birchmeier M125, Birchmeier Sprühtechnik AG, Stetten, Switzerland). The inoculation was conducted at mid-anthesis (GS 63–65). The date for this growth stage varied considerably throughout the years: For example, in 2006, this stage was reached on 12 June, whereas in 2007 it was reached on 21 May. Wheat was inoculated from both sides along the plots, directing the spray towards wheat heads. Water and suspensions with ABs (GC, RP, FA) and

TA were applied one day before the fungal inoculation and depending on weather conditions, again one or two days after the inoculation. For the application of TA, PrP and water, spray nozzles from TeeJet® XR11002 (4.0 bar, 450 L ha^{-1}) and for GC, RP and FA, spray nozzles from Floodjet® (1.5 bar, 450 L ha^{-1}) were used. Weather data were obtained from the MeteoSwiss operated weather station located at Zurich-Reckenholz.

Visual disease assessment from 4×10 randomly selected wheat heads within an individual plot was conducted in the field during two occasions between 14 and 25 days post-inoculation by counting spikelets with typical FHB symptoms and estimating the percentage of the diseased head area. Plots were combine-harvested when the cultivars reached GS 92 (caryopsis hard). Processing of the harvested samples, the procedure of a seed health test to determine the percentage incidence of FHB causing species and the preparation of samples for the analysis of toxins was done according to Vogelgsang *et al.* [42]. The method for the analysis of toxins was conducted as described above.

In 2007, there were no treatments with GC and RP in our field experiment with artificial infections and due to technical problems, the inoculation of wheat during anthesis was not possible. Hence, due to the resulting lack of orthogonally to the other years, the data of this experiment were not used for the analysis.

To determine the correlation of the field data with artificial inoculations, the results of 2007 were used but not those of 2008, because in 2008, the overall incidence by *Fusarium* species in wheat grains was close to 100% and hence, no seed health test was conducted.

3.7. Field Experiments with Semi-Natural Inoculation—Reduction of FHB and Mycotoxins in Wheat

To evaluate the performance of the ABs under conditions that mimic cropping of wheat without tillage after maize, *Fusarium* infected maize stubbles were applied in experimental field plots in 2010 after wheat emergence. The stubbles were inoculated in mid-November 2009 with FG/FCr suspensions of 1×10^6 conidia mL^{-1}, and stored in plastic boxes permeable to air in a greenhouse with an average temperature of 10 °C. The incubated stubbles were distributed in the field plots at the end of November 2009 (2–3 maize stalk pieces at about 0.5 kg m^{-2}). Based on the indication of FG infection periods by FusaProg [9], the treatments were applied on 7 June or 7 and 8 June 2010, respectively (Figure 6). With the exception of the inoculation, all procedures and assessments of parameters were conducted as in the field experiment with artificial inoculations.

3.8. Statistical Analysis

For all experiments except for the one with semi-natural inoculations, results from the experimental runs were pooled in case of equal variances. In case of a failed normality or variance tests, data were arcsine, log or square root transformed before analysis of variance (ANOVA). Apart from one-way ANOVAs analysing the effect of one treatment factor only, two-way and three-way ANOVAs were also conducted for experiments where other factors than the botanicals were important (e.g., wheat cultivar and year). When the overall effect of the tested factor was significant in ANOVA, an all-pairwise multiple comparison procedure according to Holm-Sidak ($\alpha = 0.05$) was employed. In order to specify the differences between treatments or years, a Tukey post hoc test ($\alpha = 0.05$) was used. The variances of data from field experiments with artificial inoculations were not equal due to substantial year effects. Nevertheless, we conducted a three way ANOVA and a Tukey test since our experiments were orthogonal and the *p* value for the factor treatment was <0.0001 for all four criteria. In addition, separate mean values for the cultivars "Runal" and "Levis" and the corresponding standard error of means were calculated. For these ANOVAs, the open source "R" software version 3.0.1 (16 May 2013) was utilized. For calculation of Spearman correlations coefficients and for plotting of graphs from untransformed data, SigmaPlot version 11.0 (Systat Software) was used.

Toxins **2014**, 6, 830–849

4. Conclusions

The overall aim of this study was to investigate possibilities to control FHB in wheat with substances having no negative effects on humans, animals and the environment. The obtained results prove that TA and the botanicals, GC and FA can substantially reduce FHB severity and mycotoxin contents under field conditions. In several experiments, the efficacy was even close to that observed with a synthetic fungicide.

In contrast to TA and GC, FA showed almost no fungal toxic effects *in vitro*. However, FA, being not effective *in vitro*, demonstrated great field performance in reducing the DON content in kernels by up to 71% under semi-natural inoculation conditions. The effectiveness of FA can best be explained by resistance inducing effects.

With a better understanding of the interactions between FA and wheat, such as the type and the dynamic of potentially induced compounds, the efficacy could be improved further. In addition, an optimized application strategy could be developed. The elicitor FA might be applied at the end of ear emergence. Subsequently, by using a forecasting system, which predicts FHB infection by FG and DON contamination, a second treatment with an antifungal product including TA or GC could be applied during flowering. With such an approach, it might be possible to obtain FHB and DON reductions as good as those from the best commercial fungicides.

Certainly, FHB in small-grain cereals can only be controlled with an integrated approach, employing crop rotation, tillage and proper choice of cultivars. However, data from our field experiments suggest that TA, GC and FA do have a high potential and thus could provide an excellent contribution to the production of safe small-grain cereals with acceptable toxin contents in low-input farming systems.

Acknowledgments: We are, in particular, grateful for the extensive and excellent technical contribution by Andreas Hecker in the laboratory, in climate chambers and in the field, and express to our late friend and colleague our deep thanks. We also greatly appreciate the valuable support by Keqiang Cao and Tongle Hu (Agricultural University of Hebei, Baoding, Hebei, China) and Heinz Krebs (Agroscope) in the acquisition and formulation of the botanicals. We also thank Irene Bänziger for excellent technical assistance and the field group of Agroscope at Zurich Reckenholz for great support in maintaining the field sites.

Conflicts of Interest: The authors declare no conflict of interest.

References

1. Gale, L.R. Population Biology of *Fusarium* Species Causing Head Blight of Grain Crops. In *Fusarium Head Blight of Wheat and Barley*; Leonard, K.J., Bushnell, W.R., Eds.; American Phytopathological Society (APS) Press: St. Paul, MN, USA, 2003; pp. 120–143.

2. Dill-Macky, R.; Jones, R.K. The effect of previous crop residues and tillage on Fusarium head blight of wheat. *Plant Dis.* **2000**, *84*, 71–76. [CrossRef]

3. Wegulo, S. Factors influencing deoxynivalenol accumulation in small grain cereals. *Toxins* **2012**, *4*, 1157–1180. [CrossRef]

4. European Commission. Commission Regulation (EC) No 1126/2007 of 28 September 2007 amending Regulation (EC) No 1881/2006 setting maximum levels for certain contaminants in foodstuffs as regards *Fusarium* toxins in maize and maize products. *Off. J. Eur. Union.* 2007, L255, pp. 14–17. Available online: http://eur-lex.europa.eu/LexUriServ/LexUriServ.do?uri=OJ:L:2007:255:0014:0017:EN:PDF (accessed on 20 February 2011).

5. Vogelgsang, S.; Hecker, A.; Musa, T.; Dorn, B.; Forrer, H.R. On-farm experiments over 5 years in a grain maize/winter wheat rotation: Effect of maize residue treatments on *Fusarium graminearum* infection and deoxynivalenol contamination in wheat. *Mycotoxin Res.* **2011**, *27*, 81–96. [CrossRef]

6. Zadoks, J.C.; Chang, T.T.; Konzak, C.F. A decimal code for the growth stages of cereals (maize, sorghum, forage grasses and dicotyledonous crops). *Weed Res.* **1974**, *14*, 415–421. [CrossRef]

7. Edwards, S.G.; Godley, N.P. Reduction of *Fusarium* head blight and deoxynivalenol in wheat with early fungicide applications of prothioconazole. *Food Addit. Contam. A* **2010**, *27*, 629–635. [CrossRef]

8. Forrer, H.-R.; Hecker, A.; Külling, C.; Kessler, P.; Jenny, E.; Krebs, H. Effect of fungicides on fusaria of wheat. *Agrarforschung* **2000**, *7*, 258–263. (in German).
9. Musa, T.; Hecker, A.; Vogelgsang, S.; Forrer, H.R. Forecasting of Fusarium head blight and deoxynivalenol content in winter wheat with FusaProg. *EPPO Bull.* **2007**, *37*, 283–289. [CrossRef]
10. Edwards, S.G. Fusarium mycotoxin content of uk organic and conventional wheat. *Food Addit. Contam. A* **2009**, *26*, 496–506. [CrossRef]
11. Bernhoft, A.; Clasen, P.; Kristoffersen, A.; Torp, M. Less *Fusarium* infestation and mycotoxin contamination in organic than in conventional cereals. *Food Addit. Contam. A* **2010**, *27*, 842–852. [CrossRef]
12. Birzele, B.; Meier, A.; Hindorf, H.; Krämer, J.; Dehne, H.-W. Epidemiology of *Fusarium* infection and deoxynivalenol content in winter wheat in the Rhineland, Germany. *Eur. J. Plant Pathol.* **2002**, *108*, 667–673. [CrossRef]
13. Dierauer, H.; Böhler, D. *Direct Seeding of Maize in Organic Farming*; Research Institute of Organic Agriculture: Frick, Switzerland, 2012; pp. 1–12.
14. Peigné, J. Is conservation tillage suitable for organic farming? A review. *Soil Use Manag.* **2007**, *23*, 129–124. [CrossRef]
15. Bernhoft, A.; Torp, M.; Clasen, P.E.; Løes, A.K.; Kristoffersen, A.B. Influence of agronomic and climatic factors on *Fusarium* infestation and mycotoxin contamination of cereals in Norway. *Food Addit. Contam. A* **2012**, *29*, 1129–1140. [CrossRef]
16. Dorn, B.; Musa, T.; Krebs, H.; Fried, P.M.; Forrer, H.R. Control of late blight in organic potato production: Evaluation of copper-free preparations under field, growth chamber and laboratory conditions. *Eur. J. Plant Pathol.* **2007**, *119*, 217–240. [CrossRef]
17. Hu, T.; Wang, S.; Cao, K.; Forrer, H.-R. Inhibitory effects of several Chinese medicinal herbs against *Phytophthora infestans*. *ISHS Acta Hortic.* **2009**, *834*, 205–210.
18. Bassin, S.; Forrer, H. Field screening of copper free fungicides against potato late blight. *Agrarforschung* **2001**, *8*, 124–129. (in German).
19. Krebs, H.; Musa, T.; Forrer, H.-R. Control of Potato Late Blight with Extracts and Suspensions of Buckthorn Bark. In Proceedings of the Zwischen Tradition und Globalisierung-9 Wissenschaftstagung Ökologischer Landbau, Universität Hohenheim, Stuttgart, Germany, 20–23 March 2007. (in German).
20. Vogelgsang, S.; Bänziger, I.; Krebs, H.; Legro, R.J.; Sanchez-Sava, V.; Forrer, H.-R. Control of *Microdochium majus* in winter wheat with botanicals—From laboratory to the field. *Plant Pathol.* **2013**, *62*, 1020–1029. [CrossRef]
21. Kumar, A.; Shukla, R.; Singh, P.; Prasad, C.S.; Dubey, N.K. Assessment of *Thymus vulgaris* L. essential oil as a safe botanical preservative against post harvest fungal infestation of food commodities. *Innov. Food Sci. Emerg. Technol.* **2008**, *9*, 575–580. [CrossRef]
22. Tian, F.; Li, B.; Ji, B.P.; Zhang, G.Z.; Luo, Y.C. Identification and structure-activity relationship of gallotannins separated from *Galla chinensis*. *Food Sci. Technol.* **2009**, *42*, 1289–1295.
23. Maleš, Ž.; Kremer, D.; Randić, Z.; Randić, M.; Pilepić, K.; Bojić, M. Quantitative analysis of glucofrangulins and phenolic compounds in Croatian *Rhamnus* and *Frangula* species. *Acta Biol. Cracoviensia Ser. Bot.* **2010**, *52*, 108–113.
24. Huang, W.-Y.; Cai, Y.-Z.; Zhang, Y. Natural phenolic compounds from medicinal herbs and dietary plants: Potential use for cancer prevention. *Nutr. Cancer* **2009**, *62*, 1–20.
25. Field, J.A.; Lettinga, G. Toxicity of tannic compounds to microorganisms. *Basic Life Sci.* **1992**, *59*, 673–692.
26. Ahn, Y.J.; Lee, H.S.; Oh, H.S.; Kim, H.T.; Lee, Y.H. Antifungal activity and mode of action of Galla rhois-derived phenolics against phytopathogenic fungi. *Pestic. Biochem. Physiol.* **2005**, *81*, 105–112. [CrossRef]
27. Knudson, L. Tannic acid fermentation. I. *J. Biol. Chem.* **1913**, *14*, 159–184.
28. Henis, Y.; Tagari, H.; Volcani, R. Effect of water extracts of carob pods, tannic acid, and their derivatives on the morphology and growth of microorganisms. *Appl. Microbiol. Biotechnol.* **1964**, *12*, 204–209.
29. McKeehen, J.D.; Busch, R.H.; Fulcher, R.G. Evaluation of wheat (*Triticum aestivum* L.) phenolic acids during grain development and their contribution to *Fusarium* resistance. *J. Agric. Food Chem.* **1999**, *47*, 1476–1482. [CrossRef]
30. Nicholson, R.L.; Hammerschmidt, R. Phenolic compounds and their role in disease resistance. *Annu. Rev. Phytopathol.* **1992**, *30*, 369–389. [CrossRef]

31. Bhaskara Reddy, M.V.; Arul, J.; Angers, P.; Couture, L. Chitosan treatment of wheat seeds induces resistance to *Fusarium graminearum* and improves seed quality. *J. Agric. Food Chem.* **1999**, *47*, 1208–1216. [CrossRef]

32. Rémus-Borel, W.; Menzies, J.G.; Bélanger, R.R. Silicon induces antifungal compounds in powdery mildew-infected wheat. *Physiol. Mol. Plant Pathol.* **2005**, *66*, 108–115. [CrossRef]

33. Veloz-García, R.; Marín-Martínez, R.; Veloz-Rodríguez, R.; Rodríguez-Guerra, R.; Torres-Pacheco, I.; González-Chavira, M.M.; Anaya-López, J.L.; Guevara-Olvera, L.; Feregrino-Pérez, A.A.; Loarca-Piña, G.; *et al.* Antimicrobial activities of cascalote (*Caesalpinia cacalaco*) phenolics-containing extract against fungus *Colletotrichum lindemuthianum.* *Ind. Crop. Prod.* **2010**, *31*, 134–138. [CrossRef]

34. Vogelgsang, S.; Sulyok, M.; Bänziger, I.; Krska, R.; Schuhmacher, R.; Forrer, H.R. Effect of fungal strain and cereal substrate on the *in vitro* mycotoxin production by *Fusarium poae* and *Fusarium avenaceum.* *Food Addit. Contam.* **2008**, *25*, 745–757. [CrossRef]

35. Gindro, K.G.; Godard, S.; De Groote, I.; Viret, O.; Forrer, H.-R.; Dorn, B. Is it possible to induce grapevine defence mechanisms? A new method to evaluate the potential of elicitors. *Rev. Suisse Vitic. Arboric. Hortic.* **2007**, *39*, 377–383.

36. Kessmann, H.; Staub, T.; Hofmann, C.; Maetzke, T.; Herzog, J.; Ward, E.; Uknes, S.; Ryals, J. Induction of systemic acquired disease resistance in plants by chemicals. *Annu. Rev. Phytopathol.* **1994**, *32*, 439–459. [CrossRef]

37. Siranidou, E.; Kang, Z.; Buchenauer, H. Studies on symptom development, phenolic compounds and morphological defence responses in wheat cultivars differing in resistance to *Fusarium* head blight. *J. Phytopathol.* **2002**, *150*, 200–208. [CrossRef]

38. Mackintosh, C.A.; Garvin, D.F.; Radmer, L.E.; Heinen, S.J.; Muehlbauer, G.J. A model wheat cultivar for transformation to improve resistance to Fusarium Head Blight. *Plant Cell Rep.* **2006**, *25*, 313–319. [CrossRef]

39. Chen, K.-S.; Hsiao, Y.-C.; Kuo, D.-Y.; Chou, M.-C.; Chu, S.-C.; Hsieh, Y.-S.; Lin, T.-H. Tannic acid-induced apoptosis and -enhanced sensitivity to arsenic trioxide in human leukemia HL-60 cells. *Leuk. Res.* **2009**, *33*, 297–307. [CrossRef]

40. Chen, J.C.; Ho, T.Y.; Chang, Y.S.; Wu, S.L.; Hsiang, C.Y. Anti-diarrheal effect of Galla Chinensis on the *Escherichia coli* heat-labile enterotoxin and ganglioside interaction. *J. Ethnopharmacol.* **2006**, *103*, 385–391. [CrossRef]

41. Vargas, M.H. ED50plus (v1.0): Software to Create and Analyze Dose-Response Curves, 2000. Available online: http://www.softlookup.com/display.asp?-id=2972 (accessed on 20 February 2014).

42. Vogelgsang, S.; Sulyok, M.; Hecker, A.; Jenny, E.; Krska, R.; Schuhmacher, R.; Forrer, H.R. Toxigenicity and pathogenicity of *Fusarium poae* and *Fusarium avenaceum* on wheat. *Eur. J. Plant Pathol.* **2008**, *122*, 265–276. [CrossRef]

43. Bugbee, B.; Koerner, G. Yield comparisons and unique characteristics of the dwarf wheat cultivar 'USU-Apogee'. *Adv. Space Res.* **1997**, *20*, 1891–1894. [CrossRef]

44. Pfeiffer, B.; Alt, S.; Schulz, C.; Hein, B.; Kollar, A. Investigations on Alternative Substances for Control of Apple Scab—Results from Conidia Germinating Tests and Experiments with Plant Extracts. In Proceedings of the 11th International Conference on Cultivation Technique and Phytopathological Problems in Organic Fruit-Growing, Weinsberg, Germany, 3–5 Febuary 2004; pp. 101–107.

Article

Effects of Bread Making and Wheat Germ Addition on the Natural Deoxynivalenol Content in Bread

Isabel Giménez [1], Jesús Blesa [2], Marta Herrera [1] and Agustín Ariño [1,*]

[1] Veterinary Faculty, University of Zaragoza, Zaragoza 50013, Spain; gimenezi@unizar.es (I.G.); herremar@unizar.es (M.H.)

[2] Faculty of Pharmacy, University of Valencia, Burjassot 46100, Valencia, Spain; jesus.blesa@uv.es

[*] Correspondence: aarino@unizar.es; Tel.: +34-876-554-131; Fax: +34-976-761-612

Received: 30 October 2013; in revised form: 14 January 2014; Accepted: 15 January 2014; Published: 21 January 2014

Abstract: Deoxynivalenol (DON, vomitoxin) is a type-B trichothecene mycotoxin produced by several field fungi such as *Fusarium graminearum* and *Fusarium culmorum* and known to have various toxic effects. This study investigated the effect of the bread making process on the stability of DON in common bread and wheat germ-enriched bread using naturally contaminated ingredients at the level of 560 μg/kg. The concentration of DON and its evolution during bread making were determined by immunoaffinity column cleanup followed by liquid chromatography with diode array detection (HPLC-DAD). During the bread making process, DON was reduced by 2.1% after fermentation and dropped by 7.1% after baking, reaching a maximum reduction of 19.8% in the crust as compared with a decrease of 5.6% in the crumb. The addition of 15% wheat germ to the dough did not affect DON stability during bread making, showing an apparent increase of 3.5% after fermentation and a reduction by 10.2% after baking.

Keywords: deoxynivalenol; bread making; wheat germ

---

1. Introduction

Wheat bread is a staple food prepared by baking a dough of flour and water usually leavened with yeast, which is widely consumed around the world [1]. In Spain, the mean consumption of bread accounts for 86 g/day [2]. Wheat germ is a component of wheat kernel with high nutritional value for the concentration of α-tocopherol (vitamin E), vitamins of group B, dietary fiber, polyunsaturated fats, proteins of high nutritive value, minerals and phytochemicals (*i.e.*, flavonoids). Consequently, wheat germ has been used as a flavoring ingredient in the manufacture of enriched breads available in the marketplace, increasing the nutritional value as well as extending shelf-life due to the natural content of organic acids and antifungal compounds such as lectin wheat germ agglutinin [3,4]. For all these reasons wheat germ and its derivatives are attractive and promising functional ingredients.

Deoxynivalenol (DON, vomitoxin) is a type-B trichothecene mycotoxin produced by several field fungi, including *Fusarium graminearum* and *Fusarium culmorum*, that cause a wide range of toxic effects in animal and humans [5,6]. Among the trichothecenes DON is the most frequently occurring toxin, and is found worldwide, particularly in cereal crops such as wheat and their products like flour, bread and germ [7–9]. To reduce the dietary exposure to DON, maximum limits have been set in flour (750 μg/kg) and bread (500 μg/kg) by the European legislation [10], and a temporary tolerable daily intake (TDI) of 1 μg/kg body weight was established.

The bread making process consists of three major stages: mixing, fermentation and baking. The fermentation and baking conditions vary considerably throughout the world, resulting in different effects on DON levels in final baked bread [11]. Bakery processing has been reported to reduce overall DON contamination [12–15], while others suggested that DON is highly stable during this process [16,17]. Similarly, Samar *et al.* [13] reported reductions in DON content during the fermentation phase, whereas Valle-Algarra *et al.* [14] did not observe any changes and Young *et al.* [18] even showed an increase of DON in the leavened products. These discrepancies may be due to several reasons such as the activity of baker's yeast, which may produce a reduction of DON levels attributed to mycotoxin degradation or yeast absorption [13]. Then, the addition of wheat germ to the bread dough recipe may reduce the degradation rate of DON by affecting the baker's yeast, as wheat germ is known to have natural antifungal compounds. In a recent review, it is concluded that the description of DON behavior during the bread making process is very difficult, since complex physico-chemical modifications occur during the process [19].

In summary, results of bread making studies on the stability of DON have been conflicting, and the effect of wheat germ addition on the mycotoxin level during bread making has not been studied. Therefore, the aims of the present work were to evaluate the stability of DON during bread making and to estimate the effect of wheat germ addition on the DON levels during the fermentation and baking phases of the bread making process.

2. Results and Discussion

The analytical method used for DON quantification in bread products, based on water extraction, immunoaffinity column cleanup and high performance liquid chromatography (HPLC) coupled with diode array (DAD) detection, was successfully validated down to 70 µg/kg. The method provided good recoveries for DON of 96.2%, and the study of intra-day precision in terms of repeatability obtained RSDr values of 4.5%, in accordance with the validation criteria [20].

Results of DON reduction during the different stages of bread making process are shown in Table 1. In common bread (Figure 1a), DON level in the staring material (560 µg/kg on a dry matter basis) was negligibly reduced by 2.1% during the fermentation step at 30 °C for 90 min and lowered by 7.1% after baking at 190 °C during 20 min. Evolution of DON levels in wheat germ-enriched bread (Figure 1b) showed an apparent increase by 3.5% after fermentation (from 560 to 580 µg/kg on a dry matter basis) followed by a reduction of 10.2% after baking. Therefore, the bread making process resulted in low reduction rates for DON in both bread types, which were non-significant as compared to the initial levels ($P > 0.05$). Consequently, the addition of 15% wheat germ did not exert any noticeable effect on the stability of DON during yeast fermentation and baking.

Extra care should be exercised regarding the compliance with maximum limits established by Commission Regulation (EC) No. 1881/2006, depending whether analytical results are expressed on a fresh basis or on a dry matter basis. Actually, the DON level in common bread was 356 µg/kg on a fresh basis, below the maximum limit set at 500 µg/kg, but amounted to 520 µg/kg when calculated on a dry matter basis (moisture content was 31.5%) (Table 1).

The distribution pattern of DON in the crumb and the crust also evolved in a similar manner in common and enriched bread. Thus, DON reduction in finished common bread was 5.6% and 19.8% in the crumb and the crust, respectively, amounting to 7.1% and 17.9% in enriched bread. The greatest reductions of DON in both bread types were observed in the crust, the outer part of bread which supports the highest temperatures, although the differences were not significant ($P > 0.05$).

Table 1. Effect of fermentation and baking on the reduction of deoxynivalenol (DON) levels in common bread and enriched bread with 15% wheat germ.

Sample	Common bread		Wheat-germ enriched bread	
	DON μg/kg[a]	% loss	DON μg/kg	% loss
Dough before fermentation	560 ± 54	-	560 ± 62	-
Dough after fermentation	548 ± 56	2.1	580 ± 30	+3.5
Baked bread	520 ± 10	7.1	503 ± 25	10.2
Bread crumb	529 ± 74	5.6	520 ± 26	7.1
Bread crust	449 ± 45	19.8	460 ± 4	17.9

Note: [a]: results expressed on a dry matter basis, as mean ± standard deviation (n = 2 assays).

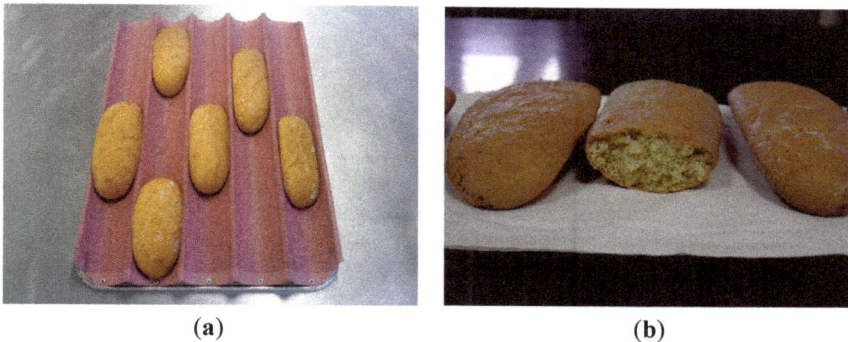

(a)   (b)

Figure 1. (a) Common bread and (b) enriched bread with 15% wheat germ.

According to the literature review, the main factors affecting the variability of the fate of DON during the bread making process include the preparation of the batter (ingredients and additives, mixing time), the fermentation step (yeast, incubation temperature and time), and the heating operation (oven-type, baking temperature and time). Some discrepancy of DON retention during baking bread may also result from uncertainty of the analytical method for DON. Thus, several studies have reported varying reductions in DON levels during bread making. Previous research showed DON overall reduction rates of 38%–44% [12], 48% [14], and 33%–58% [15]. On the other hand, other researchers reported that DON is highly stable during bread making [16,17]. These discrepancies may be due in large part to the analytical methods used, the concentration and source of toxin (natural *vs.* spiked), and the experimental conditions employed [21,22]. Thus, it has been reported that the ingredients used [23], the oven technology (commercial or home-made) [19,24] and the fermentation and baking conditions [14] influence on the reduction of DON level observed during bread making. On the other hand, Sugita-Konishi [17] reported that the DON level in flour was not reduced by bread making but that rather the biological toxicity was significantly reduced as determined by cytotoxicity bioassay.

Bakery processing has been reported to produce a thermal degradation of DON during bread making. For instance, the average reduction in DON concentration after baking (70 min at 195–235 °C) was 47.2% for bread baked in an industrial oven and 48.7% for bread baked in a log fire oven [25]. As reviewed by Kushiro [22], during baking or heating, DON is partially degraded to DON-related chemicals. Likewise, it is suggested that some DON reductions may be due to binding or the inability to extract the toxin from the matrix using current analytical techniques. Our results showed DON reductions up to 10.2% by baking at 190 °C for 20 min, which increased up to 19.8% in the crust that reached higher temperatures. Numanoglu *et al.* [26] indicated that the temperatures recorded in the crust and crumb of maize bread during baking were 100 °C and 150 °C, respectively, and thermal degradation of DON only initiated at 150 °C.

DON reduction during bread making may occur not only in the bakery due to thermal decomposition, but also during the fermentation step. Thus, yeast fermentation has been reported to produce a reduction of DON levels, which was attributed to mycotoxin degradation or yeast absorption. The fermentation stage during bread making produced DON reductions between 0% and 25% in dough fermented at 30 °C for 60 min, whereas there was a maximum 56% reduction when the dough was fermented at 50 °C [13]. In our study, yeast fermentation at 30 °C for 90 min produced minor changes of DON levels in the fermented dough, even an apparent increase of 3.5% in wheat germ-enriched bread. This is in agreement with Young *et al.* [18] who observed an increase in DON levels in yeast doughnuts, explained by the contamination of wheat with a DON precursor, which was possibly converted to DON by the active yeast.

3. Experimental Section

3.1. Bread Making and Sampling

According to a typical baker recipe, two different types of bread were manufactured at the pilot plant of the Veterinary Faculty of Zaragoza (Spain): common bread and enriched bread with 15% of wheat germ. For each type of bread (common and enriched) there were two bread making assays carried out in different times. Each assay consisted of 1.0 kg of dough that yielded 6 bread pieces of approximately 150 g. Therefore, for each bread type there were a total of two analytical samples ($n = 2$ assays) at each processing step: dough before fermentation (50 g each), dough after fermentation (50 g each), baked bread (50 g pooled from two bread pieces), bread crumb (50 g pooled) and bread crust (50 g pooled). For statistical purposes, each result in Table 1 is the mean ± standard deviation ($n = 2$ assays). The bread making process showed good repeatability between the two assays as indicated by adequate values of relative standard deviation (%RSD) calculated from Table 1. The mean %RSD was 7.3% and individual RSD values depending on sample type ranged from 0.9% to 14%.

The bread formulas were as follows: (i) common bread made with 1000 g wheat flour, 550 mL tap water, 16 g sodium chloride, and 40 g of commercial baker's yeast (*Saccharomyces cerevisiae*), and (ii) enriched bread made with 850 g wheat flour, 150 g wheat germ, 550 mL tap water, 16 g sodium chloride, and 40 g of commercial baker's yeast. A continuous high-speed mixer was used to prepare the batter by adding 450 mL pre-warmed water (37–40 °C) and mixing for 3 min, followed by the addition of baker's yeast dissolved in 100 mL pre-warmed water and mixing for another 8 min. Dough was settled at room temperature for 15 min, and then fermentation was carried out during 90 min in a camera at 30 °C and 80% relative humidity. Finally, the raised dough was baked in an oven at 190 °C for 20 min to obtain the bread.

DON levels were determined at each step of the bread making: dough before fermentation, dough after fermentation, and baked bread. For each bread several slices were cut and samples were taken from the crumb and the crust. The wheat flour and wheat germ used as main ingredients were naturally contaminated with DON and produced a concentration of 560 µg DON/kg dry matter in the starting material (dough before fermentation).

3.2. Reagents and Apparatus for DON Analysis

HPLC grade acetonitrile and methanol were purchased from Lab-Scan (Dublin, Ireland). Ultrapure water was obtained from a Milli-Q Plus apparatus from Millipore (Milford, MA, USA). The immunoaffinity columns DonStar™ were supplied by Romer Labs (Union, MO, USA). Deoxynivalenol standard solution at 100 µg/mL in acetonitrile was provided by Sigma (St. Louis, MO, USA) and stored at −21 °C. Reagents for phosphate-buffered saline solution (PBS) were obtained from Panreac (Barcelona, Spain).

The LC system consisted of an Agilent Technologies (Santa Clara, CA, USA) 1100 high performance liquid chromatograph coupled to an Agilent diode array detector (DAD) at 220 nm for the determination of DON. The LC column was Ace 5 C18, 250 mm × 4.6 mm, 5 µm particle size (Advanced Chromatography Technologies, Aberdeen, UK). The mobile phase consisted of a mixture of water/acetonitrile/methanol (90:5:5, v/v/v) at a flow rate of 1.0 mL/min.

3.3. Analysis of DON in Dough and Bread Samples

For the determination of DON in dough and bread samples, five grams were extracted with 40 mL of Milli-Q water using an Ultraturrax homogenizer for 3 min. After the extraction, the solution was filtered with Whatman #4 filter paper, and the extract collected for further cleanup by DonStar™ immunoaffinity columns according to the manufacturer's instructions. Briefly, 2 mL of the filtered extract were passed through the column at a flow-rate of 1 drop/second, followed by a washing with 5 mL PBS pH 7. DON was then eluted with 3 × 0.5 mL methanol and collected in a clean vial. The eluted extract was evaporated to dryness under nitrogen stream at 50 °C and redissolved with 400 µL of the HPLC mobile phase. One hundred µL was injected into the LC-DAD system by full loop injection system. Quantification of DON was performed by measuring peak areas at DON retention time, and comparing them with the relevant calibration curve. To facilitate the comparison of the DON levels in the different samples taken during the bread making process, results were expressed on a dry matter basis. For this purpose, a sample of 5 g was heated in an oven at 130 °C for 2 h. After cooling, the moisture content was determined by weight loss and the mycotoxin content expressed on a dry matter basis according to the formula:

$$\frac{DON\,as\,sample\,basis\,(\mu g\,/\,kg)}{\%\,sample\,dry\,matter} = \frac{DON\,as\,dry\,matter\,basis\,(\mu g\,/\,kg)}{100\%\,dry\,matter} \tag{1}$$

Average moisture contents were 41.8% for dough before fermentation, 41.7% dough after fermentation, 31.5% bread, 40.7% crumb, and 16.4% crust.

3.4. Statistical Analyses

Results from mycotoxin analyses were subjected to descriptive and comparative statistics according to Sachs [27]. DON levels determined before and after each processing step of the bread baking assays and between different sample types (bread crumb, crust) were statistically evaluated by the one-way analysis of variance (ANOVA at $P = 0.05$) procedure using the statistical software StatView SE + Graphics (Abacus Concepts, Berkeley, CA, USA). Fisher PLSD test was used when significant differences were found among means.

4. Conclusions

Mycotoxins are considered to be very stable molecules but because of their toxic effects, information about their stability in thermal processes and potential inactivation procedures is needed. This study concluded that DON was stable during the bread making process and remained stable in the enriched bread with 15% wheat germ. The fermentation step (30 °C for 90 min) and the oven baking (190 °C for 20 min) resulted in non-significant losses of DON from the initial dough. Quite different results concerning the fate of DON during bread making have been reported in the literature to date. The discrepancies between the findings reported in these different studies may be due in large part to the analytical methods employed, the differences in the experimental conditions employed and the concentration and source of toxin.

Acknowledgments: This research was supported by the Spanish MICINN (Project AGL2011-26808), the Government of Aragón (Grupo de Investigación Consolidado A01), and the European Social Fund.

Conflicts of Interest: The authors declare no conflict of interest.

References

1. Dewettinck, K.; Van Bockstaele, F.; Kühne, B.; Van de Walle, D.; Courtens, T.M.; Gellynck, X. Nutritional value of bread: Influence of processing, food interaction and consumer perception. *J. Cereal Sci.* **2008**, *48*, 243–257. [CrossRef]

2. Ministerio de Agricultura, Alimentación y Medio Ambiente (MAGRAMA). Food consumption panel. Available online: http://www.magrama.gob.es/es/alimentacion/temas/ (accessed on 11 October 2013).

3. Ciopraga, J.; Gozia, O.; Tudor, R.; Brezuica, L.; Doyle, R.J. *Fusarium* sp. growth inhibition by wheat germ agglutinin. *Biochim. Biophys. Acta* **1999**, *1428*, 424–432.

4. Rizzello, C.G.; Cassone, A.; Coda, R.; Gobbetti, M. Antifungal activity of sourdough fermented wheat germ used as an ingredient for bread making. *Food Chem.* **2011**, *127*, 952–959. [CrossRef]

5. Hussein, H.S.; Brasel, J.M. Toxicity, metabolism and impact of mycotoxins on humans and animals. *Toxicology* **2001**, *167*, 101–134. [CrossRef]

6. Maresca, M. From the gut to the brain: Journey and pathophysiological effects of the food-associated trichothecene mycotoxin deoxynivalenol. *Toxins* **2013**, *5*, 784–820. [CrossRef]

7. EU-SCOOP (Scientific Cooperation Task 3.2.10 of the European Commission). *Collection of Ocurrence Data of Fusarium Toxins in Food and Assessment of Dietary Intake by the Population of EU Member States*; Directorate General Health and Consumer Protection, European Commission: Brussels, Belgium, 2003.

8. Food and Agriculture Organization of the United Nations; World Health Organization. *Proposed Draft Maximum Levels for Deoxynivalenol (DON) and Its Acetylated Derivatives in Cereals and Cereal-Based Products*; CX/CF 11/5/6; FAO/WHO: Rome, Italy, 2011. Available online: ftp://ftp.fao.org/codex/meetings/cccf/cccf6/cf06_09e.pdf (accessed on 11 October 2013).

9. Giménez, I.; Herrera, M.; Escobar, J.; Ferruz, E.; Lorán, S.; Herrera, A.; Ariño, A. Distribution of deoxynivalenol and zearalenone in milled germ during wheat milling and analysis of toxin levels in wheat germ and wheat germ oil. *Food Control* **2013**, *34*, 268–273. [CrossRef]

10. Commission Regulation (EC). No. 1881/2006 of 19 December 2006 setting maximum levels for certain contaminants in foodstuffs. *Off. J. Eur. Union* **2006**, *L364*, 5–24.

11. Hazel, C.M.; Patel, S. Influence of processing on trichothecenes levels. *Toxicol. Lett.* **2004**, *153*, 51–59. [CrossRef]

12. Neira, M.S.; Pacin, A.M.; Martinez, E.J.; Moltó, G.; Resnik, S.L. The effects of bakery processing on natural deoxynivalenol contamination. *Int. J. Food Microbiol.* **1997**, *37*, 21–25. [CrossRef]

13. Samar, M.M.; Neira, M.S.; Resnik, S.L.; Pacin, A. Effect of fermentation on naturally occurring deoxynivalenol (DON) in Argentinean bread processing technology. *Food Addit. Contam.* **2001**, *18*, 1004–1010. [CrossRef]

14. Valle-Algarra, F.M.; Mateo, E.M.; Medina, A.; Mateo, F.; Gimeno-Adelantado, J.V.; Jiménez, M. Changes in Ochratoxin A and type B trichothecenes contained in wheat flour during dough fermentation and breadmaking. *Food Addit. Contam.* **2009**, *26*, 896–906. [CrossRef]

15. Pacin, A.; Ciancio Bovier, E.; Cano, G.; Taglieri, D.; Hernandez Pezzani, C. Effect of bread making process on wheat flour contaminated by deoxynivalenol and exposure estimate. *Food Control* **2010**, *21*, 492–495. [CrossRef]

16. Scott, P.M.; Kanhere, S.R.; Dexter, J.E.; Brennan, P.W.; Trenholm, H.L. Distribution of the trichothecene mycotoxin deoxynivalenol (vomitoxin) during the milling of naturally contaminated hard red spring wheat and its fate in baked products. *Food Addit. Contam.* **1984**, *1*, 313–323. [CrossRef]

17. Sugita-Konishi, Y.; Park, B.J.; Kobayashi-Hattori, K.; Tanaka, T.; Chonan, T.; Yoshikawa, K.; Kumagai, S. Effect of cooking process on the deoxynivalenol content and its subsequent cytotoxicity in wheat products. *Biosci. Biotech. Bioch.* **2006**, *70*, 1764–1768. [CrossRef]

18. Young, J.C.; Fulcher, R.G.; Hayhoe, J.H.; Scott, P.M.; Dexter, J.E. Effect of milling and baking on deoxynivalenol (vomitoxin) content of eastern Canadian wheats. *J. Agric. Food Chem.* **1984**, *32*, 659–664. [CrossRef]

19. Bergamini, E.; Catellani, D.; Dall'asta, C.; Galaverna, G.; Dossena, A.; Marchelli, R.; Suman, M. Fate of *Fusarium* mycotoxins in the cereal product supply chain: The deoxynivalenol (DON) case within industrial bread-making technology. *Food Addit. Contam.* **2010**, *27*, 677–687. [CrossRef]

20. Commission Regulation (EC). No. 401/2006 of 23 February 2006 laying down the methods of sampling and analysis for the official control of the levels of mycotoxins in foodstuffs. *Off. J. Eur. Union* **2006**, *L70*, 12–34.

21. Bullerman, L.B.; Bianchini, A. Stability of mycotoxins during food processing. *Int. J. Food Microbiol.* **2007**, *119*, 140–146. [CrossRef]

22. Kushiro, M. Effects of milling and cooking processes on the deoxynivalenol content in wheat. *Int. J. Mol. Sci.* **2008**, *9*, 2127–2145. [CrossRef]

23. Boyacioğlu, D.; Hettiarachchy, N.S.; Dappolonia, B.L. Additives affect deoxynivalenol (vomitoxin) flour during breadbaking. *J. Food Sci.* **1993**, *56*, 416–418.

24. Scudamore, K.A.; Hazel, C.M.; Patel, S.; Scriven, F. Deoxynivalenol and other *Fusarium* mycotoxins in bread, cake and biscuits produced from UK-grown wheat under commercial and pilot scale conditions. *Food Addit. Contam.* **2009**, *26*, 1191–1198. [CrossRef]

25. Lesnik, M.; Cencic, A.; Vajs, S.; Simoncic, A. Milling and bread making techniques significantly affect the mycotoxin (Deoxinivalenol and Nivalenol) level in bread. *Acta Aliment.* **2008**, *37*, 471–483. [CrossRef]

26. Numanoglu, E.; Gökmen, V.; Uygun, U.; Koksel, H. Thermal degradation of deoxynivalenol during maize bread baking. *Food Addit. Contam.* **2012**, *29*, 423–430.

27. Sachs, L. *Applied Statistics. A Handbook of Techniques*; Springer-Verlag New York Inc.: New York, NY, USA, 1982.

Article

Stereoselective Luche Reduction of Deoxynivalenol and Three of Its Acetylated Derivatives at C8

Philipp Fruhmann [1],*, Christian Hametner [1], Hannes Mikula [1], Gerhard Adam [2], Rudolf Krska [3] and Johannes Fröhlich [1]

[1] Institute of Applied Synthetic Chemistry, Vienna University of Technology, Getreidemarkt 9, Vienna 1060, Austria; christian.hametner@tuwien.ac.at (C.H.); hannes.mikula@tuwien.ac.at (H.M.); johannes.froehlich@tuwien.ac.at (J.F.)

[2] Department of Applied Genetics and Cell Biology, University of Natural Resources and Life Sciences, Vienna (BOKU), Konrad Lorenz Str. 24, Tulln 3430, Austria; gerhard.adam@boku.ac.at

[3] Department for Agrobiotechnology (IFA-Tulln), Center for Analytical Chemistry, University of Natural Resources and Life Sciences, Vienna (BOKU), Konrad Lorenz Str. 20, Tulln 3430, Austria; rudolf.krska@boku.ac.at

* Correspondence: philipp.fruhmann@tuwien.ac.at; Tel.: +43-1-58801-163 (ext. 737); Fax: +43-1-58801-154 (ext. 99)

Received: 6 November 2013; in revised form: 29 December 2013; Accepted: 31 December 2013; Published: 10 January 2014

Abstract: The trichothecene mycotoxin deoxynivalenol (DON) is a well known and common contaminant in food and feed. Acetylated derivatives and other biosynthetic precursors can occur together with the main toxin. A key biosynthetic step towards DON involves an oxidation of the 8-OH group of 7,8-dihydroxycalonectrin. Since analytical standards for the intermediates are not available and these intermediates are therefore rarely studied, we aimed for a synthetic method to invert this reaction, making a series of calonectrin-derived precursors accessible. We did this by developing an efficient protocol for stereoselective Luche reduction at C8. This method was used to access 3,7,8,15-tetrahydroxyscirpene, 3-deacetyl-7,8-dihydroxycalonectrin, 15-deacetyl-7,8-dihydroxycalonectrin and 7,8-dihydroxycalonectrin, which were characterized using several NMR techniques. Beside the development of a method which could basically be used for all type B trichothecenes, we opened a synthetic route towards different acetylated calonectrins.

Keywords: dihydroxycalonectrins; trichothecenes; DON; Luche reduction; scirpene

1. Introduction

Trichothecene based mycotoxins are common and widespread contaminants in food and feed. They can affect human and animal health by causing several acute and chronic symptoms after uptake [1]. Toxicity studies showed that the primary mode of action of trichothecenes is inhibition of eukaryotic protein synthesis [2–4]. When consumed in contaminated foods, trichothecenes can act as neurotoxin, immunosuppressive or nephrotoxin [5–7].

The polarity of trichothecenes depends on the number of hydroxyl groups (ranging from 1 to 5) and their esterification status. So far, more than 200 different trichothecenes have been reported, which are produced by different genera such as *Fusarium*, *Myrothecium*, *Stachybotrys*, *Cephalosporium*, *Trichoderma* and *Trichothecium* [8]. Generally, they are divided into four different groups (A–D), all containing a tricyclic 12,13-epoxytrichothec-9-ene core structure [9]. Type A toxins are compounds with at least one hydroxyl group, either no oxygen substituent at C8 or an ester functionality. In contrast, type B trichothecenes feature a carbonyl functionality at C8. The most prominent toxins of the two classes mentioned above are T-2 toxin (type A), nivalenol (NIV, type B) and deoxynivalenol

(DON, type B). From the biosynthetic point of view type A and type B trichothecenes are derived from the same precursors (Scheme 1) and most of the responsible genes are already described in the literature [10].

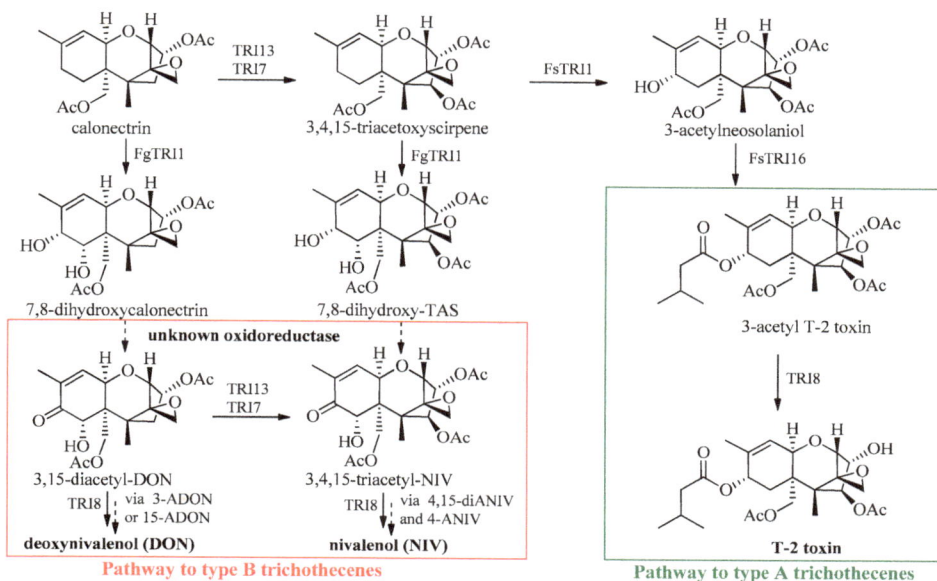

Scheme 1. .Biosynthetic pathway of Type A and Type B trichothecenes (modified from [10]). *F. graminearum* (Fg) and *F. sporotrichioides* (Fs) gene products catalyzing the reactions are indicated.

The oxidoreductase step leading to the C8 keto group is still uncharacterized. Our recent findings [11] showed the occurrence of pentahydroxyscirpene (PHS), a NIV derivative with an OH function at C8 which was isolated in substantial amounts (10%–20%) together with NIV after fermentation and also in artificially inoculated wheat. Other results showed the occurrence of 7,8-dihydroxycalonectrin [12–15] alone or in combination with 15-deacetyl-7,8-dihydroxycalonectrin [16] or 3,7,8,15-tetrahydroxyscirpene [17]. Since these compounds are all supposed to be toxin precursors, the findings suggest that there are even more acetylated forms and derivatives of trichothecene precursors that might also be present in contaminated grain, but which are not studied due to lack of standards. Therefore, we have focused on developing a reliable method to make this substance class accessible.

2. Synthetic Approach

2.1. General Aspects

The most obvious synthetic way to access 7,8-dihydroxycalonectrin derivatives and other trichothecenes with a C8 hydroxy group is the selective reduction of the C8 carbonyl function. One common characteristic of naturally occurring compounds like trichothecenes is a very well defined stereochemistry with a lot of chiral information. For example, DON has seven stereogenic centers, which influences the synthetic introduction of a new stereocenter in a very unpredictable way. Introducing a new hydroxyl group in position 8 would therefore lead to a mixture of 3,7,8,15-tetrahydroxyscirpene with its undesired isomer (Scheme 2).

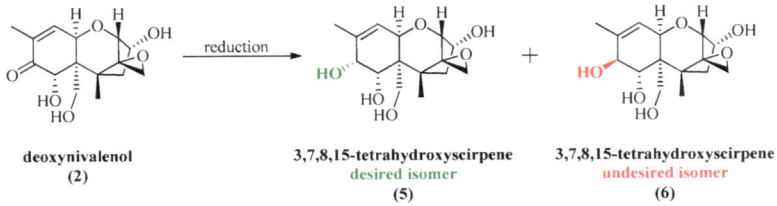

Scheme 2. Desired and undesired isomer of 3,7,8,15-tetrahydroxyscirpene via reduction of deoxynivalenol (DON).

To avoid formation of the undesired isomer and suppress side reactions of the hydride reagent, we choose to utilize the Luche reduction to achieve a very selective method for the reduction of DON.

2.2. Luche Reduction

The Luche reduction [18–20] can be used to convert α, β-unsaturated ketones into allylic alcohols using $CeCl_3$, $NaBH_4$ and methanol as solvent. The main role of cerium(III) chloride is to coordinate with the alcohol solvent, making its proton more acidic which can then be abstracted by the carbonyl oxygen of the ketone. After addition of $NaBH_4$ it also reacts with the cerium activated alcohol forming a series of alkoxyborohydrides (Scheme 3). Since alkoxyborohydrides are "hard reagents" their formation results in a selective 1,2-hydride attack on the protonated carbonyl group which leads to the desired reaction. In addition the use of $CeCl_3$ offers the possibility of coordinating [21] with the C7 hydroxy group, which results in a shielding of the backside of deoxynivalenol (Scheme 3). Due to this shielding effect, the desired frontside hydride attack should be more favored. The last point which might have an influence on the reaction, is the oxygen in the pyran ring of DON. Since this oxygen is located next to the reaction site, it is possible, that a coordination between the activated borohydride species and the oxygen is taking place (Scheme 3), which would lead to an even more targeted reduction.

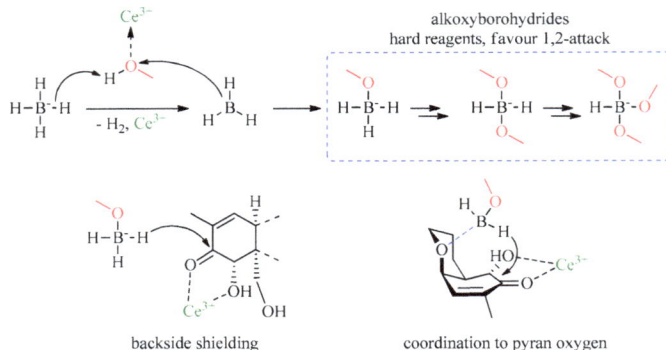

Scheme 3. Mechanism of alkoxyborohydride formation, shielding and coordination.

3. Results and Discussion

3.1. Method Development

Since DON and its acetylated derivatives are very expensive we decided to use (+)-carvone (Scheme 4) as a cheap and readily available mimic for method evaluation. Although the steric information is quite simple compared with deoxynivalenol, it provides a good model for method optimization. Therefore, the lowest possible concentration of all involved reagents as well as the estimation of possible side reactions was examined, in order to avoid needless loss of starting material. In addition **13** was used as stability test for the epoxy group, **14** as stability estimation for acetyl groups and **15** as mimic for the coordination effect in deoxynivalenol.

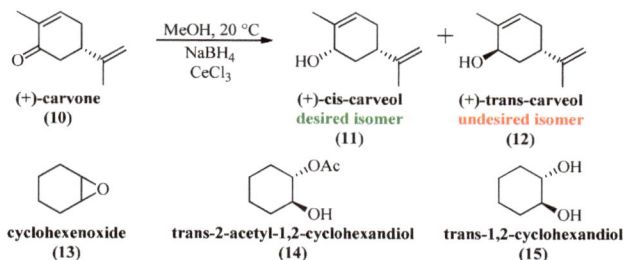

Scheme 4. (+)-Carvone and its possible reaction products as model for deoxynivalenol.

For these experiments we used CeCl$_3$·7H$_2$O as it is significantly better soluble in methanol than the anhydrous form. After some tests with varying equivalents, the reactions with 1 equivalent NaBH$_4$ and 0.5 equivalents CeCl$_3$·7H$_2$O for 1 equivalent carvone turned out to be superior to other ratios, allowing full stereoselective conversion without too much decomposition. With these conditions we were able to perform the reaction from carvone to (+)-*cis*-carveol in 30 min with an isolated yield of 92%. All attempts with lower reagent concentrations ended up with bad conversions (7%–13% of starting material left after 24 h), more byproducts (>10%) and elongated reaction times (>24 h). Stability testing regarding the epoxy group as well as the stability of the acetyl mimic under reaction conditions revealed slow decomposition over time which rose to 16% within 2 h.

3.2. Synthesis of 3-ADON, 15-ADON, 3,15-diADON and Their Reduction

3-ADON was obtained from BOKU, Dept. for Agrarbiotechnology (IFA-Tulln) and was spectroscopically pure (NMR, Figure S1) in accordance with existing literature [22]. The synthetic route towards the different acetylated DON derivatives was carried out by deprotection of 3-ADON via NaOMe/MeOH followed by Steglich esterification [23] to obtain 15-ADON and 3,15-diADON simultaneously (Scheme 5).

All synthesized DON derivatives were reduced applying an optimized Luche protocol (See *3.3. Reduction Protocol*) leading to the desired 7,8-dihydroxycalonectrin derivatives in moderate to good yields.

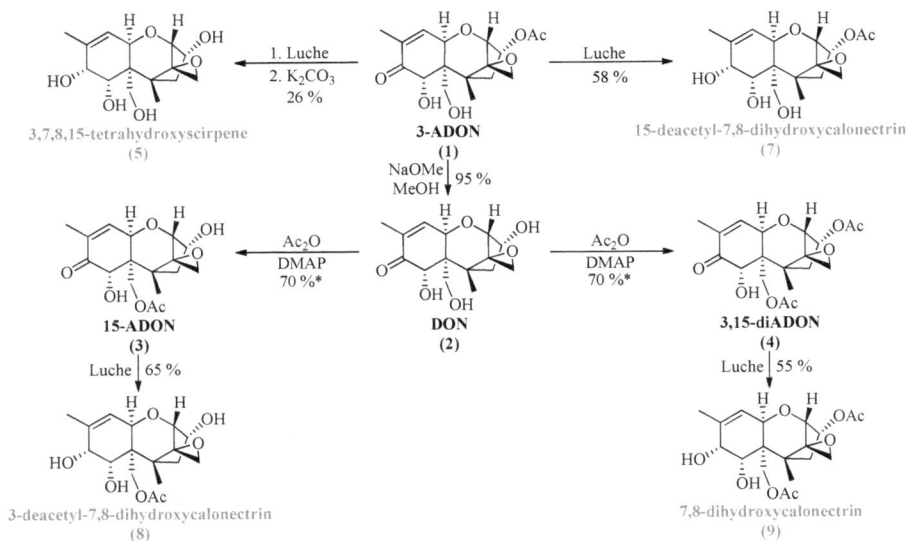

Scheme 5. Synthetic approach towards the different DON derivatives including reduction under Luche conditions (NaBH$_4$, CeCl$_3$) to the corresponding calonectrins. (* = yield as sum of both products).

3.3. Reduction Protocol

All Luche reductions towards the different derivatives were done with 0.05–0.30 mmol of DON or its corresponding acetylated form as starting material. The general procedure therefore was: Toxin (1.00 equ.) was dissolved in 1 mL MeOH and CeCl$_3$·7H$_2$O (0.50 equ. in 1 mL MeOH) was added. NaBH$_4$ (1.00 equ.) was dissolved in 1 mL MeOH and added with moderate speed (dropwise, but fast to prevent rising pressure due to H$_2$ formation). We recommend preparation of a stock solution of CeCl$_3$·7H$_2$O and NaBH$_4$ as it is easier to deal with the low substance amounts. In case of NaBH$_4$ the solution should be prepared just in time and used quickly to avoid evolving hydrogen. After addition of all reagents, the reaction was stirred at room temperature until TLC revealed conversion of the starting material. The mixture was concentrated without heating under reduced pressure to avoid decomposition and directly subjected to column chromatography.

3.4. Spectroscopic Investigation

To ensure that the desired stereochemistry of the product was achieved as well as for a full characterization including proof of stereochemistry, all products were investigated via several NMR techniques. In case of 7,8-dihydroxycalonectrin which is already well characterized in the literature, we obtained the identical spectroscopic information as published [22] by recording ^{1}H and ^{13}C spectra. In the case of the other three products we also recorded COSY, HSQC, HMBC and NOESY spectra to achieve a complete characterization (Figure 1, Table 1).

Table 1. ^{1}H NMR data (in methanol-d_4) including chemical shifts, (multiplicity) and [coupling constants] of the isolated products. Multiplicities are abbreviated as s (singlet), d (doublet), t (triplet), q (quartet), m (multiplet), and b (broad signal). R' and R are referring to the methyl signal of the attached acetyl substituent in position 3 (R') and 15 (R).

Product	2 (d)	3 (dt)	4α (dd)	4β (dd)	7β (d)	8β (d)	10 (dq)	11 (d)
(5)	3.38 [4.5]	4.31 [10.7, 4.5]	2.21 [14.6, 4.5]	1.97 [14.6, 10.7]	4.43 [4.7]	3.91 [4.7]	5.57 [5.6] (b)	4.40 [5.6]
(7)	3.68 [4.4]	5.05 [11.1, 4.4]	2.48 [15.0, 4.4]	2.04 [15.0, 11.1]	4.40 [5.0]	3.91 [5.0]	5.54 [5.5, 1.4]	4.33 [5.5]
(8)	3.41 [4.4]	4.30 [11.1, 4.4]	2.49 [14.6, 4.4]	1.97 [14.6, 11.1]	4.45 [5.6]	3.92 [5.6]	5.57 [5.8, 1.4]	4.64 [5.8]
(9)	3.72 [4.4]	5.05 [11.2, 4.4]	2.68 [15.1, 4,4]	1.95–2.15(m)	4.42 [5.2]	3.91 [5.2]	5.53 [5.9, 1.5]	4.57 [5.9]

Product	13a (d)	13b (d)	14 (s)	15a (d)	15b (d)	16 (s)	3 R' (s)	15 R (s)
(5)	3.02 [4.4]	3.16 [4.4]	1.14	3.64 [12.6]	3.88 [12.6]	1.84	—	—
(7)	3.08 [4.1]	3.20 [4.1]	1.17	3.67 [12.6]	3.90 [12.6]	1.84	—	2.10
(8)	3.04 [4.4]	3.18 [4.4]	1.15	4.34 [12.6]	4.38 [12.6]	1.85	2.04	—
(9)	3.09 [4.3]	3.22 [4.3]	1.18	4.37(s), 2H		1.85	2.04	2.10

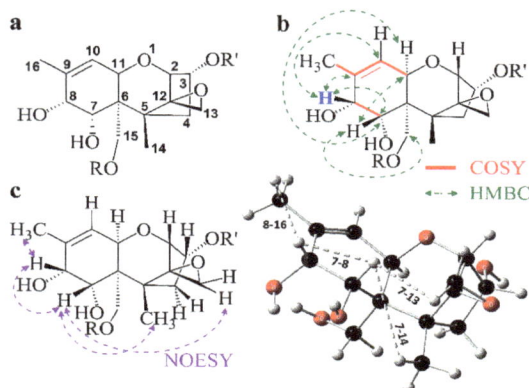

Figure 1. (a) Systematic numbering of trichothecenes; (b) Selected COSY and HMBC correlations of **5** (R = R' = H), **7** (R = H, R' = Ac) and **8** (R = Ac, R' = H); (c) Selected NOESY correlations within the 2D (left) and 3D (right, optimized geometry, Figure S30, Tables S1,2) structure of **5**.

4. Experimental Section

4.1. General

Thin layer chromatography (TLC) was performed over silica gel 60 F254 (Merck). All chromatograms were visualized by heat staining using ceric ammonium molybdate/Hanessian's stain [24] in ethanol/sulfuric acid. Chromatographic separation was done on silica gel 60 (40–63 μm, Merck, Darmstadt, Germany) using a SepacoreTM Flash System (Büchi, Switzerland). ^{1}H and ^{13}C NMR spectra were recorded on an Avance DRX-400 MHz spectrometer as well as at a Bruker DPX-200 spectrometer (Bruker, Karlsruhe, Germany). Data were recorded and evaluated using TOPSPIN 1.3 (Bruker Biospin, Karlsruhe, Germany). CD spectra were recorded using a JASCO J-815 CD spectrometer (JASCO, Easton, MD, USA), and can be found in the supporting information. All chemical shifts are given in ppm relative to tetramethylsilane. The calibration was done using residual solvent signals [25]. Multiplicities are abbreviated as s (singlet), d (doublet), t (triplet), q (quartet) and b (broad signal). 3-ADON was obtained from University of Natural Resources and Life Sciences, Vienna, Dept. for Agrobiotechnology (IFA-Tulln) and was used after ^{1}H NMR purity check [22]. All other chemicals were purchased from Sigma-Aldrich (Schnelldorf, Germany).

4.2. Deoxynivalenol (2)

3-ADON (85.6 mg, 0.25 mmol) was dissolved in 5 mL dry methanol and NaOMe (13.7 mg, 0.25 mmol) was added to the reaction. After 1.5 h TLC revealed full conversion of the starting material and the reaction mixture was concentrated to 1 mL. Finally the solution was directly purified by the use of column chromatography (CHCl$_3$:MeOH = 9:1) which yielded deoxynivalenol (79.0 mg, 95%) as white solid. The reaction product was proved to be identical to an authentic sample by TLC and thus was used for the next step.

4.3. 15-ADON (3) and 3,15-diADON (4)

DON (79.0 mg, 0.27 mmol) was dissolved in 50 mL dry dichloromethane. Pyridine (1 mL) and 4-DMAP (app. 10 mg) were added followed by the dropwise addition of acetic anhydride (27.2 mg, 0.27 mmol). The reaction was stirred overnight, treated with 20 mL HCl (2 N) and extracted 3 times with 50 mL dichloromethane. After drying with Na$_2$SO$_4$, filtration and evaporation of the solvent the remaining residue was subjected to column chromatography (CHCl$_3$:MeOH = 95:5) to yield 15-ADON (42.0 mg, 47%, Figures S2,S3) and 3,15-diADON (23.5 mg, 23%, Figures S4,S5) as white solid. Total yield = 70%, 93% conversion. 15-ADON (3): ^{1}H NMR (200 MHz, CDCl$_3$) δ = 6.61 (dq, *J* = 5.7, 1.6 Hz, 1H), 4.89 (d, *J* = 5.7 Hz, 1H), 4.83 (d, *J* = 1.6 Hz, 1H), 4.52 (dt, *J* = 10.2, 4.7 Hz, 1H), 4.24 (s, 2H), 3.78 (d, *J* = 1.8 Hz, 1H), 3.63 (d, *J* = 4.5 Hz, 1H), 3.13 (d, *J* = 4.3 Hz, 1H), 3.08 (d, *J* = 4.3 Hz, 1H), 2.22 (dd, *J* = 14.8, 4.7 Hz, 1H), 2.08 (dd, *J* = 14.7, 10.4 Hz, 1H), 1.88 (s, 3 H), 1.87 (s, 3H), 1.07 (s, 3H); ^{13}C-NMR (50 MHz, CDCl$_3$) δ = 199.6 (s), 170.3 (s), 138.8 (d), 135.6 (s), 80.7 (d), 73.5 (d), 70.1 (d), 68.9 (s), 65.5 (s), 62.2 (t), 51.4 (s), 47.4 (t), 46.3 (s), 43.3 (t), 20.7 (q), 15.4 (q), 13.8 (q). **3,15-diADON (4)**: ^{1}H NMR (200 MHz, CDCl$_3$) δ = 6.56 (dq, *J* = 5.8, 1.4 Hz, 1H), 5.20 (dt, *J* = 10.9, 4.6 Hz, 1H), 4.80 (d, *J* = 2.0 Hz, 1H), 4.69 (d, *J* = 5.8 Hz, 1H), 4.27 (d, *J* = 12.1 Hz, 1H), 4.20 (d, *J* = 12.1 Hz, 1H), 3.89 (d, *J* = 4.3 Hz, 1H), 3.80 (d, *J* = 2.0 Hz, 1H), 3.14 (d, *J* = 4.3 Hz, 1H), 3.09 (d, *J* = 4.3 Hz, 1H), 2.31 (dd, *J* = 15.2, 4.8 Hz, 1H), 2.15 (dd, *J* = 15.2, 10.9 Hz, 1H), 2.12 (s, 3H), 1.88 (s, 3 H), 1.87 (s, 3H), 1.08 (s, 3H); ^{13}C-NMR (50 MHz, CDCl$_3$) δ = 199.3 (s), 170.3 (s), 170.2 (s), 138.4 (d), 135.6 (s), 78.9 (d), 73.4 (d), 71.1 (d), 70.1 (d), 64.9 (s), 62.1 (t), 51.5 (s), 47.4 (t), 45.8 (s), 40.3 (t), 21.0 (q), 20.6 (q), 15.3 (q), 13.6 (q); ^{1}H NMR data consistent with the ones reported in literature [26,27].

4.4. (+)-cis-Carveol (11) (Large Scale Luche Reduction)

(+)-Carvone (10) (3.00 g, 20.0 mmol) and CeCl$_3$·7H$_2$O (1.86 g, 5.0 mmol) were dissolved in 150 mL MeOH and cooled to 0 °C. NaBH$_4$ (0.76 g, 20.0 mmol) was dissolved in 100 mL MeOH and added within 5 mins to the reaction solution via a dropping funnel. After complete addition of the NaBH$_4$-solution the cooling bath was removed and the reaction continued. TLC control (hexane:EtOAc = 5:1) after 30 min revealed complete conversion of the starting material and the reaction was treated with 50 mL 2N HCl and extracted three times with 100 mL Et$_2$O. The organic phase was dried over Na$_2$SO$_4$, filtered and the solvent was removed under reduced pressure. Column chromatography (hexane:EtOAc = 5:1) yielded 2.79 g (92%, Figures S6,S7) of a slightly yellow oil which was consistent with literature NMR data [28,29] for (+)-cis-carveol. ^{1}H NMR (200 MHz, CDCl$_3$) δ = 5.35 (b, 1H), 4.62 (b, 1H), 4.08 (b, 1H), 3.77 (b, 1H), 1.50 – 2.30 (m, 11H), 1.40 (dt, *J* = 12.2, 10.0 Hz, 1H); ^{13}C-NMR (50 MHz, CDCl$_3$) δ = 148.7 (s, C=), 136.6 (s, C=), 123.3 (d, =CH), 108.8 (t, =CH$_2$), 70.3 (d, CH), 40.7 (d, CH), 37.8 (t, CH$_2$), 31.0 (t, CH$_2$), 20.3 (q, CH$_3$), 19.0 (q, CH$_3$).

4.5. Conversion of 3-ADON (1) to 15-Deacetyl-7,8-dihydroxycalonectrin (7) and 3,7,8,15-Tetrahydroxyscirpene (5)

3-ADON (1.00 equ.) was dissolved in 1 mL MeOH and CeCl$_3$·7H$_2$O (0.50 equ. in 1 mL MeOH) was added. NaBH$_4$ (1.00 equ.) was dissolved in 1 mL MeOH and added with moderate speed (dropwise, but fast enough to prevent rising pressure due to H$_2$ formation). After completion the reaction was stirred until TLC revealed conversion of the starting material. The mixture was concentrated without heating under reduced pressure and subjected directly to column chromatography (DCM:MeOH = 9:1) to yield 10.6 mg (58%, Figures S13–S17) of 15-deacetyl-7,8-dihydroxycalonectrin. In order to obtain 3,7,8,15-tetrahydroxyscirpene (5), the whole reaction was repeated and evaporated to dryness. After uptake in 3 mL dry MeOH, K_2CO_3 (2.00 equ.) was added and the reaction stirred until TLC indicated the deprotection of the acetyl group in position 3. The reaction mixture was concentrated to 1 mL and purified via column chromatography ($CHCl_3$:MeOH = 9:1) to yield 13.6 mg (26% for two steps, Figures S8–S12) of 3,7,8,15-tetrahydroxyscirpene. 3,7,8,15-Tetrahydroxyscirpene (5): ^{1}H NMR (400 MHz, methanol-d_4) δ = 5.57 (bd, J = 5.6 Hz, 1H), 4.43 (d, J = 4.7 Hz, 1H), 4.40 (d, J = 5.6 Hz, 1H), 4.31 (dt, J = 10.7, 4.5 Hz, 1H), 3.91 (d, J = 4.7 Hz, 1H), 3.88 (d, J = 12.6 Hz, 1H), 3.64 (d, J = 12.6 Hz, 1H), 3.38 (d, J = 4.5 Hz, 1H), 3.16 (d, J = 4.4 Hz, 1H), 3.02 (d, J = 4.4 Hz, 1H), 2.21 (dd, J = 14.6, 4.5 Hz, 1H), 1.97 (dd, J = 14.6, 10.7 Hz, 1H), 1.84 (s, 3 H), 1.14 (s, 3H); ^{13}C-NMR (100 MHz, methanol-d_4) δ = 140.0 (s, C-9), 124.3 (d, C-10), 81.1 (d, C-2), 72.9 (d, C-11), 72.2 (d, C-8), 72.0 (d, C-7), 69.8 (d, C-3), 66.5 (s, C-12), 62.3 (t, C-15), 48.6 (s, C-6), 48.5 (s, C-13), 47.7 (s, C-5), 45.8 (t, C-4), 20.8 (q, C-16), 16.2 (q, C-14). The signal at 48.5 (C-13) is not visible in the ^{13}C spectra, but could be located in the correlated spectra. HRMS (APCI$^+$): m/z calcd for (5) [M + Na$^+$]: 321.1309; found: 321.1305. **15-Deacetyl-7,8-dihydroxycalonectrin (7):** ^{1}H NMR (400 MHz, methanol-d_4) δ = 5.54 (dq, J = 5.5, 1.4 Hz, 1H), 5.05 (dt, J = 11.1, 4.4 Hz, 1H), 4.40 (d, J = 5.0 Hz, 1H), 4.33 (d, J = 5.5 Hz, 1H), 3.91 (d, J = 5.0 Hz, 1H), 3.90 (d, J = 12.6 Hz, 1H), 3.68 (d, J = 4.4 Hz, 1H), 3.67 (d, J = 12.6 Hz, 1H), 3.20 (d, J = 4.1 Hz, 1H), 3.08 (d, J = 4.1 Hz, 1H), 2.48 (dd, J = 15.0, 4.4 Hz, 1H), 2.10 (s, 3H), 2.04 (dd, J = 15.0, 11.1 Hz, 1H), 1.84 (s, 3 H), 1.17 (s, 3H); ^{13}C-NMR (100 MHz, methanol-d_4) δ = 172.6 (s, acetyl C=O), 140.4 (s, C-9), 123.7 (d, C-10), 79.9 (d, C-2), 73.0 (d, C-3), 72.9 (d, C-11), 72.1 (d, C-8), 71.7 (d, C-7), 66.1 (s, C-12), 62.1 (t, C-15), 48.7 (t, C-13), 48.5 (s, C-6), 47.1 (s, C-5), 42.6 (t, C-4), 20.9 (q, acetyl CH_3), 20.8 (q, C-16), 15.9 (q, C-14). HRMS (APCI$^+$): m/z calcd for (7) [M + Na$^+$]: 363.1414; found: 363.1409.

4.6. Reduction of 15-ADON (3) and 3,15-diADON (4) to 3-Deacetyl-7,8-dihydroxycalonectrin (8) and 7,8-Dihydroxycalonectrin (9)

Toxin (1.00 equ.) was dissolved in 1 mL MeOH and $CeCl_3 \cdot 7H_2O$ (0.50 equ. in 1 mL MeOH) was added. $NaBH_4$ (1.00 equ.) was dissolved in 1 mL MeOH and added with moderate speed (dropwise, but fast to prevent rising pressure due to H_2 formation). After completion the reaction was stirred until TLC revealed conversion of the starting material. The mixture was concentrated without heating under reduced pressure and directly subjected to column chromatography (DCM:MeOH = 97.5:2.5 for 7,8-dihydroxycalonectrin and DCM:MeOH = 9:1 for 3-deacetyl-7,8-dihydroxycalonectrin) to yield the desired products in 55% (7,8-dihydroxycalonectrin, Figures S23,S24) and 65% (3-deacetyl-7,8-dihydroxycalonectrin, Figures S18–S22). **7,8-Dihydroxycalonectrin (9):** ^{1}H NMR (200 MHz, methanol-d_4) δ = 5.53 (dq, J = 5.9, 1.5 Hz, 1H), 5.05 (dt, J = 11.2, 4.4 Hz, 1H), 4.52 (d, J = 5.9 Hz, 1H), 4.42 (d, J = 5.3 Hz, 1H), 4.40 (s, 2H), 3.91 (d, J = 5.3 Hz, 1H), 3.72 (d, J = 4.4 Hz, 1H), 3.22 (d, J = 4.3 Hz, 1H), 3.09 (d, J = 4.2 Hz, 1H), 2.68 (dd, J = 15.1, 4.1 Hz, 1H), 1.95-2.15 (m, 1H), 2.10 (s, 3H), 2.04 (s, 3H), 1.85 (s, 3 H), 1.18 (s, 3H); ^{13}C-NMR (50 MHz, methanol-d_4) δ = 172.6 (s), 172.4 (s), 141.9 (s), 122.1 (d), 80.1 (d), 72.9 (d), 72.3 (d), 71.4 (d), 71.3 (d), 66.3 (s), 65.4 (t), 47.8 (s), 46.8 (s), 43.0 (t), 21.2 (q), 20.9 (q), 20.8 (q), 15.8 (q, C-14). 1 Signal missing due to solvent overlap. HRMS (APCI$^+$): m/z calcd for (9) [M+Na$^+$]: 405.1520; found: 405.1515. **3-Deacetyl-7,8-dihydroxycalonectrin (8):** ^{1}H NMR (400 MHz, methanol-d_4) δ = 5.57 (dq, J = 5.8, 1.4 Hz, 1H), 4.64 (d, J = 5.8 Hz, 1H), 4.45 (d, J = 5.6 Hz, 1H), 4.38 (d, J = 12.6 Hz, 1H), 4.34 (d, J = 12.6 Hz, 1H), 4.30 (dt, J = 11.1, 4.4 Hz, 1H), 3.92 (d, J = 5.6 Hz, 1H), 3.41 (d, J = 4.4 Hz, 1H), 3.18 (d, J = 4.4 Hz, 1H), 3.04 (d, J = 4.4 Hz, 1H), 2.49 (dd, J = 14.6, 4.4 Hz, 1H), 2.04 (s, 3H), 1.97 (dd, J = 14.6, 11.1 Hz, 1H), 1.85 (s, 3 H), 1.15 (s, 3H); ^{13}C-NMR (100 MHz, methanol-d_4) δ = 172.5 (s, acetyl C=O), 141.5 (s, C-9), 122.6 (d, C-10), 82.0 (d, C-2), 72.0 (d, C-11), 71.5 (2 x d, C-7, C-8), 69.6 (d, C-3), 66.8 (s, C-12), 65.4 (t, C-15), 48.8 (s, C-13), 47.8 (s, C-6), 47.3 (s, C-5), 46.0 (t, C-4), 21.2 (q, acetyl CH_3), 20.8 (q, C-16), 16.0 (q, C-14). HRMS (APCI$^+$): m/z calcd for (8) [M + Na$^+$]: 363.1414; found: 363.1411. NMR data for 7,8-dihydroxycalonectrin and 3-deacetyl-7,8-dihydroxycalonectrin were found in accordance with the ^{1}H chemical shifts reported in literature [22].

5. Conclusions

This paper presents a reliable, mild, fast and tolerant method for the stereoselective reduction of the carbonyl group in deoxynivalenol and its acetylated derivatives. Although the method was optimized for deoxynivalenol, it is likely to be applicable to nivalenol, its acetylated derivatives and other type B trichothecenes in a similar way. The isolated yields of the method were satisfying and further improvements towards better yields of the presented method might be difficult to achieve due to the instability of several functional groups (epoxy, acetyl) with prolonged reaction time. Nevertheless, we expect the method to also be applicable on masked forms of mycotoxins thereby providing a valuable tool for synthetic conversion of different mycotoxin standards.

In addition to the reduction itself, four different calonectrin derivatives were synthesized and characterized using several NMR techniques. By using NOESY (Nuclear Overhauser Enhancement SpectroscopY) we were able to prove the stereochemistry of all reaction products. The value of the compounds itself is difficult to estimate—although they are supposed to act as precursors for the biosynthesis of DON, their natural occurrence is reported rarely in literature. Nonetheless, the protocol presented can be readily applied for the synthesis of this class of compounds and therefore opens the door for their use as reference material as well as for investigations on the biologic pathway towards type B trichothecenes.

Acknowledgments: The support of the graduate school program Applied Bioscience Technology of the VUT (Vienna University of Technology) and the BOKU Vienna is gratefully acknowledged. This work was also funded by the Vienna Science and Technology Fund (WWTF LS12-021) and the Austrian Science Fund (SFB Fusarium F3702 and F3706). The authors would like to thank Franz Berthiller and Elisabeth Varga (both IFA Tulln) for providing 3-ADON. In addition we would like to thank Thiago Machado Mello De Sousa for help with CD measurements (Figures S25–S29).

Conflicts of Interest: The authors declare no conflict of interest.

References

1. D'Mello, J.P.F.; Macdonald, A.M.C. Mycotoxins. *Anim. Feed Sci. Technol.* **1997**, *69*, 155–166. [CrossRef]
2. Cundliffe, E.; Cannon, M.; Davies, J. Mechanism of inhibition of eukaryotic protein synthesis by trichothecene fungal toxins. *Proc. Natl. Acad. Sci. USA* **1974**, *71*, 30–34. [CrossRef]
3. Pestka, J. Deoxynivalenol: Mechanisms of action, human exposure, and toxicological relevance. *Arch. Toxicol.* **2010**, *84*, 663–679.
4. Arunachalam, C.; Doohan, F.M. Trichothecene toxicity in eukaryotes: Cellular and molecular mechanisms in plants and animals. *Toxicol. Lett.* **2013**, *217*, 149–158. [CrossRef]
5. Rotter, B.A. Invited review: Toxicology of deoxynivalenol (vomitoxin). *J. Toxicol. Environ. Health* **1996**, *48*, 1–34. [CrossRef]
6. Pestka, J.J. Deoxynivalenol-induced proinflammatory gene expression: Mechanisms and pathological sequelae. *Toxins* **2010**, *2*, 1300–1317. [CrossRef]
7. Maresca, M. From the gut to the brain: Journey and pathophysiological effects of the food-associated trichothecene mycotoxin deoxynivalenol. *Toxins* **2013**, *5*, 784–820. [CrossRef]
8. Ueno, Y. The toxicology of mycotoxins. *Crit. Rev. Toxicol.* **1985**, *14*, 99–132. [CrossRef]
9. Grovey, J.F. The trichothecenes and their biosynthesis. In *Progress in the Chemistry of Organic Natural Products*; Herz, W., Falk, H., Kirby, G.W., Eds.; Springer Vienna: Vienna, Austria, 2007; Volume 88, pp. 63–130.
10. McCormick, S.P.; Stanley, A.M.; Stover, N.A.; Alexander, N.J. Trichothecenes: From simple to complex mycotoxins. *Toxins* **2011**, *3*, 802–814. [CrossRef]
11. Fruhmann, P.; Mikula, H.; Wiesenberger, G.; Varga, E.; Lumpi, D.; Stöger, B.; Häubl, G.; Lemmens, M.; Berthiller, F.; Krska, R.; *et al.* Isolation and structure elucidation of pentahydroxyscirpene, a trichothecene fusarium mycotoxin. *J. Nat. Prod.* **2014**. [CrossRef]
12. Greenhalgh, R.; Levandier, D.; Adams, W.; Miller, J.D.; Blackwell, B.A.; McAlees, A.J.; Taylor, A. Production and characterization of deoxynivalenol and other secondary metabolites of fusarium culmorum (cmi 14764, hlx 1503). *J. Agric. Food Chem.* **1986**, *34*, 98–102. [CrossRef]

13. Greenhalgh, R.; Meier, R.M.; Blackwell, B.A.; Miller, J.D.; Taylor, A.; ApSimon, J.W. Minor metabolites of fusarium roseum (atcc 28114). *J. Agric. Food Chem.* **1984**, *32*, 1261–1264. [CrossRef]

14. Hanson, A. The structure of a trichothecene from fusarium roseum. *Acta Crystallogr. Sect. C* **1986**, *42*, 503–505. [CrossRef]

15. Hesketh, A.R.; Gledhill, L.; Marsh, D.C.; Bycroft, B.W.; Dewick, P.M.; Gilbert, J. Biosynthesis of trichothecene mycotoxins: Identification of isotrichodiol as a post-trichodiene intermediate. *Phytochemistry* **1991**, *30*, 2237–2243. [CrossRef]

16. Hesketh, A.R.; gledhill, L.; Bycroft, B.W.; Dewick, P.M.; Gilbert, J. Potential inhibitors of trichothecene biosynthesis in fusarium culmorum: Epoxidation of a trichodiene derivative. *Phytochemistry* **1992**, *32*, 93–104. [CrossRef]

17. Kononenko, G.P.; Soboleva, N.A.; Leonov, A.N. 3,7,8,15-tetrahydroxy-12,13-epoxytrichothec-9-en in a culture of fusarium graminearum. *Chem. Nat. Compd.* **1990**, *26*, 219–220. [CrossRef]

18. Gemal, A.L.; Luche, J.L. Lanthanoids in organic synthesis. 6. Reduction of Alpha.-enones by sodium borohydride in the presence of lanthanoid chlorides: Synthetic and mechanistic aspects. *J. Am. Chem. Soc.* **1981**, *103*, 5454–5459. [CrossRef]

19. Luche, J.L. Lanthanides in organic chemistry. 1. Selective 1,2 reductions of conjugated ketones. *J. Am. Chem. Soc.* **1978**, *100*, 2226–2227.

20. Šťastná, E.; Černý, I.; Pouzar, V.; Chodounská, H. Stereoselectivity of sodium borohydride reduction of saturated steroidal ketones utilizing conditions of luche reduction. *Steroids* **2010**, *75*, 721–725. [CrossRef]

21. Cram, D.J.; Elhafez, F.A.A. Studies in stereochemistry. X. The rule of "steric control of asymmetric induction" in the syntheses of acyclic systems. *J. Am. Chem. Soc.* **1952**, *74*, 5828–5835. [CrossRef]

22. Savard, M.E.; Blackwell, B.A.; Greenhalgh, R. An 1 H nuclear magnetic resonance study of derivatives of 3-hydroxy-12,13-epoxytrichothec-9-enes. *Can. J. Chem.* **1987**, *65*, 2254–2262. [CrossRef]

23. Grove, J.F.; McAlees, A.J.; Taylor, A. Preparation of 10-g quantities of 15-*O*-acetyl-4-deoxynivalenol. *J. Org. Chem.* **1988**, *53*, 3860–3862. [CrossRef]

24. Pirrung, M.C. Appendix 3: Recipes for TLC stains. In *The Synthetic Organic Chemist's Companion*; John Wiley & Sons, Inc.: Hoboken, NJ, USA, 2006; pp. 171–172.

25. Fulmer, G.R.; Miller, A.J.M.; Sherden, N.H.; Gottlieb, H.E.; Nudelman, A.; Stoltz, B.M.; Bercaw, J.E.; Goldberg, K.I. NMR chemical shifts of trace impurities: Common laboratory solvents, organics, and gases in deuterated solvents relevant to the organometallic chemist. *Organometallics* **2010**, *29*, 2176–2179. [CrossRef]

26. Blackwell, B.A.; Greenhalgh, R.; Bain, A.D. Carbon-13 and proton nuclear magnetic resonance spectral assignments of deoxynivalenol and other mycotoxins from fusarium graminearum. *J. Agric. Food Chem.* **1984**, *32*, 1078–1083. [CrossRef]

27. Muñoz, L.; Castro, J.L.; Cardelle, M.; Castedo, L.; Riguera, R. Acetylated mycotoxins from fusarium graminearum. *Phytochemistry* **1989**, *28*, 83–85.

28. Cravero, R.M.; González-Sierra, M.; Labadie, G.R. Convergent approaches to saudin intermediates. *Helv. Chim. Acta* **2003**, *86*, 2741–2753. [CrossRef]

29. Rafiński, Z.; Ścianowski, J. Synthesis and reactions of enantiomerically pure dialkyl diselenides from the p-menthane group. *Tetrahedron* **2008**, *19*, 1237–1244.

MDPI AG
St. Alban-Anlage 66
4052 Basel, Switzerland
Tel. +41 61 683 77 34
Fax +41 61 302 89 18
http://www.mdpi.com

Toxins Editorial Office
E-mail: toxins@mdpi.com
http://www.mdpi.com/journal/toxins

www.ingramcontent.com/pod-product-compliance
Lightning Source LLC
Chambersburg PA
CBHW051718210326
41597CB00032B/5525